A Dictionary of
CIVIL
ENGINEERING

A Dictionary of
CIVIL ENGINEERING

D.N. GHOSE
M.C.E, F.I.E, F.I.P.H.E, F.I.W.W.A.

CBS PUBLISHERS & DISTRIBUTORS PVT. LTD.
New Delhi • Bengaluru • Chennai • Kochi • Mumbai • Pune

ISBN : 81-239-1171-8

First Edition : 1991
Reprint : 2000, 2002, 2004, 2005, 2007, 2008,
          2010, 2011, 2012, 2013, 2014

Published by Satish Kumar Jain for
CBS Publishers & Distributors Pvt. Ltd.,
CBS Plaza, 4819/XI Prahlad Street, 24 Ansari Road, Daryaganj,
New Delhi - 110002, India.   Website: www.cbspd.com
e-mail: delhi@cbspd.com, cbspubs@airtelmail.in
Ph.: 23289259, 23266861, 23266867   ◦ Fax: 011-23243014

Branches:

Bengaluru: Seema House, 2975, 17th Cross, K.R. Road,
Bansankari 2nd Stage, Bengaluru - 560070
 ◦ Ph.: +91-80-26771678/79 ◦ Fax: +91-80-26771680
 ◦ E-mail: cbsbng@gmail.com, bangalore@cbspd.com
Pune: Bhuruk Prestige, Sr. No. 52/12/2+1+3/2,
Narhe, Haveli (Near Katraj-Dehu Road By-pass), Pune - 411041
 ◦ Ph.: +91-20-64704058/59, 020-32392277   ◦ E-mail: pune@cbspd.com
Kochi: 36/14, Kalluvilakam, Lissie Hospital Road,
Kochi - 682018, Kerala   Ph.: +91-484-4059061-65
 ◦ Fax: +91-484-4059065   ◦ E-mail: cochin@cbspd.com
Chennai: 20, West Park Road, Shenoy Nagar, Chennai - 600030
Ph.: +91-44-26260666, 26208620   ◦ Fax: +91-44-42032115
 ◦ E-mail: chennai@cbspd.com
Mumbai: 83-C, Ist Floor, Dr. E. Moses Road, Worli, Mumbai-400 018,
Maharashtra Ph.: +91-9833017933, 022-24902340/24902341
 ◦ E-mail: mumbai@cbspd.com

Printed at :
India Binding House, Noida (UP)

Dedicated to

SHRI SUCHIT KUMAR GHOSH

the Engineer of Civil Engineers

with my esteemed regards

PREFACE

Civil Engineering' is multi-disciplinary and ever expanding. A Civil Engineer sometimes gets lost amidst the numerous terms used in various disciplines of Civil Engineering and at times one has to put time-consuming efforts in searching out the meaning of a term.

It is an attempt made by the author to prepare "*A Dictionary of Civil Engineering*" explaining the common terms used in this field. Civil Engineering terms are arranged alphabatically with their meaning, application and self-explanatory illustrations.

It is needless to mention that various books have been consulted in preparation of this book. If the terms, which have not been included in this volume, are brought to the notice of the author, it would be highly appreciated and acknowledged.

D.N. GHOSE

This book deals with the terms used in the following disciplenes of Civil Engineering.

- Building Construction
- Docks and Harbours
- Drainage Engineering
- Geodesy
- Geology
- Hydraulics
- Hydrology
- Irrigation and Waterways
- Materials of Construction
- Railway Engineering
- Road Engineering
- Sewerage and Sewage Treatment
- Soil & Foundation Engineering
- Solid Waste Management
- Structural Engineering
- Surveying
- Water Supply Engineering

CONTENTS

A	Abacus	-	Azimuthal Projection	1
B	Bacillus Coil	-	Bypass	23
C	Cabin	-	Cylinder Test	57
D	Dabber	-	Dynamite	112
E	E	-	Eye Bolt	140
F	Fabric	-	Furring	153
G	Gabion	-	Gypsum Wall Board	179
H	Hacking	-	Hysterisis Loop	196
I	I	-	Izod Test	214
J	Jack Arch	-	Juvenile Water	223
K	Kaplan Turbine	-	Kyan's Process	227
L	Laced Column	-	Lytag	231
M	Macadam	-	Myer's Formula	248
N	Nail	-	n-value	266
O	Oak	-	Ozone	272
P	Paddle	-	Pyrometer	278
Q	Q-Trap	-	Quoin Post	310
R	Rabbet	-	Rybate	314

S	Sabin	- System Building	339
T	Table	- Tyrolean Finish	414
U	U-gauge	- U-value	440
V	Vacuum Cleaning Plant	- Vulcanite	446
W	W	- Wye	456
X	XPM	- Xylonite	473
Y	Y	- Yorky	474
Z	Z	- Zone of Saturation	476
	Addendum		479

A

ABACUS : (i) The portion of a baluster which supports the handrail.

          (ii) The upper part of a column which receives the architrave.

ABAMURUS : A buttress to support a wall.

ABATEMENT : The reduction in size of timber by sawing and planing i.e. the waste of timber.

ABATVOIX : A sounding board to deflect sound downwards.

A.B.C. PROCESS : A sewage treatment process in which precipitants used are alum, blood, and charcoal.

ABERDEEN GRANITE : Widely used building stone, grey or pink in colour.

ABNEY LEVEL : A hand level used for taking levels up steep slopes and as a clinometer. This consists of a telescopic tube and reflecting level.

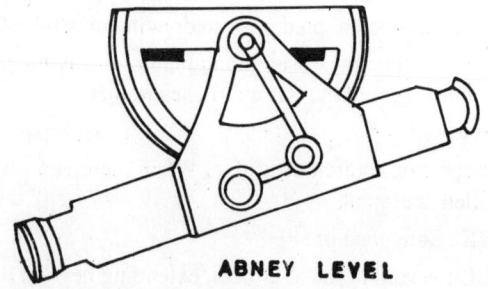

ABNEY LEVEL

ABRAMS' LAW : It states that the strength of mortar or concrete varies inversely to the water-cement ratio. This was discovered by Duff Abrams.

ABRASION RESISTANCE : It is the resistance to wear.

ABRASION TEST OF STONES : The life of stones, used as a paving material, is measured by the resistance offered by the stones against the abraiding action of the traffic. Dorry's Abrasion testing machine, sieves and a balance are required to perform test for the co-efficient of hardness of paving materials.

ABRASION TEST OF TIMBER : The specimen blocks of timber are clamped and held with a pressure against a revolving table covered with sand paper. A specimen of a standard timber is also tested at the same time to ascertain the relative wear. Abraided area of test specimen is 2" x 2". Pressure applied is 26 Ibs. and rotation is 68 r.p.m.

ABRASIVES : Hard materials used for rubbing down paint and other building materials. These may also be used for sharpening tools. Examples are glass paper, emery cloth,carborundum, garnet, diamond,emery wheels, quartz,flint sand, glass powder, etc. Red bauxite, when fused by the Lecesne process, furnishes an abrasive matter of a very good quality. Red fused bauxite has a composition approximating to emery. Red fused bauxite is not as abrasive as natural or artificial corundum at 90% of alumina.

ABREUVOIR : The mortar joint between two stones either in arch or in wall.

ABS : A kind of plastic used for making pipes.

ABSOLUTE HUMIDITY: At any condition, the actual weight of water vapour contained in 1 cu.ft. of air at that condition is the absolute humidity of air.

ABSOLUTE HYDRAULIC GRADIENT : An imaginary curve parallel to the actual hydraulic gradient line.

ABSOLUTE–REST PERCIPITATION TANK : A tank which receives a given amount of sewage for settlement without having a continuous flow.

ABSORBING WELL : It is nothing but a shaft sunk through an impermeable stratun to drain off the water to a permeable one.

ABSORPTION : It refers to the movement and storage of any liquid in the pores of any other material by the force of 'capillary action.'

ABSORPTION LOSS : Loss of water during the first filling of a reservoir in the wetting of soil.

ABSORPTIVE POWER : The relative rate of absorption of heat from a body compared with that from a black body of similar shape. This is also called 'Emissivity.'

ABSTRACTING : The method adopted prior to drawing up a bill of quantities.

ABT RACK : It is a multi-plate rack used in mountain railway to avoid slip. The teeth on different racks are staggered in relation to one another.

ABUT : To meet at one end.

ABUTMENT : A support of an arch or bridge. When there is a series of arches, the end supports are called abutments.

ABUTMENT PIECE : Sole plate or sill.

ABUTMENT WALL : A wall at the abutment, extending beyond the bridge or culvert to retain the earth behind the abutment.

ABUTMENT PIER : A pier to a wall supportng one end of a bridge or a culvert.

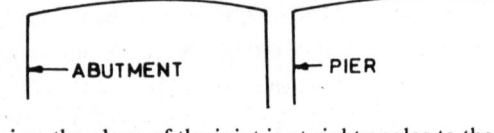

ABUTTING JOINT : A joint in which the fibres of both the joint in pieces are in the same line, i.e., the plane of the joint is at right angles to the fibres.

ABUTTING TENONS : In a single mortise, two tenons butt against each other from opposite sides.

ABYSSINIAN PUMP : A pump for the use in the abyssinian well, in which the well tube is attached to the suction tube.

ABYSSINIAN WELL : A perforated pipe (with pointed end) driven into the ground. Water can be extracted from it by pumping. It is the ancestor of common house-hold tube-well fitted with a hand pump.

ACCELERATOR : A substance which accelerates; A hardener material or catalyst which when added to a substance increases its hardening rate.

ACCELERATOR FOR CEMENT : An admixture usually calcium chloride (upto 2%) is added to cement or concrete to hasten its hardening rate by accelerating the hydration of cement.

ACCELOFILTER : Essentially a high rate Bio-filter for sewage treatment as shown in illustration.

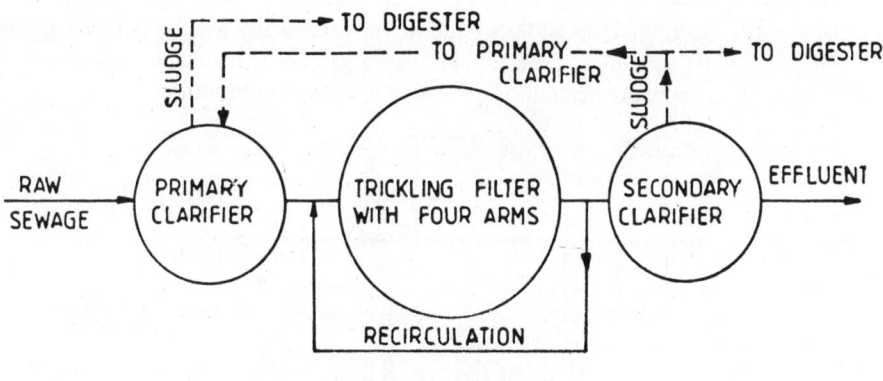

<u>ACCELO- FILTER</u>

ACCESS : An approach or entrance to a building, room, roof, etc. An access for maintenance is a detachable or removable panel or door.

ACCESS EYE : An opening in a drain which usually remains closed. When cleaning of the drain by rodding is required, it is removed or opened.

ACCESS SHAFT : It is the upper portion of a deep manhole. The minimum internal dimension of such a shaft in a deep manhole is 2′–3″ square.

ACCESS SHOE : Used to provide access for cleaning surface water or sullage drains.

ACCIDENTAL ERROR : Compensating error due to unavoidable instrumental error or due to inaccurate reading.

ACCORDION : A folding door or a folding partition.

ACCOUPLEMENT : A timber brace or tie.

ACCRINGTON NORI BRICK : Impervious brick having smooth surface and high density. Used for better class drainage purpose.

ACETONE : A highly inflammable organic solvent which is used to remove paint and varnish.

ACHROMATIC LENS : A compound lens (combination of two glasses) used to correct chromatic aberration. The double convex portion is made of crown glass and the miniscus of flint glass.

ACIDING : The process of light etching of a cast stone surface.

ACIDIC SOIL : Sandy soils, especially where rainfall is high. Salts are gererally found in tracts of poor slope.

ACIDITY OF WATER : Caused by the humic and other organic acids. This is determined by pH-Value. Water is said to be acidic, when pH-Value is 0 to 7.

ACID-RESISTING CEMENT: A compound of latex and high alumina cement, resistant to acid in high concentration, but easily broken down by alkali and injured by heat.

ACID ROCKS: Igneous rocks containing more than 66% of silica.

ACID STEEL : Steel made by a process in which the furnace is lined with silica bricks and the flux is silica.

ACID WASTE NEUTRALISATION : Industrial waste water bearing acid wastes are required to be neutralized prior to their discharge into a municipal sewer line. A two-stage acid waste neutralization system is shown in illustration.

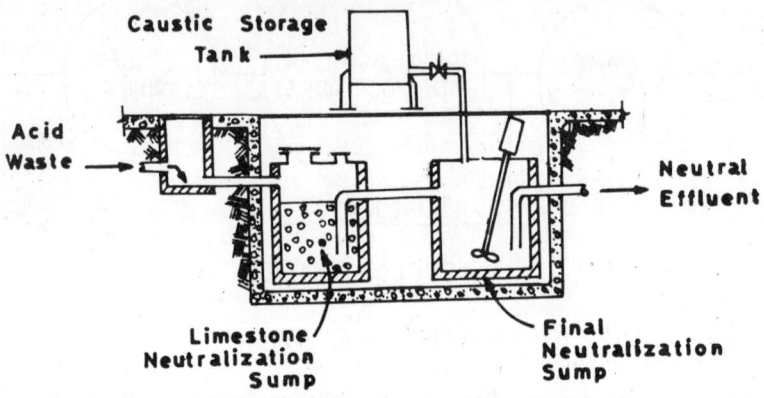

Acid Waste Neutralization System

ACISCULIS : Mason's pick with a square flat surface and a peen.

ACLINIC LINE : The line along which there is no dip of the magnetic needle. It is in the vicinity of the magnetic equator.

ACOUSTEXTILE : A material specially made for increasing the acoustic absorption.

ACOUSTICELOTEX : Thick celotex, made in form of tiles from sugar cane fibre, having perforations for increasing the acoustic absorption.

ACOUSTI-METER : Noise-meter; It helps to measure the degree of loudness of a specified sound on the phone scale.

ACOUST-TENTEST : Acoustic absorbant, made from maple wood fibre, having thick tentest mounted on edge.

ACOUSTICS : A science dealing with acoustical conditions including sound absorption and dissipation of noise.

ACOUSTIC ABSORPTION : The diminution of sound energy when it strikes on a surface of sound absorbing quality.

ACOUSTIC ABSORPTION FACTOR : The percentage of energy of sound waves which is not reflected from a plane uniform surface on the incidence of a sound wave at a specified angle.

ACOUSTIC CONCENTRATION : It is the ratio of the intensity of the directly radiated sound to the mean spherical intensity of sound for a particular distance from the radiating source of sound.

ACOUSTIC CONSTRUCTION : A construction aiming at reducing the sound entering or leaving a room i.e., to make it sound–proof.

ACOUSTIC DAZZLE : The psychological effect with high density of radiation of sound from small sources.

ACOUSTICAL FIBRE BOARD : Low-density fibre board having sound absorption quality.

ACOUSTIC PLASTER : A plaster having a high sound absorption quality. Aluminium dust may be used with a porous slag aggregate for acoustic plaster.

ACOUSTIC SHOCK : The temporary deafness due to a sudden large rush of current in a telephone receiver.

ACOUSTIC TILE : These tiles are light and often perforated. These are made of sound-absorbent materials, like cork, saw dust, straws and acoustic fibre boards.

ACRE : A unit of area : 1 acre = 10 square chains of 66 ft. chain length.
            = 4840 square yards.

ACRE-FOOT : Volume contained in 1 acre, 1 ft. deep. A unit to measure the capacity of a reservoir, which is equal to one cusec- day.

ACROTERIUM : A pedestal on a pediment over which a figure is to be built.

AC SYSTEM : Air-field soil classification system.

ACTIVATED-ALUMINA BED : To remove fluorides from wastes, the water is passed through a bed of activated-alumina. This has not been applied to public water supply.

ACTIVATED SLUDGE : Aerated sludge, which hastens the process of oxidation of the organic matters present in the sewage.

ACTIVATED-SLUDGE PLANT : A biological treatment plant for purification of sewage where activated sludge is mixed with raw sewage and aerated to hasten the process. BOD removal is about 90%.

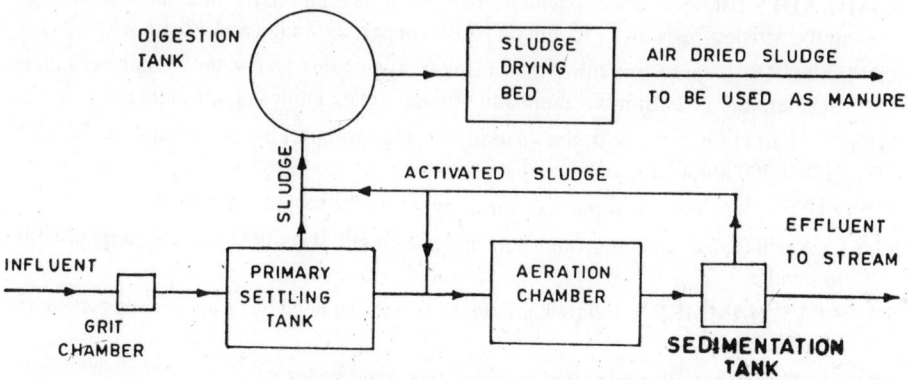

ACTIVE EARTH PRESSURE: The horizontal push from earth on to a wall.
  (1)   It acts in tension
  (2)   It helps in bulging of soil mass caused due to loading;
  (3)   It fails quickly along the shear plane.

ACTIVE LAYER: The surface layer of the ground which moves seasonally as the soil volume changes. It expands in wet weather and shrinks in summer. Foundation of a structure must always be laid below this layer.

ACTIVITY CO-EFFECIENT : In concentrated solutions, such as in sanitary engineering, the ratio of the activity to the molar concentration is called the activity co-efficient.

ADAM'S FLUSHING GULLEY : An automatic flushing tank which operates by siphon arrangement in the system. See illustration.

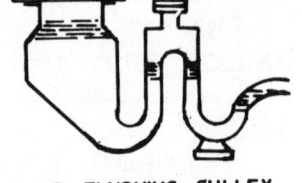

ADAM'S SEWAGE LIFT : An apparatus to lift sewage from a low level sewer to a high level sewer.

ADAMANT PLASTER : Gypsum plaster which hardens quickly.

ADAMS FLUSHING GULLEY

ADAMANTINE CLINKERS : Hard, dense and heavy kilnburnt special bricks, light yellow or pink-white in colour with smooth surface.

ADDITIVE CONSTANT : A length to be added to the product of the intercept on the staff and the multiplying constant, to get the true distance between telescope centre and staff.

ADDS : Overall quantities i.e., prior to deductions.

ADHESION : Bonding with the help of a binding material like cement, glue etc.

ADHESIVES : Cement, glue, bitumen, resin, etc.

ADJUSTING SCREW : A finely threaded screw fitted on a precision instrument to give it the final adjustment.

ADJUSTMENT : Balancing, elimination of defects and errors.

ADMIRALTY BRASS : Brass in which a part of zinc is replaced by tin to increase strength and corrosion-resistance. It contains 70% copper, 29% zinc and 1% tin.

ADMIXTURE : A subtance other than cement, aggregates and water to make concrete. Accelerator, air-entraining agent and retarder are examples of admixtures.

ADOBE : Earth reinforced with straws used to make sun-burnt blocks about 4" x 12" x 18" (100 x 300 x 450).

ADSORPTION : Condensation of a film of liquid on the surface of a solid.

ADZE : Axe like tool, used for rough surfacing of timber. Its cutting blade is perpendicular to handle.

ADZE EYE HAMMER : A claw hammer, in which the head has a socket supporting the handle firmly.

AEOLIAN SOIL : Soil transported by wind, e.g., sand dunes.

AERATED CONCRETE : Foamed calcium silicate and cellular concrete, easy to nail and saw; An insulating material.

AERATION OF WATER : Odorous gases and tastes that are present in water can be removed by the process of aeration.

AERATOR : Aerating device for the removal of volatile substances from water. There are three forms of aerators: Injection aerator, Gravity aerator, and fountain aerator.

AERATOR NOZZLES : These are provided to break up the spray into small droplets. There are varieties of nozzles of which only one is shown here.

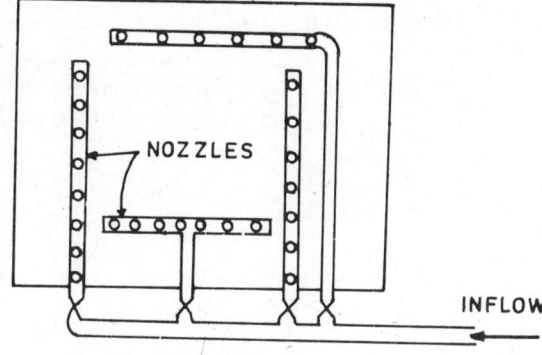

FOUNTAIN AERATOR

AERIAL ROPEWAY : Where a road, railway or bridge would be costly or slow to build, a line of towers carrying steel ropes which serve as tracks for buckets is constructed. These buckets carry materials and transport from one place to another.

AERIAL SURVEYING : Photogrammetry. Air survey. It was during the great world war of 1914-1918, that air photographs were first used for mapping purposes.

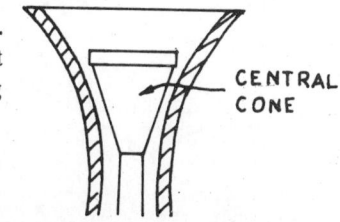

AEROBACTER AEROGENES : Non-faecal coli, usually, originate in soil and decaying vegetation. These belong to the coliform group of bacteria.

AEROBIC BACTERIA : The aerobic bacteria flourish and nourish in presence of air. With the aid of oxygen from air they produce nitrification of dead organic matters without creating any offensive odour.

AEROBIC DECOMPOSITION: Figure illustrates the cycles of Nitrogen, Carbon and sulphur in aerobic decomposition with the help of aerobes and atmospheric oxygen. The gases of decomposition go to the atmophere. Nitrogenous, Carbonaceous and Sulphurous compounds of dead organic matters are finally converted into nitrate nitrogen, Carbon dioxide, and sulphates with evolution of gases. Living plants flourish and nourish with the end products of the decomposition and the animal life consumes the plants.

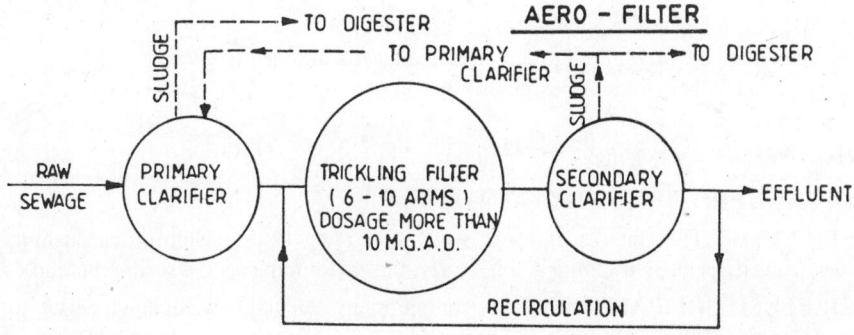

AEROFILTER : A type of high rate Trickling filter for sewage treatment as shown in illustration.

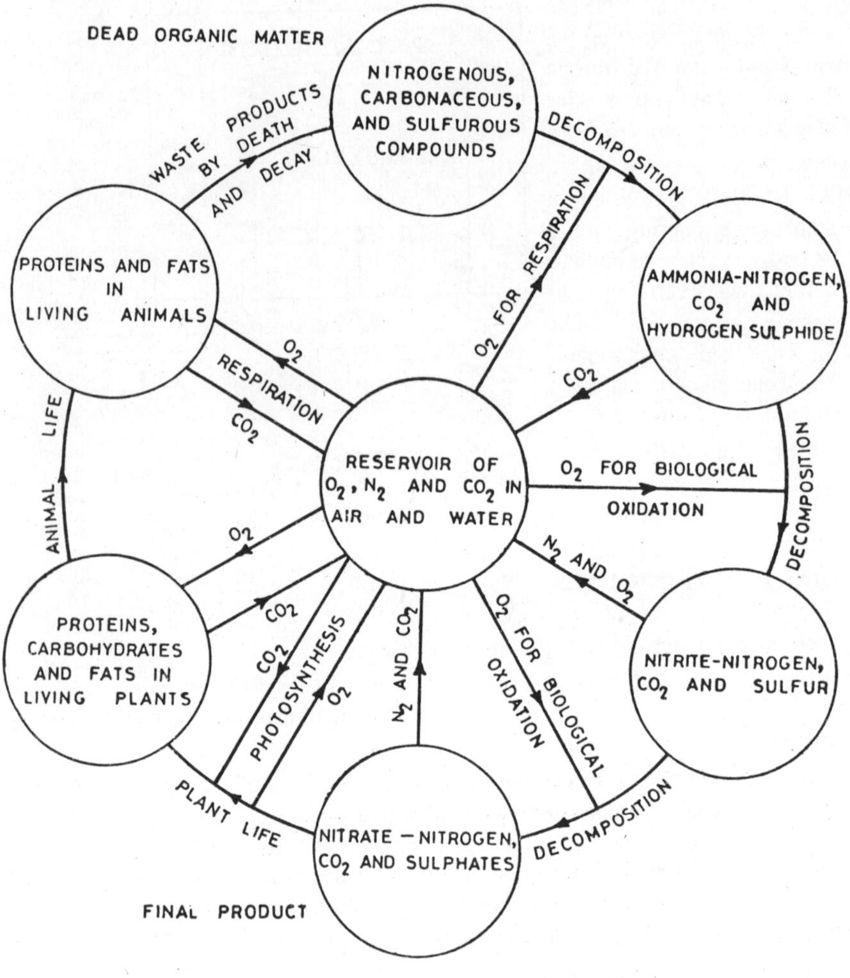

DEAD ORGANIC MATTER

NITROGENOUS, CARBONACEOUS, AND SULFUROUS COMPOUNDS

WASTE PRODUCTS BY DEATH AND DECAY

DECOMPOSITION

O_2 FOR RESPIRATION

PROTEINS AND FATS IN LIVING ANIMALS

AMMONIA-NITROGEN, CO_2 AND HYDROGEN SULPHIDE

RESPIRATION

O_2

CO_2

CO_2

RESERVOIR OF O_2, N_2 AND CO_2 IN AIR AND WATER

O_2 FOR BIOLOGICAL OXIDATION

ANIMAL LIFE

O_2

DECOMPOSITION

N_2 AND O_2

PROTEINS, CARBOHYDRATES AND FATS IN LIVING PLANTS

CO_2

CO_2

PHOTOSYNTHESIS

O_2

N_2 AND CO_2

O_2 FOR BIOLOGICAL OXIDATION

NITRITE-NITROGEN, CO_2 AND SULFUR

PLANT LIFE

NITRATE — NITROGEN, CO_2 AND SULPHATES

DECOMPOSITION

FINAL PRODUCT

AEROFOIL : The aerofoil is now used in meters for measurement of flow of a liquid. It is also required in the design of blades for turbines, centrifugal pumps and air-compressors.

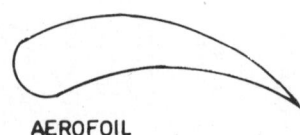

AEROFOIL

AEROGRAPH : A spray gun used for painting.

AEROPLANE PHOTOGRAPHY : Aerial Surveying.

AFTER-FLUSH : The small quantity of water remaining in the cistern after flushing water-closet pan. By trickling down slowly, this water remakes the seal in the trap.

AFTER-FLUSH COMPARTMENT : A compartment attached to wash down closet, to receive part of the flush-water and to rectify the loss of seal in the closet.

AFTER TACK : A defect of a paint film.

AGEING : Varnishes after manufacture are stored to improve their gloss and reduce crawling, pinholing, etc. This detention period is called ageing.

AGE OF TIDE : The time elapsed between the generation of spring tide and its arrival at the place. It is usually reckoned to the nearest one-fourth day and maximum of about three days.

AGGREGATE : An inert material which forms a substantial part of concrete; e.g., khowa, stone chips, gravel, cinder,slag,sand etc.

AGITATING TRUCK : A truck mixer.

AGLITE : A light-weight aggregate made from expanded clay.

AGONIC LINE : A line of zero declination.

AGRICULTURAL DRAIN : Field drain; Unglazed earthen ware or porous concrete pipes laid end to end with open joints so as to drain the subsoil water.

A-HORIZON : The uppermost of the three layers of soil science.

AIR BASE : In air-survey, it is the distance between the adjacent exposure stations.

AIR BINDING : It causes serious interference with water filtration through a filter bed. It results from negative head and from rising air bubbles escaping into the sand during back washing.

AIR BOTTLE : A vessel to collect air from high points of a hot water system.

AIR BRICK : Perforated earthen ware or cast iron brick built into a wall across an air- duct.

AIR BRUSH : A spray gun.

AIR CHANGE : Fresh air amounting the volume of a room being ventilated. Number of air-change required per hour for a room is the standard ventilation.

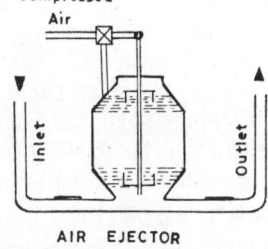

AIR EJECTOR

AIR COMPRESSOR : A machine for compressing air. Compressors are usually reciprocating or centrifugal type.

AIR CONDITIONING : The process of bringing atmospheric air to a suilable condition of temperature and humidity prior to admission of that air to a building.

AIR CONDUCTION : The flow of noise energy along an air path as contrasted with the conduction of vibrational energy.

AIR DRAIN : A cavity in the outer walls of a building to prevent the flow of damp to the interior.

AIR DUCTS : Pipes through which air is distributed throughout a building for the purpose of ventilation or air-conditioning.

AIR EJECTOR : Also known as Pneumatic ejector having non-clog characteristics. This type of sewage ejector is frequently used with economy when the quantum of sewage flow is within 1200 lpm. In the device, compressed air forces the sewage into the discharge main.

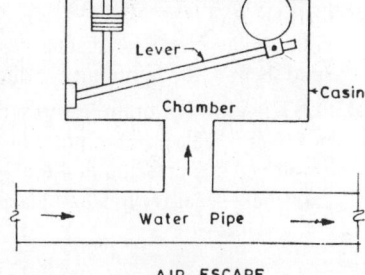

AIR ESCAPE

AIR ESCAPE : A contrivance, to discharge excess air from a water pipe, by opening the air valve.

AIR EMBOLISM: Caisson disease.

AIR-ENTRAINED CONCRETE: A Concrete having air-entraining portland cement or admixtures or air into the concrete. This concrete has high resistance to frost action and resistance to the wetting and drying salt solutions. This concrete is light in weight.

AIR ENTRAINING AGENT : To increase the frost resistance of hardened concrete and the workabillity of the wet concrete, resin (an admixture) is used as an air-entraining agent which when added to cement or concrete, entrains air in tiny bubbles.

AIRFIELD SOIL CLASSIFICATION : A classification of soil, based on sieve analysis and the consistency limits. This classification was introduced by casagrande in U.S.A. in 1948.

AIR-FLOW METER : A device (electric heater type) for measuring the flow of air through a passage by supplying the air with a known quantity of heat and measuring the rise in temperature.

AIR FLUE : A flue with an 'arnott valve' built into a chimney to remove vitiated air from a room.

AIR GRATING : A perforated iron plate built into a wall across an air duct to admit air into the building.

AIR GUN : A spray gun.

AIR LEVEL : A level tube.

AIR-LIFT PUMP : The most simple and full-proof type of pump suitable for drawing water from wells of any practicable size. There are two hanging pipes. Compressed air is injected into the larger pipe at its base from the smaller one and the pressure of water around the larger pipe forces more water into it.

AIR-LIFT BOOSTER : It is a pump or compressor inserted into a water near the discharge, so as to increase the pressure of water.

AIR LOCK : An arrangement of doors giving access to a caisson. The doors are operated by compressed air.

AIR PUMP : This pump extracts steam or air from a space so as to maintain it at a pressure below atmospheric.

AIR-RELEASE VALVE : See Air valve.

AIR SHAFT : Vertical air-passage for ventilation.

AIR SLAKING : When quick lime is exposed to air, it absorbs moisture from air and is gradually slaked into powder.

AIR SURVEYING : Aerial Surveying.

AIR TERMINATION NETWORK : A network which collects the electrical discharges from the air and send them to the lightning conductor.

AIR TEST : A kind of drain test to see whether the drain or sewer is leak-proof or not. For testing, U-gauge is employed at a pressure of 4 cm. A small U-gauge is extremely sensitive to small leaks.

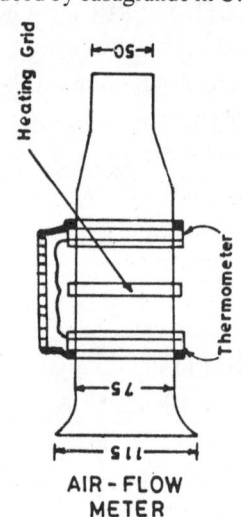

AIR - FLOW METER

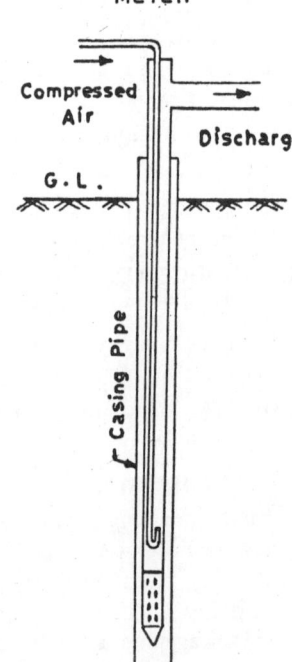

AIR LIFT PUMP

AIR-TO-AIR HEAT TRANSMISSION CO-EFFICIENT : The U-value

AIR-TO-AIR RESISTANCE : The reciprocal of air-to-air heat transmission Co-efficient. The resistasnce to the passage of heat offered by a wall.

AIR TRAP : A trap having water-seal to prevent foul air from rising from soil pipes, drains and sewers.

AIR VALVE : This is placed at every summit of mains to permit the escape of air collected at the summits of street mains. This operates automatically and it allows air only to escape and not water. This is also known as air-release or air-relief valve.

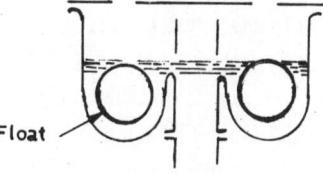

AIR WASHER: A chamber through which air passes into water sprays for air conditioning.

AIR RELIEF VALVE

AISLE : A division of a church parallel to and flanking the nave and chair. It is marked off from the walls by pillars. An aisle is found in gothic construction.

AISLE ROOF : Lean to roof; Pent or sloped roof; Pitched roof.

ALBUMINOID NITROGEN : Organic nitrogen that can be liberated by the action of alkaline permanganate upon the nitrogenous matter.

ALBURNUM : Sap wood.

ALCLAD : Aluminium coated alloy having high resistance to corrosion.

ALCLOVE : Arched or vaulted recess in a room.

ALDER : A hardwood of pale brown colour suitable for piling, joinery and for making plywood.

ALFOL : Aluminium foil used as heat insulating material.

ALGAE : A group of plants of simplest structure. The algae contain chlorophyll and often additional pigments which categorize their families and enable them to utillize radiant energy.

ALGAL-BACTERIAL SYMBIOSIS :
    See illustration.

ALGICIDES : Copper sulphate and chlorine used to kill algae.

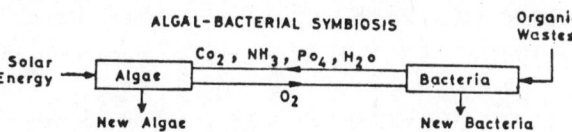

ALIDADE : Sight rule required in plane table surveying.

ALIGNMENT : Marking off points on the ground in correct line or direction for setting out a road, railway, transmission line, etc.

ALIZARIN PIGMENTS : Red, violet, blue and orange lakes obtained naturally from madder root or prepared synthetically from coal tar.

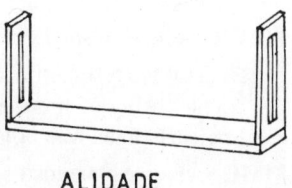

ALIDADE

ALKALINITY : The alkalinity of water is the capacity for neutralizing acids due to the presence of bicarbonate and carbonate ions.

ALKALI RESISTANCE: This is a special property of a few paints. They are durable in contact with lime, cement and soap. These paints are prepared from resins.

ALKID RESIN : A synthetic resin prepared from glycerin, with phthalic acid. This is weather-resistant and is used as binder for emulsion paint.

ALL-ROW LOCK WALL : A wall having two courses of stretchers alternating with one course of header on edge.

ALLAHABAD TILE : A special form of roofing tile.

ALLETTE : A buttress or pilaster; A wing of a building.

ALLIGATORING : Crocodiling; A defect of paint film.

ALLOY : A homogeneous mixture of two or more metals to get combined qualities.

ALLOY STEEL : Steel itself is an alloy. Alloy steel has elements that are not present in carbon steel and the percentage of manganese or silicon is more.

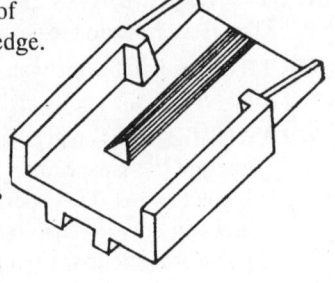

ALLAHABAD TILE

ALLOY SYSTEM : All the alloys that can be made by mixing two or more metals form an alloy system.

ALLOWANCE : Tollerance.

ALLUVIAL CLAY : Sediment of the clay grade transported by rivers from the place of its origin.

ALTAR : Pedestal, Platform; One of the steps in the stepped face of a dry-dock wall.

ALTAR TOMB : A monument or raised tomb.

ALTAZIMUTH : A large size special type of theodolite used for more precise work.

ALTITUDE : The height above sea-level; The angle of elevation of a point above the plane of horizon.

ALTITUDE LEVEL : A level fixed to the vernier arm of the vertical circle of a theodolite. This provides a horizontal datum from which altitudes may be measured.

ALUM : Aluminium Sulfate; $Al_2(SO_4)_3 \, 14H_2O$. Most commonly used chemical coagulant for coagulation process.

ALUMINIUM FOIL : Aluminium sheets thinner than 0.15 mm. which reflects visible light as well as infra-red (heat) rays. It is a good insulator.

ALUMINIUM PAINT : Corrosion–resistant paint made from aluminium powder.

ALUMINIUM POWDER : Flakes of aluminium are obtained by placing aluminium foil in a ball mill.

ALUMINOUS CEMENT : A rapid hardening cement containing 30-50% lime, 30-50% alumina, and 30-50% silica, iron oxide, etc. This is less susceptible than ordinary cement to low temperature during setting and to the action of acids and sea water.

AMBER : A fossil resin derived from extinct pine trees.

AMERICAN BOND : A bond in which every sixth course consists of headers and other courses being of stretchers. This bond can be laid very quickly. It is also known as English Garden wall bond.

AMMONAL : An explosive much used in quarries, tunnels and dry bore holes. It is a non-nitro-glycerine powder containig powdered aluminium, TNT and ammonium nitrate.

AMPHITHEATRE : A building of circular or oval shape in which spectators' seats surround the central open space (arena), where the spectacle will be presented.

AMPLITUDE : The depth or displacement of a wave from the level of calm water.

ANAEROBIC BACTERIA : All bacteria are strictly facultative anaerobes. Anaerobic bacteria do not require atmospheric air and light for their survival. Anaerobic decomposition of organic wastes takes place by the action of this type of bacteria.

ANAEROBIC DECOMPOSITION : The cycles of Nitrogen, Carbon and Sulphur in anaerobic decomposition are presented in illustration. The end products of decomposition are ammonia nitrogen, humus, carbon dioxide, methane gas and Sulphides. The gases of decomposition go to atmosphere and the end products are consumed by the living plants. The animals again consume the plants and thus the substances are reconstructed to proteins and fats. The dead plants and animals and their waste products constitute nitrogenous, carbonaceous and sulphurous matters which go on changing in anaerobic decomposition.

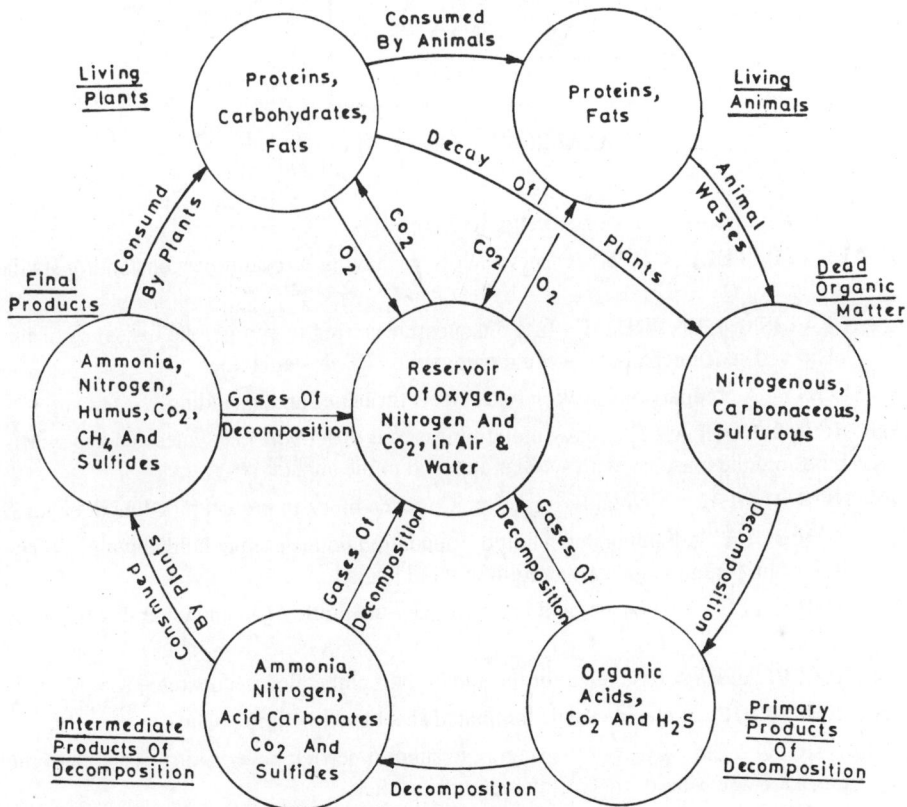

ANAEROBIC FILTER : A vertical filter bed filled with stone ballasts, gravel, cokes and plastic medium, suitable for treatment of various waste water at high loading rates and with a satisfactory treatment efficiency. But, it has the problem of clogging and short-circuiting.

ANALLATIC LENS : The additional lens in a telescope for internal focussing. This is placed between the object glass and eye piece which optically reduces the additive

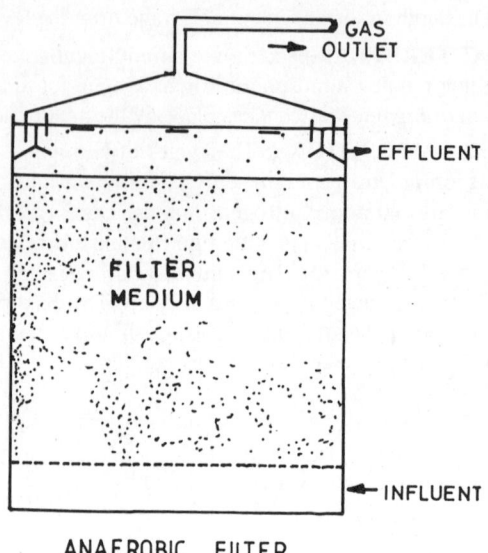

ANAEROBIC FILTER

constant to zero for a tacheometric telescope.

ANALLATIC TELESCOPE : A special telescope having zero additive constant for stadia work.

ANALLATISM, CENTRE OF : In a distance-measuring telescope, the distsance of the observed staff intercept between the upper and lower stadia lines.

ANALOGY : A comparison between two effects for better understanding.

ANCHOR & COLLAR : A heavy metal hinge used to support lock gates. The anchor is built into the masonry with a collar attached to the anchor.

ANCHOR BLOCK : A wood fixing brick; Concrete block to protect pipe line at bends.

ANCHOR BOLT : A holding down bolt or foundation bolt is used to hold down machines or building frames against vibration or wind loads.

ANCHOR GATE : A lock gate held in position by the pintle of an anchor and collar at its top.

ANCHOR PLATE : A base plate for the anchor bolt embedded in concrete.

ANCHOR STRIP : A thermoplastic laminated sheet used as lining plate.

ANCHOR TOWER : A part of the staging to support derrick tower gantry. The leg of the derrick crane is used for shifting materals on site.

ANCIENT LIGHTS : The legal right possessed by a class of windows by enjoyment of the flow of daylight for nineteen years.

ANCILLARY SHORING : The auxillary shoring (a temporary support) to relieve the wall to be underpinned. This consists of floor, roof and window strutting..

ANDIRON : Fire dog; A metal support for wood in an open fire.

ANGEL BEAM : A hammer beam, horizontal member of a mediaeval roof truss, decorated with angels curved on its exposed end.

ANGIOSPERMS : A group of flowering plants including deciduous trees which produce hardwood.

ANGLE BEAD : Corner bead; A small round moulding of hard plaster or wood or metal placed at the external angle of two walls to protect the corner from any accidental fracture.

ANGLE BLOCK : A small wooden block having a shape like a right angled triangle is fitted into the corner of a frame joint to make it stiffened and more rigid.

ANGLE BOARD : A gauge board used to plane a timber face to a definite angle.

ANGLE BOND : The bonding of brickwork at the corners of footings with special bricks or metal ties.

ANGLE BRACE : Angle tie; A bar fixed across the internal angle in a framework to stiffen it or to make it more rigid.

ANGLE BRACKET : A bracket projecting from the corner of a building and not at right angles to the face of the wall.

ANGLE CLEAT : Angle iron, to support a member in a structural framework.

ANGLE CLOSER : A specially shaped or cut brick to close the bond at the corner of a wall.

ANGLE DIVIDER : A bevel which bisects angles. Also used as a try square.

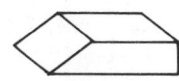

ANGLE CLOSER

ANGLE DOZER : A bulldozer with the mould board set at an angle to push earth sideways and ahead.

ANGLE-DRAFTED-MARGIN : A drafted-margin round the corner of a stone wall.

ANGLE FLOAT : A plasterer's tool to shape an internal corner of walls or any moulding.

ANGLE GAUGE : A tool to set off and test angles in Carpenter's and mason's work.

ANGLE IRON : See angle cleat. A rolled section having shape like L. The legs may be equal or unequal .

ANGLE JOINT : A joint between two pieces of timber at a corner at right angles or nearly to that.

ANGLE MULLION BRICK : A special brick. See illustration.

ANGLE OF DEPRESSION : In a surveying instrument, the vertical angle measured below the horizontal.

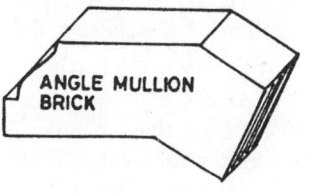

ANGLE MULLION BRICK

ANGLE OF DEVIATION : The angle by which a line or a ray is deviated from a standard line or ray.

ANGLE OF ELEVATION : The vertical angle measured above the horizontal.

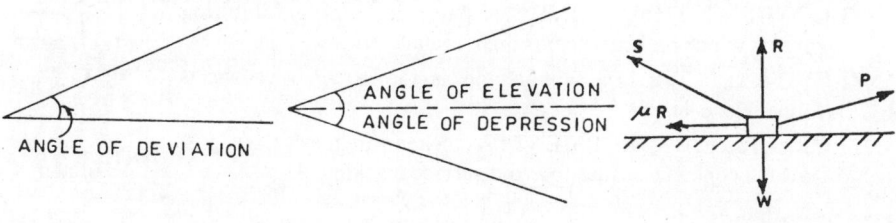

ANGLE OF DEVIATION

ANGLE OF ELEVATION
ANGLE OF DEPRESSION

ANGLE OF FRICTION: The angle between the resultant force and the perpendicular to the surface.

ANGLE OF REPOSE : Angle of rest; The steepest and stable slope to the horizontal maintained by granular material when heaped up on a platform.

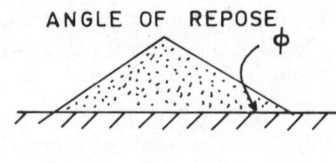

ANGLE OF SAWTOOTH : The angle between the face and the back of a saw tooth. It varies from 40° to 70°.

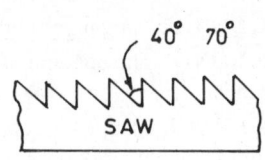

ANGLE OF SHEARING RESISTANCE : It is ϕ in coulomb's equation for cohesive soils. $S = C + P \tan \phi$.

ANGLE PLANE : A plane which shapes an internal angle.

ANGLE RAFTER : A rafter at the hip of a roof to receive jack rafters.

ANGLE RIDGE : Angle rafter.

ANGLE SECTION : See angle iron.

ANGLE SHAFT : An angle bead made of wood or pressed steel, sometimes enriched.

ANGLE STAFF : Angle shaft.

ANGLE SECTION

ANGLE STONE : A quoin stone used at the corner of a stone wall.

ANGLE TIE : A dragon tie (horizontal member) carrying one end of the Dragon beam and ties the wall plates together at a corner of a building.

ANGLE TILE : Arris tile for covering a hip or ridge.

ANGLE TROWEL : A margin trowel.

ANGLE VALVE : A valve used to change the direction of flow of a liquid.

ANIMAL BLACK : Black pigment prepared by calcining animal bones, ivory chippings, etc.

ANIMAL GLUE : Also known as scotch glue. It is the glue prepared by boiling animal bones, hides and skin, horns, and hooves. This glue has no resistance to water.

Angle
Valve

ANNEALING : This is a heat treatment given to a metal to produce machinability and cold working properties. In this process, the metal is heated to a suitable temperature for several hours after which it is gradually cooled.

ANNUAL VARIATION : The annual change in magnetic declination of a place which is more or less constant every year.

ANTI-ACTINIC GLASS : A heat-absorbing glass.

ANTI-CONDENSATION PAINT : A paint containing inhibiting pigments which prevents corrosion of metal surfaces.

ANTI-'D'-TRAP : This is a syphon trap as shown. Also known as 'Hopper Head Trap'.

ANTI-CRACK REINFORCEMENT : A chicken wire mesh placed just below the concrete surface to avoid surface cracking.

ANTI-'D' TRAP

ANTI-CREEPER : Anticreep anchors are provided to hold the rail bottom firmly with the sleeper with a view to preventing creep or advancement of rails.

ANTI FLOOD AND TIDAL VALVE : A check valve provided in a storm water drain below the high tide or flood level.

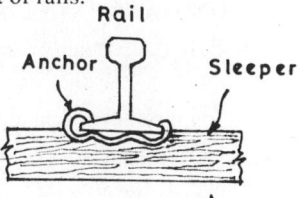

ANTICREEPER

ANTI-FOULING COMPOSITION : A special paint applied to ships' bottoms to reduce attachment of seaweed, mollascs, etc.

ANTILL'S TRAP : See illustration.

ANTIMONY OXIDE : A white pigment (Sb_2O_3) which produces a brilliant white paint with a faint pink undertone.

ANTI-NOISE PAINT : A paint that produces a rough surface like that of anti- condensation paint and has high sound absorption power.

ANTILL'S TRAP

ANTI-SAG BAR : A vertical tie bar used in a frame or structure with the objective to reduce sag or deflection.

ANTI-SIPHONAGE PIPE : A pipe that admits fresh air to the downstream side of the water seal for ventilation, the main function being prevention of sucking away of water from the water seal and passing down the soil pipe.

ANTI-SIPHON TRAP : A waste water trap whose water seal is maintained by increasing the volume of water in the trap i.e., the lower part of U is either enlarged or its depth is increased.

ANTI-SKINNING AGENT : With a view to preventing skin formation at the surface of a paint during its storage, anti-skinning agent like pine oil is added to the paint. Sometimes, the space above the paint in a container is filled with nitrogen or other inert gases which act as anti-skinning agent.

ANTI-SLIP PAINT : For finishing wooden decks and floors the surface is sometimes painted with anti-slip paint containing very fine sand, cork dusts, asbestos fibres, etc.,

APARTMENT : A dwelling unit for one family in a building.

APARTMENT HOUSE : A building having a number of apartments.

APEX STONE : The stone block forming the crown of a gable coping, a vault or a dome. Also known as 'saddle stone'.

APRON :   (i)   A panel below a window board and slightly projecting into the room.

(ii)   Vertical asphalt on a 'fascia'.

(iii)   A hard standing in an airport.

(iv)   A hard surface or pavement to the bed of a canal, stream or river to prevent scour as shown in illustration.

APRON EAVES PIECE : It is a T-shaped section which fixes the eave of the roofing sheet and acts as a flashing.

APRON FLASHING : A one-piece flashing as used at the' lower side of chimney penetrating a pitched roof.

APRON LINING : Also known as 'breast plate'.

APRON MOULDING : A moulding on a door lock rail.

APRON RAIL: A door lock rail.

APRON WALL : A spandrel in a building.

AQUEDUCT : A duct or conduit made of brick or stone masonry or concrete for conveying water over long distances.

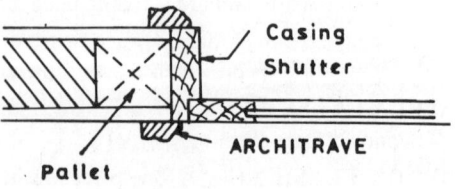

AQUEOUS ROCK : Rock that is formed in water. Usually, limestone and sandstone.

AQUICLUDE : An impermeable geological formation containing water but incapable of transmitting appreciable quantity of water.

AQUIFER : A reservoir of ground water.

AQUIFUGE : An impermeable geological formation, neither containing nor transmitting water, e.g., solid granite.

ARCADE : (i) A roofed passage with shops on both sides.

(ii) A series of arches carried on a colonnade.

ARCH : This may be considered as a beam curved in a vertical plane required to carry heavy load on a large span. Prior to introduction of steel and R.C.C. in construction works, arch construction was extensively used.

ARCH BAR : A rectangular steel bar used under a flat arch to hold the arch bricks or stones in place.

ARCH BRACE : (i) An inclined strut in a timber truss near the support.

(ii) An arch shaped (curved) timber in a wooden frame.

ARCH BRICK : An wedge shaped brick, called 'Voussoir', used for construction of an arch.

ARCH CENTERING : A timber frame required for construction of an arch. See illustration.

ARCH DAM : A dam held up by the horizontal thrust from the sides of the abutments. The arch used is curved in a horizontal plane.

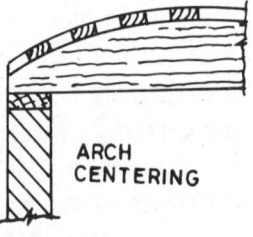

ARCHIMEDIAN SCREW : An ancient device to lift water by turning an inclined spirally threaded screw in a pipe, the lower end of the pipe being submerged in water. Screw pumps are made on this principle.

ARCH RIB : A local deepening in form of a rib of an arch used to act as a load bearing member.

ARCH RING : The load bearing ring of an arch.

ARCHITRAVE : An architrave is used to conceal joints between the door frames and the plaster at the door. It consists of two vertical and one horizontal members nailed to the plugs and edges of casings.

ARCHITRAVE BLOCK : A block at the base of an architrave into which the skirting board is fitted. Also known as 'Skirting Block' or 'Plinth Block.'

ARCHITRAVE JAMB : A moulding along the sides of a door or window opening i.e., along the jambs.

ARCH STONE : V-shaped stone, called 'Voussoir', used to build a stone arch.

ARCUATED CONSTRUCTION : An arched construction.

ARE : Metric unit of area,speaking of 100 sq. meters.100 ares make one hectare.

AREA : A space in any one plane.

ARM : A projected member; A lever arm.

ARMOURED PLYWOOD : Plymetal.

ARRIS : A sharp plastered edge; A sharp edge of a brick.

ARRIS FILLET : A tilting fillet.

ARRIS GUTTER : V-shaped yankee gutter made of wood.

ARRIS RAIL : Triangular fence rail fixed to fence post with the arris upwards.

ARRIS TILE : Angle tile.

ARRIS-WAYS : Sawing timber diagonally or laying bricks and tiles diagonally.

ARRISING TOOL : A float-like tool used in rounding off the edge of a concrete slab.

ARROW : A short piece of galvanised wire having one end pointed and other end bent is used in surveywork for marking points on ground.

ARTERIAL ROAD : A main road to which the tributary roads meet.

ARTESIAN WELL : A bore hole through which water comes out without pumping.

ARTIFICIAL CEMENTING : Strengthening loose soil by injection of certain binding materials.

ARTIFICIAL HARBOUR : A harbour formed by building breakwaters round a sea area.

ARTIFICIAL HORIZON : For measuring altitudes with the help of a 'sextant', a saucer of mercury is used to give a truly horizontal direction by reflection. This is artificial horizon.

ARTIFICIAL MARBLE : Marble made with gypsum plaster.

ARTIFICIAL STONE : High grade concrete resembling stone.

ASBESTOS : A mineral silicate comprising thin and tough fibrous crystals found in veins of rocks. This can withstand very high temperatures. This is used as a heat insulator.

ASBESTOS BLANKET : A small blanket of asbestos is usually wrapped round pipes being welded or brazed.

ASBESTOS-CEMENT : Cement mixed with asbestos fibres to make roof and wall cladding sheets.

ASBESTOS-DIATOMITE : Insulating, fire-resisting building boards made by mixing asbestos, diatomaceous earth and a binder.

ASBESTOS PLASTER : Heat insulating plaster for pipes.The plaster consists of asbestos-diatomite material.

ASBESTOS ROOFING : Plain, corrugated or patterned asbestos-cement sheets used for wall cladding or roofing.

ASBESTOS WOOD: Also known as asbestos wall board. Sheets with a higher proportion of asbestos used for lining walls.

ASH PIT : Ash pits are required to collect ashes from steam engines. These are built under the main railway tracks, just outside the platform of a station and near water column, such that a locomotive can dump ashes while taking water for its steam engine.

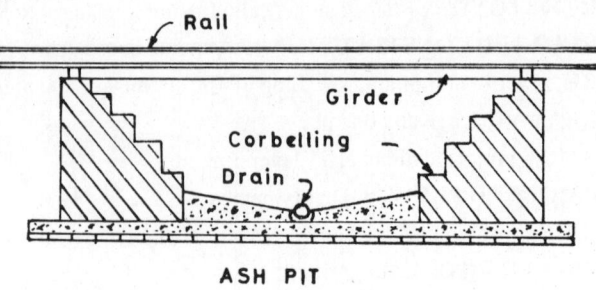

ASH PIT

ASHLAR : (i) A square-hewn stone.

(ii) Stone walls made with dressed stones.

ASHLAR FACING : Brick walls faced with stones that are rough tooled, rustic or chamfered. See illustration.

ASPHALT : Mineral hydrocarbons, black in colour, containing bituminous substances which are essentially products of petroleum decomposition. These are naturally found beneath the ground surface where petroleum deposits occur.

ASHLAR FACING

ASPHALT CEMENT : A kind of asphalt or bitumen used as a binding material.

ASPHALT ROOFING : Roofing with mastic asphalt by applying two to three coats.

ASPHALT SHINGLES : A term used for 'strip slates' in USA.

ASPHALTUM : A term used for asphalts derived from distillation of crude mineral oil.

ASPHALTIC CONCRETE : Road surfacing by using rolled asphalt.

ASTRONOMICAL EYE PIECE : The eye piece of a telescope designed to have the minimum loss of light.

ASTRONOMY : The observation and study of the sun, stars and planets and their movements. This includes precise measurements of latitude, longitude, time and geographical bearing.

ATOMIZATION : Breaking up of fluid paint, lacquer or varnish into very fine droplets as it comes out of a spray gun used for painting.This is effected by the airblast from the compressor.

ATTERBERG LIMITS : In1911, Atterberg proposed a series of tests for determining soil properties with the variation of moisture content in a clay sample. These are liquid limit,plastic limit and shrinkage limit. These are of empirical nature to some extent, but are very important to investigate the plastic characteristics of a soil.

AUDIOGRAPH : An acoustic instrument used for measuring the rate of sound absorption in a room.

AUGER : A corkscrew-like drilling tool used in boring wood and soil. Different types of augers with bits are used for different purposes.

AUTOCLAVE : A pressure vessel within which sand-lime bricks are cured at steam pressure and at high temperatures.

AUTOCLAVING : A unique method of curing green concrete, aerated concrete or sand-lime bricks by high pressure steam at 370^0F for 10 to 12 hours. The strength thus attained by concrete in one day is equivalent to the strength attained in one month by the normal curing method.

AUTOGENOUS HEALING : The disappearance of hair cracks and fissures in concrete by keeping it in damp condition.

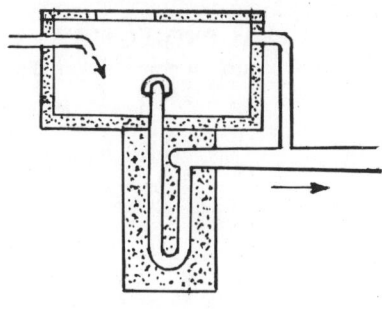

AUTOMATIC FLUSHING TANK : An automatic flushing tank is required to flush sewers laid at flat slopes, in which silt deposition is likely to occur. This is essentially a water tank with a trap. The tank gets filled with water intermittently. It works on syphonic principle and flushes the sewer at regular interval.

AUTOMATIC FLUSHING TANK

AUTOMATIC SYPHON SPILLWAY : It works on the syphonic principle and it discharges the excess flow in a combined sewer automatically to the relief sewer, when a certain level of sewage is reached in the former. For illustration, see Syphon spillway.

AUXILIARY TELESCOPE : It is used in mine survey for reading steep angles, when the line of sight of the ordinary telescope is interrupted by the instrument plate.

AVAILABLE WATER : Soil moisture available for plant growth ranging between field capacity and the wilting point.

AVENUE : A road which usually runs north-south with trees on either side.

AWL : A tool with a steel point used in carpentry for marking or scribing wood.

AXE :   (i) A tool for dressing timber roughly.
         (ii) A hammer used by masons.

AXED BRICK : A brick cut to required shape with axe.

AXED WORK : Stone facing with axe.

AXIAL FLOW PUMP : This pump is suitable for pumping a large quantum of storm water against a low head. For pumping raw sewage or sludge, this type of pump is not at all suitable. The pump consists of a multiple-bladed propeller placed in a casing with fixed guide vanes. It has no valve.

AXLE PULLEY : Also known as 'Sash pulley.' In a sash window, this is a wheel placed at the top of the window for carrying sash cords.

AXONOMETRIC PROJECTION : Axonometric projection is a special type of orthographic projection by which a three–dimensional drawing is prepared. The object is turned and tilted conveniently so that three faces of the object are displayed in one plane. Thus, the same object may be drawn at different angles. Isometric, dimetric and trimetric projections belong to axonometric projection.

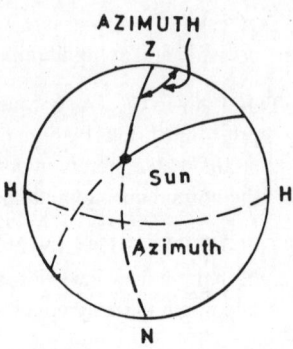

AZIMUTH : Horizontal bearing (clockwise) of a line measured from the meridian.

AZIMUTHAL PROJECTION : A map on which from the centre, all points are on their true beariings.

B

BACILLUS COLI : Sometimes found in drinking water. It is tougher than the disease-producing bacteria. It may cause bacterial infection.

BACK : (i) The extrados of an arch.

(ii) The opposite surface being the face.

(iii) The unexposed surface of a facing.

(iv) The top edge of a saw.

(v) The upper face of a slate.

BACKACTER OR DRAG SHOVEL OR TRENCH HOE : A face shovel which can dig down to a depth of 12 ft. below the tracks.

BACK BEARING : The bearing of a back station.
Back bearing = Fore bearing ± 180°

BACK BERM : Also known as Banquette; With a view to providing a minimum earth cover of 1m. above the saturation line, back berm is made at the rear side of an embankment as shown.

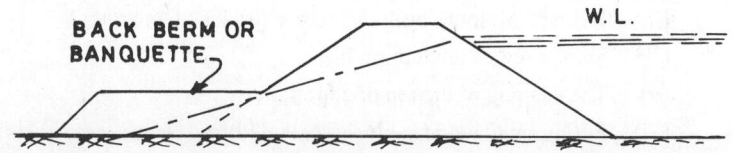

BACK BLOWING OR BACK WASHING : Reserved motion of water which is helpful to promote flow of sand particles which have clogged the passage. It is the process of cleaning a choked strainer by forcing air or water under pressure.

BACK BOILER : A boiler fitted at the back of a hearth for supplying hot water in a small house and for space heating in bedrooms.

BACK BOXING : See Back Lining.

BACK CUTTING : Additional excavation to be made to make up an embankment when it is found that excavation made elsewhere in cutting is insufficent.

BACK DROP OR DROP MANHOLE : Used to make connection from a high level sewer at a lower level. See illustration.

BACK EDGING : Cutting off a length of glazed pipe (made of earthen ware or other ceramic) by first chipping all round the pipe and then by chipping the lower portion of the pipe.

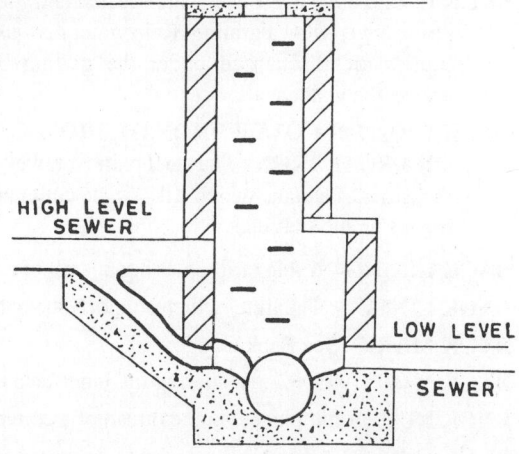

BACK DROP OR DROP MANHOLE

BACK FILL OR BACK FILLING: Earth, cinder, moorum, rubbish or stone chips used to fill the foundation trench after the foundation have been laid.

BACK FLAP OR BACK FOLD OR BACK SHUTTER : The part of a boxing shutter which can not be seen when the shutter is folded away.

BACK FLAP HINGE : A door hinge which looks like a butt hinge. It is used when the shutter is too thin to be carried by a butt.

BACK GAUGE OR BACK MARK : The distance from the back edge of a rolled section to the centre line of the rivet or bolt hole.

BACKGROUND OR BACKING : The surface on which the first coat of plaster is to be applied.

BACKGROUND HEATING : The opposite of full central heating. It is intended for room heating in cold weather.

BACK GUTTER OR CHIMNEY GUTTER : A gutter provided between a sloping roof and the uphill side of a chimney.

BACKING : (i)   The surface of a wall on which facing bricks or stones are set.

(ii)   A backup strip.

(iii) Coursed masonry built over the extrados of an arch.

(iv) Stone used as random rubble.

(v)   The shaping of the top of a hip rafter.

(vi) Fitting small wooden blocks, called furring pieces, onto joists to make a levelled floor board.

BACKING COAT : Rendering coat i.e., first coat of plaster; also called pricking-up coat when it is done on laths.

BACKINGS : Furring strips on beams or joists.

BACKING UP : Using inferior bricks for backing.

BACK INLET GULLEY : A water-sealed entry to a drain covered by a grating. Rain water or waste liquids are discharged under the grating but above the water seal.

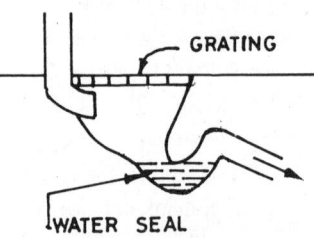

BACK INLET

BACK IRON OR COVER IRON OR IRON CAP OR BREAK IRON : The steel plate screwed to the cutter of a plane which stiffens the cutter and breaks up the shavings.

BACK LINING : A thin strip covering a jamb of a sash window frame.

BACK LINTEL : The lintel which supports the backings of a wall.

BACK MARK : See Back Gauge.

BACK MORTARING : Rendering the inner face of a connector to make tight joints.

BACK NUT : A nut placed on the thread of a connector to make tight joints.

BACK OBSERVATION : Back sight i.e., an observation taken towards the back station. This refers to readings of a levelling staff.

BACK PLASTERING : See Back Mortaring. It means rendering or plastering the part of a wall which is not seen.

BACK PROP : A raking strut used in timbering a deep trench. It helps to transfer the weight of timbering to the ground.

BACK PUTTY : Bed Putty.

BACK SAW : A saw stiffened by a folded plate at its back; e.g., a tenon saw.

BACK SAWING : Flat sawing of timber enabling the heart to be boxed.

BACK SHORE : Also known as 'jack shore'; A support to the foot of the rider shore. An outer support in a raking shore.

BACK SHUTTER : A back flap.

BACK SIGHT : See 'back observation'.

BACK UP : The backing bricks or stones of a wall.

BACKING UP : A lathing board or backing ; A narrow strip of wood fixed to a partition on which lathing is to be nailed.

BACK VENT : A fresh air inlet.

BACK WASHING : See 'back blowing'. This is required when a filter bed is choked. It is done by injecting water under pressure through the bottom of the filter bed.

BACK WATER : Water held back by an obstuction in its flow. This causes water hammering.

BACK WATER CURVE : It is the surface curve of a flowing water or sewage when backed up by an obstruction either by a dam or by a submerged flow condition. Due to the effect of back water, the depth of flow becomes increased at the upstream point.

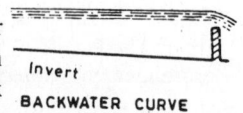

Invert

BACKWATER CURVE

BACK WATER VALVE : Essentially, a non-return valve, sometimes provided in small diameter sewers with a view to preventing the back flow from larger sewers. See illustration.

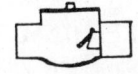

BACKWATER VALVE

BACTERIA : Unicellular micro-organisms, either aerobic or anaerobic in Saprophytic group, live on dead orgnic matters and the pathogenic group flourish and nourish in living bodies at a body temperature of 37°C.

BADGER : A tool used in bringing out the jointing mortar from inside of a drain.

BAFFLE : A small hanging plate or a thin partition placed in a chamber to deflect the flow of a liquid. Deflector vanes and guide walls are placed in flowing water or sewage for regulating velocity of flow, absorbing energy, diverting or guiding the flow path, agitating the liquid and checking the eddy currents.

BAFFLE PIER : A groyne used in diverting the flow of a stream.

BAG PLUG : An inflatable plug used in plumbing work to block the lower end of a drain length.

BAG WORK : A kind of revetment work to protect the river banks from erosion. Dry concrete or gravels sewn in bags are tamped against the bank and held together by dowel rods driven through the bags.

BAILER: (i) A 6 to 12 m. long pipe with a foot valve which is introduced through an oil-well casing for raising oil containing sand.

　　　　(ii) A sand pump.

BALANCE BAR: A large wooden beam projecting from a lock gate. When the water level is same on either side of the gate, this beam is pushed for opening the gate.

BALANCE BOX : A loaded box at the remote end of a crane from the jib for counter balancing the load.

BALANCE BRIDGE : A bascule bridge.

BALANCED CONSTRUCTION : A term given to plywood which has the same thickness of wood in both directions with odd number of plies.

BALANCED EARTHWORK : An earthwork in which the excavation equals the fills i.e., cuts and fills are same and no excess earth is left over.

BALANCED FLUE : An arrangement of air intake and flue outlet in gas fired boilers.

BALANCED SECTION : The section of a member of a structure or a frame which is designed such that its critical natural axis coincides with the centroidal axis.

BALANCED STEP : Also known as 'dancing step'. This is a winder having its narrow end narrower than the parallel treads.

BALANCE POINT : The point of intersection between a Mass-Haul curve and the datum line speaking of balanced cuts and fills.

BALK : Earth between excavations.

BALLAST : Broken stones used to form a bed for railway sleeper in a permanent way.

BALL CATCH : Also known as 'Bales Catch' or 'Bullet Catch'. An automatic door closing arrangement with a spring loaded ball.

BALL COCK : A floating copper or PVC ball used in float valves for opening or closing a flow.

BALLOON FRAMING : A typical timber house construction used in U.S.A.

BALL-PEEN HAMMER : A hammer with a hemispherical peen.

BALL TYPE JOINT : A pipe joint. See illustration.

BALSA WOOD : A hardwood which can not be nailed or screwed, but can easily be glued. It is used as a light-weight insulating coreboard.

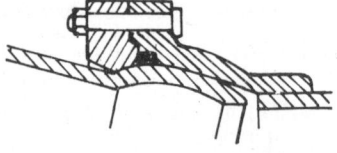
BALL TYPE JOINT

BALUSTER : A post in a balustrade of flight of stairs. Wooden balusters are made ornamental by turning in a lathe machine. These are 75 to 100 mm. in diameter.

Baluster

BALUSTRADE : Collective name given to the whole infilling from a handrail to the edge of a stair or a bridge.

BANDING : Thin strips to cover the edges of the corewood in veneering. Use of sheets for shutting stile of a flush door.

BAND CHAIN : A steel or invar tape of 100 ft. length, used to make accurate measurements. Also known as 'Steel band'.

BANDEROLLE : A ranging rod or pole used in surveying.

BAND SAW : A power-driven endless saw comprising a steel belt with teeth at one edge, running on two pulleys at two ends. This saw is used to cut timber for conversion or for cutting intricate shapes.

BAND SCREEN : An endless band or a moving screen of wire mesh for removal of solids at the point of water intake by pumping surface water.

BANISTER : A name for baluster in scotland.

BANK : An embankment for a waterbody, a road or a railway.

BANKER : (i) A mason's workbench, usually made of timber.
(ii) A board or platform on which mortar or concrete is mixed manually.

BANKING : Super elevation or cant provided on a road or railway curve.

BANK PROTECTION : Protectionary measures adopted for river or sea embankments against scour by use of revetments, groynes, mattresses, turfing with sods, etc.

BANK STORAGE : Water absorbed by the banks of a river or stream during high water level and returned to the river or stream at low water level.

BANQUETTE : (i) A back berm.
(ii) A foot bridge above road level.

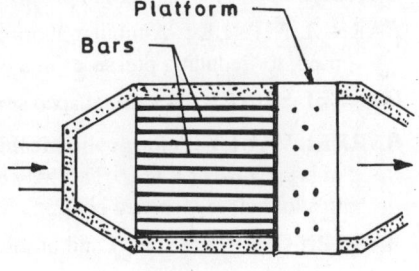

BAR : (i) A round, square or rectangular steel bar used in construction.
(ii) A glazing bar.

BARBED DOWEL PIN : A barbed steel pin used in mortise and tenon joints in woodwork.

BAR BENDING : An item of work in bending mild steel bars for reinforcement in cement concrete.

BAR DOOR : A batten timber door.

BAR SCREEN : A screen made of bars. Also known as 'Bar Rack' used in sewage pumping station to arrest floating matters and solid particles brought with the sewage flow. This is cleaned either manually or mechanically.

BAR SCREEN

BAREFACE TENON : A tenon with a shoulder on one side.

BARGE : A water vessel.

BARGE BED : A bed of mud near the edge of a river bank where a barge can moor and rest on the mud at low water level. See illustration.

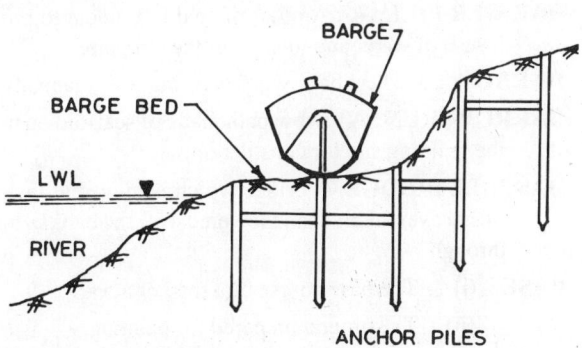

BARGE BOARD: A sloping board (usually made of timber) placed along a gable to cover the ends of roof timbers and to protect the barge courses from rain.

BARGE COUPLE : A pair of rafters used at the end of a roof, oversailing the gable.

BARGE COURSE : Also known as 'verge course'.

    (i)   A course of bricks-on-edge laid across a wall at its top to serve as a coping.

    (ii)   Slightly overhanging tiles next to the gable.

    (iii) A slightly overhanging coping on a gable wall.

BARIUM PLASTER : A plaster containing barytes aggregate with portland cement or gypsum plaster as binding material, used in plastering the inside walls of X-ray rooms with an objective to reduce radiation through the walls.

BARN DOOR HANGER : The hanging arrangement of a barn door from a set of overhead pulleys tarvelling on a rail hung from the lintel. This is required for closing or opening heavy doors.

BARNE'S FORMULA : A formula for finding out velocity of flow in slimy sewers.

    Velocity, $v = 107\, m^{0.7}\, i^{0.5}$

    where m = hydraulic mean depth, ft. and i = slope

BAROMETER : An instrument to measure atmospheric pressure.

BAROMETRIC PRESSURE : Atmospheric pressure. Air pressure measured by an aneroid barometer is used by surveyors to calculate the altitude of a place.

BARRAGE : A low height dam with sluice gates, constructed across a river to raise the water level for irrigation or navigation.

BARREL BOLT : A cylindrical bolt running in a case, used in doors and windows.

BARREL LIGHT : A sky light of curved shape built on roof.

BARREL NIPPLE : A tubular short piece of pipe with external threads at either end, used for reducing pressure in a pipeline.

BARREL SEWER : A vault-shaped sewer as shown in illustration.

BARREL VAULT : A long cylindrical vault of semi-circular arch made of brickwork or R.C.C. This may be used for making factory roofs to avoid closely spaced columns.

BARREN GROUND : (i) Land unsuitable for cultivation.

                 (ii) In mining, rock or soil met without coal or any other valuable mineral.   **BARREL SEWER**

BARRIER PILLAR : A pillar of coal left uncut to protect mine from inrush of water and to support the structure.

BARROW : A wheel barrow for carrying load manually by push or pull.

BARROW RUN : A narrow path made of scaffold boards for wheeling loaded barrows to the building site for constuction.

BASCULE BRIDGE: A bridge which is hinged at the bank of a river such that it can be raised over the river and lowered over the bank behind the hinge to allow ships to pass through.

BASE : (i)   The base course of a road or a wall

        (ii)   The ground prepared for painting.

        (iii) Widening the foot of a column.

(iv)   The part of a floor or stair below the granolithic finish.

(v)   The main ingredient (pigment) of a paint.

BASE COURSE : (i) The surfacing layer below the wearing course of a road.

(ii) The bottommost course of a masonry wall that is visible.

BASE EXCHANGE : A reversible chemical process.

(i)   In soil mechanics, a hydrogen clay having absorbed hydrogen ions can be converted into a sodium clay with absorbed sodium ions, which decreases the permeability of clay.

(ii)   In water treatment, water is softened by base-exchange. Hard water is passed through an effective zeolite bed which absorbs the salts present in water and responsible for hardness. When the zeolite bed becomes ineffective, it is regenerated by washing with brine solution.

BASE FLOW : The portion of stream flow coming from ground water discharge, which is estimated for the purpose of segregating surface runoff from the total flow of a stream.

BASE LINE : The starting line for calculations in a triangulation survey. The base line forms one side of a triangle or several triangles.

BASEMENT : A storey whose major portion is below ground level This may be used as a storage place, garage or a living room.

BASE MOULDING : The moulding at the base of a column or wall immediately above the plinth level.

BASE PLATE : (i) The plate on which a machine rests. This holds the machine and distributes load over a larger area.

(ii) The plate carrying the lower ends of the three foot screws whose upper ends are fixed to the tribrach of a theodolite. In setting up a theodolite, the base plate is screwed to the tripod head.

BASE SHOE : A quarter-round bead used at the junction of a floor board and skirting board to cover the joint.

BASE TRIM : An ornamental moulding for decoration at the base of a column or a wall.

BASIC LEAD SULPHATE : A white pigment prepared by burning galena (Pb S), the properties of the pigment being those of white lead.

BASIC REFRACTOY : A refractory lining material containing lime,magnesia or calcined dolomite with low silica content, used for lining a metallurgical furnance.

BASIC STEEL : The steel manufactured by the basic process in a furnace lined with the basic refractory. The lining combines with the phosphorus present in the metal and it is floated off in the slag.

BASTARD ASHLAR : Facing stones of a rubble wall which are dressed like ashlar.

BASTARD SAWN : Plain sawn timber.

BASTARD TUCK POINTING : A pointing work with projection from the face of the brick wall.

BAT : (i) A brick cut across its length, e.g., three-fourth bat, half-bat and quarter-bat.

(ii) A lead wedge.

BATCH: One measured box of mixed concrete or mortar.

BATCH BOX: A gauge box for measuring dry ingredients for proportioning a concrete mix.

BATCHING PLANT : A mechanical equipment for measuring different ingredients by weight or by volume.

BATCH MIXER : A concrete mixer for mixing concrete batches.

BAT FAGGOT : A fascine of 0.75 m girth and about 1.5 m long made of hazel or oak sticks.

BATTED SURFACE : An ashlar surface vertically incised with a batting tool.

BATTEN : (i) A square-sawn timber piece.

(ii) Horizontal strips 25 mm x 50 mm used for fixing slates and tiles.

BATTEN BOARD : A core board made of a core of strips glued together between the outer veneers. Each srip is less than 75 mm wide.

BATTEN DOOR : A door made by fixing vertical boards over three horizontal ledges and two diagonals, without any frame round the edges. Also known as ledged door.

BATTENING : A base for lathing walls.

BATTEN PLATE : In framing a built-up stanchion with channels, joists and angle sections, a horizontal square or rectangular plate is connected across the pairs of sections.

BATTER : An inclination or slope, expressed as one horizontal to so many verticals.

BATTER BOARD : Profile board. Horizontal boards fixed at each corner at a datum level outside the foundation trench for a building. The level of each floor and depth of foundation are measured from this level. Nails on the top edges of the boards are tied with stretched strings to indicate the line of excavation, building line, etc.

BATTER BRACE : A diagonal brace used in a truss.

BATTER LEVEL : A clinometer used to measure the slope of earth cutting and filling.

BATTER PEG : Pegs driven into the ground for setting out limitations of an earth slope.

BATTER PILE : A raking pile i.e., a pile driven at an angle to the vertical.

BAULK : A squared log of timber, hewn or sawn.

BAUXITE : An important ore of aluminium (Al_2O_3 $2H_2O$), which melts at $1600^{\circ}C$.

BAY : Uniform spacing of columns, beams and roof trusses.

BAYLIS TURBIDIMETER : See illustration. Out of the two glass tubes, one is filled with turbid water whose turbidity is to be measured and the other is filled with a standard solution of known turbidity. The bulb of 250 watts is lighted and blue light from both the tubes is observed from top. If the colour of both the tubes do not match, the procedure is repeated with another solution of known turbidity and it should be continued until both the colours are found same. This instrument can measure turbidity upto 5 ppm only.

BAY WINDOW

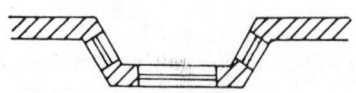

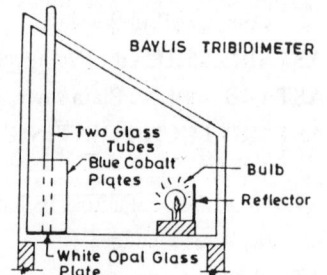

BAYLIS TRIBIDIMETER

Two Glass Tubes
Blue Cobalt Plqtes
Bulb
Reflector
White Opal Glass Plate

BAY WINDOW : A window placed in a projection of a wall which is beyond the general line of the wall, but has foundation.

BAZIN'S FORMULA : Used in Computation of flood flow.

$$C = \frac{87}{1 + M/\sqrt{R}} \quad ....\text{in Metric Units}$$

$$C = \frac{157.6}{1 + M/\sqrt{R}} \quad ....\text{in Imperial Units}$$

BEACHING : Loose stones of 75mm to 225 mm in size used in a layer of about 0.5 m to 0.7 m thickness for revetting embankments or reservoirs below the pitching level to resist against erosion.

BEAD : (i)   A strip of timber used in holding glasses in glazing.

(ii)   A mouiding used to cover a joint.

(iii)   A strip of metal or wood used in filling a narrow gap.

BEAD AND BATTEN : A framed partition wall made of battens with a bead along the edge.

BEAD AND QUIRK : A bead separated by a quirk, a very narrow groove.

BEAD AND BUTT : A thick door panelling whose one face is flush with the frame without any bead. The flushing face is decorated with mouldings such that it butts against the rails of the frame. The other face of the panel is recessed.

BEADED SECTION : An angle section, Tee section or channel section made of light alloy, whose one of the extreme ends is provided with a bead or bulb.

BEADING : Forming a bead in a member.

BEAD PLANE : A plane with a narrow cutter for making grooves and beads on wood.

BEAM : A structural member, usually horizontal, to support a floor or roof slab. The load on a slab is carried by the beam. A beam is usually made of Timber, steel and reinforced concrete.

BEAMAN STADIA ARC : A direct reading Tacheometer used in surveying.

BEAM BENDER : A machine used for bending or straightening R.S.J.

BEAM COMPASS : Used in drawing arcs of large radii which is not possible even with an adjustable compass. It is essentially a light wooden beam with adjustable heads for fixing a pin at one end and a pen at the other.

BEAM ENGINE : A steam engine with vertical cylinder used in early days.

BEAM FILLING : Brick nogging or masonry at the end supports of R.S.J. (Beam) to hold it firmly.

BEAM HANGER : A hanger in form of a stirrup strap to hold a beam.

BEAM TEST : A test to determine the 'modulus of rupture' of a test specimen concrete beam under tensile test in a Universal testing machine. The load applied and bending moment are recorded from which the maximum tensile stress in bending is computed.

BEARER : A horizontal timber or a rolled steel joist to support other timbers (burgahs) on which the roofing tiles rest in making a flat roof.

BEARING : (i)   The horizontal angle of a line made with the true North in surveying.

(ii)   The beam–supporting area at the end of a beam.

BEARING BAR: A wrought iron bar instead of a wall plate placed on a brickwork to support the beams or joists at level.

BEARING CAPACITY: The bearing power or stress of a material to withstand a load without causing any sinkage to the member from which the load is coming over it. It is given by the load per unit area.

BEARING PILE : A pile that carries load and transfers to soil either through its end or by skin friction.

BEARING PLATE : A plate over a brickwork to support a beam or joist by distributing the load over a larger area.

BEARING PRESSURE: Bearing load divided by the bearing area. Also, known as 'bearing power' or 'bearing capacity'.

BEARING STRESS : Bearing pressure or power.

BEARING WALL : A wall that bears a load.

BEARING TEST : A test, usually the load bearing test conducted to determine the carrying capacity of a soil.

BEAUFORT SCALE : A scale to measure wind velocity, which reads 0 for calm air to 12 for a hurricane (over 75 mps). The wind velocity in mph is given by $V = 1.875 \ B \ \sqrt{B}$, where $\sqrt{B}$ is the Beaufort number.

BED : The under surface of a brick, tile, stone or any other material.

BED BLOCK : A block of hard stone or concrete provided under a beam or truss end to distribute the load.

BEDDING : A layer of material acting as a bed of something to rest on it.

BEDDING STONE : A plain marble slab used by masons to check the surfaces of stones to be laid.

BED DOWEL : A dowel placed at the centre of the bed of a stone to hold it.

BED JOINT : The horizontal joint in a masonry.

BED LOAD : The quantum of sand, silt and gravel rolled in unit time along the bed of a flowing stream or river.

BED PLATE : (i)   'Sole plate' or 'Base plate' of a machine to hold it firmly to the concrete foundation by anchor bolts.

(ii)   A cast iron plate under a joist to support it and distribute the load to the masonry wall.

BED PUTTY : Also known as 'Back putty'. Glazier's putty on which a glass is bedded.

BED ROCK : A hard stratum on which a heavy foundation rests.

BED STONE : A bed block.

BEECH : A hardwood found in Europe and Asia Minor, which presents spindle–shaped markings on the surface of the flat-sawn boards.

BEESWAX : The wax obtained from honey bees, used in matt varnish and wood polish.

BEETLE : Maul or Mallet of heavy type used for driving pegs.

BEETLE HEAD : A drop hammer.

BEL : A unit of sound intensity which is equal to 10 decibels.

BELFAST TRUSS : A bow string girder made of timber, which is curved at top, the string being a horizontal tie as shown. This truss is suitable for a span upto a maximum of 15 m.

BELGIAN TRUSS : A 'Fink Truss'.

BELL AND SPIGOT : A spigot and socket joint used in cast iron pipes as shown in illustration.

BELL CAST EAVES : A term used in Scotland for Eaves with cocking pieces.

BEL DOLPHIN : A large bell-shaped concrete or steel fender supported by a group of piles, for mooring of vessels at sea. Also, known as 'Baker Bell Dolphin'.

BELLMOUTH : A bell-shaped mount provided at the suction end of a pipe for pumping out water or sewage.

BELLMOUTH OVERFLOW : An overflow from a reservior through a bellmouth provided at overflow level in a tower built up from the reservior bed.

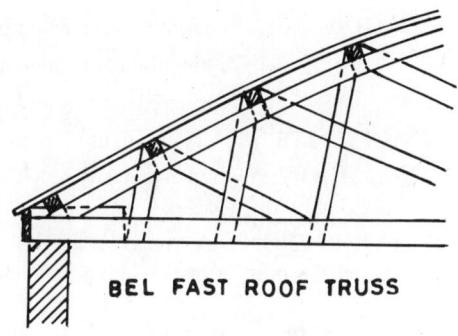

BEL FAST ROOF TRUSS

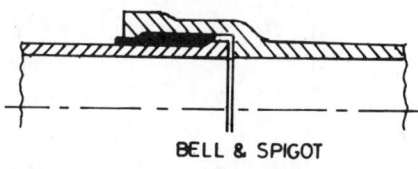

BELL & SPIGOT

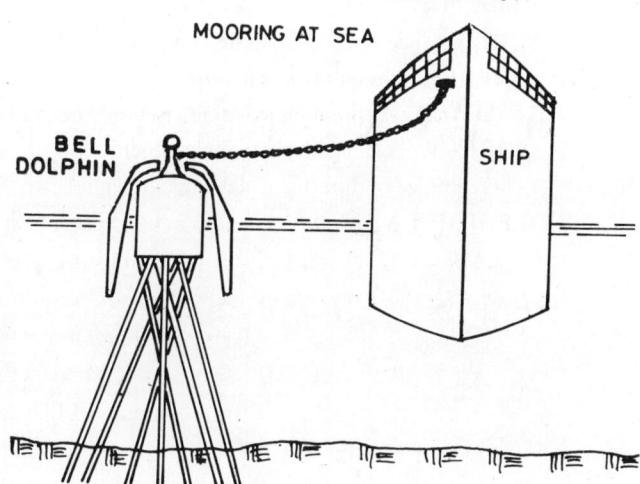

MOORING AT SEA

BELL DOLPHIN

SHIP

BELL TRAP : This trap has shallow water seal of 10 mm only and it has a tendency of getting choked. Nowadays this trap is not used.

BELL TRAP

BELLY ROD : A camber rod below a trussed beam.

BELT CONVEYER : An endless moving belt of 0.5 m to 1.5 m width placed at an angle with the horizontal for carrying loose materials in a factory. This saves manpower, time and cost.

BELT COURSE : A string course used in a building.

BENCH : (i) A long horizontal step

(ii) A berm

(iii) Worker's bench

BENCHED FOUNDATION: A stepped foundation on a sloped bearing stratum, usually required in a hill area.

BENCHING: (i) A berm between an embankment and the borrowpit.

(ii) A stepping made in quarrying minerals.

(iii) A concrete cast round a semi-circular channel in a manhole.

BENCH HOLDFAST : A cramp used to hold wood to the working bench.

BENCH HOOK : A wooden piece fitted to the working bench to protect the bench top from striking by the tools.

BENCHING IRON : A triangular steel plate with pins at corners at right angles to the plate. The pins are driven into ground at change of level or for fixing temporary bench mark.

BENCH KNIFE : A knife blade kept projected from the bench surface for steadying the timber end, while working.

BENCH MARK : A fixed point having known level with reference to the mean sea level. This point is used as a datum or reference point in level survey.

BENCH PLANE : A plane commonly used for smoothing flat pieces of timber on a bench.

BENCH SANDER : A sanding machine for finishing a wood surface with glass paper or cloth fitted to the machine.

BENCH SCREW : The threaded spindle of a bench vice.

BENCH STOP : Bench hook or Bench knife.

BENCH TRIMMER : A trimming machine used in woodwork.

BENCH WORK : Work done manually on a bench.

BEND : A curved piece of pipe, the usual angle being 90°.

BENDING FORMULA : The formula used in bending of beams.

$f/Y = M/I = E/R$ Where, f = Flexure stress allowable

or, $M = fz$ y = Depth of Neutral Axis

M = Bending Moment

I = Second Moment of Area

E = Modulus of Elasticity

R = Radius of curvature

Z = Section Modulus

= I/Y

BENDING IRON : A curved steel bar known as 'bend iron', used by plumbers for straightening lead pipes.

BENDING MOMENT : The algebraic sum of all the moments that produces the bending effect at any section of a structure due to the loads applied. This may be a sagging moment or hogging moment.

BENDING MOMENT DIAGRAM : A diagram showing the amount of bending moment at any point along a beam or any structural member subjected to a loading. From this diagram, the location of maximum bending moment, and point of contraflexure (zero bending moment at a point due to the change from positive bending moment to negative bending moment, can be found out.

BENDING MOMENT ENVELOPE : A number of bending moment diagrams due to different loadings are drawn on top of each other to represent the worst possible situation due to the combined effect of bending moments at all sections of a structure.

BENDING SCHEDULE : It is a bar bending schedule that furnishes a list of reinforcement bars with dimensioned shapes and diameters of all the bars required. The bars are bent by the bar bender by following this schedule.

BENDING SPRING : Close-packed, circular steel spring 0.75 m. long, used in bending small diameter copper or lead pipes (12 mm to 50 mm dia). With a view to maintaining the circular cross-section at the bent portion of a pipe, the spring is coiled and inserted into the right diameter of the pipe and the pipe is bent gradually. The cross-section at the bent portion remains circular due to the loading effect of the spring.

BENDING STRESS : The stress developed due to bending of a structural member. Also, known as 'Flexure stress' or 'skin stress'. This can be calculated from the bending equation $f/Y=M/I$; where, $f=flexure\ stress$, $Y=$depth of nutral axis, $M=$bending moment and $I=$ Second moment of area.

BEND TEST : A test to verify the ductility of a flat steel bar by bending the bar through 180°, when cold. If no crack is found, the piece is considered ductile.

BENT : A self-supporting two-dimensional frame, placed at right angles to the length of a structure viz. supports to a pipeline.

BENTHAL DEPOSIT : The deposits at the bottom of a slow-flowing drain or nullah which form sludge banks. Mostly, these deposits undergo anaerobic decomposition. The sludge masses are churned up along with the gases of decomposition.

BENTHAL OXYGEN DEMAND : It is the demand of oxygen by the benthal deposits for stabilization by aerobic decomposition of the top layer of the deposits.

BENTONITE : A natural soil derived from alteration of volcanic ash by chemical actions. This is composed of exclusively very active colloidal particles which swell in presence of water.

BENZENE : A highly inflammable liquid derived from coal or petroleum, used
　　　　(i) in quick drying finishing paint and
　　　　(ii) as a paint remover.

BERM : A horizontal ledge at the bottom of an embankment or at the top of a cutting.

BERNOULLI'S THEOREM : It states that the total head of each particle in a mass of flowing liquid is same. This theory is used in verturi Meter for measuring the quantum of flow through a pipe.

BERTH : A place where a ship is moored for loading and unloading.

BERTHING IMPACT : The forces on jetties and piers from the Kinetic energy of a vessel during its berthing.

BERTH JETTY : See illustration. This consists of mooring dolphin, breasting dolphin and a platform connected by catwalks. This is constructed for berthing vessels.

BESSEMER PROCESS: A process used in making steel in which excess carbon, silicon, manganese and phosphorus are removed along with slag.

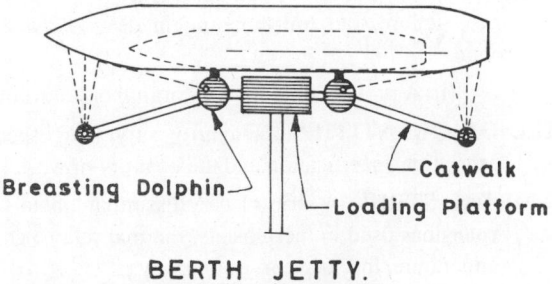

BERTH JETTY.

BETHELL'S PROCESS: A process of creosoting under pressure the railway sleepers and timbers for use in marine works.

BEVEL : (i) The meeting of two surfaces at an angle, other than right angle.

(ii) The cutting edge of a chisel or a cutter.

BEVELLED BAT   : A brick cut at an angle, as shown in illustration.This is used in making angular joints in brick work.

BEVELLED CLOSER : A brick cut longitudinally from the middle of one end to the far corner i.e., one-fourth of the brick is cut-off.

BEVELLED WASHER : A wedge-shaped steel washer, thin at one edge and thick at the other, so that it can fit well under a nut on a tapered flange.

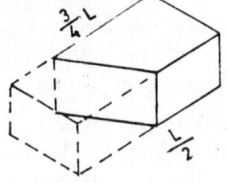

BEVELLED BAT

BEVEL SIDING : A clapboard.

BEVEL SQUARE : A tool used by carpenters for checking the bevelled surfaces. See illustration.

B-HORIZON : In soil mechanics, it is the horizon containing metal oxides and soluble materials leached from A-horizon. It is the lower part of the top soil.

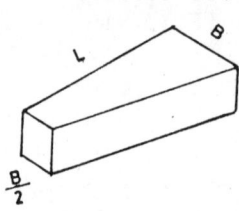

BEVELLED CLOSER

BI-CABLE ROPEWAY : An aerial ropeway with one track cable to carry loads, while a traction rope moves along it.

BEVEL SQUARE

BIB COCK : A water tap connected to a horizontal pipeline for drawing water.

BIFURCATED STAIR : A stair with a wide bottom flight and the upper flight bifurcated in two directions.

BILHAM'S FORMULA : A formula for heavy rainfall in a short period.

$$n = 1.25\ T\ (r + 0.1)^{-3.55}$$

where n = No. of occurence of storms.

T = duration of storm in hours

r = total rainfall in inches during T.

BILLET : (i) A timber piece with three sides sawn square, the fourth being made round.

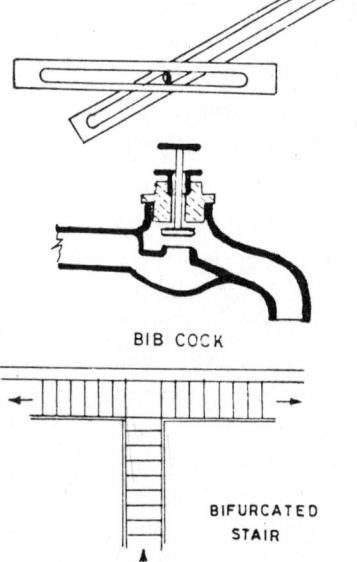

BIB COCK

BIFURCATED STAIR

(ii) A product obtained in forging or hot rolling wrought iron or steel.

BILL OF QUANTITIES : Quantity surveying, speaking of items of work with description/specification and the quantity of work involved in each item of work.

BI-METAL STRIP : A strip of two dissimilar metals having different co-efficients of expansions used in thermostats, thermal relays, gas heaters, etc., Due to change in temperature, the strip curves one way.

BINDER :   (i)   A binding material like cement, lime, gypsum plaster, bitumen, tar, etc.

(ii)   Stirrups in beams and lateral ties in columns.

(iii)   Linseed oil, size and resin used in a paint to hold the pigment in a coherant film.

(iv)   A binding beam, wooden or steel, used to support common joists.

BINDING BEAM : A binder used in a timber truss.

BINDING RAFTER : A purlin in a timber truss.

BINDING WIRE : Annealed wire (black wire) used in binding reinforcement bars.

BIOASSAY TESTS : Tests carried out to estimate the toxicity of waste waters to the biological life of a water-body into which the waste waters are discharged.

BIO-CHEMICAL OXYGEN DEMAND : Abbreviated term is BOD, most widely used parameter of organic pollution of a water or waste water. The demand of oxygen for bio-chemical oxidation of organic wastes and their stabilization in 5-days at $20^{\circ}C$ by micro-organisms.

BIO-FILTRATION : A process for treatment of sewage in which BOD is removed to a great extent by passing the settled sewage through a bed of stone blocks, brick bats, ballasts, etc.

BIO-FILTRATION PLANT : Also known as 'Trickling Filter Plant'. A treatment plant by employing bio-filtration process for treatment of sewage. The plant may be a single–stage or two-stage bio-filter and the plant layout may be made series, series-parallel or parallel. A simple layout of bio-filtration plant is shown.

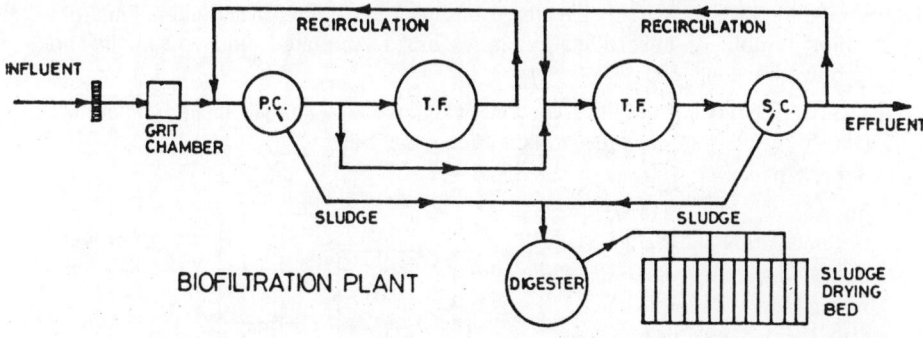

BIOFILTRATION PLANT

BIO-FILTER : A circular bed of broken stones or brick bats, the depth being normally 1.8m. The settled sewage is spread over this bed continuosly at a uniform rate by revolving distributors. As the sewage passes through the bed, the microbes (in form of slime layers) absorb the soluble organic wastes and the BOD of sewage gets reduced.

BIO-GAS PLANT: A plant used for generation of gas from human wastes and cattle dung. The illustration shows various units required in a bio-gas plant to produce gas from sewage and cattle dung slurry.

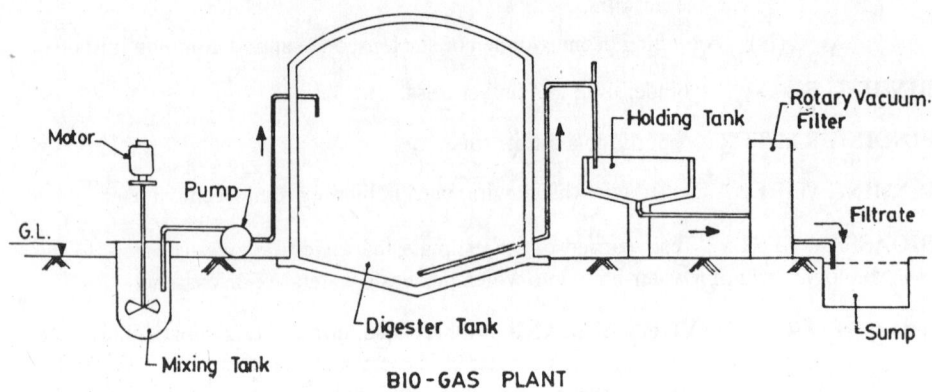

BIO-GAS PLANT

BIOLOGICAL DECOMPOSITION: Decomposition of organic wastes with the help of micro-organisms. It may be aerobic or anaerobic depending upon the enviornmental condition. Aerobic decomposition is much more faster than anaerobic decomposition.

BIOLOGICAL FILTER : 'Bio-filter' or 'Trickling filter' used in filtration of sewage for BOD removal.

BIOLOGICAL FLOCCULATION : The conversion of finely divided suspended and colloidal matters and also the dissolved matters into a film substance by contact operation in which formation of biologically active flocs or slimes of microbes is promoted under aerobic condition.

BIOLOGICAL TREATMENT : Treatment of sewage by biological activities of microbes, either aerobic or anaerobic. Examples are : Activated sludge, Bio-filtration, stabilization pond, etc.

BIOLOGICAL SHIELD : A thick concrete wall covered with lead sheet, built around a nuclear reactor to protect the workers from radiation.

BIOLYTIC TANK : A dual purpose tank (sedimentation and sludge digestion in a baffled compartments tank); similar to Travis hydrolytic tank, from which 'Imhoff Tank' has been developed.

BIO-PRECIPITATION PROCESS : A process of sewage treatment, in which the waste liquid is

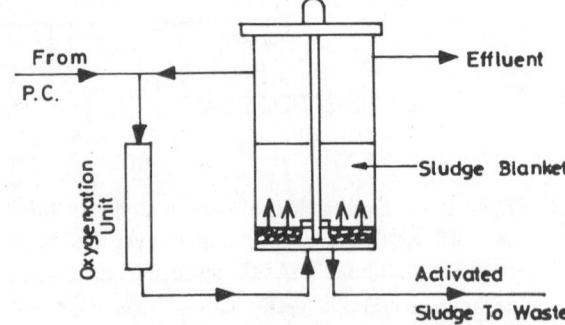

BIO –PRECIPITATION PROCESS

pretreated in a counter current oxygenation unit and is passed upward through a sludge blanket (biologically active floc).

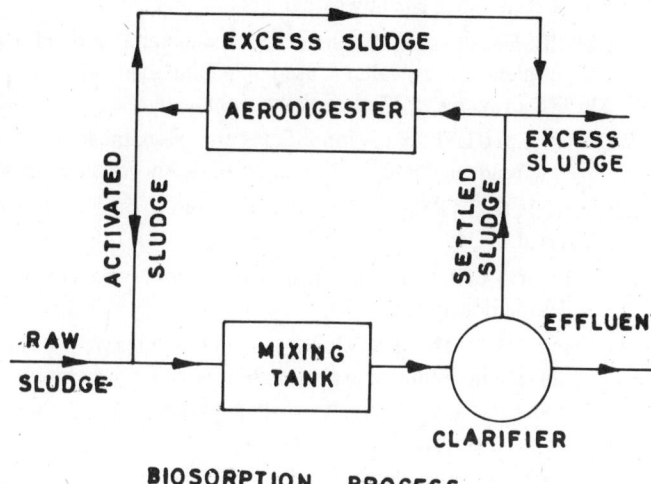

BIOSORPTION PROCESS

BIOSORPTION PROCESS : A sewage treatment process, in which sewage is preaerated with activated sludge in a mixing chamber for a period of about 30 minutes during which bio-sorption (biological adsoption) takes place. After this, the sewage is aerated in an aeration tank for about 2.5 hours. See illustration.

BIRD PECK : A small hole or a patch of discoloured and distorted grain found in a timber,

BIRD'S EYE : Small pointed depressions in annual rings of a wood show a figure like birdseye, when cut in a rotary cutting machine.

BIRD'S EYE VIEW : An oblique aerial view to produce three dimensional effect of an object

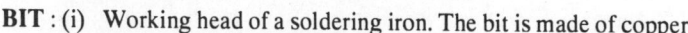

BIRDS' EYE VIEW

BIRD'S MOUTH BRICK : A special brick as shown in illustration.

BIRDSMOUTH JOINT : A joint made by making a cut at the end of a timber, such that it fits over a cross timber. As an example, fitting of a rafter end over a wall plate. See illustration.

BIT : (i) Working head of a soldering iron. The bit is made of copper.

(ii) Cutter, usually a drill bit.

BITCH : A spike made of steel, whose points are at right angles to each other.

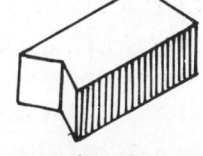

BIRD'S MOUTH BRICK

BIT EXTENSION : A steel rod being held by a brace at one end, grips a bit at the other end. This is used for drilling holes deeper than the length of the bit.

BIT GAUGE : A small piece of metal fitted to a bit by a wing nut temporarily to avoid full depth drilling.

BIRD'S MOUTH JOINT

BIT STOCK: A brace used by carpenters.

BITUMEN : Non-crystalline solid or viscous material derived from petroleum. It consists of petrolene and asphaltene, used in making road pavements. It contains 87% carbon, 11% hydrogen and 2% oxygen by weight.

BITUMEN EMULSION : A bituminous emulsion, made of water and bitumen, which is used to roads in damp cold weather. Also, known as 'bitumen road emulsion'.

BITUMINOUS CARPET : The wearing course of a road, using bitumen as binding material.

BITUMINOUS FELT : An underlining felt used for roof covering, lining damp walls and under the flooring.

BITUMINOUS PAINT : An elastic paint that adheres well to a metal surface. It is prepared by dissolving bitumen, asphalt and mineral pitch in petroleum, naptha, paraffin and various oils. It is capable of resisting actions of acids and alkalies.

BITUMINOUS PLASTIC : Plastics made from natural bitumens that can be used in making furnitures, doors, windows and electrical fittings.

BLACK BOLTS : The bolts which are covered with black iron oxide.

BLACK DIAMOND : Diamond of black or deep grey colour used for making valuable cutting edge.

BLACK GANG : Blacksmiths, fitters, riveters, welders and riggers engaged in civil engineering construction.

BLACK JAPAN : A bituminous paint of good quality or the best black varnish.

BLACK MORTAR : A low-strength, cheap-quality mortar in which ashes or coal dusts are mixed to impart colour.

BLADE GRADER : A grader used in constructional works.

BLADING BACK : Pushing a soil mass back to its position (with the help of a grader), wherefrom it came.

BLANC FIXE : Barium sulphate, an amorphous extender which is made fine by chemical precipitation and not by grinding, for use in a paint.

BLANK CARBURIZING : A process of heat treatment in which carburizing is done without carbon.

BLANK DOOR : A door which is walled-up.

BLANKET : Insulating materials made of eel grass, asbestos fibre, glass wool, cork, mineral wool, etc. backed with paper or aluminium foil.

BLANCK FLANGE : A flange bolted to the pipe flange at the dead end of a pipeline.

BLANK NITRIDING : A process of nitriding without the use of nitrogen or ammonia.

BLANK WALL : A wall without any opening.

BLAST-FURNACE CEMENT : A cement produced by grinding Blast-furnace slag. Sometimes, it is mixed with ordinary portland cement.

BLASTING : Breaking or dislodging rocks and coals in mining operation with the help of explosives.

BLEACHING : Removal of colour by chemical action.

BLEACHING POWDER : Calcium hypochlorite, a white powder of lime containing about 30% of Chlorine. This is used as a disinfectant.

BLED TIMBER : An inferior quality of timber obtained from trees that have been tapped for resin.

BLEEDER TILE : Drainage pipes built into the foundation of a building for draining out the water from the basement.

BLEEDING : (i)   Passing of vehicle through a material above it.

          (ii)   Penetration of glue through a veneer.

          (iii)   Appearance of molten bitumen on the road surface in hot climate.

BLENDER : A brush of soft hair with round blunt tip, used for removing brush marks on a freshly painted surface and also for blending colours.

BLIND AREA : A dry area.

BLIND DRAIN : A rubble drain.

BLIND FLOOR : A rough floor.

BLINDING :   (i)   Spreading of sand to fill the voids in a road wearing course.

          (ii)   A mattress of lean concrete, 50 to 100 mm thick, laid on clay to seal it and to provide a bed for putting the reinforcement steel.

BLIND HEADER : A half bat or header which does not come in view on one face.

BLIND HINGE : A concealed hinge used in joinery works.

BLIND HOLE : A drilled hole not passing through the material.

BLIND MORTISE : A mortise not passing through the timber, encloses a 'stub tenon'.

BLIND NAILING : Secret nailing used in joinery.

BLISTER FIGURE : A quilted figure in timber.

BLISTERING : A defect in painting. Bubbles appearing on a painted surface due to vaporisation of moisture under the surface.

BLOATED CLAY : Expanded clay or sintered clay used as light-weight aggreagate in concrete.

BLOCK : (i)   Building unit made of timber, clay, stone,concrete and glass.

          (ii)   A frame holding pulleys of a hoisting tackle.

BLOCK BOARD : A board made of of core strips of 25 mm width glued together between the outer veneers.

BLOCK BONDING : Bonding new brickwork into old; Bonding several courses of brickwork of one wall into a few courses of brickwork of another wall.

BLOCK BRIDGING : Solid timber bridging used in carpentry.

BLOCK FLOORING : Wood block or concrete block paving.

BLOCK-IN-COURSE : Large stone blocks laid in courses for building dock walls. The depth of these blocks is about 300 mm, the length being variable.

BLOCKING COURSE : A course of special brick or masonry placed over a stone cornice to hold it by weight.

BLOCK PLAN : A small-scale plan showing outlines of buildings or structures.

BLOCK PLANE: A small plane of about 15 cm. length without any back iron, used by carpenters for cutting end-grain of timber.

BLOCK TIN : Pure tin commercially used in plumbing work.

BLOCK PAVEMENT : Pavement with stone blocks for a road wearing surface.

BLOCK WALL : A gravity type quay wall consisting of a number of concrete blocks placed on top of each other and capped with a mass concrete wall.

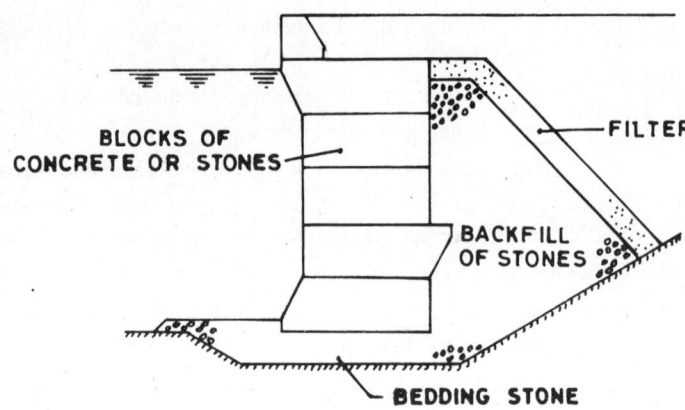

BLOCK WALL

BLOCK WORK : Use of precast concrete blocks or stone blocks in construction of breakwaters, marine structures and buildings.

BLOCK YARD : A yard, where precast concrete blocks are manufactured.

BLONDIN : A French word for a cableway.

BLOOM : (i) A forged or rolled section of wrought iron or steel, which is not finished properly.

(ii) A thin film formed on an old glossy paint or varnish, which can be removed by wiping with a cloth.

BLOW DOWN : Ejecting out boiler sediment by openig a valve at bottom of the boiler.

BLOWING : Expulsion of plaster from a surface due to the expansion of material at the back or due to the slaking of lime lump present in mortar. This forms small depressions on the plastered surface.

BLOW LAMP : A lamp that produces powerful flame required to melt solder or lead and for burning off old paint.

BLOWN JOINT : A joint used in plumbing, where one lead pipe is fitted over another of same diameter by shaping two ends like cup and cone. The cone end is introduced into the cup end and then joined with fine solder poured in. This is also known as 'Cup and Cone Joint'.

BLOWN OIL : Oxidised oil by blowing air through heated linseed oil with driers.

BLOW-OFF VALVE : A valve used in a pipeline at its lowest point for discharging sediment or emptying the line for repair/renovation work.

BLOW OUT : In working with compressed air, a sudden loss of compressed air from a caisson that may cause disaster.

BLOW TORCH : A blow lamp.

BLUE PRINT : A contact print of a drawing on tracing paper made on 'ferro- prussiate paper', which is developed in water.

BLUE STAIN : Sap stain, which is a blue fungal discolouration of sap wood.

BLUSHING : A fault in a paint, caused by moisture or lack of compatibility. A milky opalescence produced in a lacquer.

BOARD : Plank formed by sawing wood, the thickness being less than 50 mm.

BOARDING : Laying boards over rafters or studs to produce a floor of boards.

BOARDING JOISTS : Common joists used in boarding by carpenters.

BOARD LATH : (i) Wood lath,
               (ii) Gypsum plank

BOARD MEASURE : A unit of mesuring board in North America. It is 12" x 12" x 1" prior to sawing and planing. Also known as 'board foot'.

BOARD OF TRADE UNIT : Abbreviated as B.O.T. unit. It is the commercial unit of electrical energy and is one killo-watt hour.

B.O.D. : Bio-chemical oxygen demand i.e., the quantum of oxygen required for stabilisation of wastes by bio-chemical activities. It is usually measured at $20^{\circ}C$ in 5 days.

BOASTED ASHLAR : A rough finish given to ashlar by boasting.

BOASTER : A boasting chisel of about 50 to 75 mm wide, used by stone dressers for boasting.

BOASTING : Dressing stone surface with the help of a boasting chisel and a mallet. The surface is dressed with roughly parallel, oblique or vertical strokes of boaster.

BOAT SCAFFOLD : A cradle.

BODY : In painting, it is the stiffness or solidity of a dried paint film.

BOIL : Flow of water with fine sand and silt into the bottom of an excavation due to water pressure outside the excavation. A small spring of water with sand and silt is a 'boil'.

BOILED OIL : Linseed oil boiled at $500^{\circ}F$ with soluble driers is used in preparation of a paint, because a paint with boiled oil dries quickly than a paint with raw linseed oil. Boiled oil is available in pale and dark colour.

BOILER : A plant for steam generation.

BOILING HOLE : A hole of 1m x 1m x 1m made in the ground, in which a putty is prepared from magnesian quicklime. During slaking, the hole is kept covered for preservation of heat within the hole.

BOILING TUB : A tub used like a 'boiling hole'.

BOLE : Tree trunk.

BOLECTION MOULDING : A door or window panel moulding raised above the frame and rebated around.

BOLLARD: (i) A post strongly anchored into the masonry of a quay wall, used for mooring vessels.

(ii) A post suitably anchored in a road to protect a roadside wall, street island or to divert the traffic.

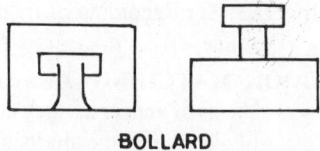

BOLLARD

BOLSTER: A support for a bridge truss on the abutment.

BOLT : A cylindrical bar with threads for screwing a nut, required in temporary fastening.

BOLTING IRON : A narrow chisel used by carpenters for mortising.

BOND : (i)   Arrangement of laying bricks according to a pattern for interface strength and breaking vertical joints.

        (ii)   Grip or adhesion between mortar or concrete and reinforcement bar.

        (iii) Placing of slates or tiles in a pattern for strength and exclusion of rain water.

BOND BREAKER : A sealant for lift slabs or a release agent.

BOND COURSE : A layer of headers as used in 'American bond'.

BONDER : A bond brick or stone.

BONDING : See 'bond'.

BONDING BRICK : A special brick used for holding together the two leaves of a cavity wall.

BONDING COMPOUND : A sealing compound, usually an oxidised bitumen applied hot to fix felt to roof in bitumen-felt roofing.

BONDING TIMBER : In early days, long timber pieces were used into brick work to strengthen it. Now-a-days, this is not practised because of the danger of the timber rotting.

BOND LENGTH : The grip length of a bar used for reinforcement.

BOND STONE : A stone laid in a wall as a header for making the bond.

BOND STRESS : The shearing stress at the surface of a reinforcement bar, which prevents its coming out from the concrete.

BONE BLACK : A black pigment made from charred bones and ivory chips. It consists of pure carbon.

BONING : In surveying, it is a method of setting out a slope with the help of boning rods.

BONING PEG : Small size hardwood cube placed at each corner of a large stone which is to be dressed.

BONING ROD : A T-shaped staff used in fixing up level or slope in laying a sewer line. This is also required in setting out foundation trenches for a building.

BONNET :   (i)   A roof over a bay window.

        (ii)   A spherical wire netting cover placed at the top of a chimney or a vent stack.

BONNET HIP TILE : A cone tile or a hip tile with a rounded top.

BOOJEE PUMP : A grouting machine used for pressure grouting with cement slurry into fissures in rock or concrete or masonry work. This essentially consists of a compressor and a pump with cement slurry container.

BONNET
HIP TILE

BOOKING : Recording of field data in a book (during surveying) for the use in drawing office.

BOOK MATCHING : Also called 'herring bone matching' in veneering. The successive sheets of veneer as they come from the knife during slicing, are placed like the pages of a book i.e., the alternate sheets are placed face up and face down.

BOOM : (i)   The horizontal members (top or bottom) of a built-up girder.

(ii)   The beam used in a lifting tackle.

(iii)   The jib of crane .

BOOSTER PUMP: A pump used in a water line with a view to increasing the pressure of water, such that the water consumer at the furthest end of the pipeline gets water at required pressure.

BOOT : (i)   A projected part of a R.C.C. beam, which carries facing brick or stone.

(ii)   The lower end of a bucket elevator used for lifting materials.

BORDER STONE : A kerb stone used on the road edge.

BORE : (i)   Internal diameter of a pipe.

(ii)   Hole in a material.

(iii)   A wave travelling with a nearly vertical front upstream.

BORED PILE : Cast-in-situ pile. This is cast at site by pouring concrete with reinforcement into a hole formed in the ground.

BOREHOLE : A hole bored into the ground for information about the geological strata, avaiability of ground water, oil, gas, sulphur, etc .

BOREHOLE LATRINE : A type of latrine used in rural areas. It consists of a borehole of 300 to 400 mm dia, provided with a concrete or wooden squatting slab. The depth is usually upto 1 m. below ground water table.

BOREHOLE PUMP : A centrifugal pump, usually of submersible type, driven by a motor on the surface, the power being transmitted through a rod. This pump is commonly used for pumping ground water from deep well.

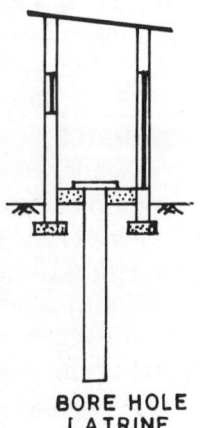

BORE HOLE LATRINE

BOREHOLE SAMPLES : Samples obtained from boreholes for the purpose of investigations on nature of strata and its strength.

BOREHOLE SURVEYING : The technique of measuring the deviation of a borehole from the vertical.

BORING : (i) Driving a borehole by auger, drill or wash–boring method.

(ii) Forming a hole in rock with the help of a percussive drill, for blasting.

BORROW : Soil or other material dug out from a place to fill other site.

BORROWED LIGHT : A window in an internal wall.

BORROW PIT : Pit formed due to excavation of soil from a place to fill other site.

BOSS :   (i)   A keystone of a dome.

(ii)   A rounded projection down from a ceiling, which is often found at the intersection of the ribs.

BOSSAGE : Rough ashlars projected from a wall.

BOSSING : Shaping sheets of lead, zinc, tin, etc for fixing to a sloped roof.

BOSSING STICK : A boxwood shaper used for sheet lead lining to a tank.

BOTTLE NOSING : A half round nosing used in wooden steps.

BOTTOMING : (i) Levelling and dressing the bottom of an excavation.

(ii) Laying stone blocks on the formation of a road.

(iii) Preparation of ballast bed for a permanent way.

BOTTOMLESS HOLE : A hole passing through a material helps in anchoring the material with a screw anchor.

BOTTOM-OPENING SKIP : A drop–bottom bucket, which opens out, when dropped at the bottom of a vertical shaft or a trench.

BOTTOM RAIL : The horizontal member at the bottom of a door or window.

BOTTOM SAMPLER : A sounding lead for picking up material from sea bed.

BOTTOM SHORE : The shore used next to a building when it is required to provide raking shores to hold up the building.

BOULEVARD : The central strip of a wide road in a city, planted with trees and shrubs.

BOUSSINESQ EQUATION : The equation developed by Boussinesq in 1885 based on his analysis of stresses in the soil under a loaded foundation. This shows that the lines of equal vertical stress under a loaded point are nearly circular.

BOW : Warping or bending of a timber at right angles to its face.

BOWDITCH'S RULE : The rule used in a closed traverse survey for adjustment of the error cripped in. The sides and angles of a closed traverse are considered equally liable to error. For an error, the correction to be applied to a line is given by.

(length of line/perimeter of traverse) x total error.

BOW DRILL : A device for boring a timber , in which the drill bit fitted is rotated by the grip of a cord of a bow. The bow is given to and fro motion by hand.

BOWLED FLOOR : A floor of a theatre or church or cummunity hall, which is given a slope of 1 in 24 or so.

BOWL SCRAPER : It consists of a bowl-shaped scraper hung from a wheeled frame. When it is towed forward, its bottom edge digs into the ground and fills the bowl. It is used in levelling and dressing the ground.

BOWL TYPE URINAL : This is a standing type urinal suitalble for installation in public places. See illustration.

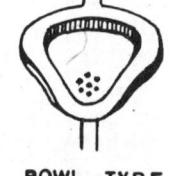

BOW SAW : A saw having a detachable blade in a wooden frame, used for sawing round curves.

BOWSTRING GIRDER : A girder shaped like a bow with strings, developed from 'belfast truss'. This may be made of timber, steel or concrete. Such a girder made of timber may be upto a span of 45 m.

BOWL TYPE URINAL

BOWSTRING TRUSS : See illustration.

BOW WINDOW : A bay window.

BOX BEAM : A box like built-up beam or girder.

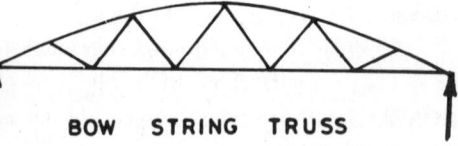

BOW STRING TRUSS

BOX CAISSON : A reinforced concrete box with open top is built on shore or bank of a river for foundation of a bridge pier or abutment. The box caisson helps to construct in the dry and ultimately the box becomes a part of the structure.

BOX CASING : Inside lining given to a foundation trench.

BOX CORNICE : A built-up wooden hollow cornice of box shape enclosed by a gutter. This is used in sloping roofs.

BOX CULVERT : A box like culvert of square or rectangular section constructed for drainage of water.

BOX DAM : A cofferdam.

BOX DRAIN : A drain of box shape, square or rectangular section constructed for drainage of water.

BOX GIRDER : See 'box beam'

BOX GUTTER : Trough gutter made of timber and lined with flexible metal, asphalt sheet or roofing felt. This is used behind parapets and in valleys formed in a sloping roof.

BOX HEAD WINDOW : A window having a wide slot at its head into which the sash can be pushed to provide opening for ventilation.

BOXING : (i)   Sheathing used in buildings.

(ii)   A frame with casing.

(iii) Hollowing out a part of window recess for receiving the boxing shutter.

(iv)   A ballast bed between rail sleepers.

BOXING SHUTTER : Folding shutters inside a window that fold away into the boxing i.e., inside the room.

BOXING UP : (i) Encasing a frame or member with timber.

(ii) Packing ballast under railway sleeper.

BOX PILE : A steel pile made with two channel sections or two sheet piles or with angle sections and R.S.J.

BOX SEXTANT : A type of sextant used in surveying for measurement of angles. It is a very handy and compact instrument.

BOX SHEAR TEST : A standard method of testing the shear strength of a soil sample in a box, split in two parts.

BOX STAIR : A closed stair.

BOX STAPLE : A metal box fitted to a door post into which the 'rim lock latch' passes, when the door shutter is closed.

BOX UP : Boxing up.

BOX WOOD : A very hard wood used for making handles of instruments and appliances.

BOXWOOD BOBBIN : An egg-shaped tool used for removing distortions in a lead pipe and also in truing the bends in lead pipes.

BOX WOOD DRESSER : A tool for straightening lead sheet.

BOXWOOD TAMPIN : A conical turning pin used for opening out the end of a lead pipe.

BRACE: (i)   A cranked tool used for turning a drill bit to make holes in wood.

(ii) A diagonal member that stiffens a structure. It is either in tension or in compression.

BRACE BLOCKS : Wooden keys that prevent sliding of the parts of a built-up beam.

BRACED FRAME : A wooden building frame with bracings between the widely spaced posts, at different heights.

BRACE EXTENSION : A bit extension used in a brace.

BRACING : A stiffener used in a structure or a member.

BRACE JAWS : Clamping device to hold the bit in a brace.

BRACKET : A projecting support in a structure.

BRACKET BALUSTER : A metal baluster bent at its base and built into the side of a stair.

BRACKETED STAIR : A stair carried on a cut string.

BRACKETING : Fixing wooden brackets to carry a plaster cornice.

BRACKET SCAFFOLD : A light scaffold made of steel–framed bracket hooked to a wall for taking up repair work.

BRAD : A tapering cut nail used for fixing floor boards.

BRADAWL : A short 'awl' with a chisel point which is pushed into wood for making holes prior to introducing nails or screws.

BRAD PUNCH : A nail punch.

BRANCH : A take-off line from a pipe.

BRANDER : A fillet of 40 to 50 mm x 25 mm nailed to the soffit of wooden joists at 30 cm c/c to facilitate fixing of ceiling laths.

BRANDERING : Making provision of branders on the soffit of the joists, such that the laths can be nailed to the branders.

BRANDING IRON : Indenting roller used for making indentation.

BRANSBY-WILLIAM'S FORMULA : A British formula to calculate the time of concentration

$$T_c = 0.88 \ L \ x \ 60/M^{0.1} \ S^{0.2}$$

Where L is length of source to given point under investigation (miles), S is average slope of length (percentage) and M is drainage basin area (square miles).

BRASS : An alloy of copper and zinc.

BRAZING : A technique of soldering brass, copper, steel and cast iron with a film of copper-zinc alloy. Also, known as hard solder.

BREAK : (i)   A change in direction of a wall.

(ii) A gap in continuity of a wall or a slab.

BREAKING DOWN : (i) Degradation.

(ii)   Conversion.

(iii) Separation of the constituents of an emulsion.

BREAK IRON : Back iron of a plane.

BREAKING STRENGTH : Rupture strength or ultimate strength.

BREAKING STRESS : The maximum stress developed in a material at the point of its breaking.

49

BREAK JOINT : A structural joint to break the continuity, with a view to giving allowance for expansion and contraction.

BREAK-POINT CHLORINATION : Chlorine residual will show a drop in a curve of chlorine residual versus chlorine applied and then the residual will increase with the increase in chlorine dosage. This is the break-point as shown in illustration. Free chlorine residual occurs beyond the break-point.

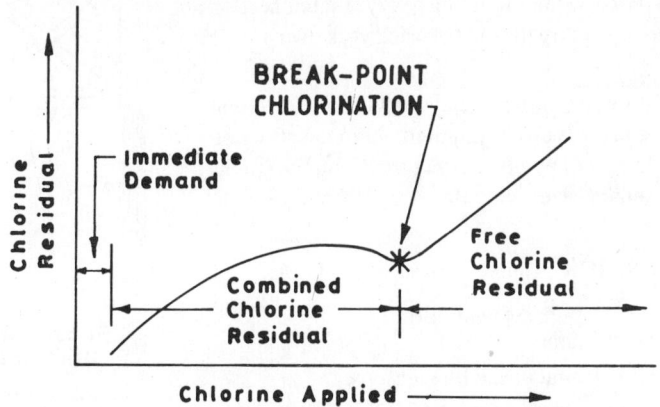

BREAK PRESSURE TANK : In hill areas, small tanks are placed at hydraulic gradient level of a gravity main, with a view to bringing down the pressure of water. Thus, the gravity main discharges water into individual break pressure tanks from which new pipelines take off.

BREAK WATER : A wall of rubble mound or blockwork constructed into the sea to protect a harbour from the wave action of sea water. See illustration.

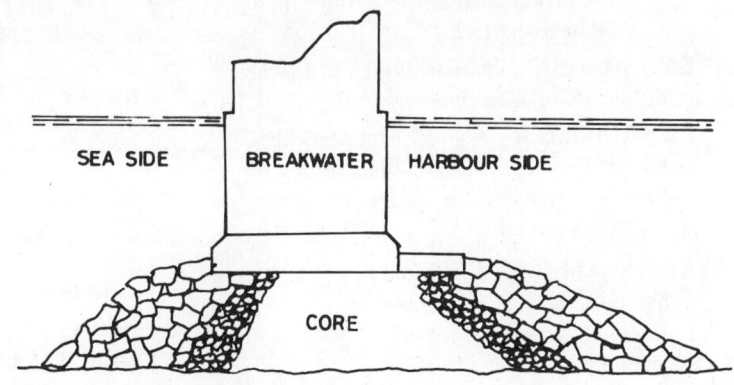

BREAST: (i)   The mould board of a dozer.

(ii)   The riser of a stair.

(iii) The wall from floor level to window sill.

(iv) The projection made inside a room for fire place.

BREASTING DOLPHIN: A dolphin that supports fenders to absorb berthing impacts and on shore wind loads on the moored vessel. These are placed widely apart in a berth jetty.

BREAST LINING : Wooden lining inside a window.

BREAST WALL : A retaining wall constructed upto a breast–high level to retain earth.

BREEZE BLOCKS : Building blocks made of clinker or coke breeze.

BRESSUMMER A long lintel of heavy timber section used earlier to carry the load of brickwork over a large opening.

BRICKS : Burnt or unburnt clay blocks of different shapes and sizes as per requirement in construction of a building or any other structure. Common bricks have standard size, which also vary from place to place.

BRICK AND BRICK : Gauged brickwork in USA.

BRICK AND STUD : Brickwork infilling between studs of a wooden frame.

BRICK AXE : A hammer used by masons.

BRICK CLAY : Brick earth.

BRICK CORE : Brickwork under the soffit of a relieving arch.

BRICK FIELD : A ground used for manufacturing bricks. See illustration.

BRICKING : (i)   Brick laying.

(ii) Forming imitation brick on a plastered surface.

BRICKLAYER : Mason; A tradesman skilled in laying bricks for different works in civil engineering.

BRICKLAYER'S HAMMER : A small hammer with a chisel end used by masons for cutting bricks to required shape and size and for hammering bricks during laying over a mortar.

BRICKLAYERS' SCAFFOLD: A scaffold made of bamboo or steel tubes and wooden planks required for laying bricks at a height.

BRICK MASONRY : Brickwork.

BRICK MATTRESS : A type of brick lining as shown in illustration.

BRICK MOULD : A rectangular box as shown in illustration is made of wood, or wrought iron. The projected sides act as handles. The mould is made 6 mm. deeper than is required for the brick size.

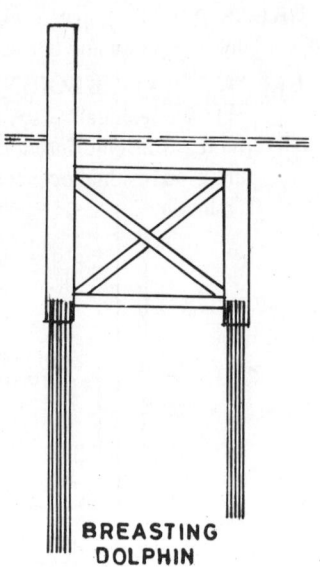

BREASTING DOLPHIN

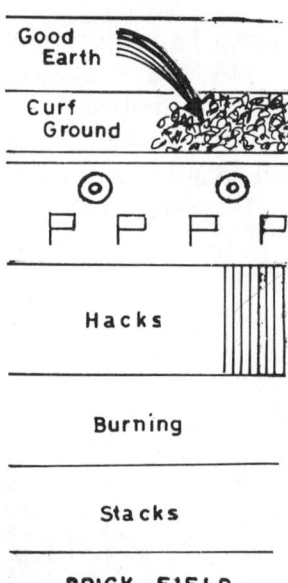

BRICK FIELD

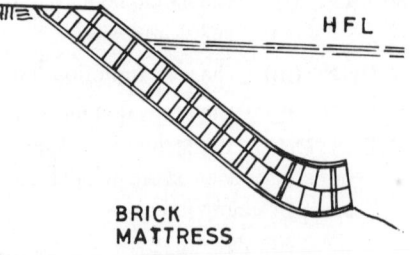

BRICK MATTRESS

BRICK NOGGING : Infilling between the studs of a framed wooden partition, with brickwork.

BRICK ON EDGE : Bricks laid on their edges at the sides of a a metalled road surface or round the periphery of a patterned brick flooring.

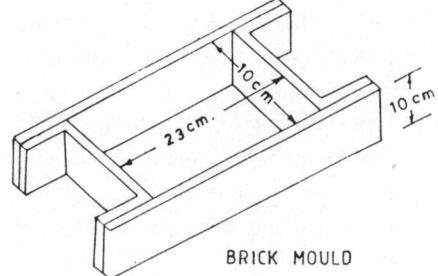

BRICK MOULD

BRICK ON EDGE SILL : A window sill made of headers laid on edge.

BRICK ON END : Laying bricks on their ends.

BRICK VENEER : An outer covering made of a half brick wall used to protect a timber house.

BRICKWORK : Masonry work with bricks.

BRICK-ON-EDGE

BRICK ON-END

BRICKYARD : Brick field, where bricks are manufactured.

BRIDGE : A structure usually horizontal, across a stream, river, canal, or railway.

BRIDGE ABUTMENT : The end support of a bridge.

BRIDGE BEARING : A support at the abutment or pier of a bridge to carry the load of the bridge. This is either fixed or provided with rocker and roller arrangement.

BRIDGE BOARD : A cut string in timber work.

BRIDGE CAP : The topmost part of a bridge pier, on which the bridge bearing is placed.

BRIDGE DECK : The bridge floor that carries the load and transfers to the bridge girder.

BRIDGE PIER : The intermediate support of a long bridge of continuous beam type.

BRIDGE RAIL JOINT : It is a joint made between two rails in a railway track as shown in illustration.

BRIDGE STONE : A long stone spanning a gap.

BRIDGE THRUST : A horizontal thrust coming at the end of an arch bridge.

BRIDGE RAIL JOINT

BRIDGE TRUSS : A truss or girder used for carrying bridge loads. This is vierendeel girder, warren girder or Pratt truss.

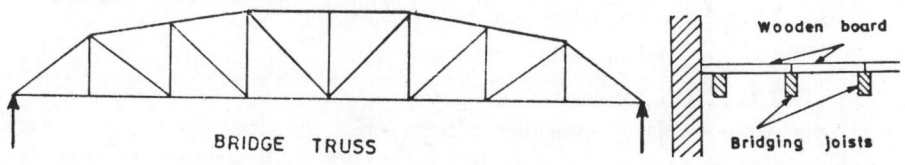

BRIDGE TRUSS

Wooden board

Bridging joists

BRIDGING: (i) Spanning across a gap.

(ii) Stiffening of adjacent joists by herring-bone strutting against the joists, required in wooden flooring.

(iii) Covering a gap in a base by a paint film.

BRIDGING FLOOR: A wooden floor supported by common joists only.

BRIDGING JOIST : A common joist that supports a wooden floor.

BRIDGING PIECE : A short piece of timber placed between or across common joists to support a partition.

BRIDLE JOINT : Birdsmouth joint developed from mortise and tenon joint. This is used in heavy wooden frame work.

BRIDLE JOINT

BRINDLED BRICK : A brick having striped surface.

BRIQUETTE : A specimen of cement-sand mortar cast in the shape of an 'hour glass' which is used in testing tensile strength of cement. See illustration.

BRISE-SOLEIL : A concrete shield that prevents intensely hot direct sunrays from entering a room. Such a shield usually consists of vertical or horizontal precast concrete strips. Also, known as sun-breaker.

BRISTLE BRUSH : A painting brush made of bristles, either of hog-hair or synthetic fibres.

BROACH :   (i)   A pin inside a lock.

             (ii)   A pointed chisel used by masons.

BRIQUETTE

BROACHED WORK : Punched work in stone dressing.

BROACHING : A method of excavating rock without using any explosive. A line of holes close to each other along the breaking line is drilled. The solid rocks between the holes are cut and the entire block is knocked out. Also, known as 'broach channelling'.

BROAD AXE : An axe having a broad bevelled cutting edge, used for dressing timber roughly.

BROAD FLANGED BEAM : A rolled steel joist having flange width almost equal to its depth. It is economical to use as a column section.

BROAD GAUGE : The widest gauge used in a railway track. It is 4' — 8 1/2" in some countries and 5'- 6" in some other countries.

Broad Flanged R S J

BROAD IRRIGATION : Disposal of sewage in farm yard by land irrigation. See illustration.

BROKEN-JOINT TILE : Single-lap tile.

BROKEN-RANGE ASHLAR : Uncoursed rubble work.

BROKEN WHITE : A white toned down to a creamy, off-white or ivory colour.

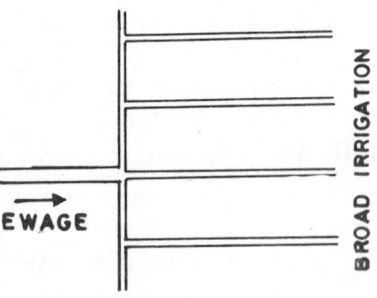

SEWAGE

BROAD IRRIGATION

BRONZE : An alloy of copper and tin.

BRONZE POWDER : Gold bronze powder used in preparation of a paint.

BRONZING FLUID : A lacquer used for applying gold bronze powder.

BROOMING : (i)   Scratching a concrete surface (when green) with a broom, to provide key for plastering.

(ii)   The crushing and spreading of fibres at the top of a wooden pile during its driving into a hard ground.

BROTCH : A spar in thatching.

BROTHERS : A two-legged sling of chain or rope.

BROWN COAT : A floating coat to serve as a base for plastering.

BROWN ROT : A decay of timber, due to which the timber becomes converted into a brown soft mass.

BRUNSWICK BLACK : A bituminous paint.

BRUNSWICK GREEN : A lead chrome green used in prepartion of a paint.

BRUSH : A tuft of animal hairs or artifical fibres held on to a handle. It is used in painting a surface.

BUBBLE TRIER : A level trier used in surveying.

BUBBLE TUBE : A level tube with an air bubble, used in surveying and construction works.

BUBBLING : A defect in a painted surface, when the paint contains highly volatile solvents with air bubbles.

BUCKET: (i)   A container of a bucket elevator.

(ii)   A kibble.

(iii)   A cup on the perimeter of a pelton wheel.

(iv)   A reverse curve given to the profile of a spillway, so that the overflow water is deflected horizontally on to the downstream apron at the base of the dam.

BUCKET ELEVATOR : An elevator consisting of an end less chain with buckets, used for raising loose material.

BUCKET-LADDER DREDGER : A dredger provided with a bucket-ladder elevator, used in digging a trench.

BUCKLE : A spar in thatching; A brotch.

BUCKLING FENDER : A buckling column-type fender consisting of a solid rubber section with one end flat and other end arched. One end of the rubber piece is bonded to the steel plate attached to the quay wall and the other end of the rubber is fixed to another steel plate supported by a pile at front.

BUCKLING LOAD : The crippling load of a column which tends to buckle it.

BUCK SCRAPER : A horse-drawn scraper used earlier. It is replaced by power-driven graders.

BUFF : Polishing terrazzo floor with mild abrasion to bring a floor finish.

BUFFER STOP: A fixture made of old rails and sleepers which is fixed at the end of a railway track to take the impact load of a moving wagon, if needed to stop it.

BUILDING BLOCKS:

Solid or hollow walling blocks made of clay, concrete, gypsum or glass, used in construction works.

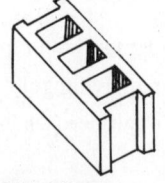

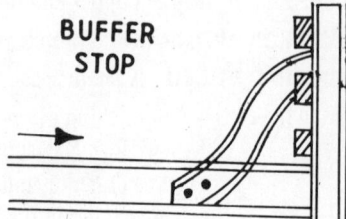

BUFFER STOP

BUILDING BOARDS : **BUILDING BLOCK**

Boards made of wood chippings, saw dust, cane fibre, gypsum plaster, etc., used as a facing to interior walls and ceilings.

BUILDING PAPER : A sheathing paper specially made for covering walls and ceilings to improve the appearance.

BUILT-UP : A member made of several pieces glued, nailed, bolted, riveted or welded together to bring the required shape and size.

BUILT-UP BOX BEAM : A box beam built-up with several sections to meet the requirement.

BUILT-UP ROOFING : Covering roof with two to three layers of bitumen felt.

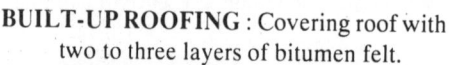

BUILT-UP BEAM

BUILT-UP SECTION : A beam or column section made of several sections as per the design.

BULB ANGLE : An angle section enlarged to a bulb at one end.

BULB OF PRESSURE : Bulb-shaped lines of equal vertical stress developed in soil below a loaded foundation. See illustration.

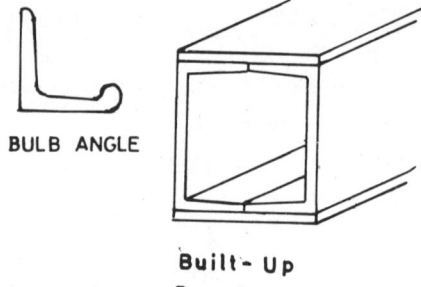

BULB ANGLE

Built- Up Box Beam

BULK BEAD : (i)   A box shaped cover over a stair well, lift shaft and water tank.

(ii)   A masonry or timber structure to retain earth along a water-front.

(iii) A partition within a ship's hull or superstructure.

BULKING :   (i)   The increase in volume of dry sand with addition of water upto a certain limit due to the formation of water film surrounding the sand grains. The increase is about 30% when 5% water is added to dry sand.

(ii)   The increase in volume of excavated loose material.

BULB OF PRESSURE

BULKING OF SLUDGE : A phenomenon that may occur in an activated sludge plant causing execessive volume of sludge without concentration.

BULK MODULUS : Ratio of normal stress to volumetric strain, denoted by 'K'

BULLDOG GRIP : A U-shaped bolt threaded at both ends.

BULLDOG PLATE : A toothed plate used as a timber connector.

BULL DOZER : A tractor with a wide blade at front which can be raised or lowered. This is used in removing material from a site by pushing.

BULL HEADER : A header brick with the projected upper arris rounded, used in construction of window sill.

BULL HEADED RAIL : A rail section with head larger than the foot, which is a modification of a double-headed rail. This rail is suitable for making points and crossings in a railway track.

BULL HEADED TEE : A tee section with a bullhead formation at the end of the leg.

B.H. RAIL

BULL-NOSE BRICK : A special brick with its one upper arris rounded as shown in illustration, used as a plinth brick.

BULL'S EYE : A small circular or oval opening in a wall.

BULL STRETCHER : A stretcher with the projected arris rounded.

BULL WHEEL : A large wheel provided at the base of a derrick mast, which rotates the mast.

BULL HEADED
TEE

BUNKER : A container used for storage of ores, coal, etc.

BUOYANCY : The decrease in weight of a body immersed in a fluid which is equal to the weight of the fluid displaced by the body.

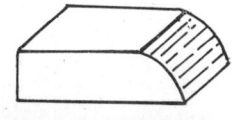

BULL NOSE BRICK

BUOYANT FOUNDATION : A raft or floating foundation in silt or mud in river estuaries. It is so designed that the total load of the foundation is nearly equal to the weight of silt, clay and water displaced by the foundation.

BURLAP : A hessian canvas used as a reinforcement for a plaster.

BURNETTIZING : A method of preserving timber by impregnating it with zinc chloride solution under pressure.

BURNT SHALE : Carbonaceous shale that has been oxidised by spontaneous combustion in a colliery tip. This can also be made by destructive distillation process.

BUSH HAMMER : A light tool of about 7 lbs. weight, used in surface dressing of stones to remove the outer skin by 0.25 inch.

BUSHING : A screwed short piece of pipe having both inner and outer threads, required to connect two pipes of different diameters, without using a tapered piece.

BUTT : To meet end to end and not overlapping.

BUTTERFLY VALVE : A circular disc hinged at two pivots inside a pipe, acting as flow-controlling valve in large pipes.

BUTTERING : Spreading of mortar on a vertical face of a brick, prior to laying.

BUTT GAUGE : A marking gauge used in timber joinery works.

BUTT HINGE : The common hinge used for hanging door and window shutters.

BUTT JOINT : A joint made between two parts by bringing them end to end i.e.,butting against each other.

BUTTRESS: Thickening pier at right angles to a wall, to facilitate the wall to resist earth or water pressure. Sometimes, high compound wall is buttressed. The thickness of buttress gradually decreases from the base to the top of a wall, i.e., a battered face is made.

BUTTRESS THREAD : A screw thread of buttress form, used to take heavy axial load in one direction only.

BUTT STRAP : A cover strap used in making a butt joint.

BUTT WELD : A weld between two pieces butting end to end.

BYATT : A horizontal member that supports docking or walk-ways in trench excavations.

BYE-CHANNEL : A diversion cut or a spillway which leads water round a reservior, when full.

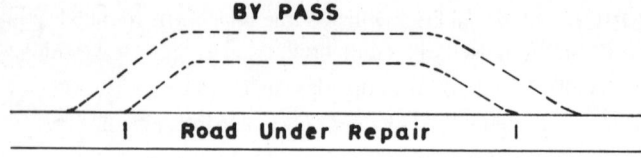

BY-PASS : An arrangement for diverting a traffic flow or the flow path of a liquid.

C

CABIN : A hut at a construction site where the supervisor keeps his working drawings and other documents.

CABINET FILE : A half round, fine, single-cut file, used in cabinet making to make a smooth finish on a joint.

CABINET FINISH : A polished or varnished finish on hard-wood.

CABINET SCRAPER : A flat steel piece used to remove plane marks by drawing over the wood surface with a view to making the surface suitable for sand papering.

CABINET PROJECTION : It is an oblique projection to show the three dimensions of an object. The lines of sight make an angle of 45° with the plane of projection i.e., the receding lines are drawn at 45° and the measurements on the receding axis are made one-half the measurements on the other axes.

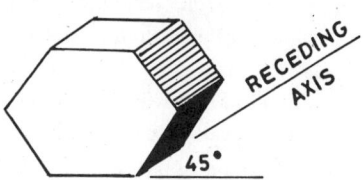

CABLE : A collection of ropes or conductors, insulated and protected where needed.

CABLE DRILL : A percussion drill (heavy drilling rig) used in drilling 100 mm to 300 mm.vertical holes in clayey soil, soapstone or hard limestone. A steel wire rope hung from the derrick (tower) of the rig which lowers the tools into the hole and raises up.

CABLE DUCT : A hole cast in concrete through which electric cables or prestressing cables are pulled.

CABLEWAY : A material handling device used in construction of bridges, dams and tall buildings, ore transportation and excavation. A heavy steel rope, called a track cable, is supported from two towers and a carriage on grooved track wheels can be pulled to the desired place on the track cable by the hoisting rope. The load hangs from the carriage by the hoisting rope.

CAISSON : A water-tight box like chamber made of timber, steel or concrete is usually sunk when the foundation of a structure is to be laid under water and the depth of water is great. Fresh water is pumped out from the chamber and the bed material is excavated within it and scooped out. Gradually the caisson goes down to the required depth of foundation and ultimately it becomes a part of the foundation.

CAKING : The setting of dense pigments into a compact mass in a paint, which can not be readily mixed by stirring.

CALCINATION : Heating of ores at a high temperature to drive off CO_2 and H_2O from ores.

CALCITE : Crystalline $CaCO_3$ found in marble stones.

CALCIUM CARBONATE : The chemical name for chalk and marble stones ($CaCO_3$).

CALCIUM CHLORIDE : It is a white deliquescent and hygroscopic salt commonly used for quickening the setting time of cement and accelerating the hardening of concrete.

CALCIUM HYDROXIDE : It is slaked lime, $Ca(OH)_2$. This is obtained when water is added to quick lime. $CaO + H_2O = Ca(OH)_2$.

CALCIUM OXIDE: It is quick lime, CaO. It is obtained by burning calcium carbonate. $CaCO_3 = CaO + CO_2$.

CALCIUM SULPHATE : $CaSO_4$, a mineral anhydrite. Also, called 'Gypsum'.

CALIFORNIA BEARING RATIO : It is a measure of shearing resistance of a soil (bearing power) to penetration under controlled density and moisture conditions.

CALIPER LOGGING : Measurement of well diameters along a well with the help of a hole caliper, which consists of four spring - extended arms and an electric resistor.

CALLUS : It is a rindgall, an extra growth over a surface wound of a tree.

Callus

CALLIPERS : A measuring instrument comprising two bent legs pivoted together just like a divider. Callipers are used to measure outside or inside diameters/dimensions.

CALORIFIC VALUE : The quantum of heat liberated by complete burning of unit weight of a fuel, expressed in heat units (BTU or K Cal) per unit weight.

CAMBER : A cross fall in a road curvature (across the road width) to permit water to run-off a road. It is also the hog provided in a girder to counter balance the effect of deflection.

CAMBER ARCH : An arch with a level extrados and about 1% rise in intrados to counteract the sagging appearance.

OUTSIDE
CALLIPERS

INSIDE
CALLIPERS

CAMBER BEAM : A beam cambered on its upper surface.

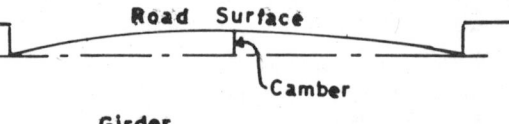
Road Surface
Camber

CAMBER BOARD : A template used in forming a camber.

CAMBER ROD : A tie rod (belly rod) in a trussed beam.

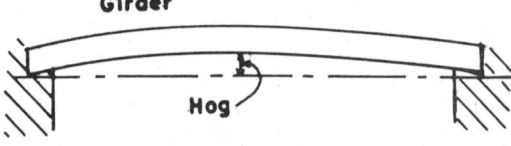
Girder
Hog

CAMBER SLIP : A timber piece, cambered on its upper surface used for forming the soffit of a camber arch.

CAMBIUM : The material just below the bark of a living tree, which gets converted into wood with its age.

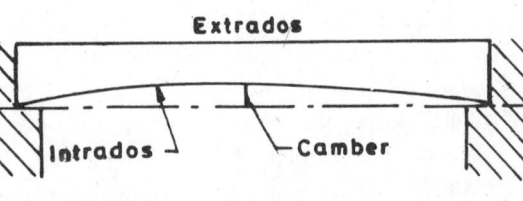
Extrados
Intrados
Camber

CAMP SHEATHING : This is a retaining wall in form of a sheath constructed to protect the river bank at a barge bed. Two rows of interconnected timber piles are placed in position, the space between (2 m to 3 m wide) being filled with puddled clay, brick bats and stone boulders.

CANAL : A channel to carry water for navigation, irrigation and other purposes.

CANALIZATION : The river course is divided into several reaches by providing dams, barrages and weirs with canal system to facilitate navigation, to control flood, to generate power and to irrigate land.

CANAL LIFT : A device for passing barges through a lock with a lift greater than 15m.

CANAL LOCK : See illustration.

CANOPY : A hood or sunshade projected from a wall.

CANT : Banking or Superelevation; It is the transverse inclination (inward tilt) given to a carriageway on a horizontal curve to minimise the overturning effect of a moving vehicle due to the centrifugal force acting on it.

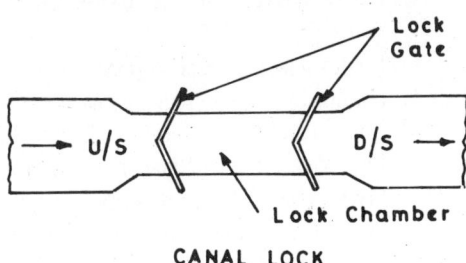

CANAL LOCK

CANT BAY : A bay window with three straight sides provided in a canted wall.

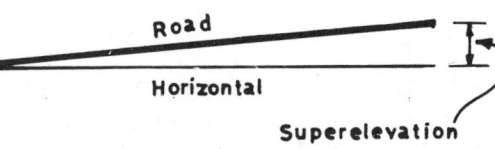

CANT BRICK : A splayed brick.

CANTED BEARING PLATE : Used in fixing rails.

CANT STRIP : A tilting fillet.

CANTED WALL : Two walls joined at an angle.

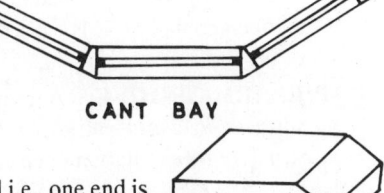

CANT BAY

CANTILEVER : An overhanging beam from a wall i.e., one end is fixed and the other end free.

CANTILEVER ARM : The arm projected from a fixed support.

CANT BRICK

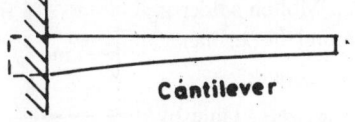

Cantilever

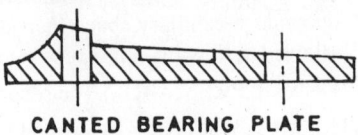

CANTED BEARING PLATE

CANTILEVER BRIDGE: In a cantilever bridge, the outer span of the cantilever arms is anchored down at either end and overhanges into the central span

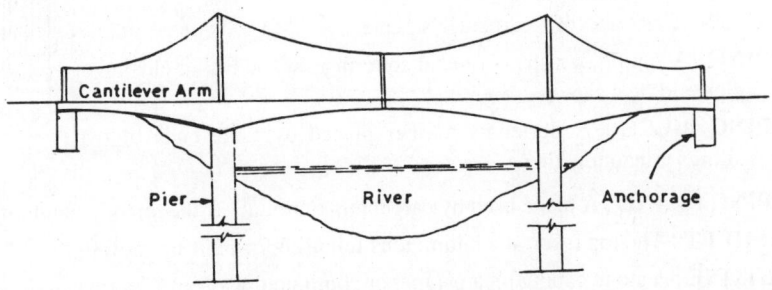

DOUBLE CANTILEVER BRIDGE

CANTILEVER CRANE: It is a transporter crane, one or both ends of which are overhanging.

CANTILEVER FORMWORK : It is a climbing formwork.

CANTING STRIP : A water table.

CANTING TABLE : A saw bench which can be tilted at any angle required for bevel cutting.

CANTILEVER FOUNDATION : A foundation of a column or a wall which does not get adequate space for a truly central base. Thus, the foundation is made cantilever and the loading becomes eccentric.

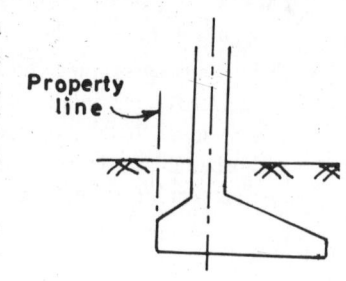

CANTILEVER WALL : A retaining wall made stable by the load of the retained material on its heel.

CAPACITY CURVE : A graph from which the volume of a tank can be read at any level of the liquid in the tank.

CAPACITY FORMULA : $Q = Av$; where Q is capacity, A is C.S. area and V is velocity of flow.

CAPILLARITY : The rising of a fluid in a very thin tube above the level of the fluid in contact with the tube in an open container. It is caused by the surface tension of the fluid.

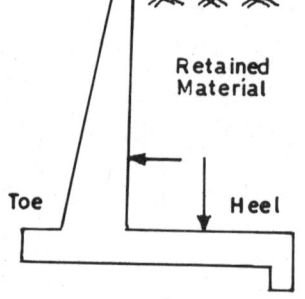

CAPILLARY GROOVE : A groove kept between two surfaces to avoid entry of water into a building or structure by capillary movement of water.

CAPILLARY JOINT : Also known as 'Sweat Joint'. This joint is made in light gauge copper tubing. The tube is inserted into a fitting slightly larger in diameter, so that the annutar space becomes a capillary space (hair diameter). Molten solder is allowed to flow into this space spreading round the perimeter.

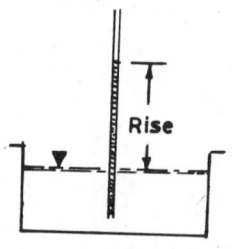

CAPILLARY PRESSURE : The seepage force of water.

CAPILLARY WATER : Water held above the water table by capillary action.

CAP : A plug used in a pipeline, See illustration.

CAP IRON : Back iron of a carpenter's plane.

CAPPING : A caping or a strip of metal covering a wood roll or closing of a pipe end by a cap.

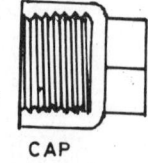

CAP

CAPPING PIECE : A horizontal timber placed over the ends of two wallings joined together.

CAPPING PLANE : A plane used by carpenters to round off the top of a handrail.

CAP SHEET : The top layer of a bituminous felt used for built-up roofing.

CAPSTONE : A stone caping on a parapet or compound wall.

CARBONATION : The natural process of slow hardening of lime mortars to convert into calcium carbonate by absorption of CO_2 from air.

CARBON BLACK : A black pigment–bone black, lamp black or vegetable black, used in paint.

CARBON CYCLE : A cycle in the decomposition of organic wastes by bio-chemical reactions. Complex organic carbonaceous matters are coverted into CO_2 and H_2O

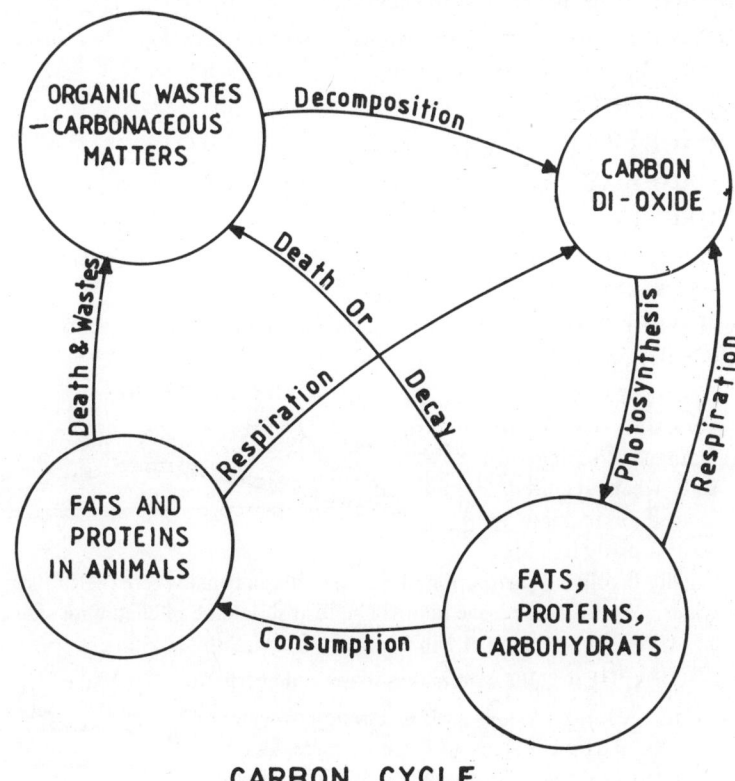

CARBON CYCLE

CARBONIZATION : The destructive distillation of coal, wood and bone for obtaining coke, charcoal and bone black with various liquids as by-products.

CARBON STEEL : Usually high carbon steel, the properties being guided by the carbon content in steel.

CARBORUNDUM : A silicon carbide harder than quartz. It is a refractory and abrasive material.

CARBURIZING : The case hardening of steel to produce a hard case of cementite with a ductile core. The method of introduction of carbon into the steel surface at a critical temperature.

CARCASE OR CARCASS : The load bearing portion of a structure without doors and windows.

CARNAUBA WAX: A hard wax used in polishing wood and stoving varnishes.

CARPENTER: A man working with wood to make wooden frames, structures, furnitures, etc.

CARPENTER'S HAMMER : A claw hammer.

CARPENTRY : Art of wood works; wood craft.

CARPET :A road surface made by laying bitumen or tar concrete to a thickness of 25 mm.

CARPET STRIP : A wood strip fixed to the floor just beneath a door to minimise the gap.

CARPORT : A car shelter close to a house.

CARRIAGE : A stair horse ; An inclined timber fitted between two strings against the underside of a wide wooden stair.

CARRIAGE BOLT : A coach bolt .

CARRIAGE PIECE : A carriage used in stairs.

CARRIAGEWAY : The part of a road over which vehicles move.

CARRIER : The container travelling on the track rope of a cableway.

CARTESIAN CO-ORDINATES : Co-ordinates measured from fixed axes that are at right angles to each other.

CARVED BRICKWOOD : Dressed brickwork with very thin joints of 1.5 mm thickness.

CART CROSSING : A communication work across a canal to facilitate crossing of carts where required, instead of making a costly bridge. This

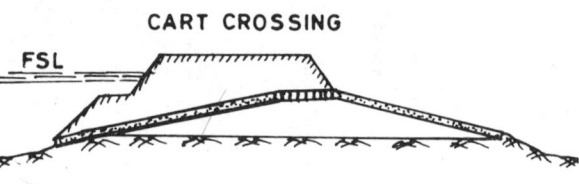

CART CROSSING

usually involves construction of a ford across a canal where the full supply depth of water does not exceed one metre. The ford should be pitched with side cut-off walls and with ramps sloping at 1 in 12 to 1 in 20.

CARTOGRAPHER : One who makes maps and charts from field data.

CASCADE : A series of steps used in aeration of water or a series of vertical steps in large diameter drain or sewer to avoid very steep slopes for flow of storm water and sewage.

CASE : The facing brick or stone of a wall.

CASE BAY : Joists enclosed between two binders.

CASED FRAME : Hollow, fixed part of a sash window, with sash weights and pulleys.

CASCADE   **AERATOR**

CASE HARDENING : Hardening of steel surface by carburizing, cyaniding or nitriding.

CASEIN GLUE : A glue made from milk, which is more resistant than animal glue. It is sometimes used as a binder for paints over cement and lime works.

CASEMENT : The hinged fanlight of a casement window.

CASEMENT DOOR : A hinged door or window, fully glazed. Also, known as French door.

CASEMENT STAY : A bar used for fixing the open casement in position.

CASING : (i)   Form-work for concrete casting.

      (ii)   Steel pipe lining to wells or bore holes.

      (iii) An enclosure made of timber or other materials on the face of a wall or ceiling to accomodate pipes and cables.

CASSAVA GLUE : A starch glue prepared from tapioca plant.

CASTELLATED BEAM : A steel beam formed by cutting a R.S.J. in a zig-zag pattern along the web and then the two halves are rearranged with the crests meeting and are welded. This results 50% increase in depth with an increase in moment of resistance, without increasing its weight.

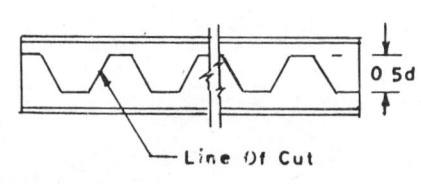

ORIGINAL   BEAM

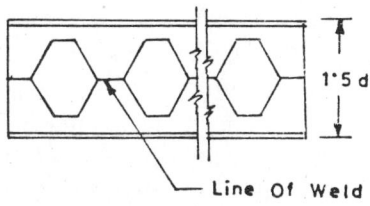

BEAM AFTER CUTTING AND WELDING

CASTING YARD : A block yard.

CAST-IN-SITU : Cast at site or cast in place

CAST IRON : This is obtained by purifying foundry pig iron. Cast iron contains 2 to 5% of carbon with other impurities. There are varieties of C.I. used for different purposes.

CAST STEEL : It is the strongest and most uniform steel used for high grade instruments. Blister steel and shear steel are reheated at a low temperature and made compact and homogeneous by hammering.

CAST RESIN : A synthetic resin, epoxide resin.

CAST STONE : Imitation stone made from a quality-controlled concrete core.

CATALLOY : Resin reinforced with glass fibres.

CATALYST : A substance that hastens a chemical reaction. This is an acce- lerator.

CATCH BASIN : A term for a 'catch pit' used in USA.

CATCH FEEDER : A ditch made for the purpose of irrigation.

CATCHMENT AREA : Drainage basin, drainage area or gathering ground to feed water to a reservoir or lake. Also, an area drained by a water course.

CATCH PIT : A pit made in a drainage system to receive storm water or sullage by arresting the detritus matters in the pit.

CATCH POINT: In a railway track on an up-grade, a point where the rails are so cut that an unhitched wagon while running back, is harmlessly derailed.

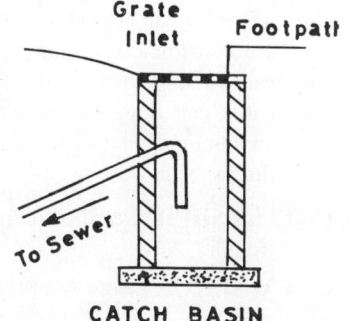

CATCH BASIN

CATENARY: A peculiar curve of varying radius formed by a rope hung between two points, quite a long distance apart.

CATENARY CORRECTION : Sag correction in surveying.

CATENARY SUSPENSION : Overhead suspension of an electric power cable by vertical links hung from a tightly-stretched steel wire rope above the cable, with a view to maintaining the same height all throughout the suspended length.

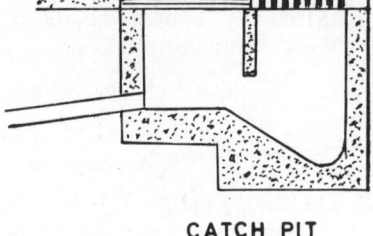

CATCH PIT

CATERPILLAR GATE : A heavy duty steel gate used to control flow of water through a spillway.

CAT EYE : A pin knot in timber.

CATHODIC PROTECTION : A method of protection of steel water mains and underground steel structures from corrosion. This is provided by using direct current to feed electrons into the metal and render it cathodic.

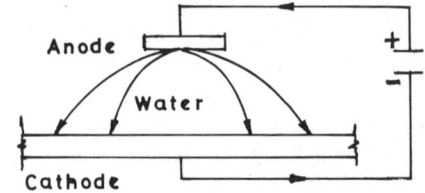

CATHODIC PROTECTION

CATION EXCHANGER : Cation or base exchanger is used to soften water or to remove iron and manganese from water.

CAT LADDER : A ladder to give access to the workmen for easy inspection and repair works. This ladder is provided with rods or cleats fixed ar regular intervels for stepping.

CATSKILL AQUEDUCT : A rock tunnel as shown in illustration, used in transportation of water from source to the distribution system.

CATTLE CROSSING : When the depth of water in a canal is more than one metre through out the year, cattle crossing is provided at a low cost instead of making a foot bridge. For cattle crossing, ramps are constructed on either bank of the canal and at such a distance apart, that cattle can cross the flowing canal by swimming from one side of the canal to the opposite side.

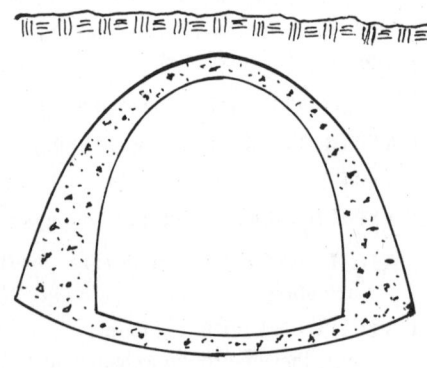

CATSKILL AQUEDUCT

CATTLE SHED : See illustration. A home for cattle with provision of manger and water trough.

CAT WALK : A gangway of restricted width provided to give access for inspection and repair.

CAUL : A hot-pressed thin aluminium sheet to protect veneered wood.

CAULKING : Blocking a gap of a pipe joint by driving in hemp rope, jute or lead wool tightly so that the joint becomes water-tight or air tight.

CAULKING GUN : A pressure gun used for sealing joints with mastic in patent glazing.

CAULKING TOOL : A blunt cold chisel having offset shape, used for caulking a joint .

CAUSEWAY : A road carried over a marshy land or low-lying area, over which drainage water flows during monsoon.

CAVITY BLOCKS : Precast concrete blocks of such

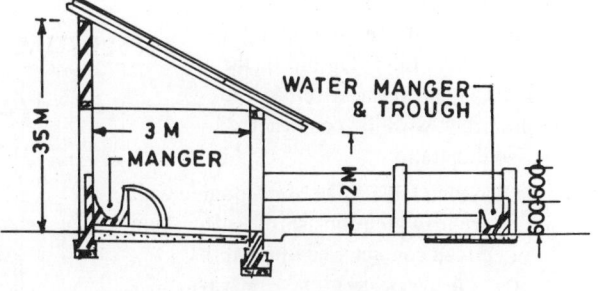

CATTLE CROSSING

CATTLE SHED

shapes that a cavity wall can be formed by laying these blocks over each other.

CAVITY FLASHING : A damp proof course crossing the gap of a cavity wall.

CAVITY WALL : A wall made of two leaves with a space in between. The leaves are connected by wall ties at intervals.

C.B.R. TEST : The test developed by the California State Highways Department for evaluation of subgrade strengths required in design of road pavements. The test can be doen on all soils.

CEILING : The soffit of a covered space or room.

CEILING JOIST : A joist that carries a ceiling beneath it, but not the floor above it.

CEILING STRAP : A wooden strip nailed to rafters or floor joists, required for suspending ceiling joists.

CELESTIAL EQUATOR : The great circle traced on the celestial sphere passing through the earth's centre and is perpendicular to the polar axis.

CELESTIAL HORIZON : The great circle traced upon the celestial sphere by a plane perpendicular to the Zenith-Nadir line and passing through the earth's centre.

CELESTIAL MERIDIAN: The great circle traced on the celestial sphere by intersection of a plane passing through the two celestial poles.

CELESTIAL POLES : The points at which the polar axis when extended, intersects the celestial sphere.

CELESTIAL SPHERE : An imaginary vast sphere with the earth as its centre. The stars appear to lie on the surface of sphere.

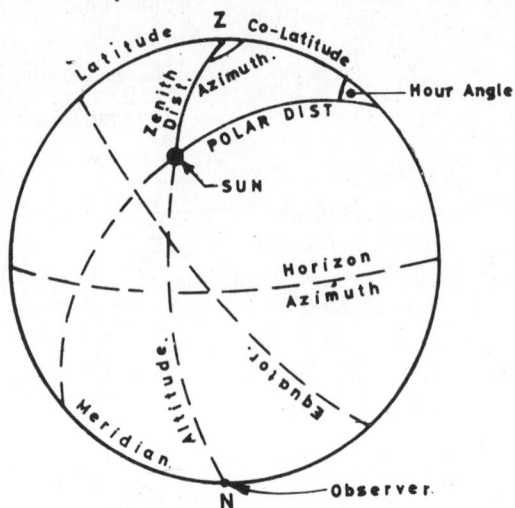

CELESTIAL SPHERE

CELLAR : A room or space whose major part is below ground level, used for the purpose of storage of materials or installation of a boiler for central heating system.

CELLULAR BULKHEAD WALL : A gravity quay wall consisting of a number of sand-filled, semicircular or circular cells constructed of steel sheet piles. On top of the cells at the front, a reinforced concrete wall is constructed. See illustration.

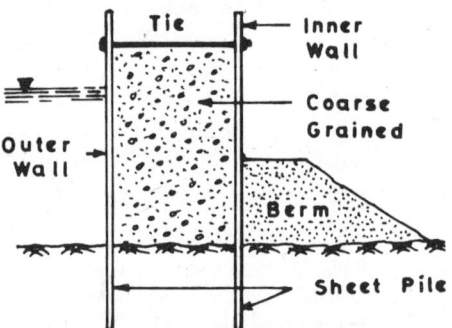

CELLULAR COFFERDAM

CELLULAR COFFERDAM : A double wall cofferdam which consists of a succession of cells in contact. See illustration.

CELLULAR CONCRETE : A term for 'aerated concrete'.

CELLULOSE ACETATE : An excellent insulating material which is non-inflammable.

CELLULOSE NITRATE : Nitrocellulose, an explosive.

CELLULOSE SHEET : A sheet made from a mixture of saw dust, wood flour, cork dust and pigments with nitro-cellulose in gelatine form on a backing of woven jute, used as a floor finish sheet.

CEMENT : A binder having cementing property which is used in preparation of a mortar or concrete. There are various types of cement used for different works.

CEMENT COLLAR JOINT : A joint made in cement concrete pipes as shown in illustration.

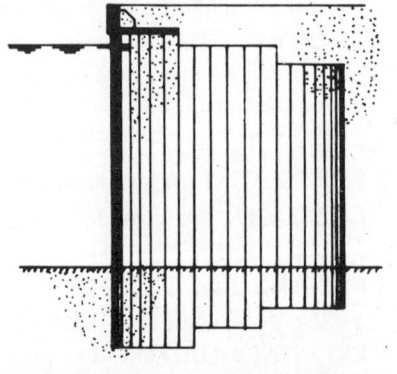

CELLULAR BULKHEAD WALL

CEMENT CONCRETE : A concrete made of cement, sand and stone chips or khowa (brick ballasts) with adequate quantum of water.

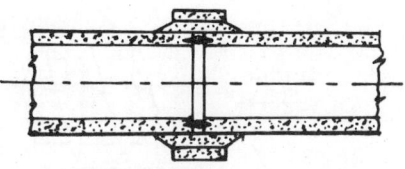

CEMENT FILLET : A weather fillet made of cement-sand mortar to fill in the corner between slates or tiles and a wall.

CEMENT COLLAR JOINT

CEMENT GROUT : A thin paste of cement and sand applied by a spray gun to fill in the cracks and crevices found in a concrete or masonry structures.

CEMENT MORTAR : A mortar prepared by mixing cement, sand and water in required proportion. This is required for brickwork, Plastering and preparation of concrete.

CEMENT PAINT : A cement slurry made of cement and water, is applied as a paint to make a surface water-proof.

CEMENT RENDERING : Treatment of a surface by applying cement-sand slurry.

CEMENT-RUBBER LATEX : Used in making a jointless flexible floor. This consists of cement, marble chips, cork and wood chips. The floor can be buffed like a terrazzo floor.

CEMENT SCREED : A strong cement mortar laid on a concrete slab.

CEMENT SLURRY : A very thin mixture of cement and water that can be used as a wash over a wall.

CEMENT-WOOD FLOOR : A jointless floor made of cement, saw dust, wood flour, sand and pigment.

CENTERING : A curved framework made of timber, required for construction of an arch or a dome.

CENTRAL HEATING : A system of heating individual rooms in a building by circulating hot air or steam through pipes, the heating arrangement being located at one place.

CENTRE OF GRAVITY : A point in a body at which its weight acts and it will remain in a balanced state, if supported at that point.

CENTRE OF MASS : Centre of gravity.

CENTRE OF PRESSURE : A point at which the total pressure of a fluid on an area is considered to act.

CENTRIFUGAL PUMP : A pump in which the impeller blades rotating at a high speed throws water outwards and discharges to the required head.

CENTRIFUGATION : A process widely used for dewatering sewage sludge, for thickening slurries and for separating liquids of different density.

CENTRIFUGE : A power-driven machine used for centrifugation. Used for dewatering of sludge continuously and discharging sludge cake and liquor separately. See illustration.

CENTROID: The centre of an area, about which the static moment of all the elements of the area is zero.

CENTRIFUGAL PUMP

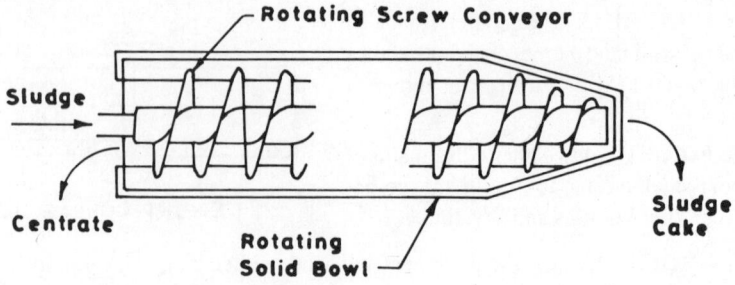

Rotating Screw Conveyor

Sludge

Centrate

Rotating
Solid Bowl

Sludge
Cake

CENTRIFUGE

CENTROIDAL AXIS: The axis passing through the centroid of an area or mass. The centroidal axis and the neutral axis of a body may not be same.

CESS : A drain along the base of a railway cutting to drain out the water.

CESSPIT : An underground pit or tank made of brick masonry or concrete, where sewage from an individual house or from a group of houses is collected and anaerobic decomposition takes place. The sewage solid is converted into humus. The gas produced escapes and the effluent passes out. The tank may be of pervious type or impervious type.

CESSPOOL : A cesspit. See illustration.

CHAIN : (i) A flexible cord made of links or loops, used in tieing or suspending a member.

(ii) A chain used for measuring lengths in surveying consists of wire links, 100 in number. An Engineers' chain is 100' and a Gunter's chain is 66' in length.

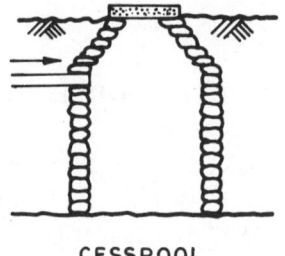

CESSPOOL

CHAINAGE : A length measured by a chain.

CHAIN BLOCK : A chain-pulley block or differential pulley block used as a hoisting tackle.

CHAIN BOOK : A field book used during chain surveying to record the chainages and measured lenghs of offsets.

CHAIN-BUCKET DREDGER : A bucket ladder dredger used in dredging operation.

CHAIN LEWIS : See illustration. A device for lifting stones, consisting of three rings & two curved steel legs. The lewis is placed into the dovetailed slot of a stone . As the crane lifts the top ring, the side rings get pulled apart and the legs grip the stone.

CHAIN LINE : A line on ground along which a chain is laid for measuring lengths and taking offsets.

CHAIN PUMP : A device to lift water for short heights by means of discs passing up a pipe on chain.

CHAIN OF LOCKS : A series of interconnected locks used in irrigation and flood control device, in which each lock gate

CHAIN LEWIS

is followed immediately by another. The head gate of a lower lock is the tail gate of the upper lock.felling trees and cross-cutting logs.

CHAIN SAW : A power-driven cross-cut saw with a projecting jib for travelling of the chain carrying cutting picks. This saw is used for felling trees and cross-cutting logs.

CHAIN SLING : A sling made of wrought iron used in handling heavy materials.

CHAIN SURVEY : The simplest form of surveying moderately small areas with chain, tape, ranging rod and offset staff. In chain surveying, a skeleton framework of straigth lines is formed.

CHAIN TIMBER : An old practice of embedding timbers in a chain into a wall with a view to strengthening the wall. In modern days this has been discarded, because the timber rots inside the wall.

CHAIN TONGUE : A heavy pipe grip used by the plumbers to hold a pipe firmly with the help of a chain linked to a toothed bar.

CHAIR BOLT : A bolt used to fix up the rail chair with the sleeper.

CHAIR RAIL : A wooden moulding or hard gypsum plaster moulding provided at the dado rail height of a wall to protect the wall from striking of the chair backs. It is hardly used now-a-days.

CHALKING : Breaking-up of pigment films on a painted surface due to the decomposition of binder by weathering action.

CHALK LINE : A chalk line made by masons on a brick surface or a plastered surface by plucking (striking) a stretched thread rubbed well with chalk, against the surface.

CHAMBER : An enclosed space with or without top cover.

CHAMBER LEVEL TUBE : A level tube provided with an air chamber to add air to the bubble by tilting, which is required for adjustment of bubble length due to variation in temperature.

CHAMFER : A bevel.

CHAMIER'S FORMULA : A British flood-intensity formula.

$$Q = 640 \, iKa^{3/4}$$

where Q is maximum flood intensity in cusecs,

i is average rate of rainfall, in./hr.

K is the co-efficient of run-off and

a is the catchment area in square miles.

CHANGE FACE : In surveying, reverse face is made by rotating the telescope vertically as well as horizontally through $180°$ to obtain the vertical circle at the opposite face.

CHANGE POINT : A turning point used in a level survey at which two readings are taken-a back sight and a fore sight.

CHANNEL : A passageway with two sides and a base.

CHANNEL BRICK : A special brick of channel shape used for construction of a narrow drain. See illustration.

CHANNEL PIPE: An open pipe of semi-circular section used for draining wate.

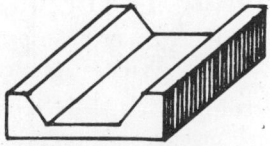

CHANNEL BRICK

CHANNEL SECTION: A rolled steel section of channel shape, used for making built-up box girder and for other structural works.

CHANNEL SECTION

CHANNEL TILE : The under tile used in Italian roof tiling.

CHARGING DOOR : A door through which materials are charged into a system.

CHARGING HOPPER : A hopper through which materials are fed into concrete mixer or into a furnace.

CHASE : A groove made into a wall or floor to receive pipe, cable or conduit.

CHASE MORTISING : Cutting a mortise in a timber member, which is in position in a frame.

CHASE WEDGE : A handled wooden wedge required in bossing lead.

CHECK : (i)   A crack formed in a converted timber across the rings and along the grains due to stresses developed in seasoning.

(ii)   A deep crack on a painted surface.

(iii)   Verification of a work.

(iv)   A structure to control the water level in a canal.

(v)   A device to control the flow of a fluid.

(vi)   A tract of land between ridges holding irrigation water.

(vii)   The side of a mortise or tenon.

CHECK CUT : A bevelled cut at the bottom of a jack rafter or hip rafter or valley rafter

CHECKED BACK : Rebated or recessed.

CHECK FILLET : An asphalt kerb provided on a roof to guide the flow of rain water.

CHECK GROUND : A rebated timber edge to receive lathing to cover an opening.

CHECK LOCK : A device to hold a door bolt in locked position.

CHECK NAILING : Double nailing of thin slates through a hole.

CHECK RAIL : A guide rail used in points and crossings and turnouts in a railway track.

CHECK THROAT : A very thin groove under a door or window sill.

CHECK VALVE : A valve having controlling device to regulate the flow of a fluid through it.

CHEESINESS : A paint film on a surface has cheesiness (soft) prior to complete drying.

CHEMICAL COAGULATION : Coagulation by the use of coagulants (chemicals like alum, ferrous sulphate and ferric chloride) in a turbid water or sewage for floc formation and clarification by settling of suspended solids. For efficient coagulation and clarification with least sludge production, selection of coagulant and adjustment of pH are essential.

CHEMICAL FLOCCULATION : Floc formation due to chemical coagulation. Flocculation of suspended matters in a water or sewage takes place within 20 to 30 minutes time. Chemically flocculated sewage solids settle rapidly.

CHEMICAL GAUGING : Also known as chemi-hydrometry. It is a device to measure the quantum of flow of water by determining the dilution of a chemical solution at the downstream side with the known strength of the solution injected at the upstream side.

CHEMICAL OXYGEN DEMAND : Abbreviated as C.O.D. Quantum of oxygen required for chemical oxidation of organics in a liquid. Under acid conditions, organic matters are converted to CO_2 and H_2O regardless of the biological assimilability of the organic substsances.

CHEMICAL PRECIPITATION : A method sometimes adopted in sewage treatment, by using chemicals like alum, lime, ferrous sulphate, ferric chloride, etc. Chemical flocculation takes place and suspended sewage solids settle rapidly.

CHEMICAL ROCKS : Sedimentary rocks precipitated by chemical activity or left by evaporation. Chemical sediments are usually microscopic.

CHEMICAL SLUDGE : Chemical precipitation removes more suspended matters as compared to plain sedimentation. Thus, more sludge is produced in chemical precipitation. The dry solids in a chemical sludge may be double than those of plain sludge. However, chemical sludge digests readily.

CHEMICAL STABILISATION : Stabilization of soil (sands and silts) by injection of chemical solutions into the voids of the soil. The chemicals impart cohesion to the soil Solidification of soils by injection of chemicals consists of successive injection of solutions of water glass and calcium chloride which react with soil and form a binder.

CHEMISORPTION : Chemical adsorption, where the adsorbate is held to the adsorbent by chemical forces (valance) instead of physical forces (vander waals).

CHEQUERED BRICK : A patterned brick as shown in illustration is used in wall decoration or in making a non-slip floor.

CHEQUERED PLATE : A steel or cast iron plate with patterns on its surface, used to make a non-slip floor in a factory or in a damp or wet place.

CHEQUERED WORK : Rubble walls built of bricks and stones in alternate squares.

CHEQUERED BRICK

CHERRY PICKER : An overhead travelling crane used to lift heavy vehicles and set it down on the neighbouring ground.

CHEVRON DRAIN : Also known as "Herring Bone Drain". Stone-filled trenches in railway cuttings laid in herring bone pattern for draining water into a buttress drain. The butress drain is laid along the steepest slope.

CHEZZY'S FORMULA : Used for determination of velocity in open channel flow.

$$V = C\,(RS)^{0.5} \quad \text{where} \quad V = \text{Velocity of flow}$$

C = A constant determined by kutter's equation.

R = Hydraulic Mean Depth

= Cross-sectional area/Wetted Peremeter

S = Bed Slope

CHICK 'S LAW: The effect of time on germicidal value of chlorination is expressed by Chick's law as

$$\log\,(N1/N2) = \text{kt}$$

Where, $N1$ = no. of organisms present initially;

$N2$ = no. of organisms present after a contact time t ;

and k = a constant.

CHIMNEY: A tall vertical structure made of brick masonry or pipes or steel sheets, containing a vertical flue.

CHIMNEY BACK : The wall behind a fireplace.

CHIMNEY BAR : A wrought iron bar built into the fireplace jambs to carry the brickwork over it.

CHIMNEY BLOCK : Precast concrete blocks used in lining the inner surface of a chimney.

CHIMNEY BOND : A brick bond with stretchers, used in making a half brick thick partition wall in a chimney.

CHIMNEY BREAST : The chimney wall being projected into a room is attached to the fireplace and the flue.

CHIMNEY CAP : An ornamental chimney hood designed to improve the draught.

CHIMNEY COWL : A revolving ventilator provided over a chimney.

CHIMNEY GUTTER : A back gutter.

CHIMNEY HOOD : A chimney cap which is usually made ornamental.

CHIMNEY LINING : A flue lining i.e., lining of the inner surface of a chimney.

CHIMNEY POT : A clay pipe provided at the top of a chimney stack to lead the smoke and flue gases clear of the masonry.

CHIMNEY SHAFT : The part of a chimney that stands freely.

CHIMNEY STACK : A chimney containing several flues projected above a roof.

CHINAMAN : A gantry-like structure with a ramp to facilitate loading a lorry below, by pushing earth or garbage or any other material with the help of a bulldozer or scraper.

CHIPPING : Removing surface defects from iron or steel by chipping chisel. Chipping is also done to remove the surface defects from concrete or to make a concrete or masonry surface rough for plastering.

CHIPPING CHISEL : A cold chisel.

CHIPPING HAMMER : A compressed-air tool used by welders to clean a welded surface.

CHIPPINGS : Crushed stone, wood or any other material.

CHISEL : A sharp bevelled-edge cutting tool of various forms with a wooden handle used by carpenters and masons.

CHISEL KNIFE : A stripping knife having a square edge.

CHLORAMINES : When chlorine is added to a water containing organic nitrogen or ammonia, chloramines are formed due to chemical reaction. These are referred to in practice as 'combined available chlorine'.

CHLORDANE : A pesticide which when applied in manholes or sewers, controls the growth of vermin and other living pests.

CHLORINATED COPPERAS : A chemical coagulant used for effective sludge conditioning in activated sludge process. It is capable of reducing BOD by 70 to 80% and suspended solids by 80 to 90%. This is also very effective in removing colour of water at a low pH-value.

CHLORINATED PARAFFIN WAX : Resinous liquid used in preparation of fire-resisting paints.

CHLORINATION : (i) Application of chlorine to a sewage treatment plant effluent to reduce bacterial count and lower BOD.

(ii) Chiefly adopted for disinfection of water. Also, used for removal of colour and odour of water.

CHLORINE-AMMONIA TREATMENT : Chlorine and ammonia are applied simultaneously to a water containing phenols, for the purpose of disinfection, without producing any disagreeable taste and odour. This treatment is also adopted for disinfection of water containing no phenolic conpounds.

CHLORINE DEMAND : As a strong oxidising agent, chlorine reacts with reducing substsances to produce so called 'chlorine demand'. Actually, it is the demand of chlorine to oxidise the reducing substances present in a water, to remove the colour of water and to kill or inactivate the organisms present in water. Thus, chlorine used up in disinfection of water is a part of the demand.

CHLORINE INJECTIOIN TOWER : A tower in which chlorine soloution is prepared for applicationto a treated water. See illustration.

CHLORINE RESIDUAL : This is either free chlorine residual in form of Hocl & Ocl or combined chlorine residual in form of chloramines in a disinfected water.

CHORD : The top or bottom flange or boom i.e., horizontal part of a girder.

C-HORIZON : The virgin soil below the top soil of A-horizon and B-horizon.

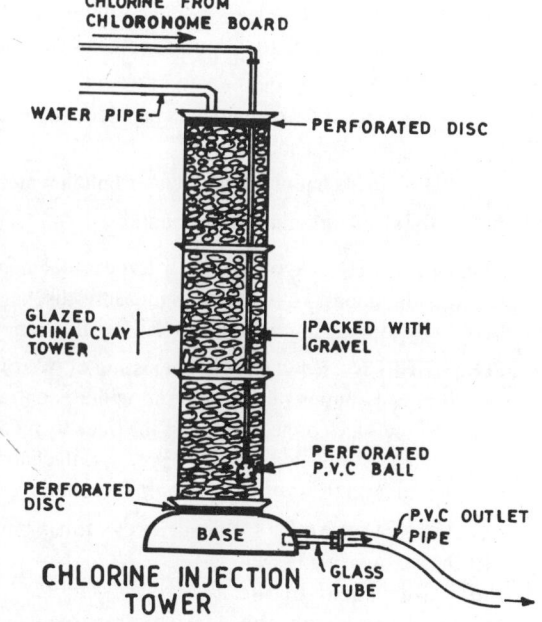

CHLORINE INJECTION TOWER

CHROMATING : (i) Providing a base coat on iron and steel surface with lead or zinc chromate.

(ii) A protective coating given to articles of magnesium alloys by dipping them in a hot solution of alkaline dichromate.

CHROME GREEN : A stable green pigment consisting chiefly of chromic oxide obtained by igniting mercurous chromate or potassium chromate and ammonium chromate.

CHROME STEEL : Used in manufacture of stainless steel. Addition of chromium increases strength and hardness of steel.

CHROME YELLOW: Lead chromate, the most valuable yellow pigment obtained from the precipitation of lead salt.

CHROMIUM PLATING: A method of preservation of iron and steel from rusting. Iron or steel to be coated with a thin film of chromium is made cathode and chromium solution as anode in the process of chromium plating.

CHROMIUM RECOVERY : It is made by an ion-exchange system. One anion bed and two beds with cation-exchange resins are used. The cation impurities are removed by

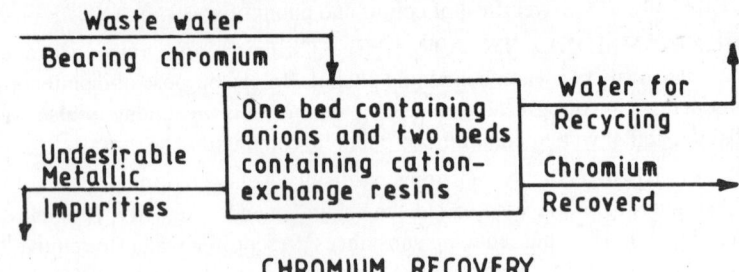

CHROMIUM RECOVERY

the resin and chromate ions are removed in anion bed.

CHRONOMETER : A valuable piece of equipment used earlier to give 'Greenwich Mean Time' especially for nautical purpose. It is still in use in conjunction with radio and other signals.

CHUTE : (i) A vertical or inclined duct provided in a multi-storeyed building for disposal of garbages/solid wastes.

(ii) A steep channel to lead water onto a water wheel for power generation.

CIMENT FONDU : High-alumina cement.

CIPOLLETTI WEIR : A weir with a trapezoidal notch having side slopes 4 : 1, used to measure discharge directly by dipping a graduated stick. See illustration.

CIPOLLETTI WEIR

C. I. POT SLEEPER : Bowl sleeper consisting of two oval shaped pots or bowls placed interted under each rail. The two bowls are connected across the track by means of a tie bar. Onthe top of each pot, a rail chair is provided to hold the rail with the help of cotters.

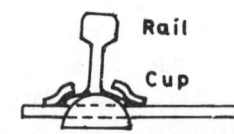

C. I. POT SLEEPER

CIRCLE OF INFLUENCE : Influence circle. It is a circle with the centre of a tubewell and radius equal to the distance to which the ground water table is affected by the pumping and consequent draw down. The radius of the circle of influence is tentatively determined by $R = (H — h)/S$

Where S is the slope of ground water table. See illustration.

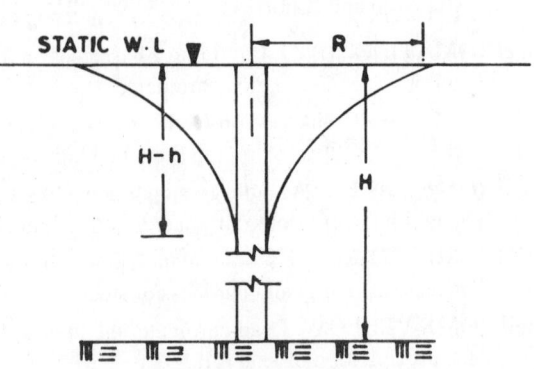

CIRCLE ON CIRCLE : Double curvature used in timber works which is curved both in plan and in elevation.

CIRCULAR ARCH : See illustration. This is used in circular opening in a building and this is usually made at a high level.

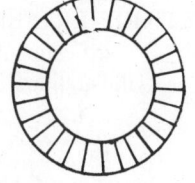

CIRCULAR ARC METHOD : Slip circle method of determining the stability of an earth slope. It is assumed that the failure occurs by shear along a circular arc.

CIRCULAR LEVEL : A level tube having the upper glass surface spherical. This is used in surveying instrument.

CIRCULAR PLANE : A compass plane used in joinery works.

CIRCULAR SAW : A circular toothed steel disc which rotates at a high speed for sawing wood and other materials.

CIRCULAR STAIR : See illustration. Used when space for a regular stair is restricted.

CIRCULATING WATER : Movement of water in a closed circuit.

CIRCUMPOLAR STARS : Stars that never set at certain latitudes on the earth. At the equator no star is visible, but at the pole all visible stars are circumpolar.

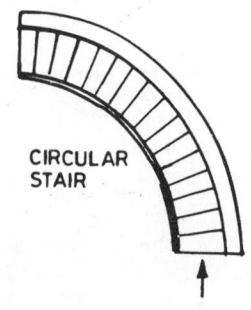

CIRCULAR STAIR

C. I. SLEEPER : A C.I. plate type sleeper is illustrated. It consists of two plates laid parallel to the rails. At the top of each plate, stiffeners are used to increase the strength of the plate. The plates are connected across the track by a tie bar.

CISTERN   : A cast iron, galvanized iron or chinaware tank with inlet, outlet, overflow pipe and flushing arrangement by pulling a chain. Used to flush a closet or a urinal after use. See illustration. A flushing cistern may be made automatic.

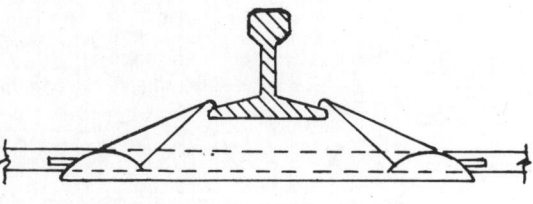

CAST IRON SLEEPER (Plate Type)

CLAMP : (i) A holding device, e.g. a cramp in joinery.

(ii) A stack of bricks for burning.

CLAMPING PLATE : A wooden connector.

CLAMPING SCREW : A screw used for clamping the vernier in a theodolite, such that the tangent screw can be used.

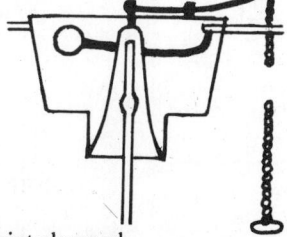

CLAMPING TIME : The time required for keeping a glued joint clamped.

CLAMP NAIL : A form of corrugated fastener used in light frames and at mitred corners.

CLAMSHELL GRAB : A grab having the shape like a clamshell.

CLAPBOARD: Weather board.

CLAPOTIS : Lapping of waves reflected on a wall higher than the water level.

CLAP SILL: A lock sill used in hydraulic structures.

CLARIFIER : Usually a circular tank, with hopper bottom for sludge accumulation. The clarification takes place through sedimentation of the suspended solids in a liquid. See illustration.

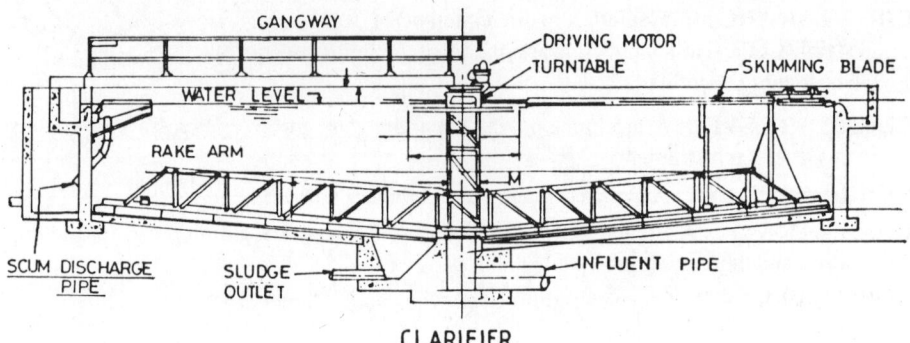

CLARIFIER

CLARIFILTER : A proprietary device combining the features of a contact aerator and a clarifier in treatment of sewage by 'Contact Aeration' process.

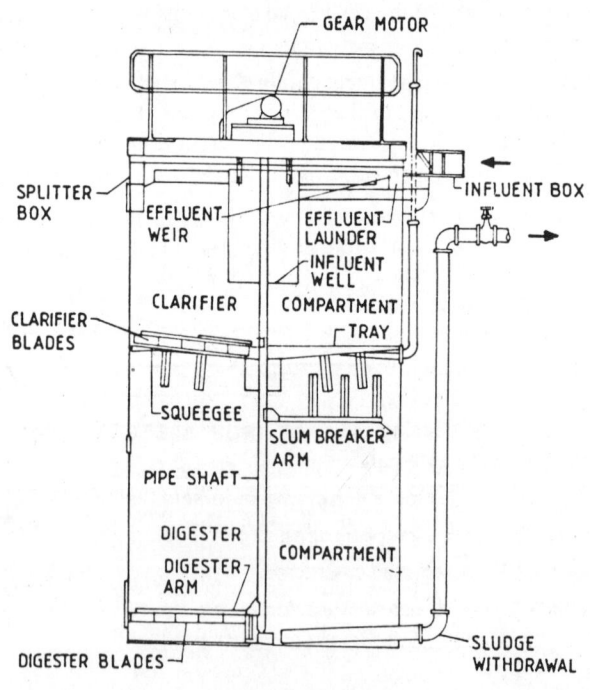

CLARIGESTER

CLARIGESTER : A sludge digestion tank equipped with a clarifier mechanism. It is a two-storeyed tank, in which the upper part is the clarifier and the digester mechanism is in the lower chamber. The side depth of most clarigesters is 6 m. This was devised to make the processes of digestion and clarification economical. See illustration.

CLARKE'S SPHEROID : For determination of shape and size of the earth, the U.S.coast and geodetic survey adopted clarke's spheroid in which 1 minute of latitude varies from 6045.95 ft. at the equator to 6107.85 ft at the pole and 1 minute of longitude varies from 6087.15 ft. at the equator to nil at the pole. Roughly, a minute of latitude or longitude at the equator corresponds to an arc of 1.15 miles.

CLASSIFICATION OF SOIL : Because of varying types of soil deposits all over the world, there are several methods of classifying soil. These are :

    (i)   by mineralogical composition

    (ii)   by origin

    (iii) by structure

    (iv) by texture

    (v)   by grain size distribution

    (vi) by cassagrande classification of airfield materials

CLASP NAIL : A cut nail of square section.

CLASTIC ROCKS : Rocks containing mineral fragments derived from pre-existing material. The texture of clastic rocks may vary from coarse grained to microscopic.

CLAW BAR : A pinch bar used by carpenters.

CLAW CHISEL : A mallet-headed chisel used by masons, in which the cutting edge has several broad nicks. This is used for shaping a stone roughly.

CLAW HAMMER : A hammer used by carpenters in which one end is split, claw shaped peen. This facilitates in drawing out nails from the work piece.

CLAW HATCHET : A shingling hatchet.

CLAW PLATE : A plate to connect timber pieces.

CLAY : One of the principal types of soil, which remains mixed with other types like silt and sand. Clay is a fine-grained soil consisting of hydrated silicate of aluminium.

CLAY MORTAR : A mortar made of clay and surki or sand or cinder with adequate quantum of water, used for kutcha-pucca houses.

CLAY PUDDLE : Puddled clay of plastic type used in cofferdam filling, in cut- off walls to dams, in dam core or in lining a ditch or pond. This is also used for water proofing, for its very low permeability.

CLAY SPADE : A grafting tool.

CLAY TILE : Roofing or paving tiles made of clay, which are well-burnt.

CLEANING EYE : An access door provided in a Y-branch or double Y-branch for inspection and cleaning purposes.

CLEAN OUT: A device for cleaning a sewer line. It is normally provided at the upper end of a lateral sewer in place of a manhole. It consists of a pipe connected to the sewer line, one end remaining at ground level as illustrated. For cleaning purpose, the top cover of the pipe is taken out and the sewer is cleaned by rodding and flushing with water.

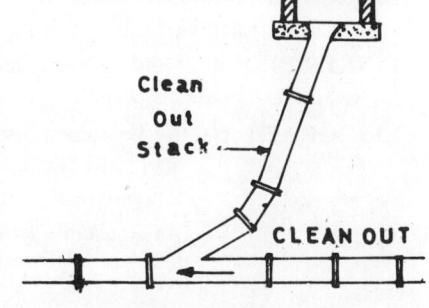

Clean Out Stack

CLEAN OUT

CLEAR SPAN: The horizontal distance of a beam or girder between one support and the other i.e., the clear opening.

CLEAR TIMBER : A timber obtained from a matured tree, close-grained, uniform colour, well-seasoned and free from any defect.

CLEAR WATER RESERVOIR : A service reservoir where filtered and disinfected water is stored for water supply.

CLEAR WATER SUMP : A clear water reservoir.

CLEAT : (i)   A wedge-shaped piece of timber used in a timber truss to support the purlin for keeping it in position.

     (ii)   An angle section made of mild steel, used in structural framework.

     (iii)   A strip of insulator across a cable to hold it in position.

     (iv)   A wooden piece plugged to a wall to carry a bracket.

     (v)   A tingle in roofing with flexible metal.

CLEAVAGE : A fracture in wood or stone or any other material.

CLEFT TIMBER : Timber split along the grain almost to the right size.

CLENCHING : A technique used in making boats and ledged and braced shutters. A nail is driven through a timber and the timber is bent over the point and on to the back of the timber.

CLENCH NAIL : A duckbill nail used for clenching.

CLERESTOREY WINDOW : A window placed above the roof of another part of a house for ventilation purpose. The window shutter can be operated with the help of a cord passing over a pulley. This type of window is found in the nave of a church which is above the roofs of the side aisles.

CLERE STOREY WINDOW

LEAN TO ROOF

CLIMBING FORMWORK : Also known as cantilever formwork, which is self-supporting, being held by hook bolts cast into the concrete. These bolts can easily be taken out on completion of the work. The soldiers (vertical members) carried by the bolts, support several lifts of formwork.

CLINK : (i)   A pointed steel bar for breaking up road surfaces or old lime terracing. One man holds the clink, while another man strikes it with a sledge hammer.

     (ii)   A seam between adjacent bays of flexible metal sheet roofing.

CLINKER : Sintered ash obtained from furnaces, which is used as a very good hard core or concrete aggregate.

CLINKER BLOCK : Building blocks made of clinker concrete which are strong enough as well as cheap.

CLINOGRAPH : (i)   An instrument used in borehole surveying. It measures the slope angle of a borehole at any point and records the same. Sometimes, it is provided with a camera and a gyroscopic orientation.

     (ii)   An adjustable set square used in a drawing office.

CLINOMETER : A very handy instrument that can be held on hand for sighting up or down an inclined plane for measuring the dip. This is used in surveying an undulating tract of land.

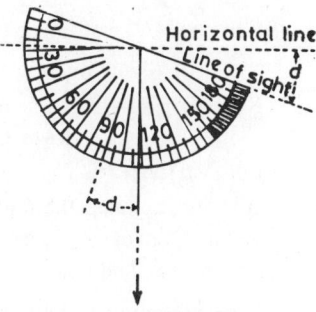

SIMPLE CLINOMETER

CLIP : (i)   A lead tack or metal strip used to fix a pipe to a wall or to fix flexible metal sheets to a sloping roof.

(ii)   A vee-shaped steel bar bolted on to traction cable in a ropeway.

CLIPPED GABLE : A jerkin-head roof.

CLIP SCREWS : An adjusting screw attached to the verniers of the vertical circle of a theodolite, by which errors can be eliminated by adjusting the verniers.

CLOSE BOARDED FENCING : A fencing usually found at the side of a railway platform, comprising vertical feather-edged boarding nailed or screwed or welded to two or three horizontal bars spanning vertical posts (made of old rails) about 3 m apart.

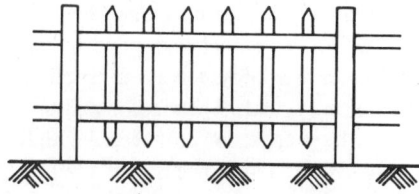

CLOSE-BOARDED FENCING

CLOSE BOARDING : Close timbering.

CLOSE CONTACT GLUE : A glue that sticks to the surfaces to be joined together, only when the gap between the surfaces does not exceed 0.125 mm.

CLOSE-COUPLE ROOF : See couple-close roof.

CLOSE-CUT HIP : Also known as 'cut and mitred hip', valley or hip formed in a sloping roof, in which the tiles or slates are cut to meet on the hip line with a metal gutter beneath them for draining out water.

CLOSED CORNICE : A box cornice made of timber or plaster mould.

CLOSED STAIR : A box stair. A stair having walls on three sides with a door at the entrance to the stair.

CLOSED TRAVERSE : In surveying, a traverse that ends at the starting point. It facilitates in checking the accuracy of a traverse. The sum of the angles turned must be 360^0 and the individual sum of eastings and latitudes must be zero.

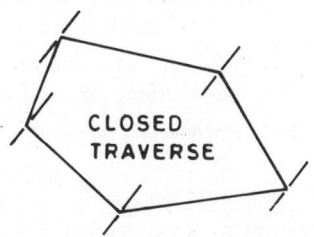

CLOSED TRAVERSE

CLOSE-GRAINED WOOD : Fine-grained wood having narrow annual rings.

CLOSE NIPPLE : A nipple used in plumbing work, which has twice the length of the standard pipe thread.

CLOSER : (i)   A brick bat; a cut brick or stone as per requirement.

(ii)   A cut sheet pile to close a coffer dam.

CLOSE STRING: Also known as 'closed stringer' in U.S.A. The outer string which has its top and bottom edges parallel. The ends of the treads and risers are housed in its face.

CLOSET: (i)   A small space or a cupboard in a room.

         (ii)   A night-soil pan used in a privy.

CLOSE TIMBERING : In a running ground, wooden planks are placed touching each other. This is also known as 'close boarding'.

CLOSING ERROR : In a closed traverse survey, this is an error due to which the finishing point does not meet the starting point. The discrepancy is due to the errors crept in the measurement of the angles and the distances. This is adjusted by distributing the error proportionately among the angles and the distances.

CLOSING STILE : The door or window stile furthest from the hinges.

CLOSING UP : Closing or upsetting a rivet by applying pressure or by hammering.

CLOTHES CHUTE : A laundry chute provided in a washerman's house.

CLOVER LEAF FLYOVER : A type of flyover junction or grade separation, having a geometrical pattern like a clover leaf, provided at the junction of two busy roads. The roads in addition to the flyover are interconnected with link roads or loops for easy circulation of traffic.

CLOUGH : A sluice gate provided in a culvert.

CLOURING : Picking.

CLOUT NAIL : A short G.I. nail with a large round flat head, required for fixing plaster board and roofing felt.

CLOVER LEAF FLYOVER

CLUB HAMMER : A mash hammer having double face, used by masons.

CLUNCH : A hard chalk.

CLUTCH : The hook-shaped edge of a sheet pile that grips the corresponding hook on the next pile, required in sheet piling work.

COACH BOLT : A round-headed bolt with an enlarged square neck to grip the timber without turning during tightening by a nut. Also, known as carriage bolt.

COACH SCREW : A gimlet-pointed screw of large size, required for making fixings to wood. This is driven into the wood by turning the square head of the screw with a spanner.

COAGULANTS : The chemicals used for coagulation, flocculation and sedimentation of particles suspended in a water or sewage. These are : aluminium sulphate, ferrous sulphate, ferric sulphate, ferric chloride and chlorinated copperas. Of these, alum, and copperas are commonly used as coagulants.

COAGULATING AGENTS : Coagulants.

COAGULATING AIDS : The chemicals that hastens the process of coagulation, when they are used in conjunction with the coagulants. These are bentonite clay, polyelectrolytes, activated silica, etc.

COAGULATION : A chemical process that involves the formation of chemical flocs which adsorb, entrap and bring together the suspended matters. It involves complex equilibria among the colloids, water or sewage, coagulant and other variables. The driving forces include lowering of zeta potential of colloids and neutralization of charges. Coagulation brings flocculation and sedimentation of dispersed particles suspended in water or sewage.

CO-ALTITUDE : The 'zenith distance' used in geodetic surveying. It is the angular distance from the object to the observer's zenith. Actually, it is 90^0 — altitude in the quadrant.

COARSE AGGREGATE : The larger size aggregates used in making concrete. These are usually stone chips and broken brick bats (khowa).

COARSE GRAINED TIMBER : A timber having wide annual rings.

COARSE STUFF : The material used in early days for the first and second coat of plaster over masonry work. Usually the mixture used, consisted of hydrated lime, sand and hairs.

COASTAL GEOMORPHOLOGY : A multifield comprising geology, geophysics, geography, hydro-dynamics, wind force, and wave current including mineralogical, chemical and biological features associated with morphological development.

COASTAL INLET : See illustration.

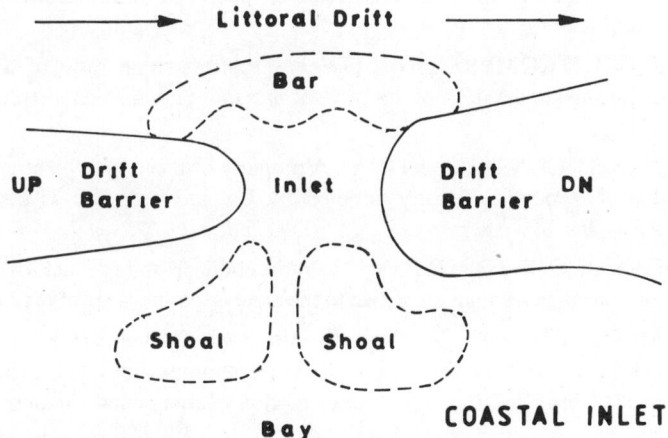

COASTAL INLET

COAT : A thin layer or film of any material over another.

COATED CHIPPINGS : Stone chippings thinly-coated with bituminous material for use in chipping carpet for the road wearing course.

COATED MACADAM : Tar macadam.

COATED ROCK BREAKWATER : A sloping breakwater in which bitumen-coated rocks are placed on the slope between the high water level and low water level. Coated rocks have great roughness which prevents them from sliding down easily.

COB :   (i)   Rammed earth construction for making a wall. Walling of damp soil mixed with cement and rammed into a formwork. This is a low-cost construction used in developing countries.

          (ii)   Unburnt brick reinforced with straws.

COBBLES: Rounded stones used as a paving material.

COBWEBBING : The ejection of spider web like threads from the nozzles of a spray gun during spray painting. This is a defect in painting.

COBWORK : Log house construction.

COCK : A valve to control the flow of a fluid through a pipeline. There are various types of cocks for specific requirements.

COCKING : Cutting a notch out of the breadth of a beam and two notches out of the wall plate, required for resting a beam on a wall plate. This facilitates in placing the beam into the notches made in the wall plate. Also, known as 'cogging'.

COCKING PIECE : With a view to providing eaves overhang, a short rafter called 'cocking piece' is nailed to each common rafter at the eaves.

COCKSCOMB : Mason's drag.

COCKSPUR FASTENER : A metal fastener for casement windows.

COD : Abbreviation for 'chemical oxygen demand'. The oxygen equivalent of the organic matter that can be oxidized. This test is performed to measure the organic matter content in a waste water.

CO-EFFICIENT OF ABSORPTION : 'Proportionality constant' denoted by 'Ks' in Henry's law of gas absorption, which states that the concentration of a gas in a liquid is directly proportional to the concentration of the gas in the atmosphere in contact with the liquid.

CO-EFFICIENT OF COMPRESSIBILITY : The stress-strain ratio of a soil. It is numerically equal to the slope of the curve of pressure versus void ratio plotted on the natural scale.

CO-EFFICIENT OF CONSOLIDATION : A property of a soil on which the degree of consolidation depends. Usually denoted by Cv and its value is evaluated by consolidation test of a soil.

CO-EFFICIENT OF CONTRACTION : The ratio of the cross-sectional area of a jet of fluid under pressure through an orifice to the cross-sectional area of the orifice.

CO-EFFICIENT OF DIFFUSION : 'Proportionality Factor' denoted by Kd in Fick's law, which states that the rate of diffusion across an area is proportional to the concentration gradient of a substance from higher concentration to lower concentration.

CO-EFFICIENT OF DISCHARGE : The ratio of actual discharge to the theoretical discharge of a fluid through a pipe, orifice or weir.

CO-EFFICIENT OF EXPANSION : The expansion per unit length of a material for each degree rise in temperature.

CO-EFFICIENT OF EXTINCTION : Rate of absorption of sunlight of a given wave length by a water.

CO-EFFICIENT OF FLUCTUATION : The ratio of standard fluctuation to the arithmetic mean of the observations in a statistical analysis. It makes possible comparisons between series of different magnitudes.

CO-EFFICIENT OF FRICTION : The ratio between the force to cause sliding of a body along a plane and the force acting normal to the plane.

CO- EFFICIENT OF GAS TRANSFER : The rate constant for gas transfer at the gas-liquid interface, where the rate of absorption of a gas is proportional to its degree of undersaturation in the absorbing liquid.

CO-EFFICIENT OF IMPERVIOUSNESS : Impermeability factor that expresses the percentrage of imperviousness of a material.

CO-EFFICIENT OF INTERNAL FRICTION : The tangent to the angle of internal frictibn between soil grains.

CO-EFFICIENT OF PERMEABILITY : The constant of proportionality between the superficial velocity and the gradient of flow through a soil.

CO-EFFICIENT OF STORAGE : 'Storage co-efficient'. Specific yield of a well.

CO-EFFICIENT OF TRANSMISSIBILITY : The value obtained by multiplying the standard co-efficient of permeability by the full saturated depth of an aquifer.

CO-EFFICIENT OF UNIFORMITY : Uniformity co-efficient to express the grain size characteristics. The ratio of 60 percentile to 10 percentile. It is denoted by 'U'.

CO-EFFICIENT OF VARIATION : In statistics, the ratio of the standard deviation of a series of values to the mean.

CO-EFFICIENT OF VELOCITY : The ratio of the actual velocity (measured) to the theoretical velocity of discharge (computed).

CO-EFFICIENT OF VOLUME CHANGE : The modulus of volume change.

COFFER : A canal lock.

COFFERDAM : A dam of temporary nature, built either by sheet piling or above the ground for exclusion of water from an area to give access to the area for work.

COGGING : Starting hot rolling of steel from the ingot.

COHESIONLESS SOIL : A type of soil that does not have cohesion in between the soil grains, e.g., sand, gravel and boulders.

COHESION OF SOIL : The shearing strength which a soil possesses by virtue of its intrinsic pressure. When loads are applied to a cohesive soil, intrinsic attraction or bond comes into action to resist the relative displacement of adjacent particless.

COHESIVE SOIL : A soil that has intergranular cohesion, e.g., silt and clay.

COIL FILTER : A type of vacuum filter, in which two layers of coil springs are placed in corduroy fashion around the filter drum. This is used for sludge dewatering.

COIL HEATING : (i)   Heating a substance by 'pyrotenex' or electric cables surrounding the substance.

(ii)   Heating a concrete slab by water pipes cast into it.

COKE BREEZE : Small cokes used for making breeze blocks.

COLECRETE : An economical method of concreting for voluminous works like dams, large foundations and roads. The method consists of spreading coarse aggregate on the ground and injecting matrix (cement, sand and water) into it.

COLD CHISEL : A chisel used by fitters for cold-cutting mild steel with the help of a hammer.

COLD DRAWING: Also known as 'wire drawing'. A process of making steel wires by drawing it through successively smaller holes.

COLD RIVETING : Forming a riveted joint by closing the rivets without heating.

COLD ROLLING : Cold bending of thin steel plates to form light structural sections.

COLD SETT : A long-handled chisel used by smithsman. During working with the chisel, a sledge hammer is used for striking the back of the chisel (sett).

COLD SETTING RESIN : Synthetic resins, which when cold, form complicated polymers on mixing with an acid accelerator. This is used in painting.

COLD WATER SERVICE : Supply of cold water through a pipeline system.

COLGROUT : A cement-sand mix with adequate quantum of water to form a thin paste used for 'colcrete'. This grout is forced into a well- packed gravel or stone chips.

COLIFORM INDEX : An index served as a basis for stating coliform concentration in a water. Also called 'MPN' (most probable number).

COLIFORMS : The group of bacteria that possesses the faculty of fermenting lactose with production of gas. This group meets the criteria for a satisfactory biological indicator of pollution. The number of these organisms in human faeces is very high.

COLLAPSE DESIGN : Plastic design of steel structures.

COLLAR : (i)   A short sleeve required in joining R.C.C. pipes.

(ii) An asphalt ring built up round a vertical pipe passing through a roof with a view to ensuring a watertight joint.

COLLAR BEAM : A horizontal tie beam of a collar beam roof.

COLLAR BEAM ROOF : A roof supported by a truss, in which common rafters are joined halfway up their length by a horizontal tie beam.

COLLAR TIE ROOF : See illustration.

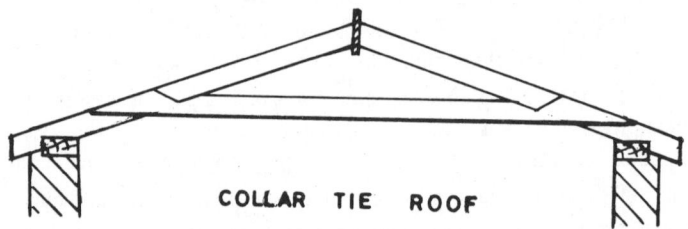

COLLAR TIE ROOF

COLLAR TIE TRUSS : See illustration

COLLECTION LINE : A house sewer or drain.

COLLECTION SYSTEM : (i) A system comprising branches and laterals in a sewer network for collection of sewage.

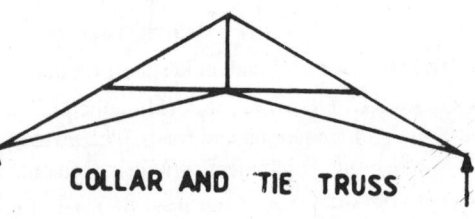

COLLAR AND TIE TRUSS

(ii)   A system adopted in collection of wastes in solid waste management.

COLLIMATION ERROR : Error in a theodolite or levelling instrument due to non-alignment of the line of sight i.e., when the line of sight is not horizontal.

COLLIMATION LINE : The line of sight in a surveying instrument. This should pass through the intersection of the cross hairs in the reticule.

COLLIMATION METHOD : In level survey, this is a method of determining the 'reduction' of the levels to some assumed datum line by taking successive back sight and fore sight with change in position of the instrument. No intermediate sight is taken. Also known as 'height of instrment' method.

COLLOIDS : Finely divided, negatively charged particles, which remain in suspension in a liquid and do not settle under normal condition.

COLLUVIAL SOIL : A soil that has been deposited primarily through the action of gravitational force, as in land-slides.

COLUMN : A post or a vertical member in a structure.

COLUMN ANALOGY : A method used in system analysis of statically indeterminate structures. This method makes use of mathematical identity which exists between the moments produced in a closed frame or beam and the fibre stresses produced when a short column is eccentrically loaded.

COLUMN HEAD : Enlargement of column top just below the slab. The construction looks like a mushroom.

COLUMN SPLICING : Two column sections, equal or unequal, joined end to end. See illustration.

COMBED JOINT : An angle joint formed by a number of tenons engaged in corresponding slots, used in woodwork. Actually, it comprises a series of dovetailed joint.

COMB GRAIN : The grains at the edge of a timber.

COMBINATION DOOR : A door having a detachable storm or winder shutter.

COMBINATION INLET : A storm water inlet which is a combination of a gutter inlet and a kerb inlet provided in roads, to receive the storm water. See illustration.

COMBINATION PLANE : A universal plane.

COMBINED STRESSES : Both bending stress and direct stress in **COLUMN SPLICING** tension or compression acting on a member of a structure.

COMBINED AVAILABLE CHLORINE : Combined chlorine residual in form of chloramines present in water.

COMBINED EXTRACT & INPUT SYSTEM : A ventilating system having two fans, for extraction and input of air in a room.

COMBINATION INLET

COMBINED SEWAGE : Waste liquid comprising sanitary sewage and storm run-off.

COMBINED SEWER : A conduit that carries both sanitary sewage and storm water.

COMBINED SYSTEM : A sewerage system in which the sewers carry both sanitary sewage and storm water.

COMBING: (i)   Laying top course of shingles by projecting above the ridge of a sloping roof to prevent entry of rain water blown by wind.

(ii)   Smoothing the face of a stone with a drag.

COMFORT STATION : A public toilet provided with wash basins.

COMMINUTOR : A stationary semi-circular screen grid with a rotating circular cutting disc, placed in a channel or in a wet well of a sewage pumping station. The larger solids intercepted by the grid are cut by the teeth on the rotating part and pass through the drum slots. See illustration.

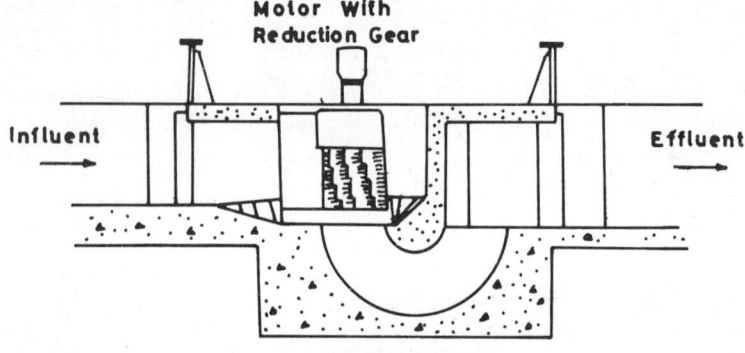

COMMINUTOR

COMMODE STEP : A riser at the foot of a stair, which is curved in plan.

COMMON ASHLAR : Hammer-dressed ashlar.

COMMON BOND : American bond.

COMMON BRICK : Locally manufactured bricks used for common purposes.

COMMON DOVETAIL : A dovetail joint, in which end grains are visible in both the members.

COMMON JOIST : Wooden joist spanning a gap between walls, onto which floor boards are directly nailed.

COMMON PARTITION : Partition made of a wooden frame consisting of a head and sill joined by vertical studs.

COMMON RAFTER : A sloping timber from ridge to eaves board supported by purlins above the principal rafters.

COMMON SEWER : A combined sewer carrying sanitary sewage and storm water.

COMMON WALL : A party wall separating two rooms.

COMMUNICATION PIPE : The portion of house service pipe for water supply from street main to the stop cock, which is under the jurisdiction of water supply authority.

COMPACTING FACTOR TEST : A laboratory test to determine the workability of freshly mixed concrete. It is more precise and sensitive than the slump test.

COMPACTION : Increasing dry density of a granular soil artificially by mechanical means.

COMPACT SOIL : A granular soil with a relative compaction of 90% or more.

COMPARATOR : (i) An instrument for comparing colour intensities.

(ii) An instrument used in photogrammetry for measuring accurately the rectangular co-ordinates of a point on a photograph.

COMPARATOR BASE : An accurately measured horizontal distance of one tape length used for checking the length of a tape in field.

COMPASS :   (i)   A very handy instrument usually called 'magnetic compass', most commonly used in compass surveying for measuring the bearings of survey lines.

(ii)   An instrument used in drawing office.

COMPASS BRICK : A 'Radial brick' used in circular brickwork.

COMPASS PLANE : A plane with an adjustable curved sole used by carpenters to smooth curved wood surfaces.

COMPASS ROOF : A roof made with curved rafters and tiles.

COMPASS SAW : A keyhole saw whose blade is tapered to a point, required to cut round sharp curves.

COMPASS BRICK

COMPASS TRAVERSE : A traverse survey in which bearings of lines are measured with the help of a compass.

COMPASS WINDOW : A bay window which is circular in plan.

COMPASS SAW

COMPENSATING DIAPHRAGM : A special fitting to a telescope in stadia work which changes the interval between stadia hairs during observation in a slope. The horizontal distance is directly calculated from the observed staff intercept.

COMPENSATING ERROR : In surveying, this error tends sometimes in one direction and sometimes in the other. Thus, the apparent result is likely to be too large or too small. The theory of probability can be applied to this class of error.

COMPENSATION WATER : The quantum of water that must pass a dam for use by the people who were enjoying the water prior to construction of the dam.

COMPENSATOR : This is provided in double wire transmission system in railway for operating signals, points and detectors. It is required to compensate the effect of expansion or contraction of wires and connecting rods due to variation in temperature. See illustration.

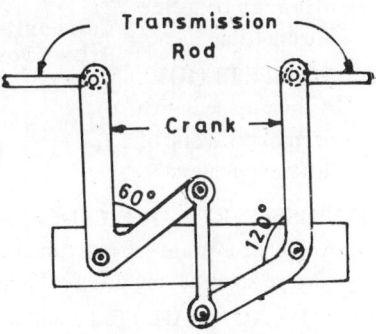

COMPENSATOR

COMPO : (i) Lead alloy for making flexible gas pipes for connection to an appliance.

(ii) A mortar of cement, lime and sand made in the ratio of 1 : 2 : 9.

COMPOSITE BOARD : (i) Hard board used for the purpose of insulation, made by compressing wood shavings, cork, asbestos, etc. soaked in a refractory binding material.

(ii) A class of plywood used in making furnitures and partition walls.

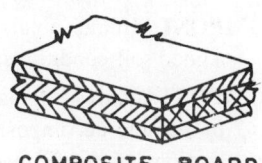

COMPOSITE BOARD

COMPOSITE BREAKWATER : A breakwater of combined structure consisting of a caisson placed on top of a rubble foundation submerged at all tidal levels.

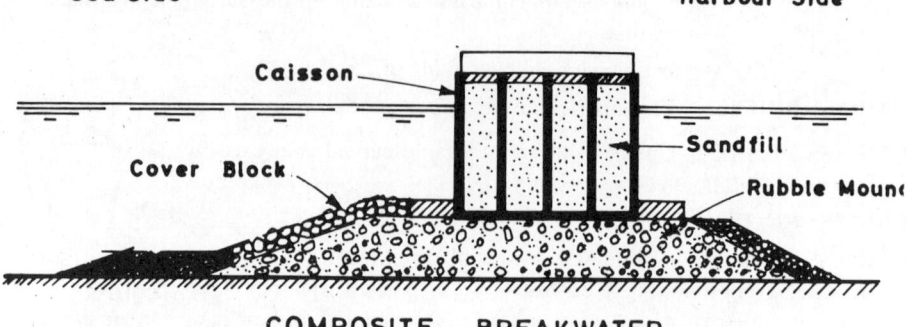

COMPOSITE   BREAKWATER

COMPOSITE CONSTRUCTION : A unit in civil construction made of different materials in conjunction.

COMPOSITE DRAW DOWN : The draw down due to the interference of the influence circles of two adjacent wells from which water is pumped simultaneously. See illustra- tion.

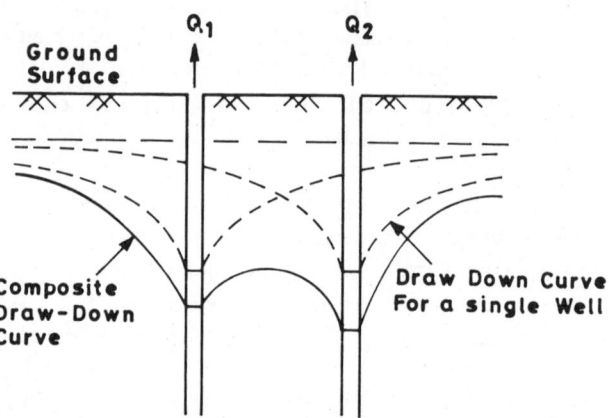

COMPOSITE FLOOR : A floor made in two or more layers of different materials.

COMPOSITE RUN-OFF CO-EFFICIENT : The run-off co-efficient calculated on weighted average of individual co-efficients of run-off for the different nature of surface areas having different imperviousness.

COMPOSITION NAIL : Brass nail used for tiling or slating in slant roofs.

COMPOSITION SHINGLES : Imitation shingles artificially made of bitumen felt for use in a sloping roof.

COMPOST : Humus obtained by decomposition of organic wastes, which can be used as a good soil-conditioner. It has the required nutrients for the growth of plants.

COMPOST PLANT : A plant to convert shredded refuse (solid wastes) into humus either by aerobic decomposition or by anaerobic fermentation. The illustration shows a sectional view of a mechanical compost plant using aerobic process of composting.

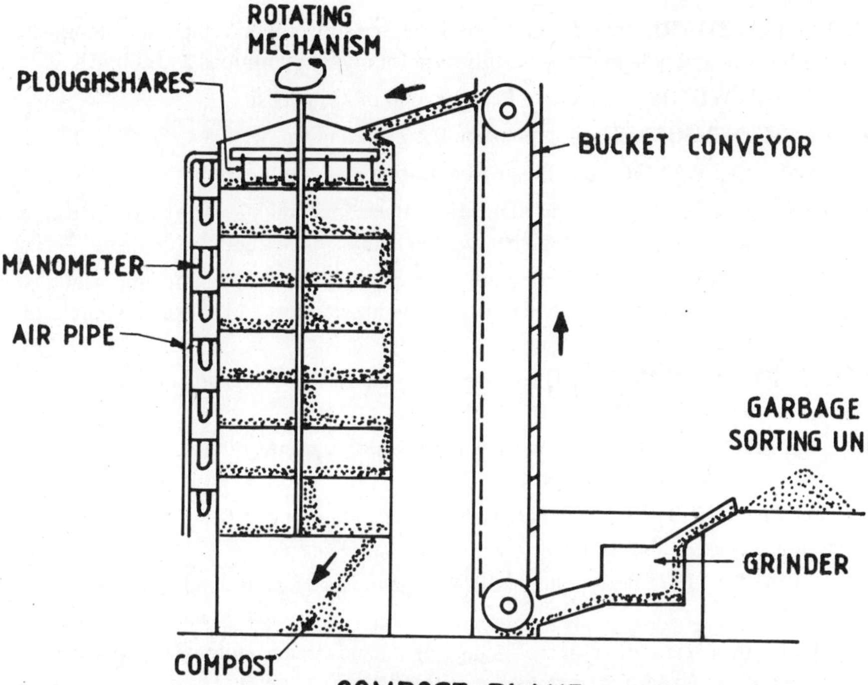

ROTATING
MECHANISM

PLOUGHSHARES

BUCKET CONVEYOR

MANOMETER

AIR PIPE

GARBAGE
SORTING UNIT

GRINDER

COMPOST

COMPOST PLANT

COMPOST STACK : Round or square based stacks, more or less haycock-shaped or dome-shaped, of compostable materials.

COMPOUND AIR LIFT : A modified air-lift pump with two air lift pipes used in places, where adequate depth of immersion is not available. The first pipe discharges into the second pipe which functions as a sump and the second pipe pumps twice the height pumped by the first.

COMPOUND BEAM : A built-up beam made of several pieces of timber nailed together.

COMPOUND CURVE : A curve comprising two or more arcs of different radii curving in the same direction and having a common tangent at the point of junction.

COMPOUND DREDGER : A bucket ladder dredger provided with clay cutter.

COMPOUND GIRDER : A plated R.S.J. in which additional plates are riveted or welded to the flanges.

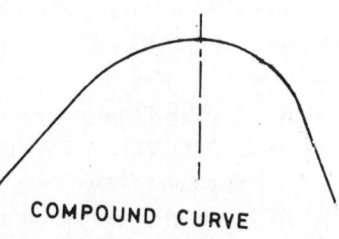

COMPOUND CURVE

COMPOUND PIPE : A pipe having different diameters in series.

COMPOUND SHAKE : Different types of shakes in combination found in a timber piece.

COMPOUND WALLING : Walls made of two or more leaves of different materials combined together.

COMPREGNATED WOOD: High-density plywood made of compressed impregnated wood.

COMPRESSED BOARD: Board made by compressing straws, wood chippings, saw dusts, corks, etc. at a very high pressure. It is used for making partitions and ceilings.

COMPRESSED WOOD : Hard board i.e., plywood of high density.

COMPRESSION BOOM : The compression flange of a beam.

COMPRESSION FLANGE : The flange of a beam under compression.

COMPRESSION JOINT : A joint used in light gauge copper tubing in which the two ends of copper tubes are screwed together end to end by a brases nut.

COMPRESSION TEST : Tests carried out to determine the crushing strength of bricks, stones, concrete, mortar, etc. Unconfined compression test and triaxial compression test are carried out for soil.

COMPRESSIVE STRENGTH : The strength of a material to withstand the compressive force acted on it.

CONCAVE JOINT : A form of pointing brickwork. A morter joint hollowed out by pressing a bar along it, while green.

CONCEALED GUTTER : A box gutter which does not come in front view, because it remains hidden by the cornice.

CONCENTRATED LOAD : A point load or knife-edge load i.e., the load is not distributed over an area.

CONCRETE : A mixture of cement, fine aggregate and coarse aggregate with adequate quantum of water to form a homogeneous mass. When set and cured, it resembles stone. This is extensively used in Civil Engineering constructions. There are various types of concrete to meet the specific needs.

CONCRETE BLOCKS : Blocks made of concrete. Hollow blocks are used in building construction. Solid blocks of heavy type are used in protection of river banks and also in construction of breakwater.

CONCRETE BREAKER : A heavy rock drill-type compressed-air tool used in breaking concrete constructions either in buildings or in roads.

CONCRETE BRICKS : Brick-shaped concrete blocks.

CONCRETE COVER BREAKWATER : A type of sloping breakwater, in which the seaward slope above the water level is protected by placing interlocking concrete tiles.

CONCRETE FINISHING MACHINE : A machine used for finishing concrete roads and runways.

CONCRETE INSERT : A metal or wooden plug built into concrete.

CONCRETE INTERLOCKING TILE : Single lap concrete tiles provided with a central nail hole at top and nibs at both ends.

CONCRETE MIXER : A concrete mixing machine comprising essentially a rotating drum into which the ingredients and water are fed for thorough mixing.

CONCRETE NAIL : A hard thick steel nail used to drive into concrete or brickwork.

CONCRETE PAVER : A concrete mixer with arrangement for spreading concrete, mounted on a crawler track, used in construction of concrete roads and pavements.

CONCRETE PILES : Precast or cast-in-situ reinforced concrete piles which are driven into ground to increase the bearing power of soil and to support a structure above ground.

CONCRETE PIPE : Prestressed concrete pipes are used for conveying water. These are pressure pipes. NP_2 and NP_3 pipes are used for conveying storm water and sewage.

CONCRETE PLACER : A device to force concrete along a pipe by means of compressed air. This device is suitable for tunnelling.

CONCRETE PUMP : A pump attached to a concrete mixer, is used to pump concrete slurry for placement of large volumes of concrete. This is used for massive concrete construction.

CONCRETE ROOF : A roof made of concrete slab according to the desired shape and design. Roofing may also be done by using precast concrete slabs to hasten the progress of construction. Shell roofs are made very thin.

CONCRETE SLAB : Slabs made of concrete as per design for flooring and roofing. Precast concrete slabs are also used for different purposes like pavement, flooring , sunshade, shelves, roofing, etc.

CONCRETE SPREADER : A machine used for spreading concrete uniformly to make a road surface or a runway.

CONCRETE VIBRATING MACHINE : A concrete vibrator used for vibrating wet concrete to bring the homogeneity in the concrete mass.

CONCRETING BOOM : A boom made of a light metal truss, on the underside of which a concreting bucket travels along a rail from one end to the other carrying wet concrete. This facilitates workmen in receiving wet mix ready at the place of concreting and also saves time and labour. The boom is made upto a span of 12 m.

CONCRETOR : A skilled workmen who works in concreting foundations, roads, runways, dams, barrages, weirs and similar such constructions.

CONDENSATE : Water that condensates from air, vapour and gases i.e., the moisture content is cooled below the dew point.

CONDENSATION : (i) Formation of water on a surface due to cooling of moisture content below the due point .

(ii) Curing of a synthetic resin .

CONDENSATION GROOVE : A lead sheet forming a channel, projected under a glazing to catch condensate for throwing it outside.

CONDENSATION GUTTER : A very narrow channel provided at the foot of a patent glazing or skylight to carry condensed water and to drain it out.

CONDENSATION WASHER : A special fitting used to raise the lower end of a glazing bar above the purlin for driving out condensed water.

CONDUCTOR : The term used for a 'down pipe' in USA.

CONDUCTOR HEAD : The term used for a 'rain water head' in USA.

CONDUIT : A pipe; An encasement for cables.

CONDUIT BOX : A distribution box.

CONE OF CONTAMINATION : Contamination of subsoil in form of a cone due to leaching of foul water from a cesspool as shown in illustration.

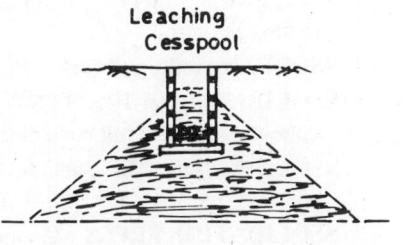

CONE OF CONTAMINATION

CONE PENETRATION TEST: A penetration test carried out on a soil by pressing a standard cone into the soil under a known load. This enables to determine the resistance to driving of bearing piles into a soil. This test is not suitable for clays or rocky soil.

CONE TILE : A bonnet tile or hip tile.

CONFINED COMPRESSION TEST : Triaxial compression test carried out for determination of strength of a soil . See 'Triaxial compression test'.

CONGE : A concave moulding at the junction of a wall and a floor.

CONGLOMERATE : Rounded stones coated with cement paste. Also, known as 'pudding stone'.

CONICAL LIGHT : A skylight of polygonal pyramid shape built-up with straight glazing bars and flat glass panes.

CONICAL ROLL : A roll-joint formed over a triangular wood roll in flexible metal roofing.

CONIFERS : Trees of gymnosperm group that produce soft wood, e.g., fir and pine trees.

CONNATE WATER : Water entrapped in the interstices of sedimentry rock during the period of its deposition.

CONNECTION PIT : A pit or a manhole to facilitate connection of incoming sewers with the outgoing sewer.

CONNECTOR : (i) A fastener used in joinery work.

           (ii) A long screw used in pipeline work.

CONSISTENCE : (i) Degree of density or stiffness

           (ii) Workability of a wet mix of mortar or c concrete determined by slump test.

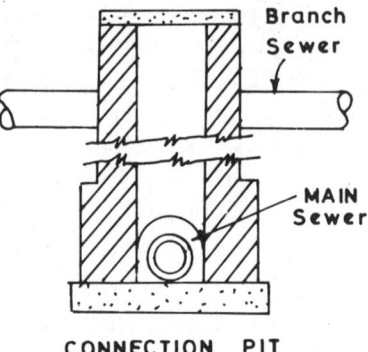

CONNECTION   PIT

CONSISTENCY INDEX : A figure to denote the stiffness of a clay. It is determined as

(liquid limit — water content)/(liquid limit — plastic limit) x 100

and expressed in percentage.

CONSISTENCY LIMITS : Atterberg limits, These are liquid limit, plastic limit and shrinkage limit.

CONSOLE : A cantilever bracket with an S-shaped scroll.

CONSOLIDATED QUICK TEST : A quick test to determine the shear strength of a cohesive soil after full consolidation under load in a laboratory.

CONSOLIDATION : The gradual compression of a soil under load. The consolidation of soil occurs when water is driven out of the voids in the soil.

CONSOLIDATION PRESS : An apparatus known as consolidometer or oedometer used in a laboratory to obtain data for plotting a curve of pressure to void ratio of a clay. This facilitates in determining the co-efficient of consolidation.

CONSOLIDATION SETTLEMENT : The settlement of a soil under load over a period of years. The settlement can be determined from the time-settlement curve.

CONSOLIDATION TEST : The soil sample is completely confined and the load is transmitted to upper and lower faces of the specimen. The pressure is increased in steps and the deformation of the sample is measured by a dial gauge. The results are plotted in form of a curve representing void ratio correspponding to each increment of pressure.

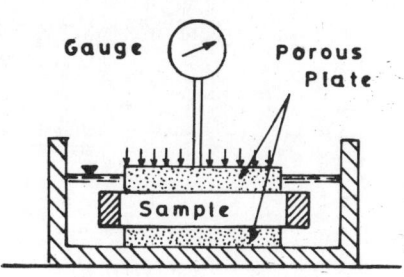

CONSOLIDOMETER : A consolidation press.

CONSOLIDATION   TEST

CONSTANT HEAD PERMEAMETER : An apparatus used for carrying out permeability test for highly permeable materials like sand and gravel. The co-efficient of permeability is computed by the equation $K = QL/hAt$. The discharge Q flowing through a sample of C.S. area A and length L under a hydraulic head h in time t is measured in the test. See illustration.

CONSTRUCTION JOINT : A narrow gap provided at intervals in a long R.C.C. wall to break the continuity with a view to keeping space for expansion and contraction due to variation in temperature. This also facilitates in avoiding cracks due to unequal settlement.

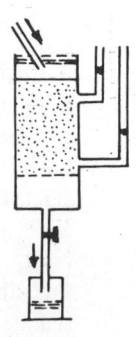

CONSTRUCTION SPANNER : A podger.

CONSTANT

CONSTRUCTION WAY : A temporary railway track for construction of a permanent way.

HEAD
PERMEAMETER

CONSUMPTIVE USE : The loss of water from a field due to transpiration and evaporation.

CONTACT AERATOR : An aeration tank in which sewage is aerated by injecting compressed air.

CONTACT BED : A bed of slime layer comprising thickly populated microbes over which sewage is passed for biosorption.

CONTACT CAPILLARITY : Where soil grains are very close to each other, the capillary water surrounding the contact points forms a miniscus and is held as contact moisture.

CONTACT FILTER : A trickling filter bed.

CONTACT
CAPILLARITY

CONTACT MOISTURE : Soil moisture held by surface tension either in smaller voids or as annular rings surrounding the points of contact between soil grains above the free water table.

CONTACT PROCESS : A anaerobic contact process is shown in illustration. It consists of a well-mixed digestion unit and a sludge settler. The essential feature is that the wash out of active anaerobic bacterial mass from the reactor is con trolled by a sludge separation and recycle system.

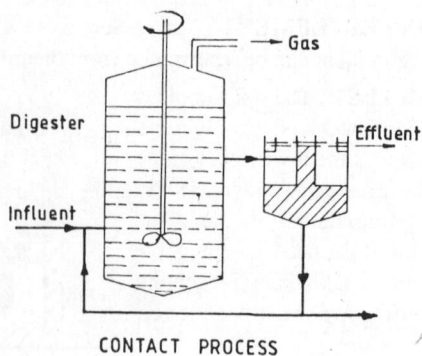

CONTACT PROCESS

CONTACT STABILIZATION PLANT: The plant process has been developed by taking the advantage of the biosorption property of activated sludge. The process is suitable for removal of BOD from the domestic wastes. The activated sludge mass is mixed with the settled sewage and aerated in a contact tank for 30 to 90 minutes, during which biosorption takes place. Then, the sludge is seperated by sedimentation and it is aerated in a sludge aeration tank for 3 to 6 hours during which absorbed organics are utilised to form new cells. The aeration volume requirement is about 50% of that of a conventional activated sludge plant.

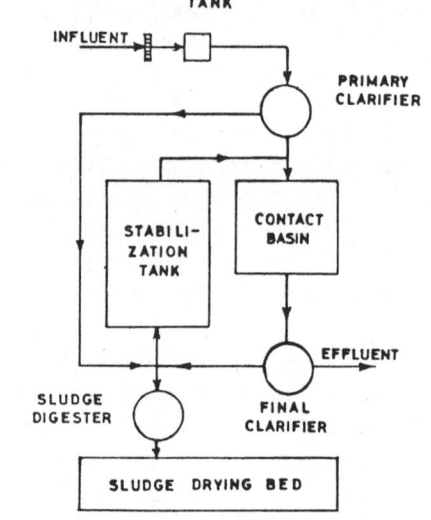

CONTACT STABILIZATION PLANT

CONTAINER BERTH : See illustration.

CONTINUOUS BEAM : A multi-span beam having intermediate supports. A continuous beam must have at least three supports.

CONTINUOUS FILTER : A bacteria bed (bio-filter) in a sewage treatment plant.

CONTINUOUS FOOTING : A wall footing.

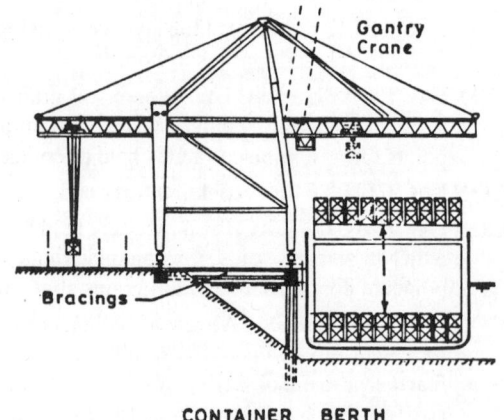

CONTAINER BERTH

CONTINUOUS HAND-RAIL : A hand-rail to a geometrical stair.

CONTINUOUS MIXER : A concrete mixer having arrangement for mixing concrete matrix continuously without any interruption. The feeding of ingredients and drawing the concrete mix are automatic.

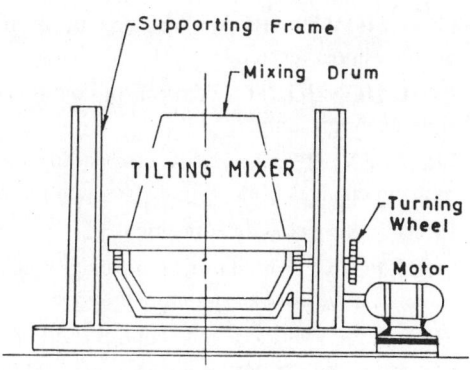

CONTINUOUS ROPEWAY : A ropeway, in which the loaded carriers travel on one side and the empty carriers return on the other side of the ropeway.

CONTINUOUS STRING : An outer string round a stairwell under a continuous handrail.

CONTOUR : A line of equal altitude upon the earth's surface.

CONTOUR CHANNEL : A channel that runs along the ridges and valleys roughly parallel to the contours.

CONTOUR CHECK : A form of terracing by making borders following the contours of the ground. Thus, the tract of land is divided into several compartments.

CONTOUR GRADIENT : A line set out at a constant slope on ground.

CONTOUR INTERVAL : The vertical distance between contour lines.

CONTOUR OF GROUND WATER : See illustration.

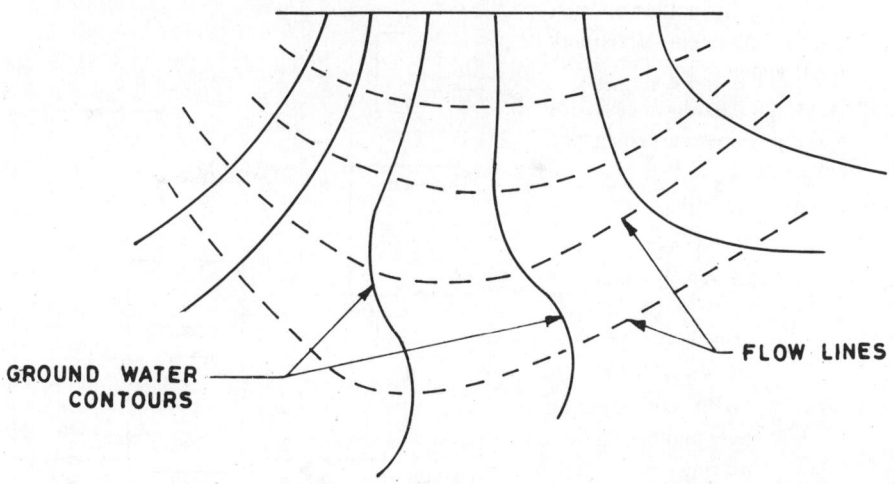

CONTOUR OF GROUND WATER

CONTOUR PLOUGHING : Soil conservation by ploughing with furrows on the slopes of road or railway cuttings. This reduces the scour of soil.

CONTRACTED WEIR : A weir having end contractions, i.e., the notch is shorter than the width of the channel.

CONTRACTION IN AREA: Necking due to reduction in cross-sectional area of a member under tension.

CONTRACTION JOINT : A break in continuity of a structure to allow shrinkage of concrete or masonry.

CONTRAFLEXURE : A point of zero bending moment i.e., the point at which the bending moment changes from hogging to sagging or vice- versa. This is a point of inflexion.

CONTROL : A check with adjustment.

CONTROL POINT : The starting point on ground of known position in plane table survey, traverse survey or photogrammetry.

CONTROL VALVE : A valve to control fluid flow through a conduit.

CONVERSION FACTOR : A number to enable conversion of units from one system to the other.

CONVERTED TIMBER : Square-sawn timber.

CONVERTER : A furnace used in making steel.

CONVEYOR : An equipment with a moving flat belt used for transportation of coal, ore, sand, etc continuously over a short distance. This is chiefly used in industries.

COOKING VAT : A steam-heated water tank into which every timber flitch is stewed for a few hours before slicing.

COOLING TOWER : A concrete tower used for cooling condenser water.

COOLING TOWER BIO-LOGICAL OXIDATION : A system as shown schematically in illustration, is used in treatment of refinery waste waters.

COPAL : Natural hard resin used for making shiny varnishes and linoleum.

COPE : (i)   A coping
(ii) To give a pro-tectionary cover to a wall with precast concrete slabs, stones, or bricks.
(iii) To fit one moulding over another without mitring.

COPED JOINT : A joint between two mouldings in which a part of one moulding is cut out to receive the other.

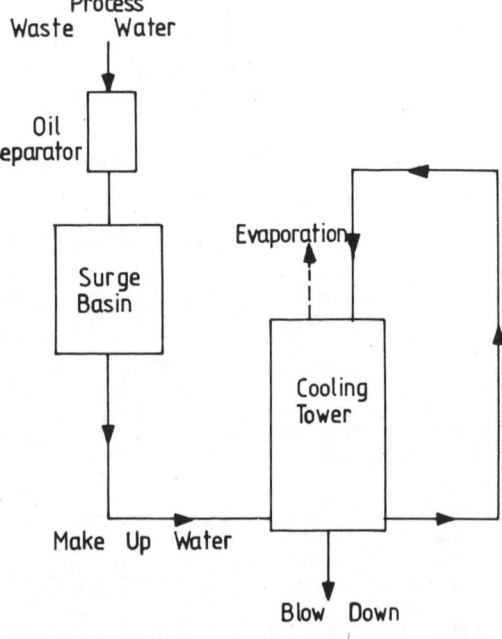

COOLING   TOWER   BIOLOGICAL   OXIDATION

COPING : A specially shaped brick, stone or concrete block placed on the top of a wall to protect it from weathering action.

COPING BRICK : A brick of special shape used in coping a wall. See illustration.

COPING SAW : A bow saw with a narrow blade used for coping out mouldings and cutting sharp curves.

COPPERAS : Ferrous sulphate, used as a coagulant for removal of turbidity of water. This acts better in the alkatine range of water.

COPPER FITTINGS : Fittings used for light-gauge copper tubing work. These are usually made of brass, copper or gunmetal. These are either compression fitting or capillary fitting.

COPING BRICK

COPPER BEARING STEEL : Steel containing 0.2 to 0.6% copper that resists corrosion.

COPPER GLAZING : Glazing with copper cames by welding.

COPPER NAILING : Nailing with copper nails where corrosion is likely to occur.

COPPER PIPE : Used in plumbing work.

COPPER PLATING : Electroplating with copper to give a protective finish to articles made of iron or steel.

COPPER ROOFING : Flexible roofing with copper sheet.

COPPER SULPHATE : Used as an algaecide to preserve water in a lake or impounding reservoir.

CORBEL : A masonry work projected from a wall surface.

CORBEL BRICK : A special brick used for corbelling.

CORBELLING : A brickwork or stone masonry with successive projections. This is chiefly used in railway platform, ashpit, building cornice, etc.

CORBEL PIECE : A bolstar.

CORBEL BRICK

CORBERL PIN : A corbelling iron (a metal plate) built into a brickwork to carry a wall plate.

CORDED WAY : A stepped path formed on a steep slope to protect from erosion.

CORDUROY ROAD : A temporary road made of 75 to 100 mm dia. saplings wired together tightly at their ends, which is laid over a liquid mud. This can be quickly laid and picked up.

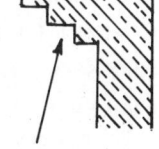

CORBELLING

CORE: (i) The innermost part of a material or a structural member. The embedded material in a composite member.

(ii) The central part of a wall or column. The core area of a wall is the middle-third of the wall. The core area of a column is the area through which the resultant of the compressive forces passes. In a spirally-reinforced concrete column, the core area is the area of concrete within the centre line of the spiral reinforcement.

(iii) The rock or soil cylinder cut out by a drill or soil sampler.

(iv)   The core wall of puddled clay or bonded rocks or concrete in a dam. A cut-off wall of clay or concrete or any other suitable material.

(v)   The brick core below a relieving arch. The wood wool within a paper-backed wood wool slab. The inner part of a veneered wood.

(vi)   The inner softer steel within a carburized steel case.

(vii) A conductor within an electric cable.

(viii) A core used in a mould for making pipes.

CORE BARREL: The length of pipe next to the cutting bit of a core drill.

COREBOARD : A hardboard made of core of strips glued together between outer veneers.

CORE CATCHER : A spring used to prevent sand from dropping out of a soil sampler.

CORE CUTTER : The cutter at the foot of the core barrel.

CORE DRIVER : A cylinder exactly of same diameter as that of a hole is used to push through the hole to clear out chips.

CORED HOLE : A hole made by leaving a core while casting a concrete piece and removing the core shortly after the concrete is set.

CORE DRILL : A power-driven diamond drill.

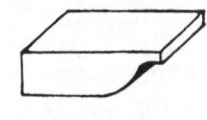

CORE LIFTER : A core cutter.

CORE WALL : See under 'Core'.

CORK : The bark of the cork oak, a tree.

CORK BOARD : A board made of compressed cork, used for insulation purpose.

CORK CARPET : An insulating carpet made of cork used for doemstic purposes. This can be rolled like linoleum sheet.

CORK-SCREW STAIR : A spiral stair .

CORK TILE : Tile made of cork board that can be used for flooring.

CORK WOOD : Balsa wood.

CORNER BEAD : A bead to protect a corner.

CORNER CHISEL : A chisel of L-shaped blade having no handle, used for cutting out mortises.

CORNER CRAMP : A cramp used for making a mitred joint by gluing.

CORNER LOCKED JOINT : A combed joint used in joinery work.

CORNICE : A moulding at the end of a roof slab projected from the outer wall to throw out the rain water.

CORNICE BRICK : A special brick used to form a moulded cornice as shown in illustration.

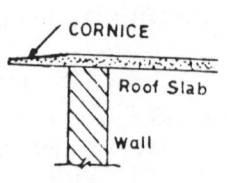

CORROSION : It is understood as a electrochemical process by which metals are eaten away or corroded and ultimately destroyed. In absence of oxygen or water no corrosion takes place. Corrosion is more rapid in acid than in alkaline solutions. Underground pipes are corroded by stray electric currents. In the process of corrosion three steps are identified

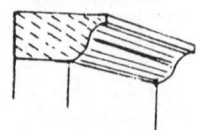

CORNICE BRICK

(i) anodic reaction by which the metal goes into solution and electrons flow through the metal to a cathode (ii) a cathodic reaction and (iii) a number of reactions of the metal ions with water.

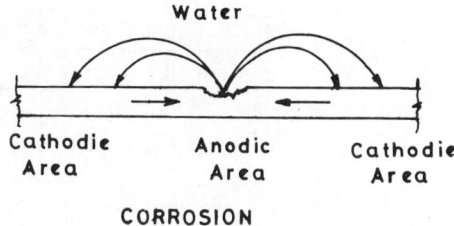

CORROSION

CORROSION FATIGUE : Weakening of iron and steel by small fatigue cracks into which water enters and corrosion takes place.

CORROSION INHIBITOR : There are a few chemicals like red oxide, chromate and sodium nitrite used in paints, which protect metals from corrosion.

CORRUGATED ALUMINIUM : Aluminium sheets with corrugations used for wall and roof cladding.

CORRUGATED ASBESTOS : Corrugated asbestos-cement sheets used for wall and roof cladding.

CORRUGATED FASTENER : This is a corrugated piece of iron or steel driven into the ends of two boards to be joined together. Also, known as wiggle nail, dog brad or mitre brad.

CORRUGATED IRON : Corrugated galvanised iron sheets used for roof cladding and forming temporary compound wall.

CORRUGATED-SHEET : Corrugated sheets of different materials like galvanised iron, asbestos cement, aluminium, pvc and glass fibre, used for wall and roof cladding.

CORRUGATED TOOTHED RING : A ring like corrugated fastener having serrations , used to connect timber pieces.

CORUNDUM : A very hard alumina (Al_2O_3) obtained as a mineral which is used as an abrasive.

COTTAGE : A village-or country-house i.e., a livable low cost house in a rural area.

COTTAGE ROOF : A sloping roof of a cottage in which the roof covering rests on purlins supported by common rafters and no principal rafter is provided.

COTTER : A steel or wooden wedge used as a fastener.

COTTERED JOINT : A joint in which a cotter is introduced normal to the long axis of the pieces to be joined together.

COULOMB'S EMPIRICAL LAW : The use of this law requires understanding of many a complex factor affecting the shearing strength of clay. The empirical law established by coulomb for shear strength of soil may be written as

$$S = C + P \tan \phi$$

where, S = shear strength,

C = effective cohesion,

P = pressure at right angles to the shearing plane,

and ϕ = angle of shearing resistance of the soil i.e., effective friction angle.

COULMB'S EQUATION: Ref. to illustration, the thrust P as given by the coulomb theory is expressed by the formula :

$$P = 0.5\ WH^2 \left[\frac{\csc \beta \sin (\beta - \phi)}{\sqrt{\sin\ (\beta + \phi')} = \sqrt{\sin\ (\phi + \phi')\sin\ (\phi + \phi')}} \frac{}{\sin (\beta - i)} \right]^2$$

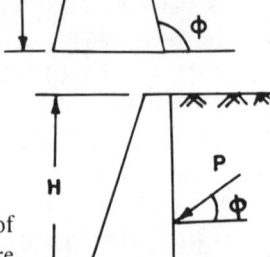

Where ϕ = friction angle

w = unit wt. of backfill

H = height of the wall and

B = vertical angle made by the wall at its base.

when B is 90° and ϕ' = i, the conditions conform to Rankine's theory.

When the wall face is vertical, the back fill surface is level and $\phi' = \phi$, the expression may be written as

$$P = 0.5\ WH^2 \frac{\cos \phi}{(1 + \sqrt{2} \sin \phi)^2}$$

COULOMB'S THEORY: The theory is based on the concept of a failure wedge bounded by the wall face and by a failure surface passing through the toe of the retaining wall. See illustration.

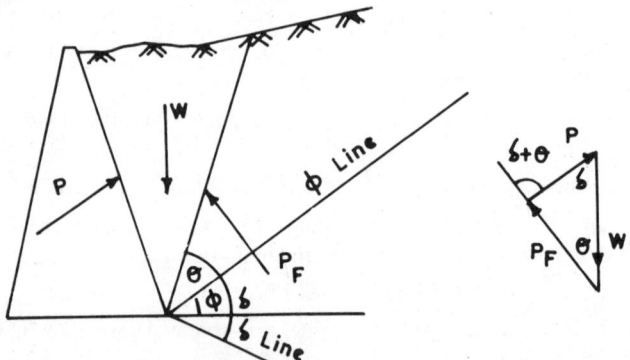

COUNTER ARCHED REVETMENT : A revetment with brick masonry provided to a cutting with arches between counter- forts.

COUNTER BATTENS : (i) Battens placed across the back of a number of boards and nailed to them for stiffening, as found in a battened door or a wooden framed floor or a drawing board.

(ii) Battens parallel to the common rafters are nailed over them on a boarded roof, over which the slating or tiling battens are fixed.

COUNTER BORING : Enlarging a hole by drilling.

COUNTER BRACING : Cross bracing or diagonal bracing provided in each panel of a truss or girder or trestle for stiffening against wind load.

COUNTER CEILLING : A false ceiling for beautification, provided under the actual ceiling counter.

COUNTER CRAMP : A batten with a number of fixed pieces of wood on to which several boards are cramped with folding wedges.

COUNTER DRAIN : A drain constructed along the base of a canal bank to drain out the seepage water or run-off to protect bank.

COUNTER FLAP HINGE : Double dovetailed-shape hinges used to fix up the lifting flap of a counter.

COUNTER FLASHING : A sheet metal flashing made at the junction of a parapet wall or a chimney and a sloping roof such that no leakage occurs at the junction.

COUNTER FLOOR : A blind floor or rough floor which is actually a sub-floor i.e., the lower set of floor boards laid diagonally in a framed wooden flooring.

COUNTERFORT : A pier built monolithic at right angles to a retaining wall with a view to strengthening it against bending and overturning.

COUNTERFORT RETAINING WALL : A retaining wall having counterforts at regular intervals as shown in illustration. This type of retaining wall is usually required when the height of the wall is great or the magnitude of superimposed load is large.

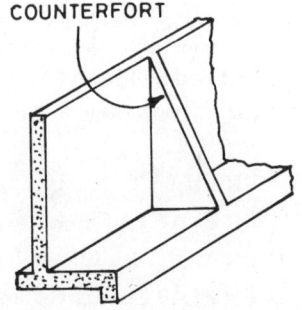

COUNTERFORT

COUNTERFORT
RETAINING WALL

COUNTER GAUGE : A mortise gauge used in joinery work.

COUNTERPOISE BRIDGE : A bascule bridge i.e., a swinging bridge.

COUNTERSUNK BOLT OR RIVET : A bolt or rivet with conical head which goes into the countersunk hole in a member i.e., there will be no projection of the head.

COUNTER WEDGING : The fixing of wooden boards by using counter cramps.

COUPLE CLOSE ROOF : A close couple roof with two rafters and a tie beam to hold the rafters in position.

COUPLE ROOF : A roofing frame with two rafters meeting at the ridge and having no tie beam. This is the simplest form of a sloping roof for a short span.

COUPLING: A coupler nut used for joining two pipes in house plumbing for water supply.

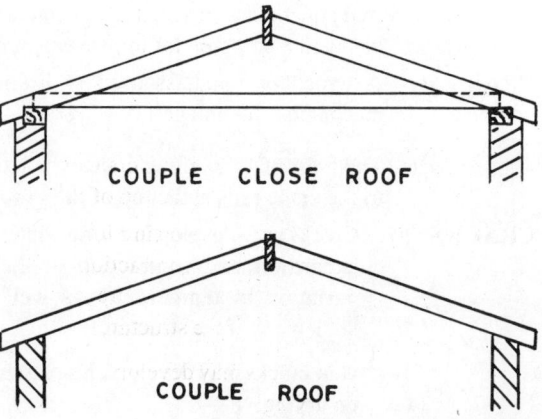

COUPLE CLOSE ROOF

COUPLE ROOF

COURSED ASHLAR: Regular ashlar work or rubble masonry.

COURSED BLOCKWORK : Laying of precast concrete blocks in horizontal courses with bonding. The coursed block work is also done in construction of break waters with heavy concrete blocks.

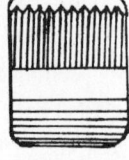

COURSED MASONRY : Coursed rubble work or blockwork.

COURSED RANDOM RUBBLE : Random rubble built in layers in construction of a wall.

COUPLING

COURSED SNECKED RUBBLE : Snecked rubble built to occasional layers i.e., not in regular courses.

COURSED SQUARE RUBBLE : Squared rubble built to occasional courses i.e., random ashlar work.

COURSING JOINT : A concentric joint in an arch separating two string courses.

COVE : Coving; A quadrant moulding joining a wall to the ceiling i.e., rounding off the joining line of the wall and the ceiling with mortar.

COVED CEILING : The ceiling curved at the junction with the walls.

COVE LIGHTING : Indirect lighting in a room from lights above a cove. The light from the cove is thrown to the ceiling, which reflects in the room and gives a pleasant diffused light with no glare.

COVER : The thickness of concrete from the reinforcement bar to the outer edge of the member.

COVER FILLET : A thin strip used to cover joints in ceiling boards or wall boards.

COVER FLAP : A flap to cover boxing shutters.

COVER MOULD : A mould to cover a joint.

COVER PLATE : A cover strap or fish plate used in a butt joint or scarf joint.

COVERING POWER : The spreading power of a paint to cover a surface.

COVING : See 'Cove'.

COW NOSE BRICK : A special brick used at plinth level or elsewhere for ornamental works in a building. Its shape is like a cow nose.

COWL : A louvred cover made of cast iron or cement asbestos fixed over a chimney or a soil pipe for improvement of draught.

COW NOSE BRICK

CRAB : (i)   A winch or windlass used to lift materials with mechanical advantage.

   (ii)   The hoist of a travelling crane which travels at right angles to the crane rails at the top of the gantry girder.

CRACKS : (i)   Cracks may develop in a long continuous structure due to expansion and contraction of the materials with the variation in temperature as well as due to unequal settlement of the structure.

COWL

   (ii)   Hair cracks may develop on a plastered surface due to shrinkage of plaster on drying.

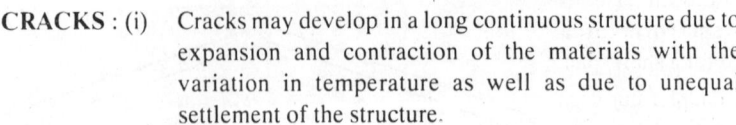

(iii)   Cracks are likely to occur in concrete due to shrinkage, poor curing, inadequate reinforcement, excessive loading, etc.

(iv)   Breaking of a paint film on exposure e.g., crocodiling, crawling and crazing.

(v)   Cracks may develop on a riveted plate due to shearing or bearing stresses.

(vi)   Fine hair cracks (not visible in bare eyes) may occur in a cast iron pipe during loading and unloading in transportation and during placement of the pipe in a trench.

(vii)   Cracks may develop due to poor bonding of a brickwork or stone masonry.

(viii)   In a wall, cracks are sometimes found under a beam due to shearing action of the load transmitted by the beam.

CRACKING SOUND : When a tin-foil is folded or bent, it produces a cracking sound. This is known as 'crying of Tin'.

CRADLE : A floating scaffold hanging on ropes used by workmen for painting outside walls of a tall building. Also known as 'boat scaffold'.

CRADLING : Fixing of rough timber around a steel joist for lathing.

CRADLING PIECE : A short beam from the wall on either side of a chimney breast to the trimmer beam to carry floor boards.

CRAIG'S FORMULA : A British Flood-intensity formula.

$$Q = 440 \, b \, N \log_e 8L^2/b$$

where, Q is flood intensity in cuseces,

L is greatest length of the catchment in miles,

b is average width of catchment in miles, and

$N = kvi$

where k is coefficient of run-off,

v is velocity of run-off in ft./sec,

i is intensity of rainfall in in/hr.

and N is a co-efficient for a particular catchment.

CRAMMING : Plugging a pipe prior to taking up a repairwork.

CRAMP : (i)   A joiners' clamp used by carpenters for gluing together timber pieces by pressure.

(ii)   A cramp iron of U-shape to hold ashlars (dressed stones) to each other.

CRANE : A power-driven lifting device with a jib commonly used in loading and unloading cargos. Also used in massive civil engineering constructions for lifting and transporting materials over a horizontal distance from the work site.

CRANE GANTRY : A gantry having rails, over which the overhead crane travels in a factory.

CRANE GIRDER : A gantry girder.

CRANE POST : The upright mast of a jib crane.

CRANE TOWER : Out of the three towers supporting a derrick, it is the king tower which supports the mast and the crane.

CRANK: A bar having bends at right angles, which is used as a lever in turning.

CRANK BRACE : A timber brace with crank used by carpenters in turning.

CRANKED SHEET : A specially bent A.C.corrugated sheet used at the junction of two sloping roofs.

CRAWLER TRACK : A tractor having an endless chain of plates instead of wheels. Also called 'Caterpillar'.

CRAWLING : A defect in a painted surface due to the shrinkage of the glassy top coat. This occurs due to the presence of grease on the base or ground.

CRAWLWAY : A duct about a metre deep through which a man can crawl.

CRAZING : (i) Cracking of a painted surface with deep checks.

(ii) Numerous intersecting cracks developed on a plastered surface.

(iii) Hair cracks formed on a concrete surface.

CREASING : Laying of one or two courses of tiles (with projection from the wall face) under a brick on edge coping with a view to throwing out rain water.

CREEP : (i) Permanent extension of a hard-drawn steel subjected to high stresses.

(ii) Increase in deformation of a metal at a high temperature, the stress remaining constant.

(iii) Advancement of rails in the direction of moving trains in a railway track.

(iv) Forward movement of a retaining wall due to the swelling of the shrinkable clay (retained at the back) during monsoon.

(v) Forward leaning movement of a sheet pile during driving into the ground.

CREEPER CRANE : A heavy duty crane used in construction of steel cantilever bridge. Such a crane travels along the top chord.

CREEP TRENCH : Similar to a 'crawlway'.

CREOSOTE : A preservative oil for timbers, derived from distillation of coal tar.

CREST GATE : A gate built into the spillway of a dam, used for adjustment of water level on the upstream of the dam.

CRETEWAY : A village road having two narrow strips of pavement running parallel to facilitate movement of vehicular traffic as shown in illustration.

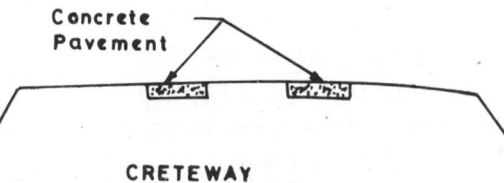

CRETEWAY

CRIB : A single layer or two layers of timber or steel laid across each other to spread a load over a large area.

CRIB DAM : A dam made of precast R.C.C. slabs laid vertically one above another with interlocking joints forming an enclosure which is filled with rock or earth.

CRIB INTAKE : See 'Intake crib'.

CRIB WALL : Similar to the construction of a 'Crib dam'.

CRIBWORK : Formation of large cells built with rectangular interlocking timber members upto the full depth of a bridge foundation and pouring concrete into the cells.

CRIMP'S FORMULA : Used in dewatering of sludge in a sewage treatment plant.

$$W_2 = (100 - P)/(100 - Q) W_1;$$

where W_1 = original weight of sludge.

W_2 = Weight of dewatered sludge.

P = % moisture of sludge before dewatering.

Q = % moisture of sludge after dewatering.

CRIMP AND BRUGE'S FORMULA : A British formula Recommended to apply to estimate flow through circular sewers and culverts.

$$Q = 3.072 \, D^{8/3} / I^{0.5} ;$$

where Q = discharge in cfm when pipe is running full;

D = Pipe diameter,in inches

I = Slope or gradient.

Also, used to determine velocity of flow in sewers.

$$V = 124 \, m^{0.67} \, i^{0.5} ;$$

where m is hydraulic mean depth and.

i is hydraulic gradient.

CRIMPER : An indenting roller used for making indentations on a concrete surface.

CRIPPLE : (i)   A bend provided at a chimney top.

(ii)   A shortened member in a frame at an opening e.g., a jack rafter in a roof.

(iii)   A bracket hooked to the ridge of a roof to hold a scaffold for laying tiles or slates.

CRIPLING LOAD : The load which when applied to a long column, the column starts bending. Also known as 'buckling load'.

CRITICAL DEPTH : With a constant quantum of flow, the velocity depends on the depth of flow. The critical depth is the depth at minimum specific energy, beyond which the specific energy increases and below which the specific energy decreases.

CRITICAL HEIGHT : The height to which a cohesive soil will stand vertically during excavation of a trench by vertical cuts ie., no timbering will be required to hold the sides of the trench upto that height. This height is proportional to the cohesion of the soil & it is nearly equal to 2 times the unconfined compressive strength of the soil divided by its density.

CRITICAL PATH METHOD : Critical path scheduling. A method adopted in developing a work progress chart. The normal time required for all the unit operations in a project is calculated and the timings are plotted in sequence. A critical path is followed to link all the operations with a view to arriving at the least time required to reach the end of the project. To achieve this, the timings of two or more operations are likely to overlap each other ie., the operations can be done simultaneously with optimal use of time.

CRITICAL VELOCITY : The velocity at which the flow will develop transverse eddies and be converted into turbulent flow. It occurs at a certain value of the Reynold's number.

CRITICAL VOID RATIO: The void ratio at which the prevention of volume change leads to no change in strength of a soil. A particular soil may have a wide range of critical void ratios under different loading conditions.

CROCODILING : Crazing of a paint film producing a pattern like the skin of a crocodile.

CROSS : A pipeline fitting looking like a cross with two short pipes crossing at right angles as shown.

CROSS BAND : In plywood, layers of veneer placed across the core to prevent cracking and shrinkage.

CROSS BRACING : Used in lattice girder, bridge, trestle and building frame for strengthening the structure.

CROSS BRIDGING : Herring bone strutting used in timber flooring.

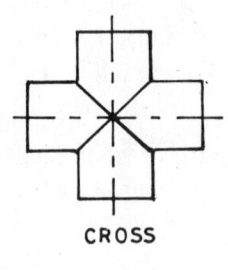

CROSS

CROSS CUT : A saw cut at right angles to the timber grains.

CROSS CUT CHISEL : A cold chisel used for hard cutting.

CROSS CUT FILE : A file having two rows of cuts intersecting each other.

CROSS CUT SAW : A saw used to cut a timber across its grain.

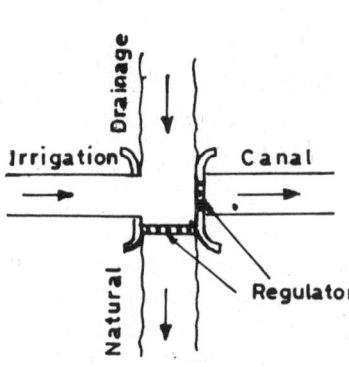

CROSS DRAINAGE : The drainage arrangement to carry the discharge of a natural drainage across a canal that intercepts the drainage. See illustration.

CROSS GRAIN : Fibres in a timber, which are not parallel with the length of timber. The fibres are diagonally or spirally interlocked in a hard wood.

CROSS HAIR : Vertical or horizontal spider line set in the reticule of a telescope of a theodolite or levelling instrument to fix up the line of sight.

CROSS JOINT : Vertical mortar joints perpendicular to the wall face.

CROSS LAP JOINT : A cross joint of two pieces of timber, both of which are halved for lapping over each other.

CROSS DRAINAGE

CROSS NOGGING : Herring bone strutting used in timber flooring.

CROSS OVER : (i) Two parallel rail tracks connected by another rail track for switching over the railway carriages from one track to the other as shown in illustration.

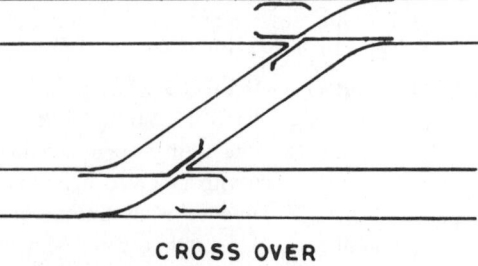

CROSS OVER

(ii) A pipe bent of U- shape to pass over another pipe.

CROSS POLING : Introducing short poling boards horizontally behind the runners to cover the gap between the runners in timbering a foundation trench.

CROSS SECTION : A transverse section of a body obtained by cutting it across its length.

CROSS SECTIONAL AREA : The area covered by the transverse section of a body.

CROSS TONGUE : A slip of wood glued into a saw-cut between two members with a view to stiffening the angle joint.

CROSS WELT : A seam between adjacent flexible roofing sheets, usually made parallel to the ridge or eaves.

CROW BAR :A Long steel bar with a claw of chisel shape at one end, used in withdrawing rail spikes, lifting up sleepers and rails, etc. in maintenance of a railway track.

CROWN : The highest part of an arch, a sewer and a road .

CROWN COURSE : A course of curved or cranked corrugated sheets laid over a ridge.

CROWN COVER : A protective cover over a circular saw or vertical fan blades.

CROWN MOULDING : A moulding at the cornice level just below a roof.

CROWN PLATE : A bolster used in joinery works.

CROWN POST : A king post or a vertical post placed centrally in a hammer beam roof truss.

CROWSFOOTING : Formation of minor wrinkles on a painted surface like the imprint of crows' feet.

CRUSHING STRENGTH : The compressive strength of a material i.e., the load at which a material fails in compression.

CRUSHING TEST : A cube test for mortar or concrete. A test to determine the compressive strength of a matetrial. A test to observe the failure of a very short column by increasing gradually the direct load over it.

CRYSTALLINE FRACTURE : (i) Cleavage fracture, (ii) a fracture in a metal.

CRYSTALLIZED FINISH : A finished painted surface with formation of wrinkles due to crystallizing of the lacquer.

CST-9 SLEEPER : A type of C.I. sleeper standardized by the Central Standard Track (CST) Committee. This is a combination of C.I. Plate, pot and box sleepers, which is extensively used in Indian Railways, See illustration .

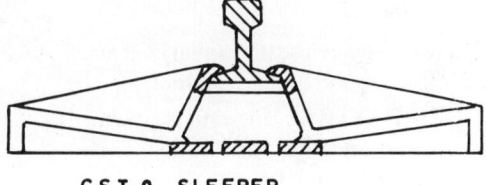

C S T-9 SLEEPER

CUBE STRENGTH : The strength of a mortar or concrete cube at the point of its crushing.

CUBE TEST : A test to determine the crushing strength of a mortar cube or concrete cube.

CUBING : Determination of the volume of a material or measurment of the volume of a work.

CULMINATION : The path followed and the time required by the sun or a star during its journey across the meridian are studied to determine the geographical meridian and the longitude. It is the meridian passage or transit.

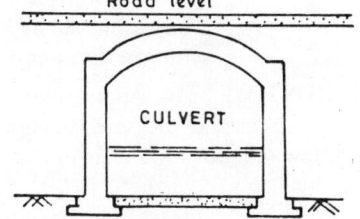

CULVERT : An underpass provided in a road or railway to facilitate the natural drainage of a land area. A small bridge over a narrow stream or watercourse for crossing it.

CUMEC : A unit for measurement of quantum of flow; One cubic metre per second.

CUMULATIVE AREA : Successive addition of areas.

CUMULATIVE ERROR : Accumulated error by summing up the individual errors developed in stages in a system.

CUP :   (i)   A hollow inverted cone fitted into a countersunk hole to withstand the thrust of a screw.

       (ii)   A form of warp commonly found in a flat-sawn timber.

CUP AND CONE FRACTURE : A form of plastic fracture in a ductile material under tension. The failure occurs in such a fashion that a cup shape is formed in one piece and a cone shape in the other.

CUP BOARD LATCH : A ball catch for securing a cupboard door.

CUP HEAD : The rounded head of a rivet.

CUP JOINT : A blown joint used in plumbing work.

CUP SHAKE : Ring shake, a defect in a timber.

CUP SQUARE BOLT : A coach bolt.

CUP TILE : A form of Italian roofing tile. See illustration.

CUP SHAKE

CUPOLA FURNACE : A vertical cylindrical furnace in which cast iron is melted for castings.

CURB : (i)   The edges of a carriageway.

     (ii)   A timber or stone upstand.

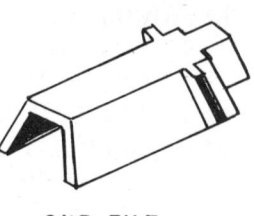

CUP TILE

CURB INLET : A street inlet provided on the road curb i.e., a vertical inlet for entry of storm water into the sewer. See illustration.

CURB JOINT : The horizontal joint made between two surfaces of a 'mansard roof'.

CURB RAFTER : The rafters provided at the upper slope of a 'mansard roof'.

CURB STRINGER : An outer string comprising a close string to support a stair, a shoe rail over which balusters are built and a facing string.

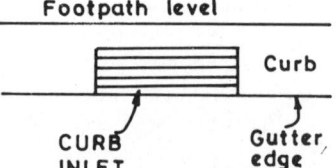

Footpath level

CURB INLET   Curb   Gutter edge

CURING : (i)   Polymerization or condensation by heating a thermo-setting resin or by adding accelerator to a cold-setting resin, so that they become strong, hard and insoluble in water.

     (ii)   A process of maturing mortar or concrete to attain its strength, by sprinkling water or keeping wet sacks over it for about seven days after setting of the mortar or concrete.

CURRENT METER : An instrument, usually a rotary meter, used to measure the current or velocity of flow of a flowing stream or river. The vane of the

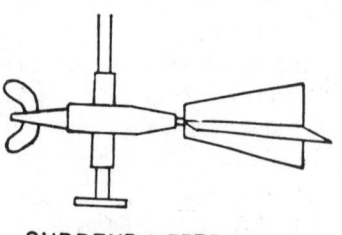

CURRENT METER

instrument rotates as the water flows through the meter and the flowing distance travelled by the water is recorded on the revolution counter of the meter.

CURTAIL STEP : Scroll steps or curved steps of spiral shape at the start of a geometrical stair.

CURTAINING : Too much of flow in paint films during painting a vertical surface which on drying produce bow-shaped ridges looking like a hanging curtain.

CURTAIN WALL : (i) A compound wall to make an enclosure.

(ii) A self-load bearing external wall between columns.

(iii) A wall serving as a screen for privacy.

CURTAIN WALLING : Wall cladding consisting of framed light-weight sheet material or fibre board for making enclosurers or cubicals on a floor.

CURTILAGE : The total land area covered by a dwelling house inclusive of its lawn and open spaces.

CURVE : A bend in a road, railway, property line, structure, etc. A curve is either horizontal or vertical . Various forms of curve are used in constructional works.

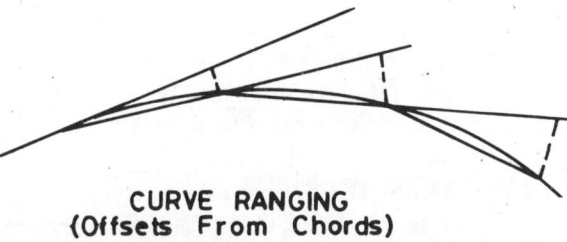

CURVE RANGING
(Offsets From Chords)

CURVE RANGING : Setting out or plotting points on ground to form a curve of desired shape.

CUSEC : A unit of quantam of flow of a fluid. One cubic foot per second.

CUSHION :   (i)   A layer of sand, gravel or any other granular material in a foundation bedding.

(ii)   A padstone over which a girder or truss end rests.

(iii) A seating made of nylon or plastic cord, asbestos or lead for the glass sheet in patent glazing.

(iv) The ballasts under sleepers in a permanent way.

(v)   An earth cover over a pipe laid underground.

(vi) A water pool over which water falls from a height.

CUSHION HEAD : A pile cap or a pile helment to protect the pile head during driving into the soil .

CUT AND COVER : A method adopted for any underground work by excavating i.e., opening out the ground and covering the trench or pit on completion of the work.

CUT AND FILL : In construction of a road or railway, the formation level is made by this method. The low areas are filled with soil by cutting the high grounds in the alignment of a road or railway.

CUT AND MITRED HIP : A hip or valley which is close-cut.

CUT AND MITRED STRING : A cut string in a timber stair in which the risers are mitred with the string.

CUT BACK EMULSION : Derived by dissolving liquid asphalt in a volatile solvent. This can be applied cold as a bitumen paint. The liquid asphalt is the viscous residue obtained by the distillation of asphaltic base crude oil to 425°C.

CUT BRICK : A brick cut to the required shape by the brick layers' axe.

CUT NAIL : A heavy duty nail of rectangular cross section made from steel plates.

CUT OFF : (i) The drainage of rain water into the soil.

(ii) A construction below foundation level or within the core of a dam to prevent the flow of seepage water.

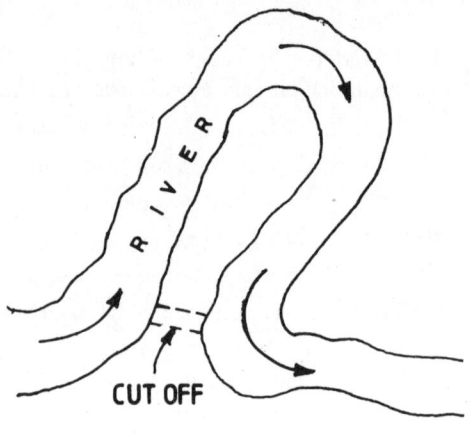

CUT OFF DEPTH : The depth of sheet pile or cut off trench below the level of a foundation.

CUT OFF TRENCH : A trench excavated below the foundation of a dam and upto the depth of an impervious stratum which is filled with puddled clay or concrete to cut off the seepage water.

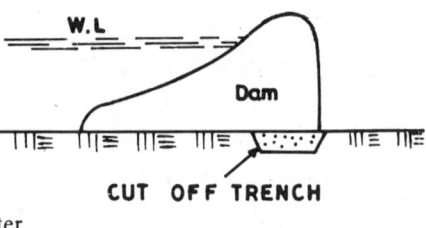

CUT OFF WALL : A wall of puddled clay or concrete built in the core of a dam to cut off seepage water. Also known as core wall.

CUT OUT : The upper part of a glazing bar cut away for flashing the glazing.

CUT RUBBLE : A rubble stone with a squared face, used in rubble walls.

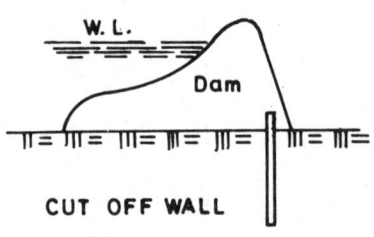

CUT STONE : A natural stone cut to desired shape.

CUT STRING : An open string or outer string in a stair, whose upper part is cut in steps such that the treads overhang it.

CUTTING : An excavation for a road or railway or for laying underground utility lines.

CUTTING GAUGE : A tool similar to the marking gauge used by carpenters. For marking, it is provided with a thin blade instead of a pin .

CUTTING IRON : The cutting blade of a plane used by carpenters.

CUTTING LIST : (i) A list of reinforcement steel bars specifying diameters and lengths prepared from the designer's bar bending shedule. Also called 'summary of Reinforcement'.

(ii) A list of timber specifying the sizes required for a job.

CUT WATER : The streamlined head of a bridge pier or any other water structure.

CYCLOPEAN : Aggregates larger than 150 mm used in concrete for massive constructions like dams, barrages, etc.

CYLINDER CAISSON : A drop shaft.

CYLINDER LOCK : A door lock which opens by key from outside and by turning the knob from inside .

CYLINDER TEST : This test is carried out to determine the compressive strength of a concrete. A concrete cylinder of 150 mm diameter and 300 mm length is tested under compression instead of using a concrete cube of 150 mm size. The cylinder test shows 0.75 times the strength of the same concrete crushed in cube test.

D

DABBER : A soft-hair brush having dome shape, used for finishing and polishing spirit varnish.

DADO : A panel of neat cement finish or any other ornamental finish around the inner walls starting from skirting level to a height of 30 cm to 60 cm.

DADO JOINT : A housed joint in timber framing.

DAM : A massive wall across a flowing stream or river to hold water at the up stream side.

DAMMAR : A natural resin of pale yellow colour obtained from exudation of trees. This is used in varnishing.

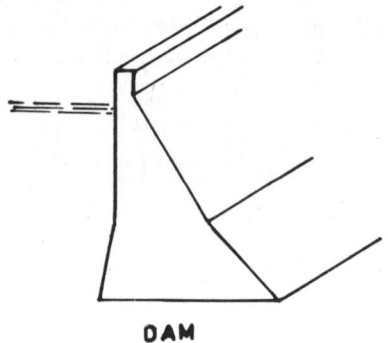

DAM

DAMPER : A metal cover across a flue that can be lifted up or let down for adjustment or closing the draft.

DAMPING : A device to reduce vibration or shocks.

DAMP PROOF COURSE : A layer of strong mix of cement concrete or any other water-repelling material laid all along the top of the walls at plinth level to prevent the ground moisture coming up the walls.

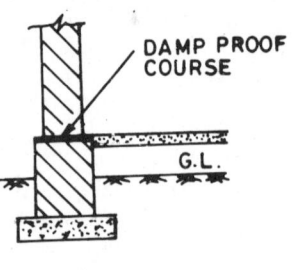

DAMP PROOF COURSE

G.L.

DANCING STEP : A balanced step with uniform rise, tread, nosing and going.

DARBY FLOAT : A long wooden float with handle on either end, required in levelling concrete or plastered surface over a long range in flooring and roofing.

DARCY'S EQUATION : The head loss in circular pipes flowing full is given by

$h_f = flv^2/d\ 2g$;   where f = friction factor,

l = length of pipe,

v = mean velocity of flow,

d = diameter of pipe,

and g = gravity constant.

DARCY'S LAW : The velocity of flow of a liquid through capillary tubes is proportional to the first power of the slope of the hydraulic gradient and it is expressed as $v=ks$; where v is approach velocity, s is the slope of hydraulic gradient and k is the co-efficient of permeability.

DATUM : A point having a level of reference which serves as a permanent bench mark from which a level surveying is carried out. Temporary bench marks, are established from the permanent bench mark.

DAYLIGHT FACTOR : (i) Sky-clearance factor.

(ii) It is the percentage illumination of the horizontal surface at a point in a room compared with the illumination that it would have from the sky having uniform bright light in a sunny day.

DAYLIGHT FACTOR PROTRACTOR : An instrument to determine the daylight factor, solid angle, slope, etc.

DAYLIGHT WIDTH : The width of an opening through which daylight enters into an enclosure.

DEAD BOLT : A door bolt of square section which enters into the recess in a door frame by turning a key in a lock. this is a locking device.

DEAD DOOR : A bricked–in door.

DEAD END : The plugged end of a pipe. Sometimes, a large size flanged C.I. pipe is blank-flanged at its end.

DEAD END SYSTEM : The free end or open end or tree system of a pipeline layout for water distribution i.e, a distribution for water supply without any loop formation.

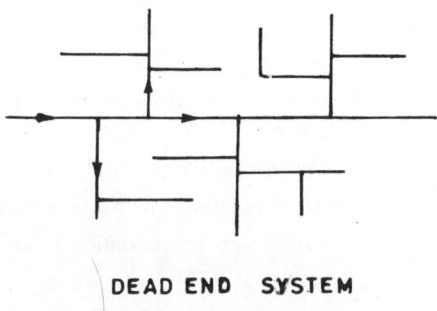

DEADENING : The pugging of floors.

DEAD KNOT : A form of knot sometimes found in a timber, which can easily be knocked out. Sometimes, it falls off forming a hole in a timber. Its fibres are not intergrown with the surrounding wood.

DEAD END SYSTEM

DEAD LEG : A hot water connection in a plumbing system in which water is kept stationary. The water cools between the draw-offs and the dead leg.

DEAD LIGHT : A fixed sash or a part of a window that does not open.

DEAD LOAD : The self weight of a structure or any non-movable permanent load fixed on it.

DEAD LOCK : A door lock which is operated by a key only from both sides without any knob or handle.

DEAD MAN : An anchor block or wall which remains buried in place by its self-weight and the passive pressure from soil. This is required to tie sheet piles or light retaining walls to keep them in position.

DEAD MILD STEEL : Soft steel containing only about 0.10% of carbon, used for pressing, drawing and bending.

DEAD SHORE : A vertical timber support to hold a needle beam to carry the weight of a wall under reconstruction. See illustration.

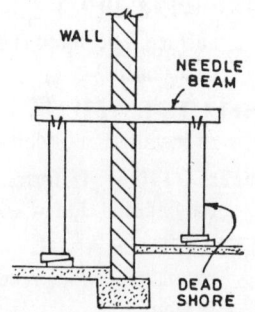

DEAD SMOOTH FILE : A file cut of finest grade.

DEAD WOOD : Timber obtained from a dead standing tree.

DEAL : A square-sawn soft wood.

DEAL FRAME: A frame-saw used by timber merchants for cutting or sawing deals.

DEATH WATCH BEETLE : A beetle that finds it way into a sap wood and burrows deeply into the timber. The adult beetle makes a ticking sound within the timber. It is difficult to kill them.

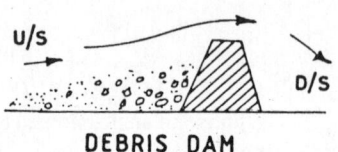

DEBRIS DAM

DEBRIS DAM : A barrier across a stream or river to store sand, gravel and pebbles.

DECANTING : A process of locking used in pneumatic caisson sinking, where accomodation in airlocks is very much limited.

DECAY : Decomposition by microbes and fungi. In timber, dry rot or wet rot is caused by fungal growth and by micro-organisms.

DECIBEL : The unit of sound intensity.

DECIDUOUS TREES : Most of the hardwoods belong to this group of trees which lose leaves every year.

DECHLORINATION : Generally required after 'Super-Chlorination'. The device to remove excess chlorine from water by keeping the residual chlorine as per requirement.

DECK : A platform, a floor without any covering at top, a bridge floor, a jetty, etc.

DECK BRIDGE : A bridge with a deck carried by the top chord of the girder.

DECLINATION : (i) The variation in angle between the magnetic north and the true north.

(ii) The angular distance of a star from the celestial equator.

DEEP BEAD : A sill bead or ventilating bead which is an upright board fixed to the board of a sash window with a rise of about 75 mm above the board. This is used for the purpose of ventilation without draught.

DEEP CUTTING : Resawing a timber parallel to one of its faces.

DEEP MANHOLE : A manhole with an access shaft above it for easy inspection.

DEEP PENETRATION TEST : See 'cone penetration test'.

DEEP SEAL TRAP : An anti-siphon trap having a water seal of 100 mm depth, normally used in one-pipe system of plumbing.

DEEP WELL : A well bored through a shallow impermeable stratum to reach a great depth to obtain greater yield of water or to draw oil, sulphur, etc.

DEEP WELL PUMP : Usually a multi-stage centrifugal pump driven by a motor at the surface. The pump at the base of the well is connected to the motor at surface by means of a shaft.

DEFECTS-LIABILITY PERIOD : A period of operation and maintenance from the date of installation or commissioning.

DEFLECTION : Deformation or deviation from the original position. In structures, it is the bending due to loading.

DEFLECTION ANGLE : In a traverse survey, it is the angle between a line and the extension of the preceeding line. This is used for curve ranging i.e., setting of curves in field.

DEFLECTION CURVE : In structural engineering, it is the elastic curve showing deflection of different points of a member subjected to loads.

DEFLECTOMETER : An instrument to measure the deflection of a structural member subjected to loads.

DEFORMATION : A term that includes deflection, permanent set and non-recoverable plastic movement of a structure.

DEFORMETER : An instrument used in drawing out the influence line of a structure by model analysis.

DEGREE-DAY VALUE : Used for calculating annual consumption of fuel in heating a building in cold climates. It is estimated by counting the number of days in a year by which the average temperature falls below $60^{o}F$. Each day with an average temperature of $59^{o}F$ is counted as 1, 58° F as 2, 57° F as 3, $56^{o}F$ as 4 and so on. Thus, a figure is obtained which is known as degree-day value.

DEGREE OF COMPACTION : In soil mechanics, it is a measure of compactness of a soil sample. It is obtained as

$$\frac{\text{void ration in looset state—void ratio of sample}}{\text{void ratio in loosest state—void ratio in densest state}}$$

It indicates the degree of density of a soil sample.

DEGREE OF CURVE : A way of describing circular curves by the degrees of angle subtended at the centre of a circle by a chord of 100 ft. length. As the radius of circle decreases, the number of degrees increases.

DEGREE OF DENSITY : Degree of compaction.

DEGREE OF SATURATION : It is given by

(Volume of voids filled with water/Total volume of voids in the sample) x 100

It speaks of the voids occupied by air in a soil sample. In other words, it is the percentage of voids in a soil sample occupied by water.

DEHOTTAY PROCESS : A ground freezing process for shaft sinking. It essentially comprises circulating of liquid carbon di-oxide in the pipes introduced into the ground.

DEHUMIDIFIER : An air-conditioning unit which reduces humidity of air by bringing down the temperature of air below dew point by spraying chilled water.

DELAMBRE'S RULE : In geodetic triangulation, for calculating lengths of sides of large triangles, this rule is used by employing spherical trigonometry. The length of a chord of the spheroid is calculated and the three spherical angles are reduced to the angles of a plain triangle formed by the chords. The two unknown sides are then calculated by plane trigonometry.

DELIQUESCENCE : The liquefying of salts present in a plaster or brickwork by absorbing moisture from air. They appear as dark patches of damps in a plastered surface.

DELIVERY PIPE : The pipeline from a reservoir to supply water or any other fluid.

DELTA :   (i)   A tract of land of triangular shape formed due to silt deposition at the junction of rivers.

(ii)   The depth of water expressed in 'acre-foot' supplied to a crop.

DELTA DEPOSIT: The silt deposits due to sudden check in velocity when a silt-laden stream enters a river or sea. This forms a land like delta.

DENSE CONCRETE : A concrete that weighs more than 1920 kg/cu.m. is usually called a dense concrete with uniform distribution of aggregates in the matrix.

DENSITY : Weight per unit volume of a material.

DENTAL : A projection like a tooth on an apron of a water- structure which breaks the forces of a flowing water.

DENTATED SILL : A notched sill in a hydraulic structure to break the forces of a flowing stream with a view to reducing the effect of scouring.

DEPARTURE : In surveying, it is the distance of a point to the East or West from the North-South reference line. Thus, a point is located by its latitude and departure.

DEPRESSANT : A bathotonic reagent used to lower the surface tension such that the finely-divided particles of gangue floating on the surface settles down.

DEPRESSED GUTTER : A street gutter is sometimes depressed where a street inlet is provided for entry of storm water into the sewer.

DEPRESSED SEWER : Synonymous with 'inverted siphon', constructed lower than adjacent sewer sections to pass beneath an obstruction. It runs full under gravity flow and at greater than atmospheric pressure, the profile being depressed below the hydraulic grade line.

DEPRESSION HEAD : The vertical difference in height between the normal ground water table (undisturbed) and the water level in a well after pumping.

DERAILING SWITCH : A trap joint i.e., a break in the continuity of the rail provided in the shunting line. The object is to derail a wagon if it tends to

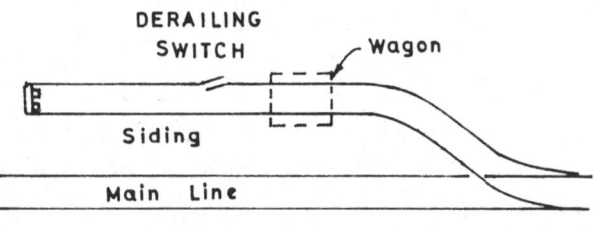

escape from the shunting line to the main running line. See illustration.

DERRICK : A lifting device, usually a stationary crane operated manually or mechanically. A 'standing derrick' is a vertically upright pole held by guy ropes and having a pulley at the top of the pole 'over which a hoisting rope passes for lifting materials. The materials are lifted up by operating a winch manually on ground. This is also known as a 'gin pole' or 'guyed mast'.

A *'guy derrick'* is a vertical mast from which a lifting jib or boom of short length is suspended which can turn a full circle. The derrick is held by guy ropes and this type of derrick is mostly used for erecting framed buildings.

The *'shears derrick'* has two poles laced together at the top from which the hoisting tackle is hung. This is also known as 'Shear legs'.

The *'three-legged derrick'* is a modified form of shear legs. It does not require any guy rope. It is used for drilling, sinking wells and driving bored piles.

The *'oil-well derrick'* is a square framed and latticed mast, which is quite tall to enable introducing the longest drill pipe into the ground and lifting it up. This has hoisting block at its top. This type of derrick is used for drilling oil wells.

DERRICK CRANE : A 'scotch derrick' of stiff legs having no guy rope, used as a stationary derrick. A permanent structure holds the mast in vertical position, the base of the mast being tied with two horizontal legs (sleepers). The legs are held down by counter weights (kentledge). The top end of the boom is hung from the top of the mast and it can swing through an angle of 240°.

DERRICK TOWER GANTRY : A steel staging consisting of one crane tower for the mast and the jib and two anchor towers for the legs and kentledge. The towers at their upper ends are tied together by derrick legs.

DESIGN LOAD : The load considered in design of a structure. The worst possible loading condition is taken into account in computing the design load. The design load must be greater than the actual load coming on a structure.

DESTRUCTIVE DISTILLATION : Heating a substance in absence of air.

DETAIL DRAWING : Working drawing.

DETAIL PAPER : Sketch paper from which a drawing is to be prepared.

DETENTION BASIN : A basin or tank or reservoir where a liquid is stored for some period. Surcharged sewer lines also serve as a detention basin.

DETENTION TIME : The time required for settlement of suspended particles in a liquid. The time varies depending upon the settling characteristics of the dispersed particles.

DETONATOR : A container having explosive mixture, used for blasting purpose.

DETOUR ROAD : An alternative road of circuitous route.

DETRITUS SLIDE : 'Creep slide' i.e., land slide by slow movement of detritus downhill clay layer or loose shales.

DETRITUS TANK : A small settling tank of short detention time for removal of heavy detritus matters without interruption to the flow of liquid. Such a tank is provided just after the screen chamber in a sewage treatment plant. The velocity of flow is restricted to 1 ft/sec.

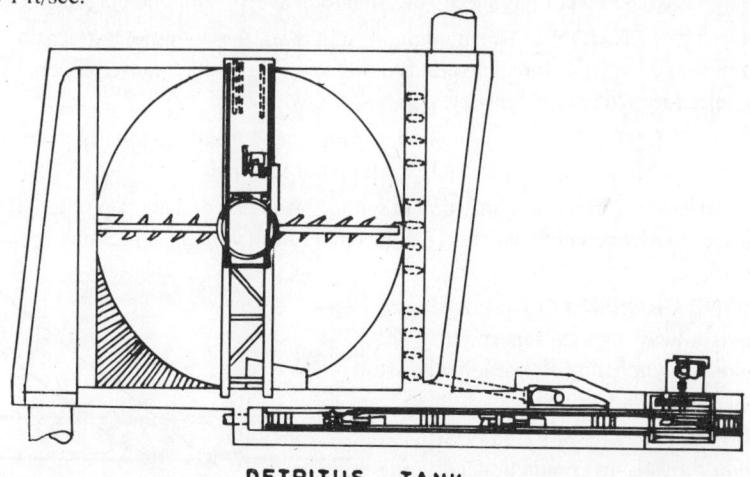

DETRITUS TANK

DETRUSION: Shearing of a timber parallel to its grain is a sort of failure, which can be prevented by using 'connectors'.

DEVIATION: (i)   Shifting from the original position.

(ii)   Deflection.

(iii)   Difference between one value of a set and the average of the set.

(iv)   Any departure from the straight.

DEVIL : (i) A stretcher carried by two labourers for loading materials.

(ii) A fire grate for heating asphalting tools.

DEVIL FLOAT : A hand float having 3 mm projected nails at either end, used by the masons for scratching a fresh plasterd surface to make a key for the next coat to be applied.

DEWATERING : (i)   Lowering ground water table by pumping.

(ii)   Bailing out water from a foundation trench or any other excavation by pumping.

DEXTRIN : A water-soluble starch-gum used as a binder in water paints and distempers.

DIAGONAL BOND : 'Raking bond' or 'herring-bone bond' See illustration. The bricks are laid at 45° to the face. It is used in flooring.

DIAGONAL BRACE : Bracings or lacings placed diagonally to strengthen a frame against wind load and horizontal forces. These members are either in tension or in compression.

DIAGONAL EYE PIECE : The eyepiece of a prismatic telescope used in surveying.

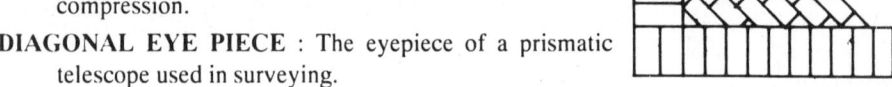

DIAGONAL GRAIN : Oblique grain when, wood fibres are at an angle to its length. This is a defect caused by faulty conversion of a timber.

DIAGONAL SLATING : Laying of A.C. diamond sheets with one diagonal horizontal.

DIAGONAL TENSION : The combined action of the longitudinal tension and the transverse shearing forces applied to a beam, which may cause diagonal cracks, is referred to as diagonal tension.

DIAGRID FLOOR : A floor having underneath a network of diagonally intersecting ribs of mild steel or R.C.C. spanning a rectangular space.

DIAL GAUGE : A measuring instrument with a graduated dial and a pointer. The pointer is free to rotate, which indicates displacement from which the pressure or flow can be read.

DIAMOND CROSSING : The crossing of two railway tracks of same or different gauges, when they cross each other at an obtuse angle. See illustration.

DIAMOND PATTERN FLYOVER : A type of fly-over junction or grade separation provided, when two busy roads having fast traffic movement cross each other. Their grades are separated by making one road to cross

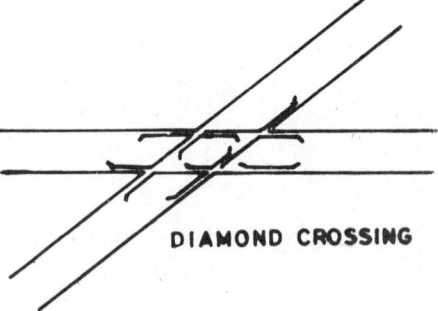

DIAMOND CROSSING

the other by means of a bridge. The roads are interconnected by link roads to facilitate traffic flow from one to the other.

DIAMOND SAW : A circular saw used to cut stone with black diamonds set in the perimeter of the saw.

DIAMOND SLATES : Square slates made of asbestos cement with two corners cut-off, so that these can be used in diagonal slating.

DIAMOND SWITCH : A special switch used in railway crossings e.g., in scissor's cross over.

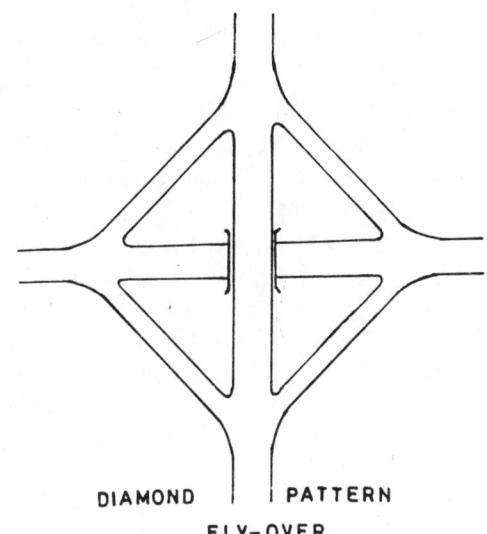

DIAMOND | PATTERN
FLY-OVER

DIAMOND WASHER : A curved washer like a 'limpet washer', used in fitting corrugated roof sheets.

DIAPER WORK : (i)   A chequered work.

(ii)   Placement of facing bricks with light and dark-shaded bricks such that it makes a diamond pattern.

DIAMOND   SLATING

DIAPHRAGM : (i) A brass fitting carrying the reticule in a surveying telescope.

(ii) A stiffening web placed across a hollow building block.

(iii)A stiffening plate between main girders of a bridge.

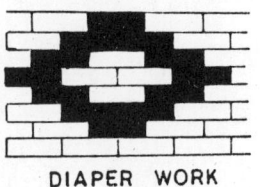

DIAPER   WORK

(iv) A diaphragm wall required during construction of a tunnel.

(v) A type of meter used to measure the flow of a fluid through a pipe

DIAPHRAGM PUMP : A type of reciprocating pump having a diaphragm of flexible leather, rubber or canvas which is moved to and fro by a rod. This type of pump is capable in handling muddy water with grits and pebbles.

DIAPHRAGM WALL: A concrete wall to keep water out of a deep excavation or to resist high earth pressure. A deep trench is excavated mechanically and is filled with bentonite clay to give support during excavation. Reinforcement rods are introduced into the mud and concrete is lowered into the trench with the help of tremie.

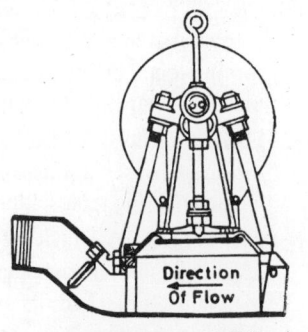

DIAPHRAGM PUMP

The construction of a diaphragm wall by this method is advantageous, because it does not cause any vibration as in driving sheet piles.

DIATOMITE: Diatomaceous earth or moler earth, a soil composed of hollow siliceous skeletons of diatoms (tiny marine organisms). This soil is so light that bricks made of it floats on water. This is used in making light-weight concrete blocks, as an absorbent for nitroglycerin in dynamite and as an extender in paints.

DIATOMITE FILTER : A schematic view of a 'Diatomite Filter' is shown. It is comparable in operating costs with the rapid sand filter. The upper limit of raw water turbidity is 30 mg/l. Only a thin layer of diatomite is on the filter element.

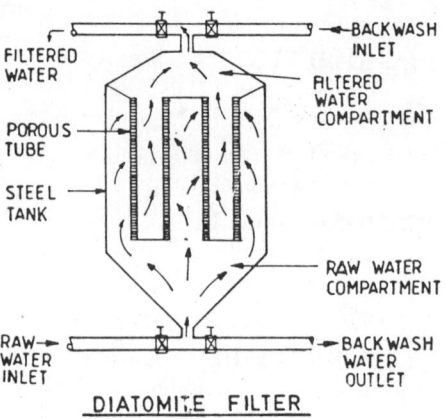

DIATOMITE FILTER

DICKEN'S FORMULA : A British empirical formula expressing flood intensity as a function of the catchment area. The formula was devised for Northern india where annual rainfall is 24" to 50"

$$Q = 825\ A^{3/4} ;$$

where Q is maximum flood intensity in cusecs and 'A' is catchment area in sq.miles.

DIE :
  (i)   A tool for making screw threads.
  (ii) A hard metal block with a hole through which a ductile metal is cold drawn to form wire.
  (iii) The enlarged square part of the upper and lower ends of a 'baluster' which meets the handrail and the steps in a stair.

DIE-FORMED STRAND : A strand for use in prestressed concrete.

DIE SQUARE : A squared timber, usually 100 mm x 100 mm.

DIE STOCK : A die-holder for cutting screw threads by hand.

DIFFERDANGE BEAM : A broad-flanged beam.

DIFFERENTIAL PULLEY BLOCK : A chain-pulley block or a hoisting tackle having an endless chain threaded over two wheels of different diameters on the same shaft turning together. It has a great mechanical advantage. This is chiefly used for lifting materials manually by winding the endless chain.

DIFFERENTIAL SETTLEMENT : Relative settlement due to unequal sinking of different parts of a structure. The worst possible differential settlement is not likely to exceed half the total settlement in a normal settlement crater.

DIFFERENTIAL WATER PRESSURE : (i) The pressure developed when the water level in a harbour basin gets lowered and remain until the ground water table behind the wall has reached the same level. The lowering of water level may occur due to tidal variations or due to short period waves or long period waves.

DIFFUSED AERATION : Aeration of water or sewage by passing compressed air through porous plates or pipes such that air bubbles coming out from the pores aerate the liquid. See illustration.

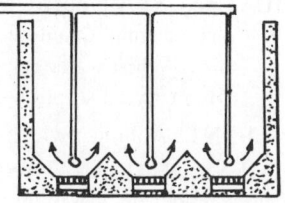

DIFFUSED POROUS WOOD : A hardwood having uniform size of pores through the annual rings.

DIFFUSER : (i)   A porous pipe or plate through which air is introduced into the sewage for aeration .

DIFFUSED AERATION

(ii)   The gradually increased cross-section of the outlet in a centrifugal pump or compressor to increase the pressure of a fluid.

DIGESTION : A process by which the organic solids of sewage sludge are converted into sludge, liquid and gas with the help of biological action.

DIGESTION CHAMBER : A chamber or tank in which the organic solids or sewage sludge is digested and the matters are converted into gases, liquid and humus by biological activities. The chamber is operated without or with heating arrangement. The microbes in the chamber are kept in anaerobic condition and they are in endogenous stage or phase.

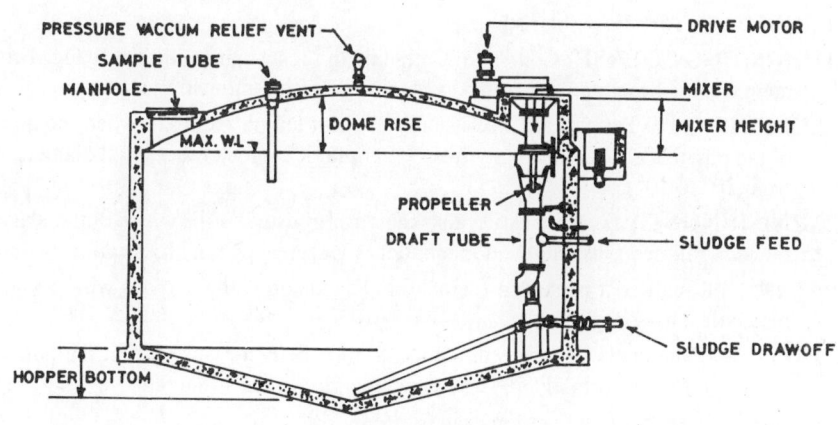

DIGESTER

DIKE : (i)   An earthen embankment built on either side of a river at a distance away from the river banks, the height being kept about 1.5 m to 2 m above the H.F.L., with a view to controlling flood.

(ii)   A stratified rock traversed by cracks.

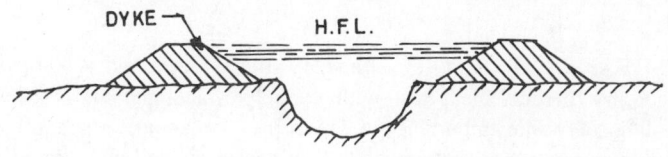

DILATANCY: A property possessed by silt which makes it distinguishable from clays. When a small quantum of wet silt is shaken in hand, it releases water and becomes shiny. When water re-enters the silt, a matt surface is produced due to increase in volume by asbsorption of water.

DILUENT : A thinner which dilutes a semi-fluid.

DIMENSIONAL ANALYSIS : The use of dynamic similarity simulating field conditions in model analysis of flow through pipelines, channels. streams, rivers, etc. and of ports, ships, break waters, weirs and other structures/members related to flow of fluids.

DIMENSIONAL CO-ORDINATION : An agreement and compromise between architects and the manufacturers of building components to suit the requirements of a modular system of building construction.

DIMENSIONAL STABILITY : A property of building materials which makes them stable by having no movement of moisture, slight temperature movement, no shrinkage or expansion due to variation in climatic condition.

DIMENSION PAPER : A standard paper having vertical columns for taking off dimensions, and making computations for record keeping.

DIMENSION SHINGLES : Shingles cut to required size of uniform width.

DIMENSION STOCK : Squared timber, sawn or hewn, having no pith. Various sizes of timber are cut out of this, according to the requirement.

DIMENSION STONE : Ashlar stone.

DIMINISHING COURSE : Courses of slates laid in such a manner that the gauge between them diminishes gradually from eaves to ridge with diminishing slate widths.

DIMINSIHING PIPE : A taper pipe usually of short length required, when the diameter of the pipeline changes abruptly from a very large diameter to a small diameter (e.g., from 20" to 10").

DIMINISHING STILE : A door stile, also known as gunstock stile, which diminishes from the lock rail upwards and the door shutter is glazed above the lock rail .

DINGING : Rough coat of stucco (cement and sand) on walls marked with a jointer to resemble a masonry work.

DIP : (i)   The angle of maximum slope of rock beds measured from the horizontal. This is normally 90^0 i.e., at right angles to the strike.

      (ii)   Angle of declination used in surveying.

DIP COMPASS : A compass having dip needle which indicates dip.

DIP NEEDLE : A magnetic needle which is pivoted horizontally in a compass so that it can swing only in vertical plane. The needle is set initially in the magnetic meridian.

DIPPER DREDGER : A dredger having a single large dredging bucket suspended from the end of a long jib.

DIRECT ACTING PUMP : A reciprocating pump in which the power cylinder (compressed air or steam cylinder) and the water cylinder are placed in opposite sides of the piston rod.

DIRECT SHEAR TEST :   The shearing strength of a soil is sometimes investigated by direct shear test with the help of an apparatus as shown in illustration. It is of prime importance to determine the shearing strength of soil, because the maximum load that can be transmitted to the soil by a foundation depends on the resistance to the shearing deformation.

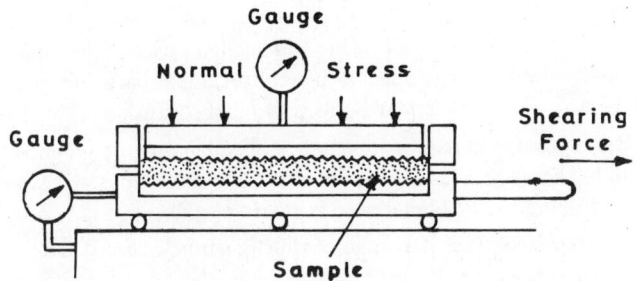

Gauge

Normal ⟳ Stress

Gauge

Shearing Force

Sample

DIRECT READING TACHOMETER : A tachometer used in tacheometric surveying, which gives direct reading of length (from the staff intercept) and the level difference between the staff and the instrument without any measurement of the vertical angle.

DIRECT STRESS : A stress which is purely either compressive or tensile without causing any moment.

DISAPPEARING STAIR : A loft ladder.

DISCHARGE : The quantum of fluid flowing through a pipe or channel per unit time. It is uaually denoted by 'Q'.

DISCHARGE CO-EFFICIENT : Co-efficient of discharge on which the actual discharge depends.

DISCHARGE CURVE : A curve related to the head of water or water level of a stream or reservoir.

DISCHARGE HEAD : The pressure of water due to its head at the point of discharge.

DISCHARGE VALVE : A control valve which regulates the discharge.

DISCHARGING ARCH : A relieving arch.

DISCONNECTING TRAP : An intercepting trap used in plumbing system.

DISCONTINUOUS CONSTRUCTION : A construction technique adopted in making a room sound-proof and heat-proof. This is done by constructing hollow walls, floating floors, etc. with the use of insulating materials.

DISINFECTION : A method of destroying or killing disease-causing microbes. For disinfection of water, disinfectants used are chlorine, chlorine-ammonia, ozone, etc.

DISPERSING AGENT : A deflocculating agent, usually sodium oxalate, to prevent quick agglomeration of the dispersed particles.

DISPERSION : Finely divided particles in suspension in a liquid. Paints and varnishes are examples of dispersion. Fine droplets in emulsion paints are also dispersion of one liquid in another.

DISPLACEMENT PUMP : Diaphragm pump or air-lift pump in which compressed air displaces water. Usuaily, this is a piston-operated or ram-operated pump.

DISPLACER : (i)   A plum in concrete.

(ii)   Any material that displaces another material.

DISSOLVED OXYGEN : The quantum of oxygen dissolved in water, which varies with the temperature of water and the pollutants present in water.

DISTANCE PIECE :   (i)   A separator used for maintaining the correct distance between the running rails and check rails in a railway track.

(ii)   A cast iron or steel separator used for maintaning the position and spacing of reinforcement bars during concreting or to keep the built-in members or formwork in position.

DISTEMPER: A sort of matt paint with appreciable quantum of pigment which is thinned with water. The washable distempers or oil-bound distempers have drying oils. In washable distempers, the binder used is casein or glue.

DISTEMPER BRUSH : A wide flat brush having long bristles used for distempering walls.

DISTRIBUTED LOAD : A load that is distributed along the length of a beam or over a slab.

DISTRIBUTION BOX : A box or chamber which gives access to the branch lines.

DISTRIBUTION CURVE : A frequency curve used in statistical analysis.

DISTRIBUTION LINE : (i)   A cable for distribution of electrical power.

(ii)   A pipeline for distribution of water or any other fluid.

DISTRIBUTION PIPE : A pipeline for distribution of a fluid from a storage tank.

DISTRIBUTION RESERVOIR : A service reservoir from which water is supplied to the consumers.

DISTRIBUTION STEEL : The subsidiary reinforcement placed at right angles to the main reinforcement bars in a reinforced concrete slab.

DISTRIBUTION TILE : Open-joint clay tiles used as agricultural drains for distribution of the sewage effluent for land irrigation.

DITCH : A narrow small channel for drainage or irrigation.

DIVERSION : A bye-pass.

DIVERSION CUT : A bye-pass channel cut to divert the flow path of a stream required during a constructional work.

DIVERSION DAM : A dam built across a stream to divert some water into a bye-pass channel.

DIVERSION OF STREAM : See 'diversion cut'.

DIVERSION TRACK : A bye-pass road or railway

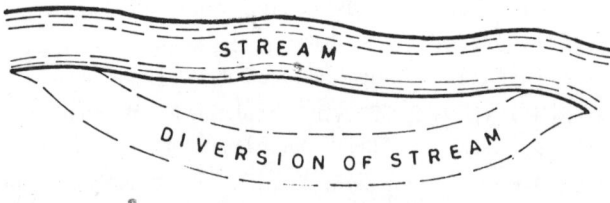

DIVERSION WORKS : Obstructions made in flow path of a stream or river to divert water into an off-taking channel. Such works are weirs, barrages and spurs.

DIVERTING WEIR : It is a device to divert the excess flow automatically from a sewerline, when the quantum of sewage flow exceeds the design flow.

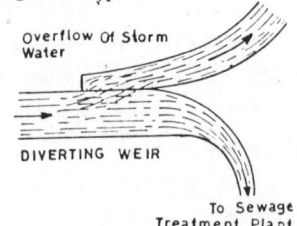

DIVIDE WALL : A wall built along a canal to facilitate diversion of flow and desilting operation.

DIVIDING DAM : It is constructed for distribution of water to different branches of a canal on the basis of the water requirement by each branch.

DIVIDING BELL : A steel chamber of bell shape which is lowered to the river bottom to facilitate working of the divers for under-water constructions and raised up. The chamber is open at the bottom. Now-a-days, this is replaced by airlock chambers.

DIVISION WALL : A fire-resisting wall built from ground level upto the roof of a building separating the apartments from fire risk.

DOCK : A basin in a river or sea, used for shipping. The basin is cut off from the action of tides by gates.

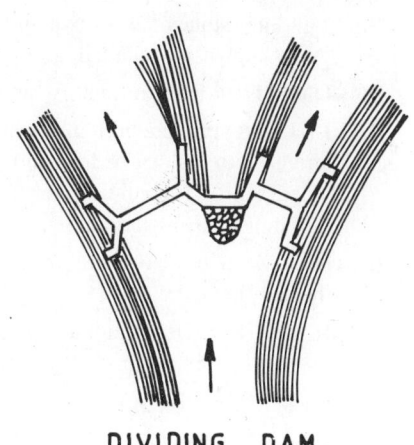

DIVIDING   DAM

DOCKING BLOCKS : The blocks of concrete or brick masonry that support the underside of the hull of a ship in a dry dock. The blocks which are provided in the central row are called 'keel blocks'.

DOCK PLATFORM : A railway platform provided in addition to passenger platform in a crossing type wayside station on a double line railway track.

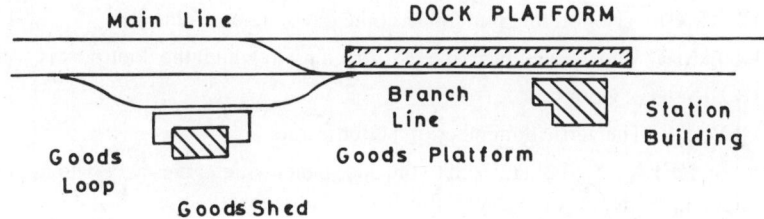

DOG : (i)   A dressing iron.

(ii)   U-shaped spike of steel used in wooden joinery works.

(iii)   A pair of hooks with a chain to lift building stones.

(iv)   A spike of square cross-section.

DOG

DOG-LEG CHISEL : A specially bent chisel used for cleaning grooves.

DOG-LEGGED STAIR : A stair having two flights without any stair well, the outer string of each flight being housed in the same newel post.

DOG-TOOTH COURSE: Course of headers laid diagonally with one corner projected to make corbelling out from a brick wall.

Dog – Legged Stair

DOG SPIKE: A dog-headed spike of square cross section driven into wooden sleeper to hold the flat-footed rail firmly.

DOG LEWIS : A type of attachment for lifting building stones as shown in illustration. Also known as 'chain dog.'

DOLOMITE : A basic refractory material.

DOLPHIN : A structure built of raking steel piles driven into the sea bed for mooring in the open sea or for guiding ships to proceed through a narrow harbour entrance.

DOME : A roof of spherical shape. A hemi-spherical vault.

DOME LIGHT : (i)   A dome made of curved glass or perspex sheet for natural light.

(ii)   A hemi-spherical vault having glazed portion for roof lighting.

DOMESTIC SEWAGE : Sanitary sewage of domestic origin.

DOOR BUCK : A door sub-frame to which the door case is fixed.

DOOR CASING : The architrave or a lining round a door frame.

DOOR CHECK : A device fixed to the door top, which closes the door automatically. Also known as 'door closer'.

DOOR FRAME : A wooden or precast concrete frame comprising two vertical posts and the head (horizontal member).

DOOR FURNITURE : Fittings and fixtures used in a door, e.g., hinges, handles, bolts, latches, locks, knobs, finger plates, etc.

DOOR HEAD : The horizontal member at the top of a door frame.

DOOR JAMB : The vertical face of a door opening to which the door post is fixed.

DOOR LINING : Door case.

DOOR POST : The vertical member of a door frame.

DOOR SCREEN : A wire net fixed to the door panel which gives access to natural air and light and prevents the entry of flies and mosquitoes.

DOOR SILL : A horizontal member of timber, concrete or brick masonry connected to the door posts at the base of the frame to keep out rain water.

DOOR STOP : A stop for the door shutter either fixed on the door jamb or set in the floor to keep the door open at a particular position.

DOOR SWITCH : An electrical switch operated by the opening and closing of a door.

DOPE : A quick-drying cellulose lacquer used for coating leather or textile.

DORMER CHECK : The upright side of a dormer window.

DORMER WINDOW : A vertical window coming through a pitched roof as shown in illustration.

DOSING CHAMBER : A dosing tank into which sewage gets accumulated upto a

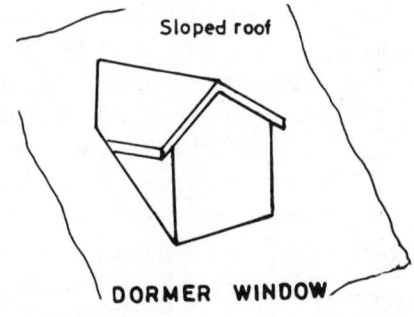

certain level, beyond which the sewage is discharged automatically to enter into a processing unit.

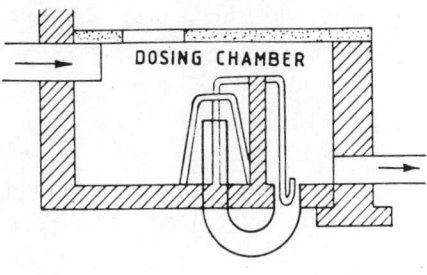

DOSING CHAMBER

DOSING SIPHON : An arrange-ment kept in a dosing chamber such that the contents of a dosing chamber get discharged automatically by siphonic action.

DOSING TANK : A dosing chamber.

DOSY TIMBER : A timber that has started decaying.

DOUBLE-ACTING HINGE : A hinge which permits a door to swing through 180^0 and facilitates in self-closing.

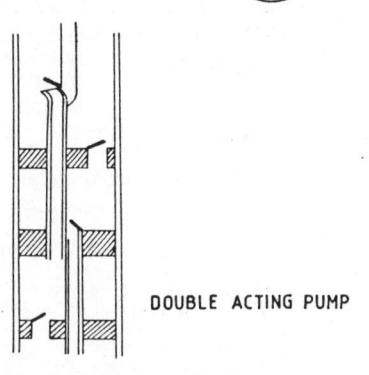

DOUBLE-ACTING PUMP :
A type of reciprocating pump in which water is

DOUBLE ACTING PUMP

discharged both by forward and return strokes of the piston or the plunger.

DOUBLE ANGLE : A mild steel section having the shape of the alphabet 'Z'.

DOUBLE BEAD : Two beads running parallel and separated by a quirk, used in joinery work.

DOUBLE BRIDGING : Two rows of herring-bone strutting used in making a wooden floor.

Double Angle Or Z - Section

DOUBLE CHLORINATION : It comprises pre-chlorination and post-chlorination for treatment of heavily polluted raw water.

DOUBLE CONNECTOR : The common term is 'long screw'. A short piece of pipe having long parallel thread at either end with a back nut and socket, which is used in connecting or disconnecting a pipeline at any location.

DOUBLE CROSSOVER : 'Scissors crossover' used in railway for interchanging tracks.

DOUBLE CUT FILE : Cross-cut file.

DOUBLE DOOR : A folding door.

DOUBLE DOOR BOLT : Espagnolette bolt.

DOUBLE DOVETAIL JOINT : A butt joint between two pieces of timber with the help of a double dovetail wooden connector.

DOUBLE EAVES COURSE : A double row of tiles or slates laid at the eaves of a sloping roof.

DOUBLE FACE HAMMER : A hammer having two striking heads, one at each end.

DOUBLE FILTRATRION: Sometimes, the filtrate of one filter is passed through another filter with a view to achieving the better quality of filtrate. This is known as 'double filtration'. In fact, the filtration of water first through a rapid sand filter (roughing filter) and then through a slow sand filter doubles the capacity of slow sand filters.

DOUBLE FLEMISH BOND: A type of brick bond in which headers and stretchers are placed alternately in rear and front elevations of a wall.

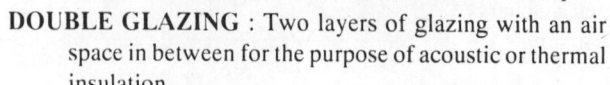

DOUBLE FLOOR : A wooden framed floor which has a counter floor to support the finished floor at its top.

DOUBLE GLAZING : Two layers of glazing with an air space in between for the purpose of acoustic or thermal insulation.

DOUBLE FLEMISH BOND

DOUBLE-HANDED SAW : A long cross-cut saw used for cutting wooden logs manually by alternate pulling at either end.

DOUBLE-HEADED NAIL : A wire nail of U-shape (hair-pin bend) which is used for fixing a formwork.

DOUBLE-HEADED RAIL : A rail section having two identical heads, so that both the heads can be used, one after another in a railway track, i.e., when one head is worn out, the other head can be used by placing the upside down.

DOUBLE HEADER : A trimmer joist made by joining two joists near the opening in a wall.

D. H. RAIL

DOUBLE HOUSE : A pair of individual houses attached to each other resembling one house.

DOUBLE HUNG WINDOW : A window having two vertically sliding sashes, each being balanced by counterweights.

DOUBLE JACK RAFTER : A jack rafter joining a valley to a hip in a sloped roofing.

DOUBLE LATHS : Laths having twice the thickness of single laths used in plaster work.

DOUBLE LOCK : Two parallel canal locks having sluice in between, to reduce loss of water during operation.

DOUBLE LOCK WELT : A cross welt.

DOUBLE MARGIN DOOR : A door shutter hinged at one side, but looks like a pair of shutters.

DOUBLE PARTITION : Two partition walls with a cavity in between for provision of a sliding door or for sound and heat insulation.

DOUBLE PIER SHAFT : A form of bridge abutment as shown in illustration.

DOUBLE PITCH ROOF : A mansard roof or any other type of roof comprising two slopes one after the other.

DOUBLE QUIRK BEAD : A return bead i.e., a bead recessed at a corner by a quirk on either side.

DOUBLE REBATED : A wide door post being rebated on both edges to facilitate opening of the door both outwards and inwards.

DOUBLE PIER SHAFT

DOUBLE RETURN STAIR : A bifurcated stair having a wide flight from one floor to the landing and two narrow flights from landing to the next upper floor.

DOUBLE ROMAN TILE : A single lap standard roofing tile made of clay with two nail holes for fixing instead of nibs.

DOUBLE ROOF : A timber roof truss in which both principal rafters and common rafters are provided, the later being placed on purlins.

DOUBLE SKIN ROOF : A specially made asbestos-cement roofing sheet whose upper skin is corrugated for draining out water and the lower is plain to form the ceiling.

DOUBLE SKIRTING : A skirting higher than the normal one and is made by rebating an upper board into the lower skirting board.

DOUBLE SLING : A two-leg chain sling.

DOUBLE SLIP : The arrangement provided in a 'diamond crossing' to facilitate the trains in changing the track, when approaching from either direction.

DOUBLE SLIP

DOUBLE STEP : A step joint made by cutting a w-shaped notch to support the principal rafter on the tie beam at eaves level.

DOUBLE-TIER PARTITION : A two-storeyed high framed-partition made of timber.

DOUBLE TURNOUT : The arrangement to take off two turnouts from different points of a main railway track. This junction is a modified form of three-throw switch. This track junction consists of two pairs of switches at two different points of a main track, three acute-angle crossings and four check rails.

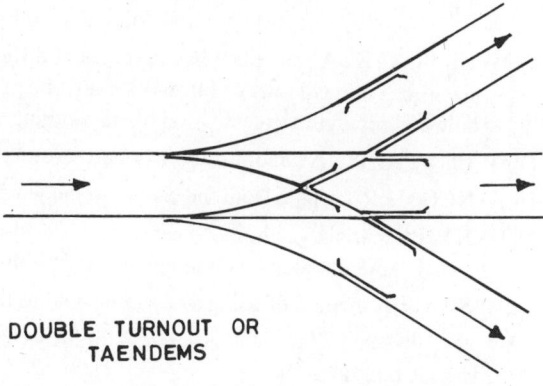

DOUBLE TURNOUT OR TAENDEMS

DOUBLE-WALL COFFER DAM : A cofferdam built with two parallel rows of sheet piles, the space in between being filled with puddled clay for water-tightness and stability.

DOUBLE WINDOW : A storm window having double shutters, one panelled and other glazed.

DOVETAIL CRAMP : A double dovetail-shaped metal cramp for stones.

DOVETAILED HOUSING JOINT: A popular joint used in timber framing. See illustration.

DOVETAILED HOUSING JOINT

DOVETAILED LATHING: A lathing either of steel or of plastic sheets bent into dovetailed shape with corrugations of 20 mm depth, which can be used as a permanent formwork for a reinforced concrete floor. This can also be plastered on both the faces, if used in a wall.

DOVETAIL FEATHER : A double dovetail key used in joinery works in timber.

DOVETAIL HALVED JOINT : A joint in which dovetails are formed in both the halved pieces of timber.

DOVETAIL JOINT : A joint used in forming interlocked corners of a wooden box or a furniture. The interlocking tenons are thinner at the root than at the end having the shape of a dove's tail, so that they can not be pulled out easily.

DOVETAIL MARGIN : A dovetail-shaped band.

DOVETAIL SAW : A back saw used in carpentry.

DOVETAIL SHEATHING : 'Dovetailed lathing'.

DOWEL : (i) A small piece of steel rod cast into a concrete floor, over which a door post is fixed.

    (ii) A metal cramp used in joining two adjacent stones in a wall.

    (iii) A short piece of hardwood rod inserted into the holes through two pieces of timber to be joined together.

DOWEL BIT : A drill bit whose cross-section is half-cylindrical. Also, known as 'Spoon Bit'.

DOWEL PIN : (i) A headless nail or pin with a barbed shank (stem) inserted into a mortise and tenon joint for making it rigid.

    (ii) A short wire nail pointed at both the ends.

DOWEL PLATE : A steel plate having holes of different diameter, used for checking the size (dia.) of the dowel or for making a wooden dowel by passing a peg through the hole and removing excess wood by hammering the peg.

DOWEL SCREW : A handrail screw (wood screw) threaded at both ends.

DOWNCOMER : A pipe from the cistern to the water closet or wash hand basin.

DOWN PIPE : A rainwater pipe to carry the surface run-off from a roof. It is normally 100 mm to 150 mm diameter and made of cast iron or cement-asbestos.

DOWSING : A method of using a 'divining rod' to locate the ground water. Also, known as 'water witching'.

DOZER : A bull dozer.

DRAFT : A strip or margin worked on a stone face, required for surface dressing.

DRAFT CHISEL : A special chisel used for drafting a stone face i.e., a stone working chisel.

DRAFT STOP : A fire stop, usually a shutter or a cover.

DRAFT TUBE : The turbine casing through which water leaves the turbine.

DRAG : (i) A steel plate used for levelling a plastered surface, while green.

    (ii) A towed implement with blades used for levelling the surface of a loose material or for scraping dirty materials from a loose surface.

DRAGLINE EXCAVATOR : An excavator used for digging purpose below the level of its tracks, which works by dragging or pulling a bucket hung from the end of a long jib.

DRAGLINE SCRAPER : An equipment having a scraper bucket controlled by ropes, used for pulling up the piled material like sand, stone chips, coal, coke and similar such loose material on to a working platform or loading platform.

DRAGON BEAM : A horizontal wooden member into which the end of a hip rafter is fixed.

DRAGON'S BLOOD : A resin obtained from a particular variety of palm tree, used for tinting varnishes. This is red in colour.

DRAGON TIE : An angle tie used in a timber truss or frame.

DRAG SHOVEL : A 'backacter'.

DRAIN : A pipe, duct or open channel used for conveying waste water or subsoil water.

DRAINAGE : A device to drain out water from a place.

DRAINAGE AREA : A catchment area or basin from which water is to be removed or drained out.

DRAINAGE BASIN : Drainage area.

DRAINAGE CHANNEL : A channel which carries the drainage water from a basin.

DRAINAGE TUNNEL : A tunnel built for drainage purpose.

DRAIN BRICK : A special brick of channel shape used in construction of small surface drains as shown in illustration.

DRAIN CHUTE : A drain pipe of special shape with the tapered enlarged end placed at the point where it enters or leaves a manhole. This facilitates in rodding the drain for cleaning purpose.

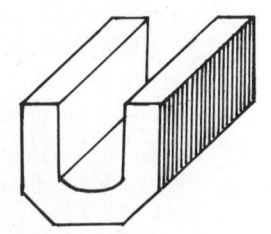

DRAIN BRICK

DRAIN COCK : A cock provided at the lowest point in a waterline system, so that water can be drained out from the system as and when required, by opening the cock.

DRAINED SHEAR TEST : The triaxial compression test carried out on a cohesive soil sample under normal load applied to the specimen with slow draining condition.

DRAIN FERRET : A thin glass bottle containing compressed strongly smelling smoke, which is broken inside a drain during carrying out smoke test for detecting leakage in the drain.

DRAIN PIPE : A conduit used for carrying drainge water. This may be made of cast iron, cement-asbestos, concrete, stoneware or earthen ware.

DRAIN ROD : A long rod made of bamboo or cane or densely coiled spring which can be moved to and fro (rodding) into a drain for removal of chokage.

DRAIN TEST : Usually smoke test is carried out after construction of a drain and prior to its commissioning with a view to detecting leakage in the drain, if any.

DRAIN TILE : Agricultural drains with open-joint clay tiles or perforated tiles.

DRAIN WELL: A well built to absorb water.

DRAUGHT : The difference of air pressure inside and outside a chimney at its base.

DRAUGHT BEAD : A deep bead used in joinery work.

DRAUGHT FILLET : A 'windguard' used in patent glazing. This is a strip required to fill the space between the lower end of the glass and the glazing bar.

DRAUGHTSMAN : The person who makes drawing.

DRAUGHT STOP : See 'Draft Stop'.

DRAW BAR : A steel bar by which the railway coaches or wagons are pulled by a locomotive engine.

DRAW BAR PULL : In a horizontal railway track, the pull exerted by a locomotive engine on its coaches or wagons.

DRAW BOLT : A barrel bolt which is operated by fingers and not by a key.

DRAW BORE : Holes are made in mortise and tenon pieces such that a tapered steel pin is inserted through the holes to have effective cramping required during gluing.

DRAW BRIDGE : A bascule bridge or swing bridge that can be moved horizontally or vertically to pass the vessels.

DRAW-DOOR WEIR : A weir with gates that can be raised or lowered.

DRAW DOWN : The vertical distance (height) by which the water level of a reservoir is lowered due to pumping of water.

DRAW DOWN CURVE : The curve showing the draw down of the ground water table due to pumping, as shown in illustration. The curve represents the slope of the ground water table due to pumping.

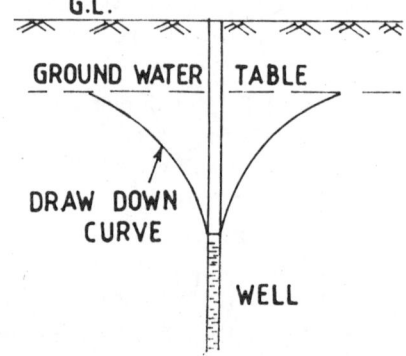

DRAW KNIFE : A U-shaped knife blade with tangs and handle at each end which can be pulled towards the carpenter by both the hands for smoothing a wooden job.

DRAWN SECTION : Architectural section made by passing through a die.

DRAW PIN : A tapered steel pin used in drawbore.

DREDGER : A vessel fitted with a bucket ladder or grab machinery for mining operation or under-water excavation.

DREDGING WELL : An access in a dredger through which the bucket ladder or suction cutter passes to the bed to be dredged.

DRENCHER SYSTEM : A manually-operated or automatic water-sprinkling system to protect the outside of a theatre or auditorium from fire.

DRESSED SIZE : The size obtained after giving final shape with dressing, planing and sand papering.

DRESSED STONE : Stone that is squared and smoothed.

DRESSED TIMBER + : A timber that is sawn, planed and sand-papered.

DRESSER COUPLING JOINT : A special type of pipe joint for plain end pipes, especially used in a pipe gallery of a water treatment plant. The pipes so joined can easily be removed as and when required.

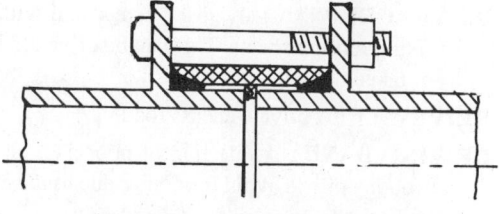

DRESSER COUPLING JOINT

DRESSING COMPOUND : Bituminous compound applied hot or cold over the roofing felt for dressing of the surface.

DRESSING IRON : A steel straight edge of about 45 cm. length with spikes at either end for fixing into a work bench. This is required in cutting slates neatly.

DRIER : Lead, cobalt and manganese compounds which hastens oxidation of drying oils in a paint or varnish.

DRIFT : The speed of a flowing body of water.

DRIFT BARRIER : A barrier of open structure built across a stream or river to catch driftwood by chains.

DRIFT BOLT : A tapered steel pin inserted into rivet holes for bringing them in line prior to riveting. This bolt is also used as a fixing between timber pieces.

DRIFT PLATE : A steel plate required for dressing one lead sheet over another.

DRIFT PLUG : A wooden plug inserted into a bent lead pipe to straighten a kink.

DRIFT TEST : A test carried on metal plates or tubes by inserting a tapered drift and forcing it into the hole until the test piece cracks.

DRILL BOW : The bow of a bow-drill used in joinery work.

DRILL CARRIAGE : A movable stage carrying rock drills, used in tunnelling.

DRIP :   (i)   A groove or throat under the edge of a cornice, coping or moulding to prevent the rain water flowing back to the wall, by throwing it off.

          (ii)   A drop apron

          (iii)   A step formed in flexible metal roofing in a flat roof, at right angles to the fall.

DRIP CAP : A projected small strip over a door or window opening, with a drip underneath.

DRIP CHANNEL : A narrow groove or throat under a cornine or coping.

DRIP COURSE : A narrow band all along the outside wall of a building at a level above the doors and windows or at the roof level with a small groove or throat or channel underneath.

DRIP EDGE : The free end or edge of a flexible metal roof which drips into a gutter.

DRIPPING EAVE : An eave with no gutter ie., the eave which drips into the open .

DRIP SINK : A shallow sink or basin provided at floor level to receive drips from a water tap.

DRIP TRAP : See illustration. A trap having scrubbing arrangement.

DRIVEN PILE : A steel, wooden or R.C.C. pile that is driven into the ground by hammering.

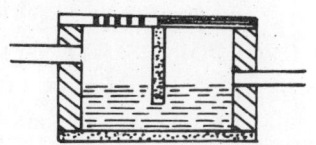

DRIP TRAP

DRIVE SCREW : A galvanised screw nail with a steep screw thread round it, used for fixing roofing sheets. This can be driven in by hammering, but can not be withdrawn or taken out without turning.

DRIVEWAY : A private access road.

DRIVING BAND : A pile ring or hoop i.e., a steel band provided round the head of a wooden pile to avoid brooming due to the blow of a pile hammer.

DRIVING CAP : The pile helmet i.e., a cap provided over the head of a steel pile to prevent the damage due to blow of a hammer.

DROP : (i)   A step down

(ii)   The thicker part of a R.C.C. mushroom slab surrounding the column head.

(iii)   The steep portion of a channel.

DROP INNUNCIATOR : An innunciator, in which a signal drops to indicate the room number wherefrom a signal is given. This is used for the purpose of communication.

DROP APRON : A metal strip fixed vertically to the eaves or verges in flexible metal roofing.

DROP ARROW : In measuring the horizontal distance in a sloping ground, surveying chains are used by the process of stepping and to mark the points vertically, a special arrow of plumb bob type (called drop arrow) is used.

DROP BOTTOM BUCKET : A container or skip whose bottom can be opened to drop concrete, when the skip reaches the bottom of a shaft.

DROP CEILING : A false ceiling.

DROP CONNECTION : See 'Drop Manhole'.

DROP ELBOW : A small elbow with lugs (ears) for screwing it to a wall which is sometimes required in plumbing.

DROP ESCUTCHEON : A drop keyplate of metal pivoted just above the keyhole of a lock matching the escutcheon.

DROP HAMMER : A metal block called 'monkey' which drops over a pilehead through a guide rod, required in pile driving.

DROP MANHOLE : When there is a difference in level (height) between inflow and outflow sewers more than 60 cm. in a junction of sewer lines, a drop manhole is constructed as shown.

DROP MOULDING : Moulding in a door or window panel below the surface of the frame.

DROP ON : A portable rail crossing which can be installed on top of two parallel tracks, set at a distance equal to the gauge of the railway track. This facilitates in transferring wagons from one track to the other.

DROP POINT SLATING : Diagonal slating in roofing.

DROP PENETRATION TEST : A dynamic penetration test carried out on soil, in which the penetrator is driven into the soil.

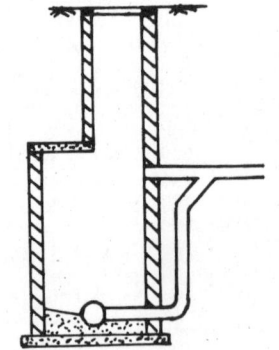

DROP MANHOLE

DROP SIDING : Weather boarding., Also known as 'Rustic siding'.

DROP SHAFT : A'cylinder caison' or 'open caison' which consists of a massive structure of brick masonry or concrete or steel . It sinks gradually into a soft ground by its own weight, when the soil is removed by grab within the drop shaft. This method is adopted in excavating deep foundation for bridge pier or abutment.

DROP SYSTEM : A heating system in which the rising pipe reaches directly to the highest point from where it feeds vertically downward branches and returns back to the water.main.

DROP WINDOW : A sash window which when required is dropped into the groove beneath the sill, keeping the window space completely open for ventilation.

DROWNING PIPE : An inlet pipe to a tank or cistern below the water level in the tank, i.e., a submerged inlet.

DRUM : (i)   A circular stone or brick block used in a column.

(ii)   A circular wall carrying a dome.

DRUM CURB : A curb used for cutting.

DRUM GATE : A spillway gate having shape like a sector of a circle, provided in dams. The opening and closing of this gate requires admittance or release of water through valves.

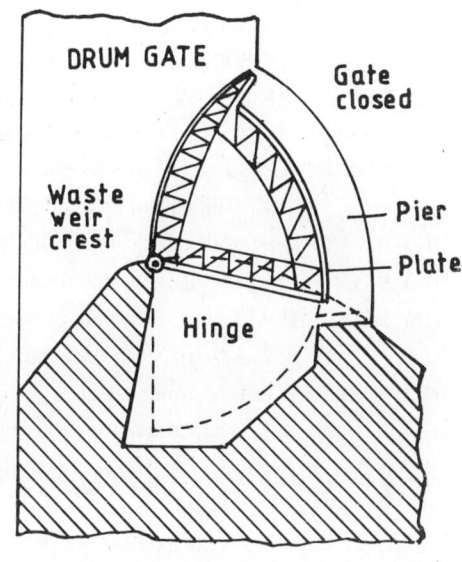

DRUM TRAP : A trap used in waste water pipe connection as shown in illustration.

DRUMMER : A striker used by blacksmith.

DRUM SCREEN : A fine screen of drum type having non-ferrous wire mesh. As the drum (partly submerged in waste water) rotates, the fine particles are arrested and discharged into a through by flushing with high pressure jets of water

DRUNKEN SAW : A circular saw set slightly oblique to its shaft with a view to making a wide cut.

DRY CONSTRUCTION : The construction of a building with prefabricated concrete blocks and without use of mortar or water. This facilitates quick construction.

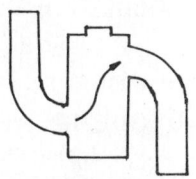

DRUM TRAP

DRY DENSITY : The weight per unit volume of a dry material.

DRY DOCK : A graving dock into which a ship to be repaired/overhauled is taken in, lock gate is closed and the water is pumped out. The ship rests on the docking blocks.

DRY GALVANIGING : A process of galvanizing steel or iron by making it fluxed with hot ammonium chloride solution, drying it and then passing it through a molten bath of zinc.

DRY HYDRATE : Hydrated lime powder in its dry state.

DRYING OIL: An oil of vegetable or animal origin which is processed for using in paints so that it produces a hard film by oxidation, when exposed to air.

DRY JOINT : A plane of contact between two parts of a structure, which is kept to permit relative movement of the parts due to unequal settlement or shrinkage or expansion. The plane of contact is made by introducing building paper into the space left between the two parts. Actually, the parts are not joined together.

DRY MASONRY : A masonry work without using mortar i.e, the bricks, stones or precast concrete blocks are laid by joggled joints or by using dowel pins to form a wall.

DRY PACK : Dry mortar or concrete mix is moistened (damp condition) and is placed in between the two load bearing structures to fill the space. The moist mix is then rammed to bring compactness. It has less shrinkage compared to thin grout and it can be loaded immediately. It is used in the space between head of cast-in situ piles and the structure above it.

DRY PRESS : A method of forming cast stones with moistened dry mix of concrete.

DRY PRESS BRICK : Bricks made by compressing moist clay (almost in dry condition). The moisture content is about 5 to 10% . The pressure required to mould the bricks is about 750 to 1500 psi.

DRY WEATHER FLOW : The waste water flowing through a sewer in dry weather condition. Normally, it comprises sanitary sewage, sullage and wash water.

DRY WELL : (i)   A term for soak -pit.

          (ii)   The part of a sump of a pump house, where the pumps are installed.

DRY ROT : A type of decay caused in timber due to dampness.

DRY STOCK : Dry wood or seasoned timber containing about 10 to 20% moisture.

DRY STONE WALL : (i)   Rubble masonry without use of any mortar.

          (ii)   Stone or brick masonry by using dowels or by joggled joints.

DRY WALLING : Dry stone wall .

DRY TRAP : A 'D'-shaped trap. See illustration.

DUAL FUEL SYSTEM : A heating system in which two different types of fuel can be used.

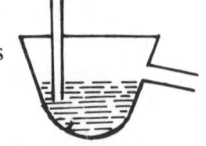

D - TRAP

DUAL SYSTEM : A two-pipe system of plumbing.

DUBBING OUT : Filling holes in a wall surface with coarse stuff prior to plastering.

DUB OFF : Removal of arrises from a tenon to enable its entry into the mortise.

DUCHEMIN'S FORMULA : A formula used to find out the pressure of wind normal to a sloping roof, when the wind pressure 'P' on a vertical surface and 'a', the angle of slope of the roof with horizontal are known.

$$\text{Normal pressure, } N = 2 P \sin a / 1 + \sin^2 a$$

DUCK : A strong and tough cotton material used for conveyer belt or belt for a power drive.

DUCK BILL BIT : A dowel bit used in making holes in timber.

DUCK BILL NAIL : A chisel-pointed nail used in joinery work.

DUCK BOARD : A cat ladder.

DUCK FOOT BEND : A 90° bend often used to support a motor-pump or a vertical pipe.

DUCT : A conduit, hollow shaft, chute, crawlway or subway.

DUCTILE : A material is said to be ductile, which can be cold- drawn.

DUCTILITY : A property of a metal like mild steel, wrought iron, copper, lead and some light alloys, to undergo cold plastic deformation.

DUCTUBE : With a view to forming cable ducts in concrete, inflated tubes are tied to the reinforcement steel and after casting and initial setting of concrete, the tubes are deflated and taken out.

DUFF ABRAM'S LAW : When concrete is fully compacted, its strength may be taken to be inversely proportional to the water-cement ratio. For concrete work without vibration, the water-cement ratio should be at least 0.45.

DUG WELL LATRINE : A toilet suitable for economically weaker section of people living in remote rural areas. This is simply a dug well, sometimes fitted with rings. Soakage takes place through its peripherial soil. See illustration.

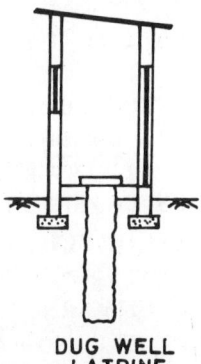

DUG WELL
LATRINE

DUMB BARGE : A self-powered and self-discharging hopper barge that carries dredged material from a dredge.

DUMB WAITER : An elevator which raises and lowers food stuff and crockery from one floor to the other in a hotel.

DUMMY JOINT : A recess or groove cut through the upper half of a concrete slab and filled with bitumen-impregnated fibrous material. This is to allow concrete to crack in line, if at all occurs.

DUMPER : A four-wheeled, rubber-tyred vehicle with a hopper or container at its front, the driver's seat being at the rear side. This type of vehicle is used for transferring materials at construction sites and is also used to carry refuse from one place to the other.

DUMPING : A mass of ground left untouched with excavation on its sides in a long excavation like tunnelling, railway cutting or dry dock. It acts as an abutment for timbering the sides of a trench or excavation.

DUMPY LEVEL : The commonest form of a levelling instrument used in surveying. The telescope and the level tube are rigidly fixed to the vertical spindle of the instrument.

DUNE SAND : Piles of fine sand in form of dunes made by rolling of sand grains swept by wind in a seashore or desert.

DUNNAGE : Waste timber.

DUODECIMAL SYSTEM : A system of units in which a large unit is subdivided into twelve small units, e.g, feet and inches system. This is used in taking measurements and quantity surveying.

DUPLEX APARTMENT : A maisonette.

DUPLEX DWELLING : A dwelling unit which has provision for accomodation of two families, one above the other.

DUPLEX ENGINE: An engine having two steam-driven pistons to drive a pump or compressor.

DUPLEX HEADED NAIL : A double-headed spike.

DUPLEX SLEEPER : A form of railway sleeper made of cast iron and used at rail joints, where additional strength is required. It consists of two plates connected across the track by means of a tie bar as shown.

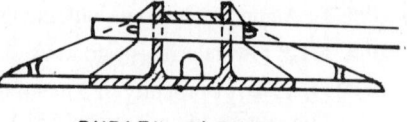

DUPLEX SLEEPER

DURALUMIN : A light alloy of aluminium with small percentages of copper, magnesium, silicon and manganese, which is corrosion- proof and is used in light structures.

DURAMEN : The heartwood of a timber.

DURATION CURVE : A rating curve to indicate the quantum of flow through a stream in a period for power generation. The area under the curve respresents the total flow in a period.

DUST DRY : The drying stage of a painted surface, at which no dust particles will adhere to it.

DUSTING : The dust producing surface of a concrete due to disintegration. This may occur due to lack of curing of concrete or due to too much of water content in a mix with dirty sand.

DUSTING BRUSH : A flat brush used for dusting a surface prior to painting.

DUTCH ARCH : A flat arch, in which the central part is formed with wedge-shaped bricks or stones.

DUTCH BARN : A building of steel frame and a curved roof without any wall.

DUTCH BOND : A brick bond in which each layer is formed by placing a stretcher and a header alternately. See illllustration. It is used in English Cross bond or in Flemish bond.

DUTCH DOOR : A stable door.

DUTCHMAN : A piece of material used to cover a defect either in carpentry or in other trades.

DUTCH BOND

DUTCH MATTRESS : A mattress of reed or timber used to protect the scour of a river bed.

DUTCH PENETROMETER : A penetromter having a cone-tipped rod through a 20 mm. dia. pipe and a pressure gauge and handle at top, used to determine the consistency of cohesive soil or the relative density of cohesionless soil by penetrating it into the soil.

DWANG : (i)   A crow bar.

(ii)   Timber struts between floor joists.

DWARF PARTITION : A partition wall of small height, i.e, not upto the ceiling height.

DWARF WALL : A low-height wall to support the ground floor joists.

DYE : A colouring material with which a solution is prepared that penetrates into another material to colour it.

DYKE : (i)   An earth bundh (mound of earth) built at a distance along the river bank with a view to retaining flood water.

(ii)   A dry wall made of stones.

(iii)   A large ditch

DYNAMIC PENETRATION TEST : The test carried out by driving a penetrometer into a soil to determine the consistency and relative density of cohesionless soil or hard deposits.

DYNAMIC PILE FORMULA : A pile formula that expresses resistance to penetration of piles during driving. The formula is based on the field tests.

DYNAMIC SIMILARITY : If a model of a hydraulic structure operates at a speed simulating the full-size project under field conditions, then the resistances R, densities P, lengths L and velocities V can be expressed by the relationship -

$$R_1/R_2 = (P_1/P_2) . (V_1^2/V_2^2) . (L_1^2/L_2^2)$$

DYNAMIC STRENGTH : The strength of a material or a member to resist suddenly applied loads.

DYNAMITE: A very powerful explosive contaning nitroglycerine absorbed in diatomite. This is chiefly employed in blasting works.

E

E : Denotes 'modulus of elasticity' of a material, speaking of its stiffness.

E = Stress / Strain = load per unit area/change in length per unit length.

EARTH BORER : A drilling rig mounted on a truck required in boring earth.

EARTHEN DAM : A dam made of compacted earth with a core of puddled clay or any other impervious material.

EARTHEN WARE : Pottery from brick earth; Earthen ware pipes are also manufactured for use in low-cost land-drainage. Earthen wares may be made salt-glazed.

EARTH MOVING PLANT : Machinery like bull-dozers,excavators, loading shovels, graders, etc., required for shifting mucks and levelling a surface by removing spoils.

EARTH PLATE : A copper plate sunk in damp ground and connected to lightning arrester for the purpose of earthing.

EARTH PRESSURE : Pressure or thrust given by a retained earth. Active earth pressure tends to overturn a retaining wall and passive earth pressure is the resistance of an earth surface against deformation by other forces.

EARTHQUAKE : cracks formed in the earth's crust with vibration due to settlement of earth's surface. Seismographs are used to measure earthquake intensities.

EARTH TABLE : Ground table or Grass table.

EARTHWORK : Digging earth or raising the ground with soil.

EASEMENT : The right of a person over another person's land to walk over it or to lay a pipeline through it. This is required to have access to the backplot.

EASEMENT CURVE : Transition curve of varying radius from straight to a circular curve.

EASTING : In surveying it is an eastward departure.

EAVES : The lowest part of a sloping roof, projected from the wall.

EAVES BOARD : A tilting fillet or a board fixed to the edge of the eaves.

EAVES COURSE : The first layer of tiles or slates at the eaves of a sloping roof.

EAVES FASCIA : A board on the edge of a sloping roof nailed to the rafters, to carry the eaves gutter or to act as a tilting fillet.

EAVES FLASHING : A drop apron provided in a sloping asphalt roof at the eaves.

EAVES GUTTER : A rainwater gutter fitted to the eaves fascia.

EAVES PLATE : A plate over the posts at eaves to support the tail ends of the rafters, when there is no end wall.

EAVES POLE : Eaves board or tilting fillet.

EAVES TILE : A special tile of short length used in first course at eaves of a sloping roof.

EAVES TROUGH : Eaves gutter.

EBB CHANNEL : A channel formed in an estuary made by a river at low water level. It is usually distinguished by a 'S' curve. In an ebb channel, sand bars have seaward direction.

ECCENTRICITY : In structural engineering, it is the distance between the point of application of a direct load to a member and the centroid of the member.

ECCENTRIC LOAD : The direct load applied to a column or to a member away from the centroid of the column or member. This load causes moment due to eccentricity.

ECHO SOUNDER : An instrument used in determining the depth of water by measurment of time required for a sound to be echoed back from the bed of sea.

ECONOMIC RATIO : In design of R.C.C. beams, economic ratio means that both steel and concrete have attained their maximum stresses.

ECONOMY BRICK : A special modular brick of 3.5" x 3.5" x 7.5" size used in USA.

ECONOMY WALL : A 4" (100 mm) thick brick wall stiffened at intervals with 8" (200 mm) thick pillars (piers) carrying the roof truss. The piers are to be built also at either side of the door and window openings.

EDDY FLOW : Turbulent flow in a stream or channel.

EDDY LOSS : Loss of energy due to eddy flow.

EDDY'S THEOREM : In structural engineering, bending moment in an arch at any point is equal to the product of the horizontal thrust and the vertical distance from the arch centre to the line of thrust.

EDGE GRAIN : A grain seen in a quarter sawn wood with growth rings nearly at $45°$ to the face of the piece.

EDGE ISOLATION : Expansion strip in timber joint.

EDGE JOINT : A joint between two veneers in the direction of the grain.

EDGE NAILING : Joining floor boards by secret nailing.

EDGE RUNNER : A grinding mill having circular rolls driven round in a steel bowl of circular shape containing mortar to be ground.

EDGE SHOT BOARD : A timber board with a sawn and planed edge.

EDGE TOOLS : Tools with a sharp cutting edge as found in chisel, gouge, plane, and hatchet.

EDGE TRIMMER : A special plane with a recessed sole for making an edge of small timber, square to its face.

EDGING STRIP : A band over the edge of a flush door.

EDGING TROWEL : A rectangular trowel with one edge turned down with a view to trimming the edges of kerbs.

EEL GRASS : Sea weed (plant) when dried, packed and formed into a blanket, serves as a sound absorbent material.

EFFECTIVE AREA OF AN ORIFICE : Actual cross-sectional area of an orifice multiplied by its co-efficient of discharge.

EFFECTIVE DEPTH OF BEAM: In reinforced concrete beams, the effective depth is the vertical distance from the outer face of the compression flange (top surface of concrete) to the centre of the steel rods for reinforcement. Thus effective depth = overall depth — (cover + 1/2 dia. of steel rod).

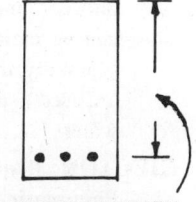

EFFECTIVE DEPTH

EFFECTIVE HEIGHT OF COLUMN: In design of a column, slenderness ratio is found out from the end conditions. The effective height varies from 0.70 to 2 times the actual column height depending upon the end conditions. The effective height, L, is shown in illustration.

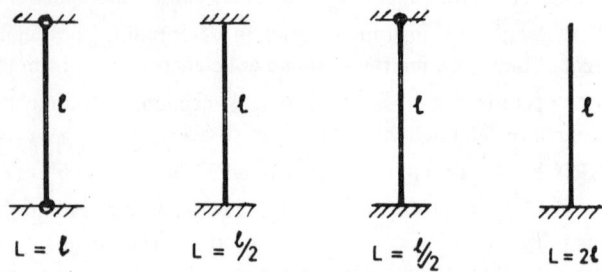

$$L = l \qquad L = \tfrac{l}{2} \qquad L = \tfrac{l}{2} \qquad L = 2l$$

EFFECTIVE LENGTH OF COLUMN : See effective height of a column.

EFFECTIVE PRESSURE : The pressure between the points of contact of soil grains i.e., intergranular pressure. In an equilibrium soil system, this is equal to the total pressure minus the neutral pressure of water in the pore space.

EFFECTIVE SIZE : According to Hazen's definition, it is the diameter of soil grain size which is larger than 10% by weight of soil particles as obtained from the grading curve of the soil particles. Hence, 10% of the particles are finer and 90% are coarser than the effective size.

EFFECTIVE SPAN : The horizontal distance between the centres of the supports of a beam or a slab. Obviously, the effective span is larger than the clear span (opening between supports).

EFFECTIVE STRESS : Effective pressure.

EFFECTIVE THICKNESS OF A WALL : Actual thickness of a solid brick or stone masonry wall, but it is two-third the thickness of two leaves added together in case of a cavity wall. This is required to determine the slenderness ratio.

EFFICIENCY : (i) For a pump-motor, it is power output divided by the power input. 100% efficiency can never be achieved.

(ii) The efficiency of a sewage treatment process is given by its BOD removal efficiency i.e., BOD removed divided by influent BOD.

EFFLORESCENCE : Powdery white salts appear on the brick or plastered surface as it dries out. This happens when the salt present in brick earth comes out at the surface.

EFFLUENT : The liquid coming out from a system i.e., the discharge from an unit.

EGG-SHAPED SEWER : A sewer section of egg shape, laid underground with its small end down so that required velocity can be maintained during small quantam of dry weather flow. Normally, these sewers are built with brick masonry. These sewers are used to carry the combined flow of sanitary sewage and storm water.

EGG SHAPED SEWER

EJECTOR : Required to eject sewage from a low level to a high level. Pneumatic ejector is in use to raise sewage by injecting compressed air. For illustration, see 'Pneumatic Ejector'.

ELASTIC : A material is said to be elastic when it expands or stretches due to application of a load and regains its original shape on release of the load.

ELASTIC CONSTANTS : Modulus of elasticity, shear modulus and bulk modulus are called elastic constants.

ELASTIC CURVE : Also known as 'Deflection curve'; The deflection shape of the neutral surface of a beam bent due to the load applied over it. Euler

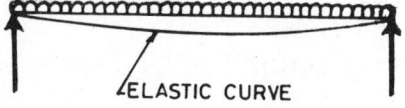

ELASTIC CURVE

established that the curve is a part of a circle of radius equal to EI/M. This is obtained from the relationship $M/I = E/R$.

ELASTIC DESIGN : Design of structures with a working stress of about half to two-third of the elastic limit.

ELASTIC LIMIT : The limit within which stress is proportional to strain. When this limit is exceeded it causes a permanent set. Elastic limit and limit of proportionality are same for most of the materials.

ELASTIC MODULUS : Modulus of elasticity denoted by E and expressed as stress/strain within elastic limit.

ELASTIC RAIL SPIKE : A rail fastening provided with a steel spring and a specially shaped head. See illustration. This provides a better grip with the foot of the rail.

ELASTIC STRAIN : A strain (Change per unit length) produced by a load within the elastic limit, which vanishes on release of the load.

ELBOW : A sharp bend in a pipeline. A fixture used in plumbing work.

ELBOW BOARD : A window board.

ELBOW LINING : Panelling over a window jamb.

ELECTRICAL RESISTANCE STRAIN GAUGE : Used to measure strain in a structure. It consists of a flat coil of very fine wire wound round a thin insulating board or plate placed

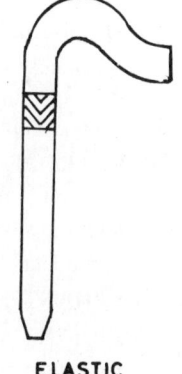

ELASTIC SPIKE

between two sheets of insulating papers. It measures 1/2" x 1". This gauge is glued to the surface of a structure and the ends of the wire are connected to a very sensitive electrical resistance measuring instrument which indicates the strain developed in the structure.

ELECTRIC ARC WELDING : Required for joining metal pieces by forming an arc by current flowing across a gap in the circuit.

ELECTRIC DRILLS : Power-operated drills used for drilling holes quickly.

ELECTRIC EYE : It functions by photo-electric effect. It is required for automatic opening of doors when a person approach them and afterwards closing the doors. It is also used in counting process.

ELECTRIC LOG : Most commonly used to determine the proper place to set well screens and to provide a basis for selecting required lengths of screens.

ELECTRIC SCREW DRIVER: On mass production work, screws are driven speedily with the same force by electrically operated screw driver.

ELECTRIC TRACTION: Running trains by electric motor, the power being supplied from overhead cable.

ELECTRIC WELDING : Arc welding or resistance welding for joining metal pieces.

ELECTRODE : A conductor leading electric current in an electrolytic cell or electric furnace.

ELECTRODE BOILER : A boiler in which water is heated by passing current through electrode. There is no chance of overheating or short-circuiting, when the power is switched on and the boiler is empty.

ELECTRO FILTRATION : It results from the flow of drilling mud into surrounding formations or of water through porous media.

ELECTROLIER : A hanging arrangment of electric light fitting used in USA.

ELECTROLYSIS : Flow of electric current through electrolytes, which deposites metal on cathode and oxygen or acid radicals are liberated at anode.

ELECTROLYTE : A liquid or solution through which conduction of electricity takes place. Acids, alkalies and salt solutions are examples of electrolyte.

ELECTROMAGNET : A soft iron bar can be made a magnet by passing direct current through a coil wound round it.

ELECTRO-OSMOSIS : A process of lowering ground water table in silty clay, which hastens the natural drainage by forcing the water to flow away from an excavation. Direct current is passed through the wet soil, by which water flows to the cathodic area. This process is expensive, but it increases the strength of soil to a great extent.

ELECTROPLATING : The deposition of a thin film of a noble metal like nickel, chromium, copper or cadmium on another metal by the process of electrolysis.

ELEMI : An oleo-resin required in preparation of siprit lacquers and nitro-cellulose products.

ELEPHANT SPINDLE : A special form of spindle moulder.

ELEPHANT'S TRUNK : A hydraulic ejector used in civil engineering works.

ELEVATED RAILWAY : A railroad carried on a bridge supported on columns above the road level.

ELEVATING GRADER : A grader having a plough collector and belt elevator placed at right angles to its direction of travel, used for digging dry soil and loose materials and transfering them to a higher level. This machine helps in fast digging and transfering the spoil away from site.

ELEVATION : Normally it is the front view of an object. However, it may be a rear elevation or end elevation. This is required to show the object in orthographic views.

ELEVATION HEAD : Potential energy. It is given by the product of the density of a fluid and its height above a point. In case of water it is expressed as feet of water.

ELEVATOR : A mechanism by which passengers or goods are elevated from a lower level to a higher level. A lift used in a multi-storied building. Lifting bucket used in mines and in dredging operation.

ELLIPSE OF STRESS : An ellipse constructed graphically proportional to the principal stresses in a plane at a point, which gives the magnitude of the resultant stress at any angle through that point.

ELLIPTICAL STAIR : A stair with an elliptical well.

ELM : A brown hardwood with twisted fibres. It is used for piles, weather boards, etc.,

ELONGATION : The extension i.e., increase in length of a structural member under tensile test.

ELUTRIATION : A unit operation of washing sewage sludge in which chemical components, organic or inorganic, that interfere with chemical conditioning and filtration, are removed. The operations may be done in a single tank or a multiple tank.

ELUTRIATOR : A device used for mechanical analysis of soil and for classification of minerals by washing them with upward flow of water. This is a classifier working on the principle that large grains settle faster than the small grains in a liquid. In industrial elutriators, the small grains in a material rises up and washed away first due to the upward flow.

ELUVIUM : Dense mineral deposits are eluvial at the place of their origin due to the disintegration and washing away of the surrounding lighter minerals.

EMBANKMENT : A mound of earth, rock or composite material forming a trapezoidal section used for a roadway or railway. This is also built along the banks of a river or stream to protect the surrounding areas from flood due to high water level in the river/stream.

EMBANKMENT WALL : A type of retaining wall built at the base of a wall to prevent it from overturning or sliding.

EMERY : A mixture of corundum and hematite or magnetite used in preparation of abrasive-paper, cloth and-wheel.

EMINENTLY HYDRAULIC LIME : Hydraulic lime obtained by burning limestones containing more than 25% of aluminium silicates which brings the hydraulicity

EMPIRICAL FORMULA : A formula or thumb rule developed from various observations and findings but without any theoretical support or mathematical proof.

EMPTY-CELL PROCESS : A process of preservation of timber by pressure creosoting without driving out the air remaining imprisoned in the timber. Therefore, the absorption of creosote oil by the timber is low.

EMSCHER TANK : Imhoff Tank.

EMULSIFIER : A chemical agent added to a mixture of fluids to form an emulsion.

EMULSIFIER SYSTEM : A system comprising a high pressure spray directed on to the oil which is emulsified by the jet action.

EMULSION : A more or less stable suspension of liquids minutely dispersed through another liquid in which they are not soluble.

EMULSION INJECTION : Artifical cementing for soil stabilization by injecting bituminous emulsion into soil.

EMULSION PAINT: When an emulsion is made stable by adding colloids like casein, glue, etc., this may be used as a paint. Bituminous emulsions, oil-bound distemper and latex emulsions are examples of emulsion paint. These paints harden by evaporation of water in them and not by oxidation.

ENAMEL: (i)   Vitreous enamel is a glass surface attached to Cast Iron or steel articles by firing. Although this is more resistant to wear than enamel paint, it cracks and chips when struck.

          (ii)   A hard high-gloss paint available in different colours are used for decorative woodwork. This paint is prepared by mixing ground pigments with varnish vehicle. For hard enamel coats, copal and water resins are used.

ENAMELLED BRICK : A glazed brick.

ENCASE : Cover with a case or lining or embedding in a material.

ENCASED BEAM : A R.S.J. beam embedded in concrete.

ENCASED KNOT : A dead knot in a timber.

ENCASTRE : A beam is said to be encastred, when its ends are built-in i.e., fixed, e.g., a beam end is built in a concrete wall or piller.

ENCAUSTIC DECORATION : Ornamental decorations burnt on to porcelain and glass articles and also on to bricks and tiles.

ENCLOSED KNOT : A knot that does not appear on the surface of a timber member.

ENCLOSED STAIR : A closed stair in an enclosure.

ENCLOSURE : A space covered by walls or fence.

ENCLOSURE WALL : An exterior wall which is a non-load bearing wall in skeleton construction. The wall need not be carried by the skeleton frame.

END BEARING : Beam support at its ends.

END BEARING PILE : A bearing pile that transfers load to a hard stratum through its tip.

END CONNECTION : In steel structures, it is the connection of a beam or girder with the stanchion or column. This may also be a connection between a girder and a secondary beam.

END CONTRACTION : The reduction in area of flowing water due to the effect of a notch.

END GRAIN : The timber grains shown by a surface, when a tree trunk is cross cut.

END JOINT : In carpentry, it is a butt joint.

END LAP JOINT : An angle joint made between two pieces of timber by halving each piece for a length equal to the width.

ENDLESS SAW : A band saw used in sawing timbers.

END SPAN : The last span at either end of a continuous beam or slab with intermediate supports.

END THRUST : A thrust from the end of a member in a structure, e.g., thrust at the ends of an arch.

ENDURANCE LIMIT : The maximum stress obtained from the fatigue test of a material, below which fractures will not occur in the material due to loading.

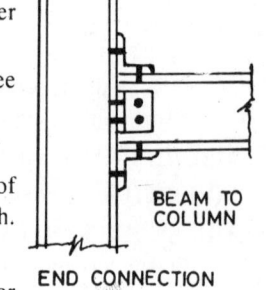

BEAM TO
COLUMN

END CONNECTION

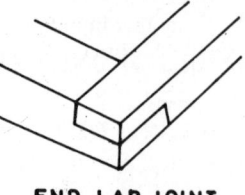

END LAP JOINT

ENGINEERED BRICK : In USA, it is the brick that measures with its mortar joint 3.2" x 4" x 8".

ENGINEERING BRICK : A brick having uniform colour, shape & size with standard crushing strength and other properties suitable for constructional works.

ENGINEERING NEWS FORMULA : A popular pile driving formula.

$$Q = 2WH/(S + C),$$

where Q = safe design load on pile in lbs.,

W = weight of ram of pile driver, in lbs.,

S = penetration of the pile in inches under the last blow.

C = a constant, 1 for a drop hammer and 0.1 for a steam hammer.

ENGINEER'S CHAIN : A chain of 100 ft. length having 100 links used extensively in chain surveying.

ENGINEER'S HAMMER : A hammer with a striking face at one end and a ball peen at the other, weighing upto 3 lbs.,

ENGINEER'S LEVEL : A levelling instrument, usually called a 'dumpy level' which has a telescope with a level tube attached to it. This is commonly used by engineers for level surveying.

ENGINEER'S TRANSIT : This is essentially a transit theodolite having a graduated vertical circle and a telescope bubble, which facilitates in reading vertical angles in addition to horizontal angles.

ENGLISH BOND : A brick bond having alternate courses of stretchers and headers, which is commonly used in brick-work.

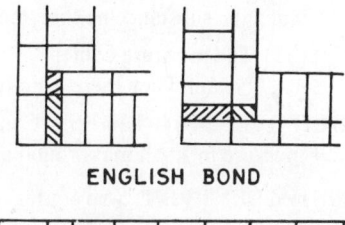

ENGLISH BOND

ENGLISH CROSS BOND : Also known as saint Andrew's cross bond. It is an English bond with the exception that in alternate stretcher courses a header is laid just after the Quoin stretcher. This brings the break in vertical joints of alternate stretcher courses which are displaced by a half brick from each other i.e., the same vertical line of stretchers as in English bond is not followed. The decorative effect of this bond is more than the English bond.

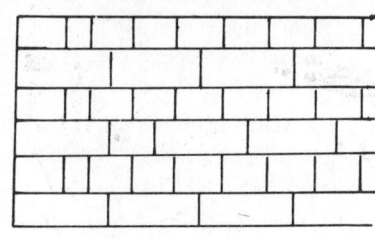

ENGLISH CROSS BOND

ENGLISH GARDEN WALL BOND : American Bond.

ENGLISH ROOFING TILE : A single lap roofing tile made of clay whose sides overlap within its thickness.

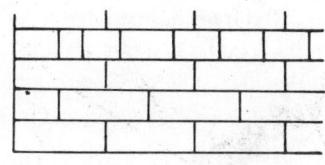

ENGLISH GARDEN
WALL BOND

ENROCKMENT : Rip-rap.

ENTRANCE HEAD: The head with which a liquid enters into a conduit and flows.

ENTRANCE LOCK : A lock through which vessels get access to a dock having water level different from the outside water level.

ENTRANCE LOSS : The loss of head at the entry point of a conduit due to eddies and friction.

ENTRY BORER : A machine used for drilling horizontal holes in a thick coal seam.

EPOXIDE RESIN : A synthetic resin used for gluing metal parts and also for house repairing works.

EQUAL ANGLE : An angle section having equal legs, used in making a structural frame.

EQUALIZATION OF BOUNDARIES : A method adopted in calculation of an area bounded by irregular lines. For simplification, the irregular lines are replaced by straight lines such that the area deducted is equal to the area added. The total area is then calculated by adding the individual areas of the triangles formed during replacement.

EQUALIZING BED : A bed of concrete or a cushioning material formed at the bottom of a trench, on which a pipeline is laid. This is required to avoid unequal settlement of a pipeline.

EQUILATERAL ARCH : A two-centred arch, the curves being struck from the two ends of the springing line as shown. The voussoirs are laid radial from the corresponding centre.

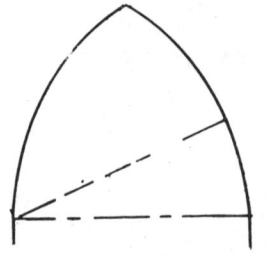

EQUILIBRIUM MOISTURE CONTENT : (i)The moisture content of any substance at a particular environment, when the substance neither looses nor gains any moisture.

(ii) The moisture content of a soil is said to be in equilibrium, when there is no moisture movement.

EQUIPOTENTIAL LINES : Contour lines of same water pressure in a soil mass round a water retaining structure.

EQUILATERAL ARCH

EQUIVALENT PIPE : An equivalent pipe is one, in which the loss of head for a specified flow is the same as the loss of head of the system that it replaces. Thus, the method is the replacement of a series of pipes of varying diameter with one equivalent pipe i.e., a single line of equivalent capacity.

EQUIVALENT TEMPERATURE : It is the temperature of a room in which the rate of cooling effect of air flow is same as that of radiation of heat from the walls. But, it does not take into account the humidity in air.

ERECTING SHOP : A fabrication shop in a large open space, where framed structures are fabricated and made ready for erection at site.

ERECTION : Placement, positioning and fixing of precast concrete frames or fabricated steel frames.

ERGONOMICS : Interactions between work and people relating to the design of machines, chairs, tables, almirahs, shelves, stairs, doors, windows, etc., to suit a person to work with least movement and no fatigue.

EROSION : (i) ‒ Scour

(ii) Wearing away of a surface due to abrasive action.

ESCAPE : In hydraulics, it is a wasteway to discharge the entire flow of a stream with a view to protecting the surrounding areas from flooding.

ESCAPE STAIR : A stair conveniently located in a multistoried building for escape of persons in the event of fire.

ESCUTCHEON : A key plate round a key hole.

ESPAGNOLETTE : A vertical bolt of special type used in casement doors, which runs the full height of the window.

ESTABLISHMENT CHARGE : Overhead cost.

ESTIMATING : Quantity surveying i.e., measurement of volume of work to be done and its tentative cost.

ETCHING : Designs made on a glass surface by cutting the surface with hydro-fluoric acid.

ETEISPHON SHELL ROOF : A corrugated shell roof of special shape as shown in illustration is used for low-cost houses. Usually, the depth of corrugation is 200 mm for a span of 9 m. The depth of corrugation determines the rigidity of the arch ring. Multispan corrugated shell roofs of this type may be used to cover large areas.

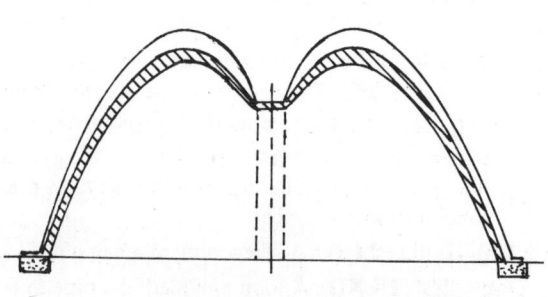

ETEISPHON SHELL ROOF

ETHANE : C_2H_6, a natural gas from oil wells.

EULER CRIPPLING LOAD : The buckling load or critical load applied to a column axially, at which the column tends to buckle. The safe working load on a column is calculated by dividing the crippling load by the factor of safety.

EULER CRIPPLING STRESS : The stress obtained by dividing the crippling load by the cross-sectional area of the column. The safe working stress can be calculated by dividing this stress by the factor of safety.

EVAPO-TRANSPIRATION LOSS : Combined loss of moisture content from a soil due to evaporation and plant transpiration.

EVEN GRAIN : Almost uniform distribution of grains in a timber surface or a mosaic surface.

EVEN TEXTURE : Uniform surface in brick, timber, concrete and steel.

EXCAVATING CABLE WAY : A cableway fitted with a clamshell bucket used for digging and removing the spoil.

EXCAVATION : Digging and removing earth.

EXCAVATOR : A power-driven excavating machine mounted on tracks, used for quick excavation in soil or rock.

EXCELSION : Wood shavings for making wood wool or packing material.

EXCITER: A material that sets when mixed with water. Lime,alkali or sulphate when added to powdered blast furnace slag, cause it to set with water. Portland cement acts as an exciter.

EXFOLIATED VERMICULITE : Vermiculite expanded to 10 to 15 times its original volume by heating it.

EXFOLIATION : Scaling of stones by weathering action.

EXHAUST SHAFT : A Passage to remove hot air from a room for ventilation

EXPANDED CLAY : Bloated clay. In USA, it is called 'Haydite'. Burnt vitreous cellular clay pellets which are hard and have air- filled cells.

EXPANDED METAL : Diamond-shaped sturdy steel or other metal mesh formed by slotting metal sheets, used for metal lathing (a base for plaster) or reinforcing concrete, or used as low-cost grills.

EXPANDED POLYSTYRENE : An insulating material having a thermal conductivity of 0.22. It is available as blocks, sheets or loose fill material.

EXPANDING BIT : A drilling bit having no twist, but with a cutter which can be adjusted to varying radii. Thus, holes can be made from smaller to a larger one.

EXPANDING CEMENT : A hydraulic cement first developed in France, by mixing ground cement clinkers & blast furnace slag with gypsum. This cement expands during its setting and first hardening, which results in prestressing after casting a reinforced concrete with mild steel.

EXPANDING PLUG : A screw plug or a bag plug.

EXPANSION BEND : A loop provided in a pipe to take up the expansion or contraction due to temperature change, without causing any damage to the pipe.

EXPANSION BOLT : An anchor bolt comprising a split cone is inserted into the hole in a masonry, the bolt between the split halves being projected outside the hole. The nut remains inside between the split halves. When the bolt head is turned, the nut presses the sides of the cone against the hole wall and thus it gets tightened.

EXPANSION JOINT : (i)   A joint used in a pipeline as shown in illustration

(ii) A rail joint with a gap for expansion of rails.

(iii) A joint provided in a structure to prevent cracks due to expansion.

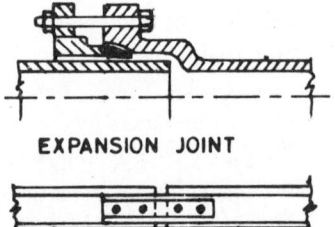

EXPANSION JOINT

EXPANSION JOINT

EXPANSION PIPE : In hot water supply system it is the pipe leading from the hot water tank to the point of discharge over the cold water cistern. This facilitates in discharging boiling water safely into the cistern.

EXPANSION ROLLER : The roller support at one end of a bridge girder or truss to permit movement due to change in temperature.

EXPANSION SLEEVE : A metal, asbestos or PVC pipe built into concrete or masonry through which a pipeline carrying the liquid passes. The sleeve allows the expansion of the pipe line if any, without causing any damage.

EXPANSION STRIP : Also known as 'isolation strip', 'insulation strip' and 'edge isolation'.

    (i)   A strip of material used for breaking the continuity of a long construction with a view to allowing expansion.

    (ii)   An insulating material used at the junction of a masonry wall and a glass wall.

    (iii)   Isolation of machine foundation by pouring loose material like sand in between two machine foundations for absorption of vibration.

    (iv)   A strip of bituminous felt used at the junction of a main wall and a partition wall.

EXPENDING BEACH : A beach designed to use the energy of sea waves.

EXPLOSIVES : Materials like detonator, dynamite, ammonal and gelatine explosives used for blasting rocks, coal seams and for demolition of very old structures prone to collapse.

EXTENDED AERATION PLANT : A sewage treatment plant of small capacity (upto 1 mgd flow) suitable for a small colony or town. The organic loading is kept low with long aeration time to avoid sludge handling problem. 'Oxidation Ditch' is an extended aeration plant.

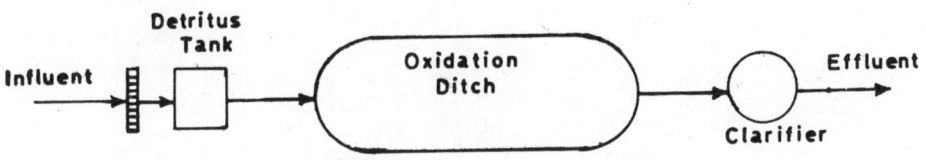

Extended Aeration Plant

EXTENDER : (i)   An inert pigment of inorganic material available in form of white powder which is added to a paint for adjustment of film formation and improvement of working properties. These are barytes, kaolin, diatomite, asbestine, mica, silica, whiting etc.

      (ii)   Wood flour added to an expensive glue for its dilution and increasing its spreading capacity.

EXTENDING LADDER : A telescopic ladder which can be extended to increase its length.

EXTENSION BOLT : Also known as 'monkey-tail bolt' which is a long barrel bolt with a long handle such that it can be pushed home or released easily.

EXTENSION RULE : A two-fold or four-fold wooden rule used by carpenters for taking interior measurements of a door or window opening.

EXTERIOR PANEL : A panel of a slab whose one end has dis-continuous support.

EXTERIOR TRIM : Wooden mouldings used in exterior parts of a building. These are barge boards, eaves boards, eaves gutter, cornice mouldings, etc.

EXTERNAL GLAZING : Glass panelling on the outside wall of a building.

EXTERNAL VIBRATOR : A vibrator fixed to the formwork for placing concrete. This is rarely used, since the formwork has to be made extra strong to withstand the vibration.

EXTERNAL WALL : An outside wall of a building whose one face is subjected to direct atmospheric actions.

EXTRACT SYSTEM : A ventilation system in which a fan sucks air from a room and blows it outside the room through a duct at a high level.

EXTRADOS : The upper surface of an arch.

EXTRAPOLATE : To extend a curve following its trend or nature drawn on the basis of the obtained data.

EXTRUDED SECTION : Commonly used light alloy structural sections formed by extrusion.

EYE : An access or door provided in a pipe for easy inspection and cleaning.

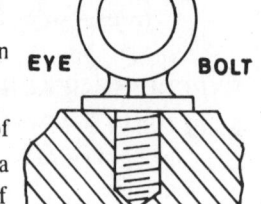

EYE BOLT : A bolt with a steel ring or loop at one end in place of a head, the stem being- threaded. The stem is screwed to a heavy machine and the machine is lifted with the help of eye.

F

FABRIC : A building carcase i.e., the structure of a building.

FABRICATION : Preparation of members of a structural frame in workshop such that the framed structure can be built easily in a short time by making assembly of the parts at site. The operations involved in fabrication are cutting to shape and size, bending, notching, squaring, welding, drilling, flame cutting, forging, planning, etc.

FACE : The front surface of a wall which is exposed.

FACE BEDDED : Stones laid with their natural bed vertical.

FACE BRICK : Facing brick which makes the front surface of a wall.

FACED PLYWOOD : Plywood faced with plastic or sheet metal.

FACED WALL : A wall faced with facing bricks or stones.

FACE EDGE : The working edge of a member from which other edges are measured.

FACE HAMMER : A hammer with a striking face and a cutting peen, used by masons during brick work.

FACE JOINT : The part of a joint that comes in view on the face of a wall.

FACE LEFT : The position of a theodolite with its vertical circle to the left of the observer.

FACE MARK : The broad surface of a square-sawn and planed timber marked by a pencil with a view to truing other edges with respect to the marked face.

FACE MEASURE : Superficial measure.

FACE MIX : A mixture of cement and fine stone chippings with water to form a paste which is placed on the inner surface of the mould with a backing of ordinary concrete.

FACE MOULD : A templet applied to the face of a stone.

FACE PIECE : A face waling.

FACE PLATE : The plate of a marking gauge which presses against a face during marking.

FACE PUTTY : The triangular fillet of putty put along the edges of a glass in a door or window.

FACE RIGHT : The position of a theodoilte with its vertical circle to the right of the observer.

FACE SHOVEL : A forward shovel attached to an excavator for digging.

FACE STRING : An outer string in a wooden staircase.

FACE VENEER : A veneer for decoration purpose.

FACE WALING : A face piece across the end of a trench.

FACING : (i) Making a surface by using a good quality material on the exposed surface of an object.

(ii) A protective or decorative coating given to an object.

(iii) Lining.

FACING BOND : A bond in which mostly stretchers are exposed.

FACING BRICKS : Bricks of pleasing appearance having good texture and uniform colour used at the face of a wall.

FACING HAMMER: A hammer having a rectangular notched head, used for dressing stones.

FACING WALL : A wall built with precast concrete slabs to protect the sides of a deep trench or excavation instead of using timber planks.

FACTOR OF SAFETY : The ratio of ultimate strength of a material to the maximum allowable strength (i.e., working strength) with which a design is made. The factor of safety to be adopted in a design depends upon the quality of material, workmanship, the type of structure and nature of loading.

FACULTATIVE LAGOON : A lagoon used for treatment of liquid organic wastes, in which aerobic action takes place in the upper layer, facultative in the mid-layer and anaerobic at the bottom of the lagoon. The interactions in a facultative lagoon are presented in illustration.

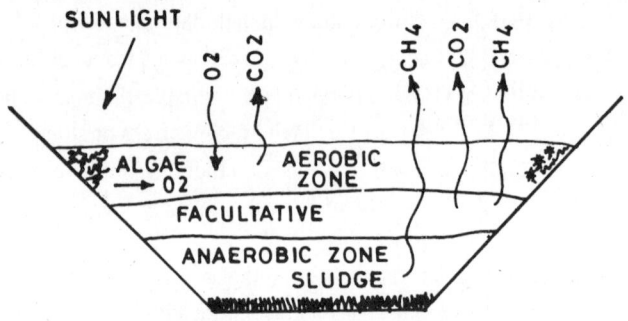

FACULTATIVE LAGOON

FAGGOT : (i)   A term for facing brick.

(ii)   A fascine;

FAIENCE : Double-fired glazed terra-cotta and wall tiles.

FAIENCE MOSAIC : Mosaic floor made with glazed plastic tiles.

FAGGOTING : Making revetment with thorn bat faggots.

FAIR AND HATCH FORMULA : A formula for determining the specific permeability. It is given by

$$K = 1/[\ m[(1 - \alpha)^2 /\alpha^3 . (\theta / 100\ \Sigma\ P/d_m)^2\}]$$

where α = Porosity, m = Packing factor,

θ = Sand shape factor,

P = Percentage of sand held between adjacent sieves and

d_m = Geometric mean of rated sizes of adjacent sieves.

FAIR-FACED BRICKWORK : A neat and smoothly finished surface of brickwork.

FAIRLEAD : A fixture (usually of metal) to a quay wall and ship deck to guide the ropes during berthing.

FALL : (i)   The slope or gradient of a road, railway, canal, stream, etc.

(ii)   A drop from a level surface.

FALL BLOCK : A pulley block in a lifting tackle, which rises and falls with a load.

FALLING APRON : A revetment on the sloping bank of a river or canal.

FALLING HEAD PERMEAMETER : More suitable for permeability tests on materials of low permeability, by adjustments and measurements of head and time accurately over a wide range of the co- efficient of permeability. The apparatus is shown in illustration.

FALLING STILE : The shutting stile of a gate which is held by a cocked hinge.

FALL PIPE : A downpipe in plumbing a building.

FALSE BODY : High viscosity of a paint which gets reduced by stirring.

FALSE CEILING : A decorative or pleasant ceiling built under a roof with a gap in between. It also provides space for running cables and pipes.

FALSE HEADER : A half-brick not used as a header in a brickwork.

FALSE HEARTWOOD : A wood looking like a heartwood due to growth of fungus and other causes.

Sample

FALLING HEAD PERMEAMETER

FALSE LEADER : A framed mast held upright on ground, used for holding a pile hammer and for guiding the pile during driving.

FALSE MEMBER : A member provided in a structure, not as a structural member but to facilitate erection.

FALSE RAFTER : A cocking piece of wood used in roofing.

FALSE TENON : A hardwood wedge or pin inserted into the mortise, when the tenon is found to be weak.

FALSE WORK : Any temporary work required during a construction.

FAN : A platform of scaffold boards or wooden planks projecting out of a wall and sloped upwards over the road, so that the falling materials during construction or demolition work hit the platform and are deflected towards the wall and not fallen on the road. This is a safety measure.

FANG : The end of a metal railing built into a wall.

FANLIGHT : A glazed frame above a door frame to facilitate entry of light.

FANNING'S FLOOD FLOW FORMULA : The formula for empirical evaluation of flood flow from drainage basin characteristics and hydrological factors.

$$Q = CA^{5/6},$$

where $C = 200$ and A is area in sq. miles.

FASCIA BOARD : A wide board fixed vertically to the ends of rafters, which carries the eaves gutter.

FASCINE : Bundles of brushwood tied in cylindrical shapes used as a protective facing to river banks.

FASTENER : Nails, screws, bolt-nut, rivets, dowel pins, spikes & dogs, are used as fasteners required in different types of jobs.

FAST SHEET : A dead light

FAST TO LIGHT: A particular colour of a paint which remains unaffected by a certain light.

FAT BOARD: A wooden board used by masons to hold the mortar in one hand during pointing work.

FAT CLAY : A clay that possesses high values of liquid limit and plasticity Index. This is also known as highly plastic soil.

FAT EDGE : A ridge of wet paint formed at the lower edge of a freshly painted vertical surface due to flow of paint downwards.

FATIGUE : A state in a member reached when it fails at a much lower stress than its usual stress, because of repeated reversals of stress.

FAT LIME : A lime having high calcium content.

FAT MIX : A rich mix for mortar or concrete in which more quantum of cement or lime is used than the usual one.

FAT MORTAR : A mortar in which more cement is used and it becomes too sticky to the trowel.

FANNING FRICTION FACTOR : A factor 'f' against Reynold's number used in ground water hydraulics is given by

$$f = d \, \Delta p \, / \, 2 \, DLv^2$$

where d = average grain diameter

p = pressure difference over a length L of porous media

D = density of fluid.

v = velocity of flow.

FATTENING : Increase in viscosity.

FATTENING UP : Increasing the plasticity of a lime putty or a soil.

FAUCET : (i) A tap provided in a household water filter or a water cooler.

(ii) The socket end of a pipe.

FAUCET EAR : Lug or ear-like projection at the socket end of a pipe which facilitates in fixing the pipe onto a wall.

FEATHER : (i) A slip tongue or spline required to join match boards.

(ii) The pendulum slip that separates the sash weights in a sash window.

FEATHER EDGE : A rule having feather edge used in plastering.

FEATHER EDGE BRICK : A compass brick.

FEATHER-EDGED BOARD : A tapered board used in close-boarded fencing and also in weather boarding.

FEATHER EDGED COPING : A coping brick or stone or concrete moulding whose one edge is thicker than the other.

FEATHER JOINT : A ploughed and tongued joint which is made with a cross tongue introduced between ploughed edges.

FEATHER TONGUE : A cross tongue used in a feather joint.

FEEBLY HYDRAULIC LIME : Lime that contains clay lumps and that has poor hydraulic property.

FEED CISTERN : A cistern used for storage of cold water to feed a boiler. It is provided with a ball cock for control of water flow.

FEEDER : A channel to feed water to canal or a reservoir.

FEEDING MAIN STSTEM : A system of water mains to feed the individual tanks at different levels from a water treatment plant as shown in illustration for a hilly area.

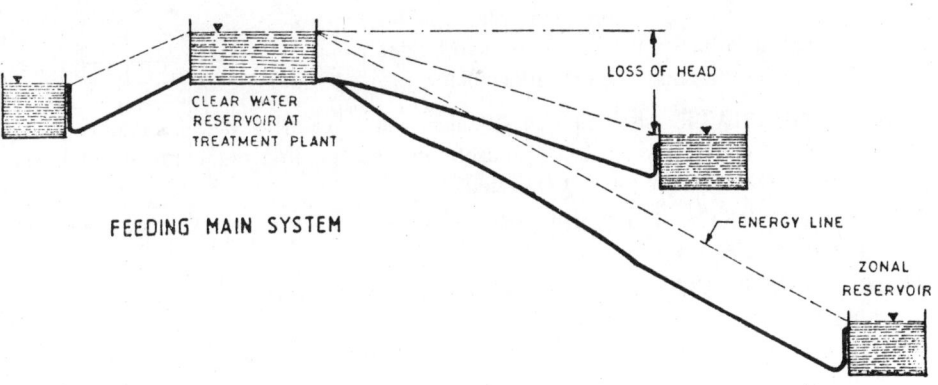

FEEDING MAIN SYSTEM

FEED PUMP : A pump used for feeding water to a boiler.

FEED WATER : The water that is treated, pre-heated to boiler-temperature and pumped for feeding a boiler.

FEINT : Bent edge of flashings or cappings in a flexible metal sheet roofing, with a view to forming a capillary break.

FELLOE : A segment of a wooden wheel rim or hub.

FELT AND GRAVEL ROOF : A roof covering of bitumen felt with pea-size gravel.

FELT NAIL : A clout nail.

FELT PAPER : A type of building paper.

FENCE : (i) A guide for timber required in sawing or cutting in a machine.

(ii) A guard round a machine for safety of the workmen.

FENDER : (i) Wooden block with rope mat or rubber block or old rubber tyre fastened to a wall or piles to protect a water vessel from impact i.e., to absorb the shock load.

(ii) A baulk placed on ground close to a street to protect the standards of a scaffold from striking by wheels of vehicles moving on the street.

FENDER PILE : A vertical wooden pile to absorb the impact of vessels with a view to protecting the berth.

FENDER POST : A bollard or a guard post acting as a fender.

FENDER WALL: (i) A dwarf wall to carry the hearth slab for a fireplace.

(ii) A wall provided with fenders and timber rubbing strip to absorb the impact of vessels during berthing as shown in illustration.

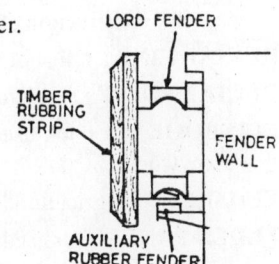

FENESTRATION: The array of windows and openings in the wall of a building to produce an architectural effect.

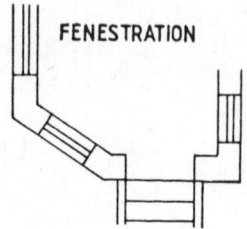

FENESTRATION

FERRIC CHLORIDE : Successfully used as a coagulant in water Treatment plants for removal of turbidity. It forms ferric hydroxide floc which settles down in a clariflocculator.

FERRIC SULPHATE : Used as a coagulant in water treatment. It reacts readily with natural alkalinity in water. It is also useful in removing iron and hydrogen sulphide from well waters. It produces dense floc.

FERRO-CONCRETE : Reinforced concrete in which steel or iron bars are used for reinforcement.

FERROUS SULPHATE : Also known as 'Copperas'. It acts as a coagulant in a better way in the alkaline range of water for removal of turbidity. It should be used after adding lime to a turbid water. It is cheaper than alum. Flocks formed are heavy, which settle dawn rapidly.

FERRULE : A one-way valve used for making water-connection from a street water main. It is fitted to the water main either at top or at mid-depth of the pipe. The arrangement is made such that water flows in one direction only and there is no scope for back flow. See illustration.

FETCH : The free distance through which wind can travel in raising waves upto a sea coast.

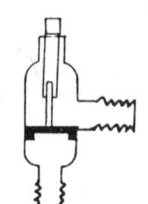

FERRULE

FIBRE BOARD : Insulation boards and decorative boards made of wood fibres, straws and vegetable fibers. These are light in weight.

FIBRE ROPE : Hemp rope or manila rope. Rope made of vegetable fibres.

FIBRE SATURATION POINT : At about 30% moisture content, the dimension and strength of a timber remain unchanged, above or below which both dimension and strength of a timber change considerably.

FIBROUS CONCRETE : A concrete made of fibrous matrix (mortar) i.e., with sawdust, asbestos and other fibres mixed with cement, sand and stone chips.

FIBROUS PLASTER : A plaster made by reinforcing with canvas, wood laths, wood fibres and wire nets.

FIDDLE BACK : A mottle figure in timber.

FIDDLE DRILL : A bow drill used by carpenters.

FIDLER'S GEAR : Lifting tackle used in laying large blocks of stones or concrete under water for construction of water structures.

FIDUCIAL LINE : A line of reference used in surveying.

FIDUCIAL POINT : A reference point used in survey work.

FIELD BOOK : A book used by surveyors for recording field measurements during a surveywork.

FIELD DRAIN : Agricultural drain.

FIELDED PANEL : A raised panel in a door or window.

FIELD MOISTURE EQUIVALENT : The optimum moisture content of a soil at which if a drop of water is put on the smoothed surface of the soil, the water drop will penetrate but will spread over the surface with a shiny appearance.

FIELD RIVETS : Rivets used in rivetting at site.

FIELD TILE : Field drain or agricultural drain tile.

FILL : Earthwork in filling.

FILLER : (i)   Finely powdered minerals added to bitumen and tar for making them stiff for use in road pavement.

    (ii)   A paste containing extenders mixed with turpentine or white spirit and gold size used for filling up indentations in a surface painted with a primary coat. Sometimes, glazier's putty is used for painted surfaces and litharge and glue for varnished surfaces.

    (iii)   A material called 'extender' used in glue.

    (iv)   Carbon black, slate dust, barytes, mica, quartz, cotton, asbestos, paper, bentonite and wood flour are the fillers added to various plastics to vary the property.

FILLER JOIST FLOOR : A floor made of filler concrete slab in which joists of small sections are closely spaced resting over large beams and the spaces between small joists are filled with concrete.

FILLER ROD : A welding rod i.e., an electrode of filler metal used in welding.

FILLET : (i)   A narrow strip of wood of triangular section used along the edge where two surfaces meet at an angle.

    (ii)   A triangular mortar strip used to replace lead sheet flashings at walls or under verges in a sloping roof.

FILLET WELD : A triangular weld between two pieces of metals meeting at right angles.

FILLING KNIFE : A knife used for laying on paste filler to fill the indentations prior to applying paint on a surface.

FILLING PIECE : A wood piece planted on another and planed to make a smooth surface.

FILLISTER : A rebate cut in a glazing bar for fixing glass.

FILM GLUE : A thin paper-like fine solid sheet of phenol formaldehyde resin is laid between a decorative face veneer and the strong backing veneer for joining together by thermo-setting.

FILTER : A strainer or a straining medium to arrest fine flocs and bacteria present in a water.

FILTER BED : A bed having a filtering media of fine sand, coarse sand, pea-size gravel, medium size gravel and coarse gravel as shown in illustration. This type of bed is used in filtering water.

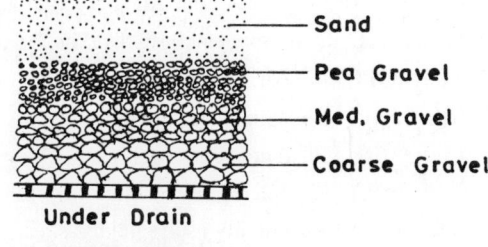

FILTER BLOCKS: Blocks of stone, concrete or bricks used in a trickling filter for bio-filtration of sewage.

FILTER MEDIUM: Graded sand and gravel are used as filter medium for filtration of water.

FILTER UNDERDRAIN : An under drainage system provided at the bottom of a filter bed to collect the filtered water as shown in illustration.

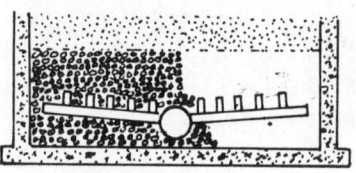

FILTER WELL : Wells constructed for drawal of ground water with a view to lowering the ground water table.

FILTER UNDERDRAIN

FILTRATION OF WATER : A process of removal of fine flocs and to some extent bacteria present in water by employing the art of filtration of water through soil as found in nature.

FINAL GRADE : The formation level of a roadway.

FINAL SETTING TIME : The time required by a cement paste to set finally. This time differs with the quality and type of cement, other factors remaining same.

FINE ADJUSTMENT SCREW : A tangent screw provided in a thedolite for fine adjustment.

FINE AGGREGATE : Sand, crushed stone, cinder, etc. are called fine aggregates which are used in making concrete or mortar.

FINE COLD ASPHALT : A road wearing course made of fine aggregates and bitumen, which is spread evenly and rolled when cold.

FINENESS MODULUS : A number indicating the fineness of a material like cement, sand, pigment, etc. The number for a particular material is obtained by conducting test by passing the material through a number of standard sieves placed one above the other. The percentage residues on each sieve are summed up and divided by 100, the quotient being the fineness modulus.

FINES : The finer particles in a sieve analysis.

FINE SOLDER : A solder having the lowest melting point of all solders, which is used in making a blown joint. It is an alloy of 67% tin and 33% lead.

FINE STUFF : The fine material used in finishing coat.

FINE-TEXTURED WOOD : A wood having fine texture which does not require the use of filler prior to varnishing.

FINGER PLATE : A plate fixed with a door latch to protect the door surface from finger marks.

FINGER SLIP : A curved hone for sharpening the inner face of gouges.

FINIAL : A pointed ornamental work at the top of a newel or gable or pinnacle.

FINING OFF : Rendering the finishing coat in plaster work.

FINISH : (i)   Finishing coat of paint.

          (ii)   Final coat of plaster.

          (iii) Completed work with final touch up.

FINK TRUSS : Commonly used steel truss.

FIRE BACK : The backwall of a fire place.

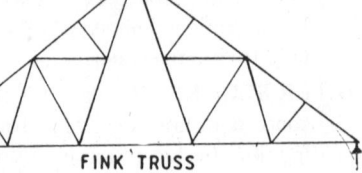

FINK TRUSS

FIRE BARS : Cast iron bars forming a grate of a furnace over which solid fuels are placed.

FIRE BLOCK : Solid bridging used as a fire stop.

FIRE BREAKS : Division wall, fire-proof door, closed stair, etc. are the devices to reduce the fire risk in a building.

FIRE BRICKS : Refractory bricks used in lining furnaces. These bricks can withstand a very high temperature.

FIRE CEMENT : High-alumina cement or refractory cement, such as fire clay used in jointing fire bricks in lining a furnace.

FIRE CLAY : A clay containing high percentage of silica and alumina with which fire bricks are made.

FIRE CRACKS : Cracks formed on a plastered surface or a concrete surface due to excessive heat by exposure to sun in hot climates.

FIRE DEMAND : Demand of water to fight fire.

FIRE DIVISION WALL : A wall raised to separate a building to protect against spreading of fire. The construction of such a wall is essential in fire-prone areas.

FIRE DOOR : A door to a furnace for feeding fuels, firing and inspection.

FIRE ESCAPE STAIR : An emergency stair provided in a commercial building and multistoried apartments or residential buildings, which should be easy approachable from all parts of the building.

FIRE EXTINGUISHER : Fire extinguishing foam sprayers, gas gun, emulsifiers, water sprinklers, drenchers and hydrants.

FIRE GRADING : Gradation of walls, columns, beams, roofs, floor, doors and windows according to their resistance to fire.

FIRE HAZARD : Danger of fire arising out of internal or external causes. The internal causes may be due to much use of fire catching materials like timber, perspex sheets, etc., leakage in gas line and burner in kitchen, storage of volatile liquids and explosives,bare electric wiring, defects in wiring, fire place, use of open-flame lantern, etc.

FIRE HOSE : A hose pipe usually of 65mm diameter used in fire fighting. One end of the hose is fitted into the fire hydrant and the other end is provided with a nozzle.

FIRE HYDRANT : An outlet from a water main into which a hose pipe is fitted for drawing water during fighting.

FIRE LOAD : The quantum of heat produced per sft. of a building floor when the combustible materials are completely burnt out during occurence of a fire. A low fire load is less than one lakh BTU/sft. A medium fire load lies between one lakh and 2 lakhs BTU/sft. and a high fire load ranges from 2 lakhs to 4 lakhs BTU/sft.

FIRE PLACE: In cold climate, in absence of central heating arrangement, a fire place is built in individual room as shown in illustration. Coal or fire wood is used as fuel for heating the room.

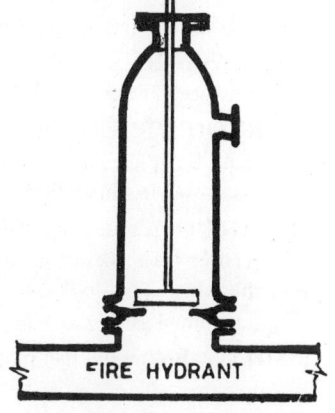

FIRE HYDRANT

FIRE POINT: The temperature at which a substance ignites from a flame.

FIRE-PRONE AREA : An area which is prone to fire risks due to presence of combustible substances, working with fire and other commercial activities with fire.

FIRE PROOF : A material capable of withstanding fire and does not ignite when a flame is put to it.

FIRE PROTECTION : Protectionary measures taken against fire risk. See 'Fire hazard' and 'Fire extinguisher'.

FIRE RESISTING FINISH : Covering the surfaces by applying a fire- retardant paint which is based on silicones, chlorinated waxes, casein, polyvinyl chloride, urea formaldehyde, borax and similar such incombustible substances.

FIRE STOP : A wall made of brick, stone or any other incombustible material, acting as a barrier to fire.

FIRE TOWER : A fire-escape stair or spiral ramp well with fireproof door at every floor of a tall building.

FIRE WALL : A fire-resisting dividing wall.

FIR FIXED : Dressed but unplaned timber fixed by nailing.

FIR FRAMED : Unplaned timbers framed by making joints.

FIRM CLAY : A clay that can be taken out by digging and can be moulded by pressing in hand.

FIRMER CHISEL : A carpenter's chisel which is not struck by a mallet. It is an ordinary chisel mostly used by joiners in carpentry.

FIRMER GOUGE : A chisel having a curved blade.

FIRST FLOOR : The floor just above the ground floor. However, when there is a basement, the floor at ground level is called the first floor.

FIRST MOMENT : Moment of an area or a force.

FIRE PLACE

FIRST STOREY : The space between the first floor and the second floor.

FIRST WEIGHT : The first lowering or removing the roof of a coal seam by blasting at the face. The coal pillrs are thus broken down.

FISHED JOINT : A joint made with fish plates and bolts as shown in illustration.

FISH GLUE : A glue prepared from fish skins and bladders. This looks like an animal glue.

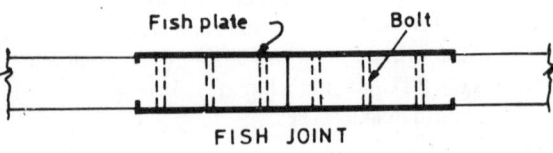

FISH JOINT

FISHING : Bolting up fish plates to rails or to form a fish joint.

FISHING TOOLS : Recovery tools used in exploratory oil-well drilling to take out broken tools.

FISH LADDER : A suitable pass for fish to travel from up-stream to down stream side of a weir. It is provided with baffles having openings to allow water to flow and fish to pass.

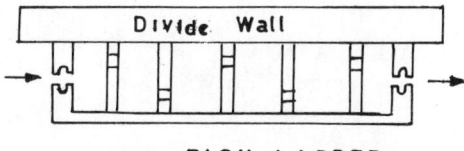

FISH LADDER

FISH PASS : A fish ladder.

FISH PLATE : (i)   A steel plate of special shape used for joining the ends of two rails by means of bolts-nuts as shown in illustration.

           (ii)   Rectangular plates used in making structural connections like extension of a stanchion or column.

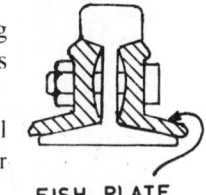

FISH PLATE

           (iii)   Cover strap or splice plate or flitch piece of steel used for making a fish joint in timber.

FISH SCREEN : A screen provided at the entry point of a channel or canal to act as a barrier to the access of fishes.

FISHTAIL BIT : A bit used in rotary drilling.

FISHTAIL BOLT : An anchor bolt with its tail split like a fishtail which is cast into concrete.

FISHWAY : A fish ladder for access of fish.

FISSURED CLAY : A clay having a network of fissures seen in dry weather condition.

FISSURES : Very fine hair cracks.

FITCH : A long-handled small brush used in painting inaccessible areas.

FITTINGS : Associated parts.

FITMENT : Fixtures.

FIT UP : A specially made formwork that can be used repeatedly.

FIXED BEAM : A beam with its ends fixed.

FIXED END : A rigid end of a beam or column.

FIXED LIGHT : A dead light i.e., a fixed sash.

FIXED RETAINING WALL : A retaining wall used in construction of a basement, the top and bottom of the wall being rigidly fixed.

FIXED SASH : Fixed light.

FIXED MOMENT : The bending moment developed at the fixed end of a structural member.

FIXING BRICK : A brick made of saw dust or wood wool and clay or wood that can be used for fixing joinery by nailing.

FIXING FILLET : A piece of timber inserted into a masonry joint to be used as a fixing for joinery.

FIXING SLIP: A fixing fillet for joinery.

FIXING STRIP: A strip of metal or other material used for fixing sheets to framed partitions or for fixing cladding to a wall frame.

FIXTURES : Fitments used in a building.

FLAG STATION : A wayside railway station similar to a halt station, but it is provided with a station building. No arrangement of permanent signals are kept. The

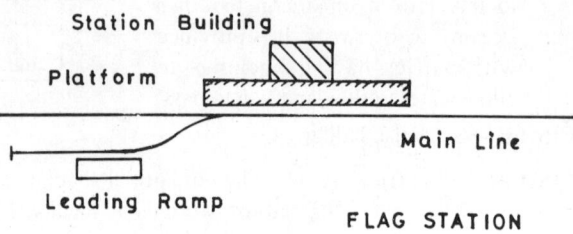

movement of trains is controlled by showing flag signals.

FLAG STONE : Thin sheets of sandstones or slabs of other stones and concrete used in paving lawns and footways.

FLAKING : (i) Dislodging of films in form of flakes from a white- washed or colour-washed or a painted surface.

(ii) A network of reeds woven over rafters of a sloping roof to facilitate laying of thatch without the use of battens.

FLAMBOYANT FINISH : A glossy transparent finish produced by applying varnish or lacquer over a surface.

FLAME CLEANING : Cleaning a structural steel prior to painting by removing mill scale, rusts, oil, grease and water with the help of a hot flame.

FLAME CUTTING : Cutting steel and other metals by means of oxy-acetylene or oxy-hydrogen, or oxy-coal gas flame.

FLAME RETARDANT PAINT : A special paint applied on timber, commercial boards or plywood with a view to retarding the spreading of a fire flame.

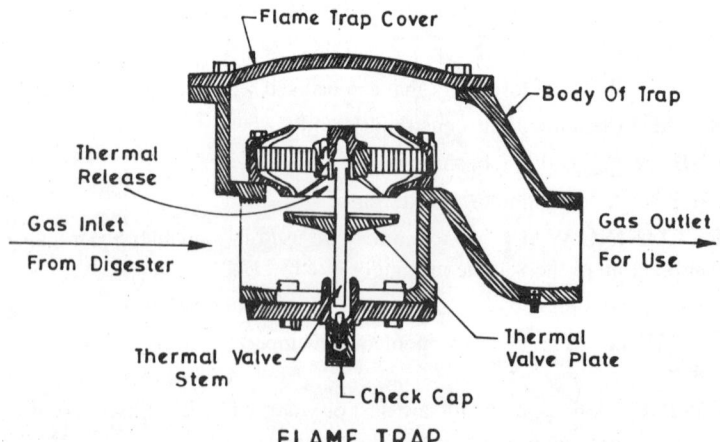

FLAME TRAP

FLAME TRAP : A flame arrester as shown in illustration, is used to prevent the passing of a flame through a gas pipe. It operates on the principle similar to a 'Miners lamp', which cool the flame below ignition point. The trap should be provided near the point of combustion and not at a distance more than 10m. from the point of combustion or fire. The flame arresting elements should be corrosion-resistant and the trap should have adequate area and depth to stop the flame without too much of resistance in its way.

FLANGE : (i)   A circular plate or disc forged or cast onto the end of a pipe to facilitate jointing with another flanged pipe by means of bolts and nuts.

(ii)   The horizontal wide strips or plates of a R.S.J. beam or girder.

(iii)   A disc attached to a shaft for jointing with another flanged shaft by means of bolt-nut.

(iv)   The base of a flat-footed rail.

(v)   The projecting disc of a railway wheel to hold it onto the rail and to prevent derailing.

FLANGED JOINT : A joint made by fastening the flanges of two pipes or shafts by means of bolts and nuts. See illustration.

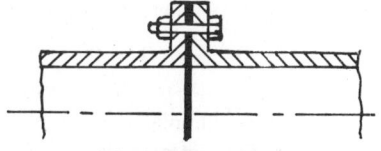

FLANGED PIPE : A pipe having flanges at its ends, required for making a flanged joint.

FLANGED RAIL : A flat-footed rail of inverted Tee-section used in railway tracks.

FLANGED JOINT

FLANGED SHAFT : A shaft having flanges at its ends for joining it with another flanged shaft.

FLANKING WINDOW : A window attached to an outside door for entry of more light.

FLANKS : (i)   The sides of a metalled road.

(ii)   The intrados of an arch at its springing points.

FLANK WALL : A wall built at one side or end of a building.

FLAP TRAP : A trap provided with a flap used in plumbing for drainage so that no back flow occurs through the trap.

FLAP VALVE : A valve provided with a hinged flap or shutter which allows flow of a fluid through it only in one direction. The shutter automatically closes when there is a tendency of backflow from the other end of the valve. This is used in a gravity outfall for drainage.

FLARED COLUMN HEAD : The widened part of a column head just below the slab, in form of an inverted cone.

FLARED HEADER : A brick with one end having dark colour due to its closeness to fire during burning. Such bricks are used in 'diaper work' which produces a pattern in the face of a brickwork.

FLASH BOARD : A stop log or stop board.

FLASH DRYING : Rapid drying of a painted surface by applying radiation of heat.

FLASHING: (i)   A piece of flexible sheet metal or any other impervious material used to make a weathered joint i.e., to throw out rain water from a junction between a roof covering and a chimney or another surface. Normally, the upper end of a flashing is rigidly held in mortar joints.

(ii) Finishing a surface with glossy patches at joints.

(iii) Formation of different colours in bricks and tiles during burning alternately with too much of air and controlled air.

FLASH JOINT: A weathered joint.

FLASH MIXER : An arrangement for mixing chemicals intimately with a liquid in a short time as shown in illustration. This is commonly used in water treatment plant for thorough mixing of chemicals with the raw water.

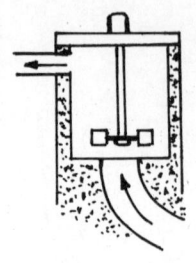

FLASH POINT : The minimum temperature needed by a material to ignite momentarily when a flame is put to it.

FLASH SET : Unusually rapid setting of a cement .

FLAT : (i) An apartment.

(ii) A matt surface produced by painting.

(iii) A level surface.

FLASH MIXER

FLAT ARCH : A straight arch provided above doors and windows in place of lintels. Its soffit and extrados are horizontal. Also known as 'Jack arch' or 'French arch'.

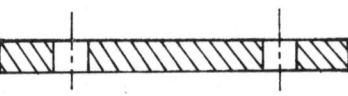

FLAT ARCH

FLAT BEARING PLATE : A sole plate to distribute the load of a machine over a large area as shown in illustration.

FLAT BOTTOMED RAIL : Flat-footed or flanged-rail used in a railway track.

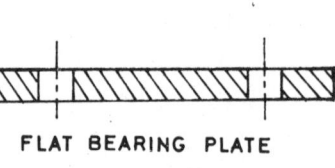

FLAT BEARING PLATE

FLAT BRUSH : A brush of black bristles used for painting.

FLAT COAT : A filler coat applied to a surface prior to painting for filling up indentations.

FLAT COST : Cost of material and labour for a work.

FLAT CUTTING : Resawing a timber making parallel to one of its edges which is level.

FLAT DRAWN GLASS : A sheet glass for use in glazing windows.

FLAT-FOOTED RAIL : A flat-bottomed rail in a railway track, as shown in illustration.

FLAT GRAIN : Grains found in a flat-sawn timber with the annual rings at angles less than 45° with the timber face.

FLAT JOINT : A mortar joint flushed with the surface of the brickwork.

FLAT POINTING : Pointing of brickwork with flat joints.

F. F. RAIL

FLAT ROOF : A roof having a slope less than 10° to the horizontal.

FLAT SAWING : Sawing a log by parallel cutting for conversion of timber with minimum wastage.

FLAT-SAWN TIMBER : See 'Flat Grain' and 'Flat Sawing'.

FLAT SLAB : A level slab made of concrete for flooring or roofing or making a deck.

FLATTING DOWN : Rubbing a surface with abrasives like glass paper, emery cloth, powdered pumice stone and felt, etc. for preparation of a surface prior to painting.

FLATTING VARNISH : Application of a varnish containing hard resin on a surface as an undercoat for varnishing.

FLAT VARNISH : A varnish whose gloss is reduced by adding filler, soap, wax or pigment.

FLAT WALL BRUSH : A brush similar to a distemper brush.

FLAUNCHING : Placing a mortar fillet surrounding the top of a chimney stack for throwing off rain water.

FLEMISH BOND : A bond used in masonry work in which stretchers and headers are placed alternately in each layer. See illustration.

FLEMISH DIAGONAL BOND : In this bond, stretchers and headers are used in alternate layers.

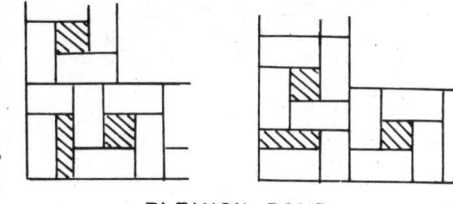

FLEMISH BOND

FLEMISH GARDEN WALL BOND : Also known as 'Sussex Garden wall Bond'. A bond, in which there is a sequence of three stretchers and one header in each layer and it appears on both faces of a 250 mm thick wall.

FLEMISH GARDEN WALL BOND

FLEXIBLE JOINT : A type of pipe joint as shown in illustration. This is used in places subject to settlement such as river beds, newly filled areas, rafts, etc.

FLEXIBLE METAL ROOFING : Roofing with sheet metals like flexible sheets of tin, zinc, lead, copper and aluminium. Most commonly used roof coverings are tin sheets and aluminium sheets, because sheets of zinc,lead and copper are costly.

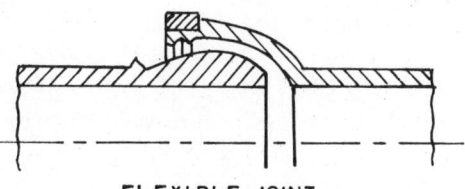

FLEXIBLE JOINT

FLEXIBLE PAVEMENT : A pavement for roads, air strips or runways made of a waterproof bituminous wearing course over a base course.

FLEXIBLE WALL : A reinforced concrete cantilever type retaining wall used as a wall for basement and underground or overground reservoir.

FLEXIMER : A seamless flexible flooring or roofing made of cement-rubber latex.

FLEXURAL RIGIDITY : Rigidity flexure which is the EI value of a structural member. It is obtained by multiplying the modulus of elasticity of the structure material by the second moment of area of the structural member.

FLEXURAL STRESS : Flexure stress or bending stress developed in a structural menber due to its bending.

FLIER : (i)   A flying shore.

(ii)   A rectangular tread of step.

FLIGHT : A series of steps joining a floor and a landing or between two floors without any landing and changing direction.

FLIGHT SEWER: It comprises a flight of steps as shown in illustration, permitting a steep slope in the sewer. The steps break the velocity of flow of the sewage. Its use may obviate the necessity for a drop manhole.

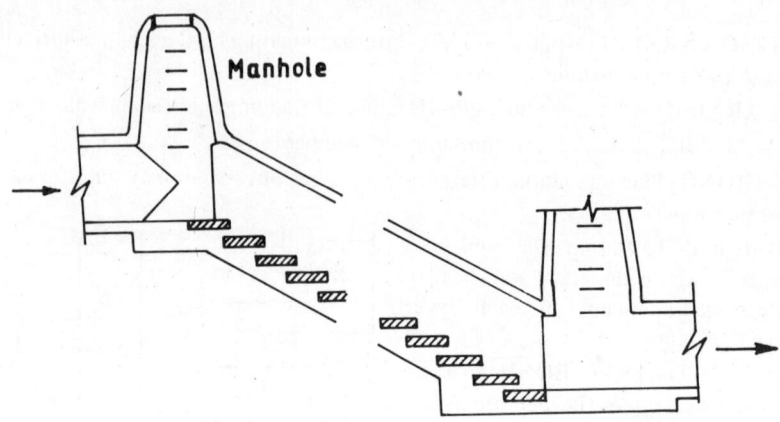

FLIGHT SEWER

FLINTS: Silica nodules, a form of stone.

FLINT WALL: A wall made of brick or concrete is faced with flints for decoration purpose.

FLITCH : A timber from which veneers are cut or a pile of veneers kept in order as it was in the timber prior to slicing.

FLITCHED BEAM : A sandwich beam made of two pieces of wooden beams placed parallel sandwiching a steel plate in between. These are bolted together to form a single beam as shown in illustration

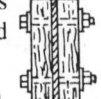

FLITCH PIECE: The steel plate used as a sandwich material in a flitched beam.

FLOAT : (i)   A cross-grained wooden tool used in plastering.

FLITCHED BEAM

(ii)   A cast iron pipe hanging from a floor level is required to drain out the waste water.

(iii) An appliance for spreading mortar or hot asphalt over a surface.

(iv) A small floating object used to indicate the direction of flow and to measure the velocity of flow of a fluid.

(v)   A ball cock used in a water tank to open or close a valve with the fall and rise of the water level in the tank and to control overflow. This is commonly found in water reserviors and flushing cisterns.

FLOAT-ACTUATED METER : See illustration. A float-controlled device for measurement of flow of a liquid through a channel. The float is hung from the integrating flow recorder and it is kept in the still

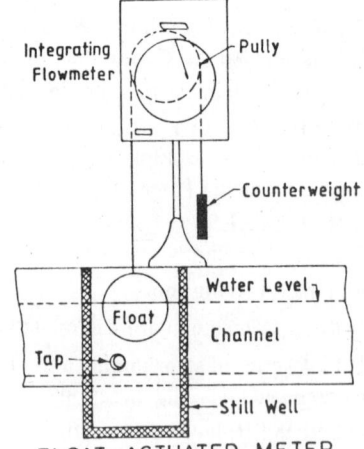

FLOAT- ACTUATED METER

169

FLOCCULANT

well. The meter records continuously the flow with the rise and fall of the liquid level.

FLOAT GLASS : A thick glass sheet made by floating molten glass over a molten metal. This produces a smooth surface of the glass sheet.

FLOATING : (i) Levelling a plastered surface by a floating rule.

(ii) Separation of pigment grains in a paint film at the surface when two dissimilar pigments are used in manufacturing a paint. This is aimed at in metal paints, especially in aluminium paints.

FLOATING BERTHS : Berths having floating arrangement used in small craft harbours. Floats are made of wood, fibre glass, polyurethane and light-weight concrete.

FLOATING CAISSON : A caisson sunk by filling it with water so that it remains hydrostatically stable at each stage. The flotation is achieved by producing caissons with sufficient buoyancy and hydrostatic stability.

FLOATING COAT : The topping coat of a two-coat plaster which is levelled by a float. Also known as 'Browning coat'.

FLOATING CRANE : A crane placed on a pontoon (a large barge) in a large port for loading and unloading.

FLOATING DOCK : A floating dry dock that consists of a floating steel structure which sinks beneath a water vessel (required for under water repairs) and makes itself buoyant when the water ballast is pumped out. Thus, the vessel is raised above the water level and is repaired in the dry condition.

FLOATING FLOOR : A floor having its wearing surface separated from the load bearing slab. This is made by placing a glass wool quilt over the rough concrete slab and wooden battens are laid over it. Then polished floor boards are nailed to the battens.

FLOATING FOUNDATION : A R.C.C. raft or mat foundation which is to remain in a buoyant condition in a soft soil.

FLOATING HARBOUR : A breakwater made of large barges (pontoons) connected end to end and to serve as a harbour.

FLOATING PIPELINE : A flexible pipeline supported on pontoons for intake of water for a water supply scheme or for removal of dredged silt and sand (taken out by a suction dredger and pumped into the pipeline).

FLOATING RULE : A long wooden rule used for leveling the floating coat of plaster.

FLOATSTONE : A porous opal used for rubbing a gauged brick.

FLOAT VALVE : A valve actuated by a ball cock.

FLOC : A fine fluffy mass formed by aggregation of very fine suspended particles in a liquid.

FLOCCULANT: A coagulant added to a turbid water for flocculation. Commonly used flocculants are alum and copperas. 'Ferrifloc' and 'Ferrichlor' are also used. Sometimes, auxiliary chemicals are added to water for hastening the process of flocculation.

FLOCCULATION: The process of forming floccules or wooly masses by bringing together the finely-divided suspended particles like colloids and fluffy organic matters in water. This is achieved by adding coagulants or precipitants in the turbid water. The particles come in contact with each other, cling together, form a large mass and settle rapidly.

FLOCK SPRAYING : Blowing soft fluffy fibres or tissues of silk, rayon, cotton, nylon, etc. onto a glued surface to bring the textile effect. This is used for various decorative works and articles of use.

FLOGGING : Smoothening a wooden floor manually by rubbing it with stones.

FLOOD CHANNEL : A channel with branches formed in an estuary due to flood water during high tide. It is wide at the mouth i.e., at the seaward end, gradually narrowing to the upstream end meeting a bar.

FLOOR : A deck or platform.

FLOOR ARCH : A term for 'Jack Arch'.

FLOOR BOARD : Wooden board or plank used in flooring.

FLOOR CLIP : A strip of steel sheet anchored to the surface of a concrete floor by casting, which is bent in form of U-shape on setting and hardening of concrete. This is required to fix battens for a timber floor.

FLOOR CRAMP : A cramp used to force several floorboards together prior to nailing them to the battens.

FLOOR DRAIN : See illustration.

FLOOR GUIDE : A groove in a floor to guide the movement of a sliding door.

FLOOR JOIST : A common joist used in timber flooring.

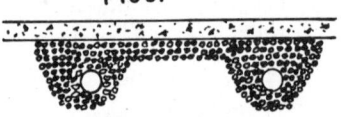

Floor

FLOOR DRAIN

FLOOR LINE : A mark given on a wall to indicate the finished floor level.

FLOOR LINING : Covering a rough concrete floor with linoleum mat or commercial board or plywood.

FLOOR PLAN : A plan showing the layout of all rooms with floor area.

FLOOR SLAB : A slab to be used as a floor between two stories.

FLOOR STOP : A door stop anchored in the floor.

FLOOR STRUTTING : Solid strutting between floor joists in making a timber floor.

FLOOR TILES : Tiles used in making a finished floor. Such tiles are made of various materials.

FLOOR VARNISH : A quick-drying, abrasion-resistant varnish used in finishing a timber floor.

FLOURY SOIL : A fine-grained soil composed of silt or rock flour.

FLOW CONTROLLER : An automatic device to control the rate of filtration through a filter bed. The principle of Venturi tube with a movable floating valve attached to a flexible diaphragm has been adopted as shown in illustration. Also known as 'Rate of Flow Controller.'

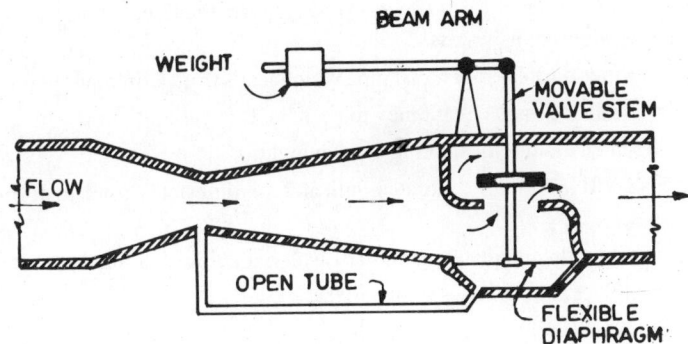

FLOW CONTROLLER

FLOW CURVE : In soil mechanics, a curve obtained by plotting the data of a 'liquid limit test' of a soil sample. It is a straight line, when values of water content are plotted on the vertical arithmetic scale and the number of blows on the horizontal log scale.

FLOW INDEX : The slope of a flow curve.

FLOW LINES : Lines in a 'flow net' indicating the flow path of water through a soil mass, which intersect the equipotential lines at right angles. See illustration of a 'flow net'.

FLOW METER : An instrument to measure the quantum of fluid flowing per unit time. There are various forms of such a meter. Integrating flow meter is used as a water meter. Commomly used flow meters are magnetic flow meter, disc type rotary meter, float-actuated meter, electrically operated meter, etc.

FLOW NET : A network of flow lines and equipotential lines expressing the relationship between water flow and potential (head) through a soil mass. The illustration shows these lines intersecting at right angles in a dam section.

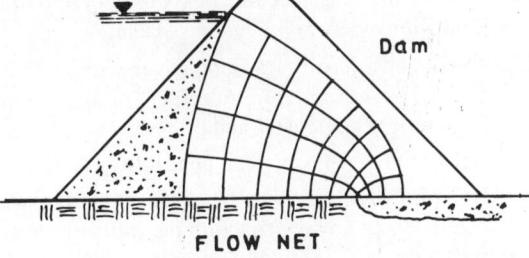

FLOW NET

FLOW SLIDE : A slide of a mass of silt or loose sand caused when it is in a semi-fluid state. On sliding, the material spreads out to a flat slope. Also known as 'mud flow' or 'Earth Flow'.

FLOW TABLE TEST : A test similar to slump test to measure the stiffness of a concrete mix.

FLOW THROUGH CHAMBER : The upper compartment of a two-storied sedimentation tank.

FLUE : The passage of smoke in a chimney.

FLUE BLOCK : Precast hollow blocks which are sometimes used to form a flue.

FLUE GAS: The smoke from a furnace containing CO, CO_2 and N_2.

FLUE GATHERING: The flue at the base of a chimney.

FLUE LINING : Pargetting or use of fireclay pipes to form the flue. Linings are sometimes made of aluminium or asbestos.

FLUE PIPE : A cement-asbestos or metal pipe which leads from a fireplace to the chimney.

FLUE WAY : The space inside a chimney to pass the flue gases.

FLUME : An open channel with sluicing arrangement.

FLUORESCENT PIGMENTS : Borates, silicates or tungstates used to form brilliant paints.

FLUSH : (i)   To pour an adequate quantum of water at a time at a high velocity.

   (ii)   To place two surfaces in one plane.

FLUSH BEAD : A bead with a quirk on each side.

FLUSH DOOR : A door having a smooth surface without any panel and built of commercial boards either solid or with a hollow core.

FLUSHED JOINT : A flat joint.

FLUSHING : Hydraulic filling.

FLUSHING TROUGH : A long cistern or water tank to flush a series of water closets at a regular interval.

FLUSHING VALVE : A valve to supply a definite quantum of water required for flushing a water closet without the use of a flushing cistern

FLUSH PANEL : A panel which is flush with its frame.

FLUSH SOFFIT : The level and smooth undersurface of spandrel steps.

FLUSH TANK : A flushing cistern or trough.

FLUSH VALVE : See 'Flushing valve'.

FLUTES : Semi-circular vertical grooves at intervels in a pillar or column.

FLUX : A fusible substance like borax used in soldering, brazing and welding to cover the joint with a view to preventing oxidation.

FLY-ASH : The ash particles from pulverised coal that pass through the chimney. This can be used as an admixture to cement or as pozzolana. Fly-ash is also used in making light-weight aggregates and bricks.

FLYING BOND : An American bond or Monk bond.

FLYING BUTTRESS : A buttress that supports a wall and acts as a stiffener at a high level without any contact with the wall at its lower level.

FLY-OFF : Evaporation of rain water from a surface.

FLYING SCAFFOLD : A scaffold hung from top.

FLYING SHORE : A horizontal strut provided between two walls above ground level as shown in illustration to keep the walls vertical when the intermediate construction has been removed.

FLY RAFTER : A barge board.

FLY WIRE : A fine wire mesh.

FOAMED CONCRETE : Aerated concrete made of foamed cement and light-weight aggregates.

FOAMED SLAG : Foam-textured blast-furnace slag which is used as a thermal insulator and light-weight aggregate.

FOIL : A very thin sheet of a material.

FOLDED FLOORING : A flooring made by pushing floorboards forcibly into place instead of using a floor cramp.

FOLDING CASEMENT : A pair of casements with meeting stiles hinged to the frame without any mullion.

FOLDING DOOR : A door, with two or more leaves (shutters) ninged to each other and hung from the frame.

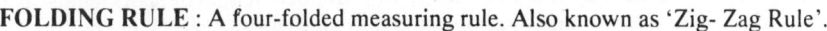

Straining Piece

FLYING SHORE

FOLDING RULE : A four-folded measuring rule. Also known as 'Zig- Zag Rule'.

FOLDING SHUTTER : Boxing shutter i.e., a shutter made of several leaves hinged to each other that can be boxed by folding.

FOLDING STAIR : A ladder which can be folded.

FOLDING WEDGE : Hardwood wedges used in pairs to tighten up or loosen supports like dead shores, raking shores and flying shores.

FOLIATED STRUCTURE : The most obvious characteristic of metamorphic rocks which refers to the parallel arrangement of needle like minerals like sericite, hornblende, biotite and chlorite.

FONDU : A term used for high-alumina cement.

FOOT BLOCK : (i)   A timber block used as a base for a post.

(ii)   An architrave block.

FOOT BOLT : A vertical tower bolt.

FOOT CUT : A horizontal saw-cut at the foot of a common rafter to make a birdsmouth joint.

FOOTING : A wall or column foundation which is widened to distribute the load to a large area.

FOOTING DRAIN : A drain consisting of open-jointed tiles laid in back-filled trenches close to the foot of a basement wall to permit removal of seepage water by gravity flow into sewers or ditches.

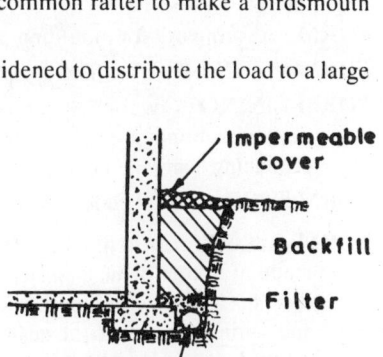

Impermeable cover

Backfill

Filter

FOOTING DRAIN

FOOT IRON : Iron step or step iron used for accessibility of the operation & maintenance staff into a small chamber.

FOOT PLATE : A base plate or sole plate.

FOOTPRINTS : An adjustable wrench with serrated jaws used in plumbing work by pipe fitters. Also known as 'Combination pliers' or 'Pipe tongue'

FOOT SCREWS: Three screws attached to the tribrach of a theodolite, required for levelling the instrument. These are also known as 'Levelling screws' or 'Plate screws'.

FOOT STONE: A stone block used as a gable springer.

FOOT SUPER : Square foot.

FOOT VALVE : A check valve used at the end of a suction pipe in pumping water. This is a one-way valve fitted at the lower end of the section pipe so that the suction pipe remains full with water all the time and no priming is required.

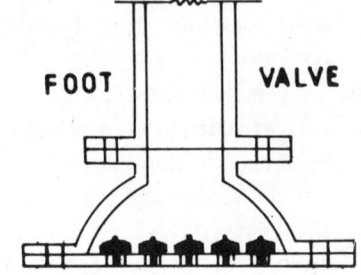

FOOTWAY : The pathway or footpath for pedestrians.

FORCE CUP : A simple and cheap tool for taking out slush mucks (wastes) from a drain. It consists of a rubber cup fitted to the end of a wooden handle of short length. It is pushed up and down over the waste plug and thus the wastes are taken out by suction.

FORCED CIRCULATION : A circulation of fluid by means of pumping.

FORCED DROP SHAFT : A method of sinking cylindrical shafts in water-logged land by pushing down a series of telescopic C.I. cylinders with the help of hydraulic jacks inside a R.C.C. well curb. The soil is excavated by dredging and driving a trepan (a rotary boring machine) and the excavated material is taken out with the help of an air-lift pump.

FORCED DRYING : Drying a painted surface by radiation of heat in a short period.

FORE SIGHT : A fore observation made by means of a surveying instrument, particularly during levelling. The observation or reading is taken for the next survey point or station in the forward direction of the survey work.

FORGE WELDING : A form of pressure welding.

FORK LIFT TRUCK : A power-driven truck provided with a projected steel fork at its front by which materials are picked up from ground level, lifted up and transported to the desired location.

FORM : A formwork for moulding concrete or plaster.

FORMATION LEVEL : The dressed level surface of a ground or an embankment.

FORM LINING : Covering the inner surface of a formwork (into which concrete will be cast) with bitumen-impregnated paper, cardboards, waxed paper, etc. with a view to producing a smooth concrete surface.

FORM STOP : A dead end.

FORM WORK : A temporary framework made of timber planks, struts and posts or of steel plates with supports for casting and setting concrete to the desired shape. Also known as 'Shuttering'. See illustration.

FORSTNER BIT : A drill bit with sharp ring round it, used by carpenters for making blind holes.

FOSSIL RESIN : Copal resin which has become hard over years underground.

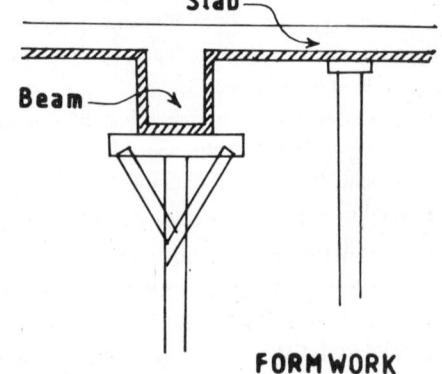

FOUL-AIR FLUE : A ventilating duct for driving out foul air.

FOUL WATER : Sewage and sullage.

FOUNDATION : The sub-structure or the part of a structure built underground for transfer of load to soil and to hold the structure. Various types of foundation are adopted depending upon the type of structure and the soil condition.

FOUNDATION BOLT : Holding down bolt or anchor bolt required to fix the base of a structure or a machine into concrete or brickwork. These are Rag bolt, Lewis bolt, Tee bolt and Eye bolt.

FOXY TIMBER : A timber that has started decaying.

FRAME : A two-dimensional or three-dimensional structure in skeleton form made by joining different members. A frame is usually made of timber, steel or concrete.

FRAMED DAM : A dam built of timber or steel frame.

FRAMED AND BRACED DOOR : A framed door with vertical boards in one face and three ledges and two diagonal braces in the other face.

FRAMED AND LEDGED DOOR : A framed door with vertical boards and ledges without any bracing.

FRAMED DOOR : A door having a rigid frame, three or four rails, hanging stile and shutting stile.

FRAMED FLOOR : A floor made of wooden boards supported by a wooden frame.

FRAMED HOUSE ROOF : A roof truss with a prop at the middle of the span as shown in illustration.

FRAMED PARTITION : A partition made by making a frame of timber or steel sections.

FRAME HOUSE : A house made of timber frame and commercial boards and weather boards.

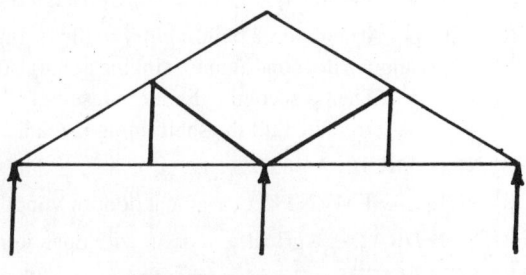

FRAMED HOUSE ROOF

FRAME SAW : A power-driven heavy duty saw used for cutting wood or stone.

FRAME WORK : A skeleton structure of timber , steel or concrete for a building.

FRAMING SQUARE : A steel square used by carpenters.

FRANCIS FORMULA : The most commonly used equation for estimating the flow over rectangular weirs.

$$Q = 3.33 (L-0.1nh)h^{3/2}$$

Where Q is discharge, L is crest length of weir, n is no.of end contractions and h is observed head upon crest of weir. This formula neglects the approach velocity in contracted weirs.

FRANCIS TURBINE : A low-head water turbine with a vertical shaft used for generation of hydro-electric power.The water enters radially and passes out downwards.

FRANKI PILE: A cast-in-situ concrete pile with a bulbous toe. A rough shaft is formed by ramming successive quantities of concrete while progressively raising the steel pipe.

FREE BOARD: The depth or height between normal water level and the top of the reservoir or channel.

FREE END : The beam end having no support or a column end having no fixity.

FREE FALL : Free flow.

FREE FLOW : A flow of water through a pipe end which remains unaffected by the tail water level.

FREE HAUL : The maximum distance upto which excavated material can be transported from a site without any additional cost.

FREE RETAINING WALL : A retaining wall that tilts or slides slightly such that the shifting of the wall top is about 0.005 of the height of the wall.

FREE STANDING : The part of a chimney which does not touch other structure for its support.

FREE STONE : Limestone or fine-grained sandstone which can be carved and worked in any direction for building construction.

FREE STUFF : A perfectly clear timber.

FREE VIBRATION : The vibration at the natural frequency of a structure. This occurs when a structure is displaced within the permissible limit and then released.

FREE WATER : Held water or gravity water.

FREE WAY : A roadway for fast moving through traffic.

FREEZING : A method of building dry walls in fine-grained water-logged soil for safe excavation in dry condition for sinking a shaft. At first, cased holes in a circular pattern are sunk. Then a second cylinder is inserted in each hole and cold brine solution at -16^{0}C is circulated till the shaft lining is made.

FRENCH ARCH : A dutch arch.

FRENCH CASEMENT : A casement door or window.

FRENCH DRAIN : Agricultural drain with open joints and surrounded with gravel.

FRENCH FLIERS : The fliers round an open well stair.

FRENCHMAN : A knife with its end bent over, used for trimming mortar joints.

FRENCH POLISH : It consists of white shellac, sandarac, methylated spirit, benzoin, gamboge and olibanum.

FRENCH ROOF : A mansard roof.

FRENCH STUC : A plasterwork resembling an imitation stone.

FRENCH TILE : Interlocking clay tile.

FRENCH TRUSS : A fink truss.

FRENCH VARNISH : A varnish prepared with shellac and methylated spirit.

FRENCH WINDOW : A term used for a casement door.

FREQUENCY CURVE : A curve representing frequency distribution in statistical analysis.

FREQUENCY DIAGRAM : A histogram showing frequency distribution such that the area under the diagram corresponds to the frequency.

FREQUENCY DISTRIBUTION : It speaks of the relation between the magnitude of a variable characteristic and its frequency of occurence.

FRESCO : Painting with alkali-resistant water paint on a surface of wet lime plaster.

FRESH AIR INLET : An inlet arrangement with a hinged mica flap provided at the top of a drain pipe with a view to admitting fresh air to the house drain for ventilation.

FRESH WATER LENS : Ground water recharge with fresh water in the neighbourhood of a sea so that a fresh water lens is produced underground by shifting the salt water as shown in illustration.

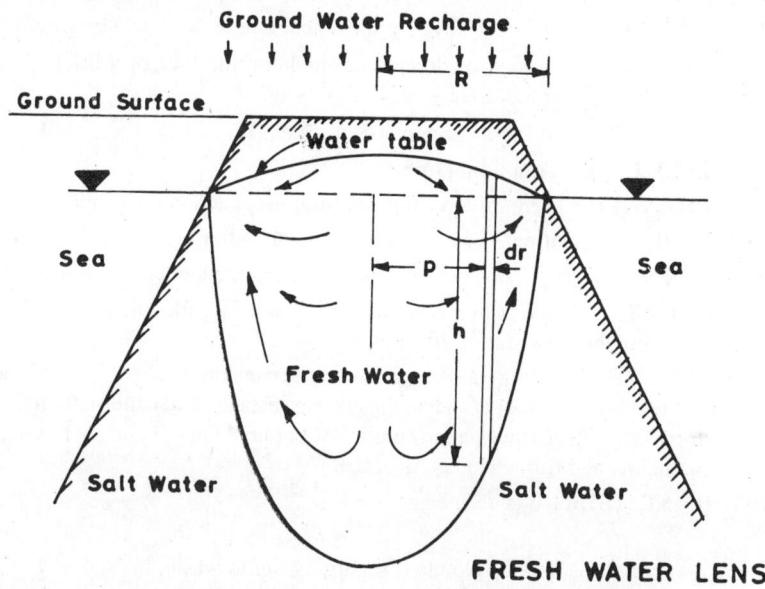

FRESH WATER LENS

FRET SAW : A saw for cutting sharp curves required for joinery works. The blades are adjustable and replaceable. The saw may be hand-operated or power-driven.

FRETTED LEAD : Lead cames of H-shape used for leaded lights.

FRETTING : Breaking away of road metal on a road surface due to heavy vehicular load.

FRET WORK : Work done by fret sawing.

FRICTION : A force that opposes a motion.

FRICTIONAL SOIL : A soil of clean silt,sand or gravel, whose shearing strength is measured by the friction between the grains. It is a cohesionless soil.

FRICTION CIRCLE : A circle of radius $R \sin \phi$ concentric with the circular rupture arc of radius R as shown in illustration, required in determining the stability of ground slopes.

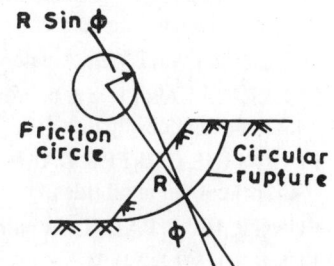

FRICTION HEAD : The energy lost by a flowing water due to friction in a pipe.

FRICTION LATCH : A spring catch mortised into a door edge.

FRICTION PILE: The pile that develops the load carrying capacity by friction on the sides of the piles. The load carrying capacity of such a pile sometimes results from a combination of point resistance and skin friction.

FRINGE WATER: A term for 'held water' above the ground water table.

FRIEZE : The portion of a room wall above the picture rail.

FRIEZE PANEL : The topmost panel in a door leaf.

FRIEZE RAIL : The rail provided in a door shutter, below the frieze panel

FRIG BOB SAW : A long hand saw for cutting bath stones.

FRIT : Finely ground and sieved sand, glass or flint to be used in manufacturing ceramics.

FROG : (i)   A Vee-shaped groove for the wheel flanges at a rail crossing.

  (ii)   An indentation or depression made on the face of a brick for grip during laying with mortar.

FRONTAGE : The road-facing dimension of a site.

FRONTAGE LINE : A building line.

FRONT HEARTH : The part of a fireplace floor projected into a room.

FROST BOIL : The softness of a soil due to thawing after heaving up of frost on it.

FROST CRACK : Cracks in wood caused by frost at the tender age of the tree.

FROST HEART : A defect in a heartwood in form of a dark colour caused by the action of frost on the tree during its growth.

FROST HEAVE : Swelling of silty soil due to expansion of water in frost and formation of several layers of ice parallel to the ground surface. When the water in frost expands, it forces the silt particles apart, increases the pore spaces and more water comes up from below and forms the layers of ice.

FROUDE NUMBER : It is defined as $F = V/\left(\dfrac{S'}{S} \ g D\right)^{0.5}$

  where V is the jet velocity discharging liquid waste,

    S' is the difference in specific gravity between the waste and the surrounding water,

    S is the specific gravity of liquid waste,

    g is acceleration due to gravity and

    D is discharge jet diameter.

The expression given is used in determining inital dilution by turbulent jet mixing.

FUGITIVE PIGMENT : A pigment that fades away rapidly.

FULL CELL PROCESS : A pressure-creosoting process for preservation of wood. Prior to pressure-creosoting, the moisture and air are withdrawn from the wood by keeping it in a vacuum chamber.

FULL-WAY VALVE : A gate valve which does not impede the flow of water.

FULLER'S EARTH : A bentonite clay used for absorbing the fats from wool and also as an extender in paints.

FULL-TIDE COFFER DAM : A coffer dam in an estuary is built quite high so that water is kept out at all tides.

FUNGICIDAL PAINT : A paint that resists the growth of fungus.

FUNICULAR RAILWAY : A ropeway for transport of passengers.

FURRING:(i)   Forming a cavity in an outer wall.

  (ii) Leaving an air space between brickwork and plaster.

G

GABION : A small cellular cofferdam or wire basket to hold soil. Originally, it was the name for a bottomless basket placed with others in a row on the edge of a trench to protect the sodiers from rifle fire.

GABBART SCAFFOLD : A strong scaffold made of squared timber in which the 'standards' consisting of three deals are bolted together and the 'ledger' rests on the middle deal which is cut off at the platform level. This type of scaffold is practised in Scotland.

GABLE : The triangular part of an end wall supporting a sloping roof. Also, known as 'gable end'. A gable may be made of brick, stone, weather board or hanging tiles.

GABLE BOARD : Barge board.

GABLE COPING : A coping to a gable wall projected above the sloping roof.

GABLE END : A 'gable'.

GABLE POST : A short post placed at the apex of a gable to house the barge boards.

GABLE ROOF : A sloping roof with gables at its ends.

GABLE SHOULDER : The projection made by a gable springer at the base of a gable coping.

GABLE SPRINGER : The overhanging brick or stone at the base of a gable coping.

GABLET : A small ornamental gable as made over a 'dormer' or 'gambrel' roof .

GABLE ROOF

GABLE WALL : A wall with a gable at its top.

GABLE WINDOW : A window built into a gable.

GABOON : An african Mahogany tree.

GAD : A pointed steel bar of short length used in wedging out coal in mining.

GAFFER : A ganger or a man in charge of a work.

GABLE WINDOW

GALE : A wind blowing at 64 km. per hour at a height of 9 m. above the ground. This exerts a pressure of about 25 kg/sq.m. on a vertical face.

GAIN : In carpentry, it is a notch or mortise in a timber to receive another timber.

GALLERY : In mining, it is a tunnel or a mine roadway to collect water in rock.

GALLERY APARTMENT HOUSE : In USA, it is an apartment house having access to the dwelling units from an open corridor.

GALLET : A chip of rock.

GALLETING : (i)   Embedding gallets or spalls on a plastered surface for decoration or for filling joints in a rough masonry work.

(ii) Setting of small size plain tiles in the mortar bed on the top of single lap tiles with a view to providing a level surface for the ridge tiles.

GALLON: A measuring unit for the volume of a liquid. One Imperial gallon contains 10 ibs. of water, while one American gallon of water weighs 8.33 lbs.

GALLOWS BRACKET : A triangular framed wooden bracket projected from a wall.

GALVANIZED IRON : Iron or steel, coated with zinc either by dipping it into molten zinc or by the process of electrolysis.

GALVANIZED PIPE : Mild steel pipes coated with zinc are used mostly in domestic installation of pipelines for carrying water.

GAMBREL ROOF : A sloping roof having a gablet near the ridge, the lower part being hipped. Also, known as 'half-hipped roof' See illustration.

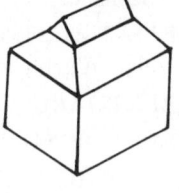

GAMMA PROTEIN : Obtained from soyabean and used as extender for casein in dstemper or water paint.

GANG : A group of workmen/labourers.

GANG BOARDING : A cat ladder used by workmen for going up and coming down.

GAMBREL ROOF

GANGER : Gangman who supervises the work of a gang.

GANG MOULD : A mould that produces simultaneously a number of similar concrete units.

GANG SAW : A framed saw having reciprocating movement.

GANGWAY : A narrow approachway for men to walk on for inspection and repair work.

GANGWAY

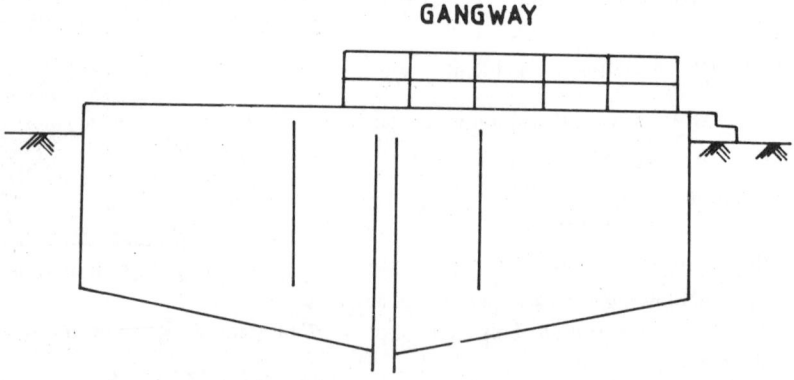

GANGUE : Impurities and other minerals associated with an ore.

GANTRY : (i) A temporary staging built of steel joists or timbers for carrying heavy loads.

  (ii) Crane gantry is a permanent gantry carrying rails for the overhead travelling crane.

GANTRY CRANE : Portal crane.

GAP-FILLING GLUE : A glue used for joining surfaces which can not be brought close to each other. The gap between the surfaces is filled with this glue.

GAP-GRADED AGGREGATE : Two sizes of aggregate (instead of using various grades of aggregate) used in making a dense concrete by addition of grits, stones, pebbles or sand to fill the gaps. Gap grading is simple and easy in practice.

GARCHEY SINK : A specially made domestic sink for disposal of small size organic and inert matters.

GARDEN WALL BOND : A brick bond of one brick thickness used in making garden walls which carry only their self loads. These are either English garden wall bond or Flemish garden wall bond.

GARDEROBE : A small room in which clothes and other domestic articles are kept.

GARNET : A gallet.

GARNET HINGE : A cross garnet hinge used in carpentry.

GARNET PAPER : Abrasive paper having finely powdered garnet.

GAS CARBURIZING : A process of carburizing steel by heating it in a current of hydro-carbon gases.

GAS CONCRETE : Aerated concrete or a light-weight concrete foamed by mixing aluminium powder with water which forms tiny hydrogen bubbles during the preparation of concrete.

GAS CRAZING : When a paint or varnish containing vegetable drying oils, dry in burnt coal gas, webbing, frosting, crows-footing or crystallizing takes place on the painted or varnished surface.

GASKET : Asbestos sheets, thin copper sheets or impregnated papers are used as gaskets for making gas-tight or water-tight joints. Sometimes, rubber sheets are used for this purpose.

GAS PLIERS : Pliers with serrated concave jaws to grip gas pipes.

GAS-PROOF PAINT : A paint film that is not affected during drying by the burnt coal gas.

GAS WELDING : Welding of metals by using oxy-acetylene flame. This is used at sites where electrical power is not available.

GATE : In hydraulics, it is a barrier put across a flow. The flow can be regulated by partly opening the gate.

GATE CHAMBER : A recess in a lock wall to receive a ship caisson when it is open.

GATE HOOK : A metal bar in form of a gudgeon pin built into the gate pier or driven into the wooden post, on which the gate hinge is dropped so as to hang the gate.

GATE PIER : A gate post made of brick, stone or concrete.

GATE POST : A gate pier on which a gate hinge is fixed or a gate shuts.

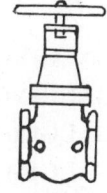

GATE VALVE : A valve used in a pipeline to close or control the flow of a liquid by closing or partly opening the gate, which is a circular disc attached to the threaded spindle of the valve. The gate can be lowered down or brought upward by turning the spindle.

Gate Valve

GATHERING GROUND : A catchment area.

GAUGE: In civil engineering it is a measuring device as follows :

(i)   Measuring sheet thickness or wire diameter by a number.

(ii)   Measuring device to indicate the quantum of rainfall.

(iii)   Device to measure flow of a stream.

(iv)   Water level measuring devive.

(v)   Measurement of ingredients for a mix proportion.

GAUGE BOARD: (i) A mortar board of 1 metre square for carrying plaster and tools used in plastering.

(ii) A pitch board used in carpentry.

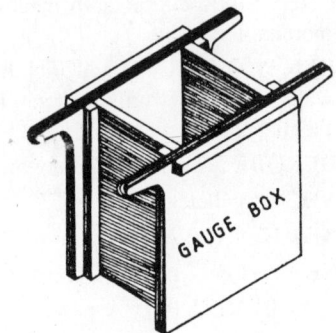

GAUGE BOX : A batch for measuring the quantity of cement and aggregates for concrete mix.

GAUGED ARCH : An arch built with gauged bricks i.e., soft bricks made to shape either by rubbing on a hard surface or by cutting the bricks. The joints are made very thin either with lime mortar or with lime putty.

GAUGED BRICKS : These are soft bricks made to the required shape by rubbing or cutting them. These bricks were used in earlier days.

GAUGED MORTAR : Cement-lime mortar.

GAUGED STUFF : Gauged plaster or putty used for finishing coat of interior ceilings, mouldings, etc.

GAUGE LENGTH : The length of a metal piece under tensile test.

GAUGE POT : A watering pot with a measured quantum of water.

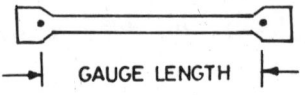

GAUGE ROD : A storey rod.

GAUGING : (i) Marking timber with a mortise gauge or marking gauge.

(ii) Sizing a soft brick by rubbing or cutting it.

(iii) Adding a measured quantum of cement, gypsum plaster or any other material to keep the proportion of mix.

(iv) Measuring stream flow

(v) Measuring thickness of sheet or diameter of wire.

(vi) Measuring the quantum of precipitation.

GAUGING BOARD : A board for mixing mortar, plaster, etc.,

GAUGING BOX : A batch box.

GAUGING STATION : A station point where a stream gauge or a rain gauge is installed.

GAUGING TROWEL : A plasterer's tool.

GAUNTLET TRACK : A railway track, when a double line of same or different gauge is narrowed over a short distance. This is used when two tracks of different gauges have to cross a river through a bridge or when one of the two tracks is under repair. This arrangement is very economical.

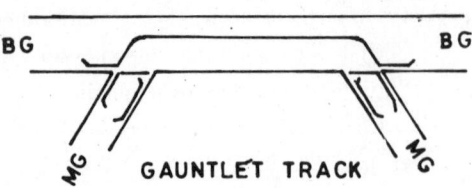

GAUSSIAN CURVE : Normal curve; A type of frequency curve which fits a large number of frequency distribution, whose probability of occurence is most.

GATHERING LINES : The arrangement when a number of railway lines (parallel tracks) are branched off from a common diagonal or inclined turnout track. Also, known as 'ladder track'.

G-CRAMP : A 'G'-shaped screw cramp made of steel which is commonly used by carpenters for jointing timber pieces with glue.

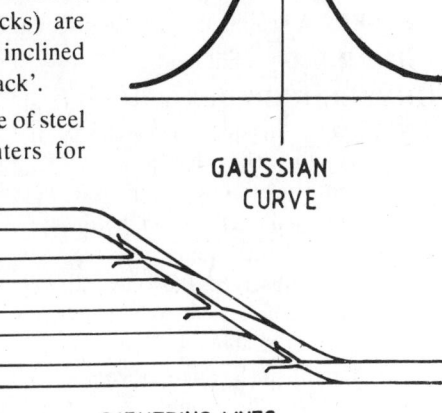

GAUSSIAN CURVE

GELATIN EXPLOSIVES : Explosives containing no gelatine have a jelly-like texture. These contain nitro-glycerine or ammonium nitrate that can be used in wet boreholes for blasting rocks.

GATHERING LINES

GELATIN MOULDING : Making jelly moulds for undercut fibrous plaster castings.

GELLING : Conversion of liquid to jelly required in paint preparation.

GEODESY : Geodetic surveying.

GEODETIC CONSTRUCTION : Stressed skin construction of structures.

GEODETIC SURVEYING : A survey conducted with high degree of accuracy to determine data concerning the shape and size of earth.

GEOID : The figure of earth defined by the imaginary mean sea level is very irregular. It has the distinctive property that its surface at any point lies in a plane tangential to the direction of gravity at that point.

GEOLOGICAL MAP : A map showing the geological formation and underground strata of an area on earth.

GEOMETRICAL STAIR : A stair without any newel post and with a continuous string round a semi-circular or elliptical well. This stair may not have any landing in between floors.

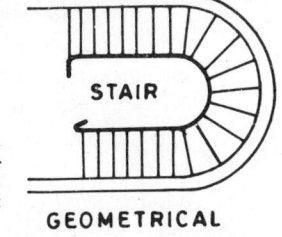

STAIR

GEOPHYSICAL SURVEY : The survey conducted for search of mineral deposits. Maps are prepared with variations of elastic properties of earth, gravitational field, magnetic field, radio activity, etc.

GEOMETRICAL

GEOPHYSICS : The study of earth from the information on meteorological details, oceanography, geography, geology, geodesy electromagnetism, etc.

GEOTECHNICAL PROCESSES : The processes which change the soil properties like electro-osmosis, ground water lowering, vibroflotation, injection of material for soil stabilization, soil compaction, etc.

GEORGIAN GLASS: Thick glass reinforced with steel wire mesh with a view to protecting damage due to external causes.

GEOTHERMAL GRADIENT : It is normally 1°C for each 30 m depth of a well. Any deviation from this gradient gives information on circulation or geologic conditions in a well. Abnormally cold temperatures may indicate the presence of gas.

GESSO : A shining white composition of glue and whiting or plaster of paris and size used on wood or plaster as a background for painted designs on it.

GEYSER : A heating unit for supply of hot water in a bath room.

GHAT ROAD : A hill road.

GHYBEN-HERZBERG RELATION : An equation derived to explain the hydrostatic equilibrium between the salt water and fresh water below ground surface adjacent to sea. It was observed that the salt water occured at a depth below sea level of about 40 times the depth of fresh water above sea level. See illustration .

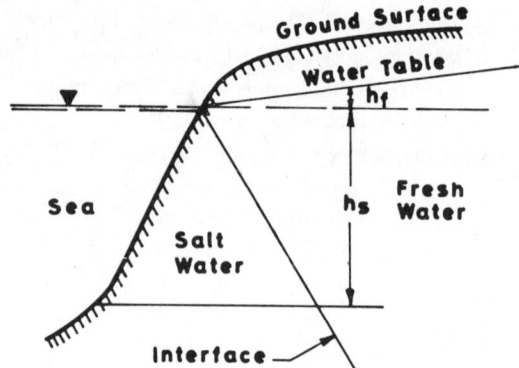

$$h_s = 40\ h_f.$$

GHOONDIE : The rounded surface at the junction of roof and parapet wall.

GIB : A metal piece used with a cotter or wedge, which clasps together the parts of a cottered joint.

GIBBS MODULE : A modular ir- rigation outlet of rigid type, consisting of an inlet pipe, Eddy Chamber, and a discharge pipe. The inlet pipe is laid under the bank of the distributary.

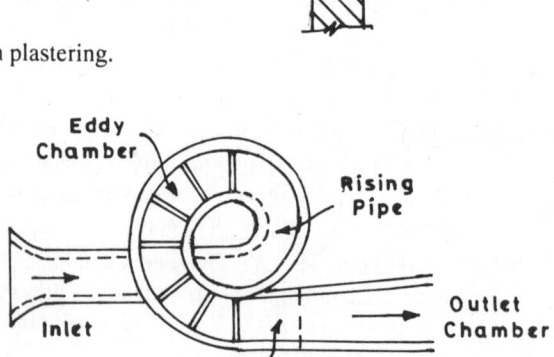

GIG STICK : A radius rod used in plastering.

GIMLET : Ancient Greeks used this small tool with a handle at right angles to the spindle having a sharp pointed end. This was used for boring hole upto 6 mm diameter in wood.

GIMLET POINT : The point of wood screw or coach screw.

GIN : A simple lifting tackle with a tripod and a gin block.

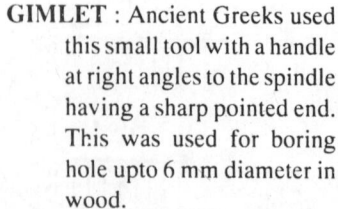

GIBBS MODULE

GIN BLOCK : A steel frame to which a single rope pulley is hooked.

GIRDER : A large size beam of timber, steel or concrete to which the secondary beams may be connected.

GIRDER BRIDGE : A bridge supported by girders.

GIRDER CASING : Sometimes, the portion of a girder below the ceiling is provided with a casing to make it fire-proof.

GIRT : An intermediate beam in a timber-framed building carrying floor joists. This may be called a small girder.

GIRTH : The circumference of a round timber.

GIRT STRIP : A ribbon board.

GIVE AND TAKE LINES : In calculation of a land area of irregular shape, the straight lines used for the equalization of boundaries.

GLAND : A compressible copper, brass or gun-metal ring used in a compression joint under the screwed fitting. This makes the joint watertight.

GLAND JOINT : A joint made on a copper tube or pipe which allows temperature movement.

GLASS BLOCK : Hollow blocks made of transparent or translucent glass with patterns on both surfaces. These are used in making decorative partition walls.

GLASS CONCRETE CONSTRUCTION : Concrete pavement lights or floor or roof slabs with glass lenses cast in, with a view to have ornamental and hidden lighting arrangement.

GLASS CUTTER : A tool with sharp diamond cutting edge.

GLASS FIBR REINFORCEDRESIN : Synthetic resinrein forced by glass fibre, chiefly used for roofing sheets, sports-car bodies, boat hulls and caravan roofs. This may also be used for repair of C.I. rainwater gutter, jointing of metal to wood and water-proof non-rusting welds.

GLASS PAPER : Also known as sand paper, emery paper and garnet paper. This is an abrasive paper made from powdered glass, sand, flint, garnet and corundum glued to paper. The quality of such papers depends on the degree of fineness.

GLASS SIZE : The glazing size.

GLASS SLATE : Also called glass tile. Glass pieces are made of same size as that of a slate or tile and are used in attic or other places for natural light.

GLASS STOP : A device provided at the bottom of the glazing bar to prevent glass panes sliding down.

GLASS WOOL : Flexible, soft and silky fibres of glass used for insulation of heat and sound. Also, known as 'glass silk'.

GLAUCONITE : A zeolite which is natural green sand used in cation-exchange process for softening of hard water.

GLAZE : (i)   Glaze coat used in painting.

       (ii)   Use of glass for lighting.

       (iii)   Glass like glossy surface made on pottery, bricks and tiles by glazing.

GLAZE COAT : Almost transparent, very thin coloured coat can be provided on a surface by the use of a special paint.

GLAZED BRICK : Smooth and shiny bricks can be produced by fire glazing.

GLAZED TILE: Wall tiles made of earthenware may be glazed with decorations for interior use.

GLAZED WARES: Stoneware pipes and fittings are made salt glazed.

GLAZIERS' PUTTY : A plastic material used for bedding glass panes and making weather proof fillets on the outside to hold the glass to the frame. This is prepared by mixing linseed oil with whiting.

GLAZING : Glass fitting for light.

GLAZING BAR : A sash bar of rebated wood or metal to hold the glass panes in a window or door frame.

GLAZING BEAD : A glass stop made of timber bead to hold the glass pane instead of using putty.

GLAZING SIZE : The glass size for glazing. The size of glass should be 1/16 inch smaller on all sides so as to have clearance between glass and the window frame.

GLAZING SPRIG : Headless nail called 'brad' remains buried in the face putty round the glass pane.

GLOBE VALVE : A valve of bulbous appearance has a circular metal disc inside, which is screwed down to the valve seat for closing the flow through it. For opening the valve, the spindle is rotated and the disc rises up. This valve gives more resistance to flow compared to a gate valve.

GLOSS : The reflection of light by a painted, polished, varnished or enamelled surface.

GLUE : A sticky liquid or semi-liquid used in joinery works. Various glues used are animal glue, casein glue, cassava glue, soya glue and synthetic resins.

GLUE BLOCK : An angle block.

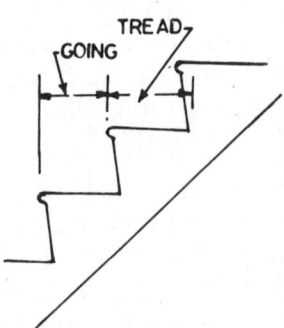

Globe Valve

GLUE KETTLE : A cast iron pot with an outside water jacket, used for heating animal glue.

GLUE LINE : A thin film of glue between two parts to be joined together.

GLYCEROL : An alcohol mixed with water in required proportion for use in preparing natural and synthetic resins for paints and varnishes.

GO-DEVIL : For cleaning a concrete pump (used for spreading concrete slurry), a ball made of paper, sack or any other material is introcduced into the pump end and is driven out through the pipeline by compressed air, which forces out the concrete remainings adhered to the pipeline.

GOING : The horizontal distance between two successive nosings of treads. The going of a flight is the sum of the goings of the treads.

GOING ROD : A rod used for setting out the goings of a flight of stair.

GOLD BRONZE : A powder of copper alloy used for bronzing.

GOLD SIZE : An oleo-resinous varnish (that dries quickly but hardens slowly) is used in fixing gold leaf to a surface. This is also used for making 'filler' in paints.

GOLD STOVING VARNISH : A transparent varnish used to form a yellow film on a white or silvery plate.

GOLIATH CRANE : A crane crab which travels along the beam at the top of a four-legged heavy portal frame having wheels to run on rails. This crane is used for erection of heavy steel structures or for lifting heavy loads in harbours or for construction of nuclear power station and similar such works.

GO-OUT : A sluice provided in a tidal embankment which impounds water during high tide and passes out the water during low tide.

GORE : A lune to cover a dome.

GORGE : A throat.

GOUGE : (i) A chisel having a curved cutting edge required for hollowing wood. Gouges of different curvatures are used for wood carving.

   (ii) A mason's tool for carving stone.

GOUGE BIT : A drilling bit with curved end.

GOUGE SLIP : Oilstone slip for sharpening gouges.

GOW CAISSON : Also known as 'Caisson pile' or 'Boston caisson'. A technique for sinking small hollow shafts through soft clay, taking out the excavated soil within it and introducing another shaft of smaller diameter within the first one and so on. Afterwards, the bore-hole is filled with concrete and the cylinders are removed.

GRAB : A split and hinged grab bucket fitted with curved jaws or teeth hung from a crane. The bucket is dropped down to excavate the soil and is raised up with the spoil inside the bucket. Different grabs are used for different types of soil to be excavated.

GRAB SAMPLING : Random sampling to have an idea about the characteristics of a material.

GRAD : (i) Gradient (ii) formation (iii) degree.

GRADED AGGREGATE : Classified aggregate having different particle sizes.

GRADED FILTER : A filter, in which the filtering medium is so arranged that layers of coarse gravel, medium gravel, fine gravel, coarse sand and medium sand are placed one above the other and when water flows through one layer to the other, the material of one layer is not carried to the other to cause clogging.

GRADED SAND : A sand sample containing coarse, medium and fine sand.

GRADE LEVEL : Formation level.

GRADIENT : The rise or fall per unit length of horizontal distance i.e., the slope. It is expressed either in degrees or in per cent.

GRADIENTER : A micrometer fitted to the vertical circle of a theodolite or levelling instrument so that the instrument can function as a tacheometer without stadia hairs and also as a grading instrument required in surveying.

GRADIENT POST : In railway, a short post is set beside a track at each change of gradient with an indication of length and degree of gradient.

GRADING : (i) Classification of different grain sizes by percentage in a sample.

   (ii) Shaping a ground surface by earth moving plant, called grader.

GRADING CURVE : A curve plotted with grain size of a sample on a horizontal logarithmic scale and percentages on a vertical arithmetic scale. Any point on this curve speaks of what percentage by weight of particles in the sample is smaller in size than the given point.

GRADING INSTRUMENT : A levelling instrument with a telescope that can be raised or lowered to set out a correct gradient.

GRADIOGRAPH : A levelling rule used for verifying the drain slopes.

GRADIOMETER : Grading Instrument.

GRADUATED COURSER : Courses that are diminishing.

GRADUATION : Scaled markings required for easy measurement.

GRAFFITO : Also known as 'Scratch Work'. Patterns made on a green plastered surface and on drying, multi-colour effects may be produced by painting the ornamental surface.

GRAFTING TOOL : A narrow spade for digging out hard clay manually.

GRAIN : Wood fibre.

GRAINER : A painter who brings imitation wood grain with knots or marble veins by the technique of his painting.

GRAINING : Painting a surface to look like wood grain or marble.

GRAIN-SIZE CLASSIFICATION : Classification of soil by its grain size.

GRANITE : An igneous rock having large crystals, the principal constituent being silica. It is hard, heavy and durable. This stone is used in bridge piers, marine works and ornamental building columns.

GRANITIC FINISH : A concrete mix can be so prepared that it will resemble granite.

GRANOLITHIC : A screed of stone chippings, cement and sand over concrete floors to produce a smooth hard wearing surface (1 inch thick). This is Indian patent stone flooring. Coloured stone chips may be used for beauty.

GRANULAR STABILISATION : Soil stabilization.

GRANULATOR : A stone breaker to break large stones into small aggregates.

GRATE INLET : A type of storm water inlet provided on the roadside gutter. This inlet is not suitable for silt-laden flow.

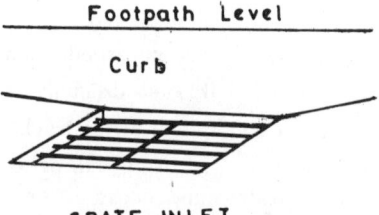

Footpath Level

Curb

GRATE INLET

GRATICULE : Reticule of a telescope used in surveying.

GRAPPLER : Wedge shaped, eyed spike driven hard into brick masonry for the top-end of a bracket scaffold.

GRASS TABLE : Earth table.

GRAVATT'S TEST : A three-peg test to indicate whether the cross-hairs of the telescope of a theodolite are on the optical axis or not.

GRAVEL : Smooth, rounded, naturally obtained building aggregate. Due to its smooth rounded surface, it is not selected for making a good concrete.

GRAVEL BOARD : A horizontal board fitted to the underside of the vertical boards of a fencing so that the vertical boards are kept above the ground.

GRAVEL PUMP : A specially made centrifugal pump for raising hydraulically loosened small gravels.

GRAVEL ROOF : A roof made water-proof by covering it with roofing felt, over which small size gravels are bonded with bitumen to protect the surface from atmospheric actions.

GRAVING DOCK : It is essentially a dry dock where bottoms of ships are scraped and smeared with graves, the dregs of tallow.

GRAVITATIONAL WATER : (i) The water above the standing water level.

(ii) Excess soil water that drains through soil under the influence of gravity.

GRAVITY ARCH DAM : A dam which resists the thrust of water retained by the arch action as well as its self weight.

GRAVITY CORRECTION : A tape correction that is needed for precision measurement at places of different gravity. For this, a tension to the tape is applied by weight on a cord passing over a pulley

GRAVITY DAM : A dam of massive construction which prevents its overturning by its self-weight alone. This type of dam is to be built quite heavy and high enough.

GRAVITY FILTER : A rapid sand filter through which settled water is filtered at the rate of 2 to 5 gpm/sft of filter bed. See 'Rapid Sand Filter.'

GRAVITY MAIN : A pipeline which carries liquid due to the action of gravity, which is imparted by providing adequate slope. The flow occurs only due to the difference in levels.

GRAVITY RETAINING WALL : A retaining wall similar to a gravity dam, which prevents its failure due to overturning by the action of its self weight. This is also a massive construction with brick or stone masonry or mass concrete. For stability of this wall, the resultant force of the thrust of the retained material and the self weight must pass through the middle-third at its base and the wall should be designed to take no tension.

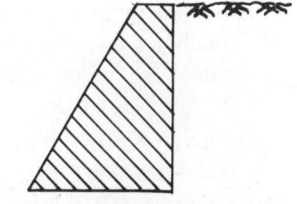

GRAVITY RETAINING WALL

GRAVITY SCHEME : A water supply or sewage treatment scheme in which no pumping is required at all and all flows are by gravity.

GRAVITY WATER : (i) Water flow by gravity

(ii) Gravitational water

GREASE TRAP : A trap provided in a house drainage or industrial effluent line or sewerage system to arrest oil and grease with a view to preventing chokage in the pipeline system. The device is very simple and oil and grease can easily be removed, since these float on the surface of the waste water.

GREASE TRAP

GREASINESS : A greasy surface developed on a paint film due to the lack of compatibility.

GREAT HEAD SHIELD : A device to protect the workmen engaged in tunnelling work in soft ground.

GREASE TRAP

GREEN BRICK: A brick in the process of drying, prior to burning.

GREEN CONCRETE: Concrete after initial set and prior to hardening.

GREEN HEART : A hardwood in British Guiana, which is very strong, hard and durable. It has a high modulus of elasticity of about 3×10^6 psi at 10% moisture content. It is used for timber works in docks.

GREYSTONE LIME : Hydraulic lime obtained by burning chalk having adequate quantum of alumina and silica.

GRID : (i)   Square or rectangular layout of straight lines which can be used in plotting points on a plan.

(ii)   An open frame of wooden or steel beams or old rails built on the foreshore which facilitates in receiving floating vessels during high tide, mooring and repairing the vessels during low tide hours, when the water leaves the shore.

GRID BEARING : The angle between the north-south line in the grid and the desired direction.

GRID IRON SYSTEM : A system layout used in water distribution pipelines as shown.

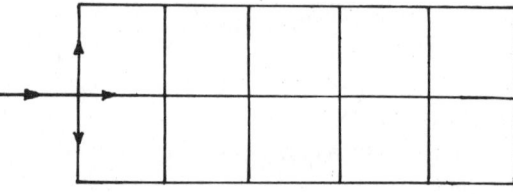

GRID IRON SYSTEM

GRID NORTH : The vertical lines of the grid in a map are directed to grid North and usually at the centre of the grid system, the meridian at that place coincides with the vertical line so that along this line grid North coincides with true North.

GRID PLAN : A plan in which major walls and building components coincide with the grid lines. Grid plan is used in designing prefabricated buildings.

GRILLAGE FOUNDATION : A footing to support a heavy column/ stanchion consisting of a frame of steel joists. Usually placed in two tiers as shown. Grillage foundation with hard timbers and old rails was in practice in early days.

GRILLE : (i)   A grating through which air passes for ventilation.

(ii)   A decorative grating used in window openings for protection and circulation of air.

GRINDER : An abrasive wheel rotated manually or mechanically or by electrical power for sharpening tools or for reducing the size of a metal piece.

GRINDING SLIP : An oilstone slip used for sharpening tools in carpentry.

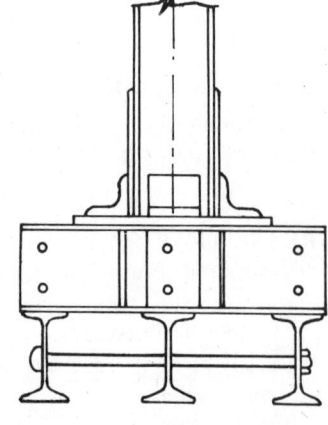

GRILLAGE FOUNDATION

GRINDSTONE : An abrasive wheel of sandstone. See 'Grinder'.

GRIP :   (i)   A small 'catch water drain' cut on the uphill side of an excavation to drive out the rainwater from it.

(ii)   A bond between two surfaces.

(iii)   A shallow drain dug on the road verge to allow the rainwater to flow from the road to the road-side drain or nullah.

GRIP LENGTH : Bond length. The length of straight reinforcement bar expressed in diameter of the bar required to anchor it in concrete.

GRIT CHAMBER : A detritus tank with a detention time of about one minute used in a sewage treatment plant for separation of sand, grits, pebbles, etc., from the sewage.

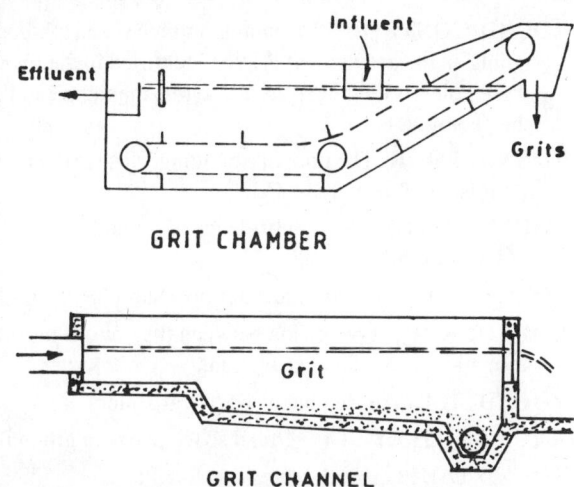

GRIT CHAMBER

GRIT CHANNEL : A long narrow channel used in a sewage treatment plant for removal of grits, sand and other inert matters. The removal takes place by sedimentation of particles under the controlled velocity of flow of sewage in the channel.

GRIT CHANNEL

GRITTER : A self-propelled machine used for spreading stone chippings over a road surface required in surface dressing.

GRITTING : Blinding on a road surface.

GROG : Broken pottery used in making refractory bricks.

GROIN : (i)   Groyne. This may be of 'fending', 'repelling' or 'attracting' type as shown.

(ii) The curved line at which the soffits of two vaults intersect.

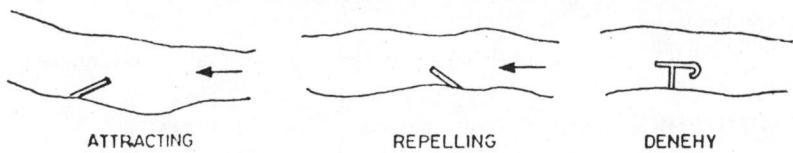

ATTRACTING   REPELLING   DENEHY

GROMMET : A hemp washer impregnated in jointing compound and used in plumbing work to make a tight joint.

GROSS ERROR : A figure showing an absurd value which is far from the actual measurement in surveying. The error can be detected by taking three other measurements & may be rectified by discarding the absurd value and taking the average of others.

GROSS FEATURES : Prior to subjecting a timber to stress the gross features of the timber to be checked are knots, shakes, splits, slope of grain, etc.

GROUND: (i)   The surface to be painted

(ii)   The first coat of paint.

GROUND BEAM : A reinforced concrete beam placed near ground level over piles which acts as a foundation of walls. This beam may also act as a strip foundation.

GROUND BRUSH : Round or oval-shaped wire-bound paint brush required for painting a large area.

GROUND CASING : A blind casing used in joinery .

GROUND COAT : The first opaque coat of paint used as a base for glaze coat.

GROUND CONTROL : In photogrammetry i.e., photographic surveying, marking of points on the ground that is to be identified in the photograph and surveying to locate these points with their altitudes so that the rader can fix the position of aircraft to take the photograph.

GROUND FLOOR : The floor of a building, close to the ground level. Normally, the ground floor is 2 to 3 feet above G.L.

GROUND MOVEMENT : Subsidence of ground surface in a mine area due to withdrawal of coal or other minerals.

GROUND PLAN : The ground floor plan showing the building layout and foundation.

GROUND PROP : A puncheon between the bottom-most frame and the foot block on the formation level of an excavation.

GROUND SILL : A sole plate used by carpenters.

GROUND STOREY : The vertical space between ground floor and first floor.

GROUND TABLE : Earth table.

GROUND WATER : Water contained under ground or under rocks, below the standing water level.

GROUND WATER LOWERIN : Ground water may be lowered by making a number of well points outside an excavation and pumping out water so that the excavated trench may be kept dry for construction purpose. It has been observed that water level can not be lowered in soils having the effective size below 0.05 mm.

GROUND WATER RECESSION : The variation of base flow with time during periods of no rainfall over a basin. It is a measure of drainage rate of ground water storage from a basin.

GROUND WORK : The work consisting of fixing battens over rafters to form a base for tiling.

GROUNDED WORK : Joinery fixed to a ground.

GROUT : Cement-sand slurry made of equal volumes which is injected into the joints of brick work, stone masonry or fissures in rocks.

GROUT BOX : An expanded metal box of conical shape, cast into concrete with an anchor plate at the bottom for securing an anchor bolt.

GROUT CURTAIN : A series of vertical holes are drilled downwards under the cut-off walls beneath a dam on both upstream and downstream sides and are filled with grout under pressure to fill fissures in the rocks and reducing the possibility of breaking up.

GROUTED MACADAM : A macadam road built with coarse aggregates, the voids being filled with cement grout or bituminous grout.

GROWTH RING : Annual rings found in a timber, each being formed in one year of growth.

GROYNE : See 'groin'.

GRUB AXE : An adze like tool for pulling up the roots of cut trees.

GRUB SAW : A special saw for cutting stones manually.

GRUB SCREW : A short screw used to hold the fittings tightly in joinery works without any projection of its slotted head.

GUARD BEAD : A bead provided round the inner edge of a sash window to guard it against swinging into the room.

GUARD BOARD : A scaffold board placed at the outer edge of a gantry to avoid objects falling below, which may cause injury to people.

GUARD LOCK : A lock guarding a dock from tidal water.

GUARD POST : A bollard.

GUARD RAIL : A check rail used in rail track.

GUDGEON : (i)   A gate hook.

          (ii)   A pin

          (iii) A metal dowel for locking adjacent stones

GUIDE BANK : Also known as 'Bell Bundh' named after J.R.Bell. It is constructed at a bridge site to protect the bridge ends by guiding the water flow in a stream.

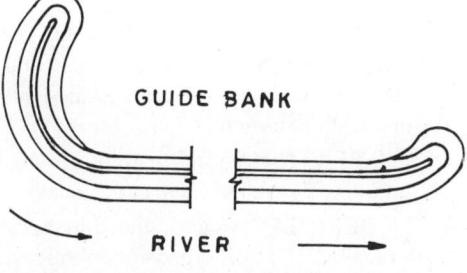

GUIDE BANK

RIVER

GUIDE BEAD : A guard bead.

GUIDE COAT : Prior to painting, a thin coat of loosely-bound paint is applied over a surface and is removed by rubbing when it is dried. This gives an indication of the undulation (high and low places) of the surface, if any.

GUIDE PILE : A vertical square timber is driven close to the sheet piles in a trench to carry the walings (horizontal members) and the sheet piles. Thus, it takes the entire earth pressure of the trench wall.

GUIDE RAIL : A check rail.

GUIDE RUNNER : A runner driven ahead of other runners to guide them.

GULLET : (i)   A narrow trench dug to the formation level in earth cutting to lay a track for movement of wagons carrying the spoil.

         (ii)   A channel in a rapid sand filter for entry of settled water and exit of wash water.

GULLEY : A pit with grating on the roadside gutter for draining of the water from road surface. The pit is provided with a silt trap from which silt is removed periodically.

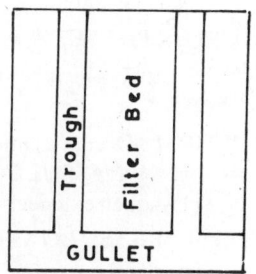

Trough

Filter Bed

GULLET

GULLEY SUCKER : A truck carrying a large tank fitted with a vaccum pump and a hose. The hose is introduced in the pit to suck silt and slush mucks from it.

GULLEY TRAP : A trap with a seal of water provided in a gulley to prevent foul gases coming out from the drain. Silt gets deposited at the bottom which is removed periodically.

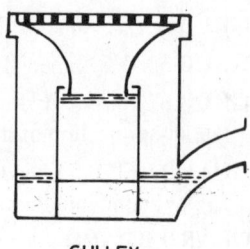

GUM ARABIC : A fine white powder obtained from 'Acacia' tree, used for joining wood and paper. This glue is also used in making transparent paints.

GUMMY : A drag on a brush from a sticky paint during painting. This occurs due to evaporation of the solvent in the paint. This can be made good by adding thinner to the paint.

GULLEY

GUM VEIN : Accumulation of resin as a streak found in some hard woods.

GUN COTTON : Nitro-cellulose used as explosive.

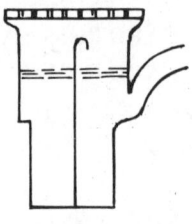

GUNITE : A cement-sand (1 : 1) paste thrown to formwork for making a dense concrete wall. It is extensively used in repairing wearing surfaces of old concrete structures, walls of mines and water tunnels.

GUN METAL : An alloy made of copper, tin, lead and zinc.

GULLY TRAP

GUN STOCK STILE : A diminishing stile used in joinery.

GUNTER'S CHAIN : A surveyor's chain of 66 ft. length having 100 links used earlier for land survey. One chain length is one-tenth of a furlong and one square chain measures one-tenth of an acre.

GUSSET PLATE : A steel plate of required shape (square, rectangular, triangular or trapezoidal) used to connect the members of a steel frame, truss or girder.

GUTTA PERCHA : An unsaturated hydrocarbon of colloidal nature, obtained from the exudation of a tree. It becomes plastic when heated and tough and hard when cold. It has good insulation property.

GUTTER : A channel along the eave of a sloping roof or along the edge of a road to carry rain water. In a sloping roof, this may be a parapet gutter, eaves gutter or valley gutter.

GUTTER BEARER : Short pieces of timber to carry the gutter.

GUTTER BED : A flexible metal sheet laid over the tilting fillet at the back of an eaves gutter.

GUTTER BOARD : A gutter bearer.

GUTTER BRICK : A specially made brick used in forming the roadside gutter.

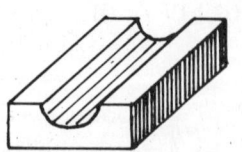

GUTTER PLATE : A wall plate below a gutter.

GUY : A rope which holds a mast, chimney derrick, shear legs, etc.

GUTTER BRICK

GYPSUM : Occurs as minerals viz. alabaster, selenite and satin spar. It is $CaSO_4. 2H_2O$. It is the raw material for gypsum plaster, an extender in distemper and water paints.

GYPSUM BASE BOARD : Square plaster board used for plastering with gypsum.

GYPSUM BLOCKS : Hollow blocks made of gypsum plaster used in building construction.

GYPSUM INSULATION : Loose flocculent gypsum of thermal conductivity of 0.45.

GYPSUM LATH : Gypsum plasterboard in form of sheets are nailed to wall or ceiling.

GYPSUM PLASTER : Plaster made from gypsum by driving out the water content i.e., anhydrous gypsum plaster ($CaSO_4$), plaster of paris (hemihydrate). It expands on setting and therefore it does not form any crack on the surface, provided the plaster backing is not defective.

GYPSUM PLASTER BOARD : A building board made of two heavy sheets of paper boards with a core of gypsum or anhydrite plaster in between. This is used as an insulating plasterboard.

GYPSUM WALL BOARD OR CEILING BOARD : Self-finished decorative gypsum plasterboards for interior decoration.

H

HACKING : (i)   Rubble walling with alternate courses of stones of different depths.

            (ii)   Placement of green bricks in such a manner that the bottom edge of each gets dried by circulation of air.

HACKING KNIFE : A knife used for removing old putty from a glass pane prior to reglazing .

HACK SAW : A hand saw having its blade fitted into a steel frame, used for cutting metals.

HAIR CRACKS : Fine hair-like random cracks that appear on surface of a material and do not penetrate the surface.

HAIRED MORTAR : A mortar prepaed by mixing hairs with the coarse stuff to reduce cracking due to shrinkage . The hairs act as reinforcment. Sometimes, fibres of jute, asbestos and manila are used in place of hairs.

HAIR HOOK : A hoe-like tool with two bent tines used for mixing hairs into coarse stuff.

HAIR PIN BEND : A bend like ladies hair pin is sometimes provided in hill roads due to the site condition. see illustration.

HALF BAT : A snap header which is made by cutting a half brick into two across its length i.e. a quarter of a brick.

HALF BRICK : A brick is halved by cutting across its length.

HALF BRICK WALL : A wall having a thickness of half brick.

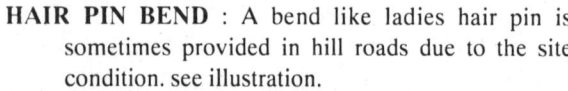

HALF HATCHET : A hatchet with a notch for drawing nails, used by carpenters.

HALF JOIST : A joist cut into two pieces along the web to form a Tee section .

HALF LANDING : An intermediate platform between two flights of a stair. The platform is built at half the height between two floors.

HALF-LAP JONT : A lap joint formed by halving both the pieces to be joined.

HALF-LATTICE GIRDER : A warren girder.

HALF-SOCKET PIPE : Drain pipes socketed only in the lower half, used for sub-soil drainage.

HALF-TIDE COFFERDAM : A cofferdam built in the sea, with a height lower than the high tide level i.e., it allows entry of water during high tide, which is bailed out afterwards.

HALF PRINCIPAL : A principal rafter not reaching the ridge in a wooden truss

HALF-RIP SAW : A close-toothed rip saw used in woodwork.

HALF-ROUND VENEER : Veneer cut from a flitch in a lathe machine, which is almost semi-circular.

HALF-SPACE LANDING : Half-landing in a stair. The length of landing is the sum of the widths of the two flights plus horizontal space if any, between the flights.

HALF-SPAN ROOF : A pent roof or lean-to roof.

HALVED JOINT : A wooden joint made by halving both the pieces to be joined.

HALT STATION : This is a type of wayside railway station having a passengers platform only with no other facilities like building, staff, crossing arrangement, etc. See illustration.

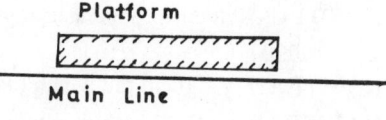

HALT STATION

HAMMER BEAM : A short horizontal member provided at wall plate level in a hammer beam roof truss.

HAMMER BEAM ROOF TRUSS :
See illustration. This type of truss was used to be constructed in early days. This gives greater headroom in the central part of a room. Timber and wrought iron were used to form such a truss.

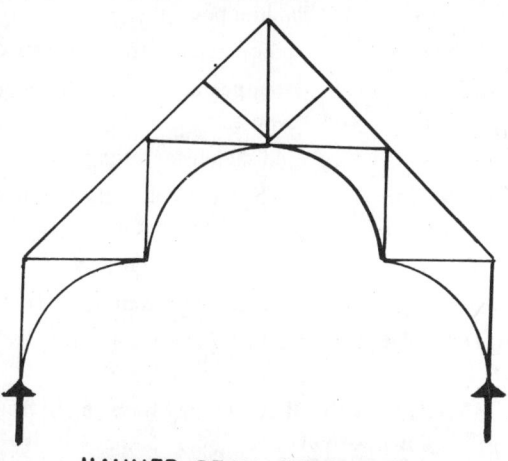

HAMMER BEAM ROOF TRUSS

HAMMER DRESSED STONE : Roughly dressed stone face after quarrying and seasoning a stone. This produces a good architectural effect.

HAMMER FINISH : A finish like a hammered metal obtained in a painted surface by spraying coloured enamel paint containing metal powder.

HAMMER-HEADED CHISEL : A mason's chisel having a conical head which is struck by a hammer .

HAMMER-HEADED KEY: (i) A metal cramp used to lock stones together, in laying stone blocks.

(ii) A double dovetail key used in joinery works.

HAMMER DRESSED

HAMMER POST : A post resting on the hammer beam in a hammer beam roof truss.

HAND BORING : Drilling holes in earth for sampling of soil with the help of shell and auger.

HAND BRACE : A brace used by carpenters.

HAND DRILL : A small, portable hand-operated boring tool. In present days, power-driven hand drills are used .

HAND FINISHER : A screed rail or similar tool used by concrete workers for shaping and levelling a compacted concrete.

HAND FLOAT : A wooden tool used by masons to finish the green plastered surface.

HAND LEAD : A sounding lead which is attached to a lead line used in hydrography survey.

HAND LEVEL: A hand-held, small, compact levelling instrument with a spirit level, used in surveying, e.g. abney level. This can be employed in carrying out contour survey upto a distance of 100m. from a point of known level.

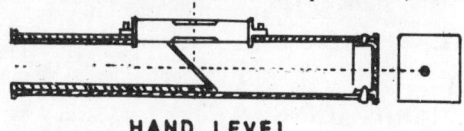

HAND LEVEL

HAND PUMP: A hand-operated pump for withdrawal of ground water for rural water supply.

HAND RAIL : A guard rail which forms the top of a balustrade in a stair or balcony.

HAND RAIL BOLT : A bolt having threads at both ends, used in fixing handrails. Also, known as 'joint bolt.'

HAND RAIL PUNCH : A tool introduced into the mortise under a handrail for tightening a nut on the handrail bolt.

HAND SAW : A hand-operated carpenter's saw commonly used in woodwork.

HAND SCREW : A carpenter's cramp with wooden jaws and a threaded spindle . Also known as 'screw clamp'.

HANGAR : A covered shelter for aircrafts.

HANGER : (i)   The bars used at the top of a R.C.C. beam from whcih the reinforcement bars are hung or held in position with the help of stirrups.

(ii)   The stirrup strap

(iii)   A member, from which some materials or parts are hung.

HANGING GUTTER : An eaves gutter usually of sheet metal fastened to the ends of the rafters.

HANGING LEADER : A steel frame hung from the top of a crane to guide a pile to its downward path .

HANGING POST : A vertical member from which a gate or saloon door is hung .

HANGING SHINGLING : Fixing shingles to vertical . Also known as 'weather shingling'.

HANGING STILE :The door or window stile to which hinges are fixed.

HARBOUR : A water area close to a land for giving shelter to ships, loading and unloading materials and men.

HARBOUR MODEL : Prior to construction of a harbour, it is the usual pactice to make a model of the harbour simulating field conditions for studying wave actions and solving problems of siltation and scour.

HARBOUR OF REFUGE : A harbour only to shelter ships during storms and not for loading and unloading.

HARD BOARD : A fibreboard which is manufactured under pressure and is used in making partitions, decorative works and furnitures.

HARD BURNT : Bricks and tiles that are burnt at high temperatures for greater strength and durability with low absorptive power.

HARD CORE : Hard materials like lumps of stones, bricks and old concrete introduced into a soft ground in foundation or used for filling a soft soil for making a road.

HARD DRY : A state, when a painted surface gets dried throughout the depth of the paint film, so that the next coat can be applied.

HARDENER : An accelerator used in synthetic resins, so that the paint when applied on a surface, gets hard quickly .

HARDENING : (i)   Setting and acquiring strength by concrete against compression .

(ii)   Steel hardened by quenching and tempering .

HARD FACING : Welding onto steel a hard surface of tungsten carbide with a view to forming an abrasion-resistant cutting edge to a drilling or cutting tool.

HARD GLOSS PAINT : An oil paint having hard glossy finish like enamel.

HARDNESS OF WATER : It is caused by the carbonates, bicarbonates, chlorides and sulphates of calcium, magnesium, iron, manganese and aluminium in water. carbonate hardness is called temporary hardness and non-carbonate hardness is the permanent hardness. Total hardness of a water = carbonate hardness + non-carbonate hardness. Carbonate hardness is known as alkalinity of water.

HARD PLASTER : A hard finish in plastering, the materrals being cement and sand mix, keene's cement,gypsum plaster. etc.

HARD PUTTY :Hard stopping used in filling cracks and holes in a timber article prior to painting .

HARD WATER : A water that makes food tasteless , forms scales in boiler tubes, causes corosion and incrustation in pipes, affects the dyeing process and causes excessive consumption of soap for cleaning purpose. Water having hardness above 100 mg/l is called hard water.

HARD WOOD : Timber obtained from deciduous (broad-leaved) trees which belong to angiosperm group.

HARDY-CROSS METHOD : A controlled trial and error method of analysis of water distribution networks for hydraulic balance. Data initially assumed are corrected by successive corrections. This is either balancing heads by corecting assumed flows or balancing flows by correcting assumed heads.

HASP AND STAPLE : A fixture for wooden boxes, gates and doors, in which the hasp is locked over a staple.

HATCHET : An axe used by wood workers for dressing wood.

HATCHET IRON : A hatchet-like soldering iron used by plumbers.

HATCHING : Drawing parallel lines to in dicate different materials, landuses or the parts sectioned.

HATCH LINES : Section lines or parallel lines for shading.

HAULAGE ROPE : The traction rope or cable in a rope way.

HAULING PLANT : Bull dozers, wheeled tractors attached to bowl scrapers crawler tractors, draglines, tipping lorries, tractor- trailers and dumpers are hauling plants. These are employed for transportation or haulage of materials from one place to the other.

HAUNCH : (i)　　The full width given to a tenon at the point wherefrom it projects.

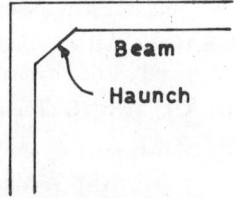

Beam

Haunch

(ii)　　The increased depth of a beam at its ends near the supports. This is also called 'Hauncheon'.

(iii)　　The part of an arch near its springing point .

(iv)　　The flank of a road.

HAUNCHED TENON : A tenon having wider part at the point of its projection and narrower part at the tip.

HAUNCHEON : See 'haunch'.

HAUNCHING: See 'haunch'.

HAY BAND: Ropes made of hays and straws left into the cavity of a cavity wall for collection of mortar droppings during construction.

HAYDITE :Expanded clay used as an aggregate in light-weight concrete.

HAYFORD'S SPHEROID : A spheroid to give the dimensions of the earth in which the equatorial radius is 6378.388 km, and the polar radius is 6356.912 km the semi-major and semi-minor axes respectively.

HAY'S PROCESS : A practical application of contact aeration, which comprises primary sedimentation,first stage contact aeration, intermediate sedimentation, second stage contact aeration and final sedimentation.

HAZEN'S LAW : It states that the permeability of a soil can be approximated based on the paricle size of the soil and it is nearly equal to D_{10} size x 100 cm/sec.

HAZEN WILLIAM'S FORMULA : The most widely used pipe-flow formula whcih is written as

$$V = 1.318\, C\, R^{0.63}\, S^{0.54}$$

where $V =$ Velocity of flow through pipe

     $C =$ Friction co-efficient

     $R =$ Hydraulic Radius and

     $S =$ Hydraulic gradient i.e. slope.

This formula is also written in the form

$$H = KQ^n$$

where $H =$ Headloss in pipe flow,

     $K =$ Pipe constant for a particular pipe,

     $Q =$ Quantity of flow through the pipe and

     $n =$ Exponent, an index.

HEAD : (i)   The pressure of a fluid is expressed as 'head'. It is the potential energy per unit weight of the fluid above a level.

     (ii)   The larger end of a hammer.

     (iii)   The head of a bolt, rivet, screw and nail.

     (iv)   The widened part of a column at its top.

     (v)   The upper horizontal member of a door or window frame or a partition.

HEAD BLOCK : A wooden block bolted to the end of a tie beam in a timber truss to bear the thrust of the rafter.

HEAD BOARD : A horizontal piece of timber (board) provided at the roof of a heading and carried by head trees at either end.

HEAD CASING :The portion of an architrave outside a door and also over the door.

HEADER : (i)   A brick which is laid across the length of a wall for bonding .

     (ii)   A conduit which distributes fluid to other pipes or a conduit which receives fluid from a number of pipes.

     (iii)   The part of a boiler to which the boiler tubes are connected.

HEADER JOIST : A trimmer joist used in construction of a floor or a frame.

HEAD FLASHING : A flashing in form of a gutter round the projected edge through a roof.

HEAD GATE : The lock gate at the upstream of a conduit.

HEAD GUARD : A metal gutter that crosses the cavity of a cavity wall to protect the head of a wooden frame from water.

HEADING : A pilot tunnel required in formation of a large tunnel or for any other construction work.

HEADING BOND : A brick bond in which only headers are used. This is required in construction of wall footings and curved walls.

HEADING COURSE : A course or layer of headers used in a brick bond .

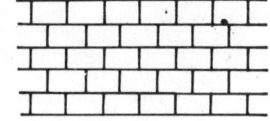

HEADING JOINT : A cross joint.

HEADING BOND

HEAD JAMB : The head of a door or window.

HEAD JOINT : Same as heading joint.

HEAD MOULDING : A moulding at the top of an opening.

HEAD NAILING : The nailing of slates near their heads in a sloping roof. The nails get covered by the overlapping part of the slates.

HEAD RACE : The channel that brings water from the forebay to a turbine.

HEAD ROOM : The unobstructed height from floor to ceiling or the headway in a stair.

HEAD TREE : A horizontal member on either side of a heading to support the bead board.

HEAD WALL : A retaining wall provided at the end of a culvert.

HEAD WATER : The water at the upstream of a river.

HEAD WEATHER MOULDING : A weathering board or a moulding framed into the head of a window or door frame to throw out rain water.

HEAD WORKS :   (i)   The structure built at the head of a channel for diversion of water into it.

            (ii)   The structure for intake of water from a river for the purpose of water supply or irrigation.

HEART BOND : A bond for making thick walls, in which the joint between two headers in the wall core is covered by another header.

HEART CENTRE : The pith in the core of a tree.

HEARTING : The infilling of a thick wall which is faced with good bricks or stones.

HEART PLANK : A plank made from the heartwood.

HEART SHAKE : A radial shake originating from the pith of a tree as shown. This is a defect in timber.

HEART WOOD : The tinber obtained from the core of a tree. Also known as 'Duramen'

HEAT INSULATION : A material used for resisting transmission of heat.

HEAT-RESISTANT PAINT: (i) Enamel paint that can be stoved   HEART SHAKE
at high temperatures.

          (ii) A paint containing silicon resin, used in articles which are subjected to heat.

HEAVY-BODIED PAINT: A viscous paint that forms a strong film.

HEAVY SOIL : A soil mostly containing clay.

HEDLEY DIAL : It consists of a large compass box having a long magnetic needle balanced upon a pivot in the centre. There is a circular attachment fixed on the side of the dial to indicate the vertical angles which are shown by a pointer travelling over a graduated horizontal arc inside the box.

HEEL : (i)   The rear end of the base of a retaining wall i.e. at the back of the wall retaining material .

(ii)   The lower end at the hinged side of a hanging door stile.

(iii)   The part of a beam resting on a support.

(iv)   The back end of a plane.

(v)   The hinge of 'points' used in a railway track at points & crossings.

HEEL POST : The post of a lock gate at its corner. Also known as 'Quoin Post'.

HEEL STRAP : A steel strap of U-shape used to join the tie-beam with the principal rafter of a timber truss at its supports. The strap passing over the principal rafter is bolted to the tie beam. Thus, the thrust of the rafter is transmitted to the tie beam.

HEIGHT BOARD : A storey rod used in carpentry.

HEIGHT MONEY : Extra money paid to an worker for working at a height more than 12 m. from ground level.

HEIGHT OF INSTRUMENT METHOD : Line of collimation method used in level survey.

HELD WATER : Capillary water that is held in soil above the standing water level by surface tension.

HELICAL HINGE : A hinge used for swing door.

HELICAL REINFORCEMENT : Steel rods bent spirally for reinforcement in R.C.C. columns.

HELICAL STAIR : Spiral stair.

HELIOGRAPH : A surveying instrument having arrangement for flashing sunrays by reflection with a view to bringing visibility of a distant survey station.

HELIOSTAT : A device to reflect sunrays in the required direction. Thus, the object to be bisected when viewed through a telescope appears like a bright star.

HELIOTROPE : Similar device as that of a 'Heliostat' used in geodetic survey.

HELIUM : A rare gas having radio-activity, often found in uranium mines. It is a natural gas . This can also be manufactured commercially.

HELIUM DIVING BELL : A diving bell used in caisson sinking, in which the workmen breathe oxygen-helium mixture instead of air having oxygen-nitrogen mixture. This arrangement is made to prevent 'caisson desease'. Because , helium is less soluble in blood than nitrogen and it diffuses out of blood more rapidly.

HEMIHYDRATE PLASTER : $CaSO_4 . \frac{1}{2} H_2O$, which is obtained by driving out moisture from gypsum $(CaSO_4 . 2H_2O)$ by heating . Also known as 'Plaster of Paris', which is quick-setting plaster.

HEMP : Plant tissues and fibres used in making ropes.

HERRING BONE BOND : See 'Diagonal Bond'.

HERRING BONE DRAIN : A 'Chevron Drain'

HERRING BONE MATCHING : Book Matching in timber .

HERRING BONE STRUTTING : Stiffening of floor joists at their midspan by cross bridging.

HESSIAN : Also known as 'burlap'. Coarse material woven from jute or hemp for making sacks .

HEW :To shape a timber from a freshly cut wood by axe.

HEWN STONE : A well-dressed stone .

H-HINGE : Parliament hinge used for flushing the door or window shutters with the wall surface while opened.

HICKEY : A portable tool used for bending tubes and rods .

HICKORY : A strong timber available in North America.

HIDING POWER : The covering power or the power of opacity of a paint.

HIGH ALUMINA CEMENT : A cement having higher alumina content than the normal one for portland cement. It is resistant to attacks by sulphates and acids.

HIGH CALCIUM LIME : A fat lime or rich lime which has high calcium content. It is relatively a pure lime that can be mixed with cement.

HIGH CARBON STEEL : Carbon steel containing high percentage of carbon than that in mild steel.

HIGH DENSITY PLYWOOD : A plywood having density more than twice that of an ordinary plywood. Its mechanical strength is very high and it can not be bent or nailed. It is chiefly used in flooring and in aircraft.

HIGH EARLY-STRENGTH CEMENT : Rapid-hardening cement which attains its strength at an early stage compared to that of an ordinary cement .

HIGH EXPLOSIVE : An explosive containing chemicals that undergo detonation .

HIGH PRESSURE HEATING : An arrangement in a central heating system by circulating hot water under high pressure through small diameter pipes .

HIGH PRESSURE STEAM CURING : Autoclaving for special bricks and concrete blocks,which occurs in a chamber containing steam under high pressure.

HIGHWAY : A roadway of standard width, meant for all sorts of traffic.

HILEY'S FORMULA: A dynamic pile formula to determine the driving resistance of a pile.

It is given by

$$R = w\,h\,n/(s + 0.5c)$$

where R is ultimate driving resistance in tons.

      w is weight of drop hammer in tons.

      h is height of drop in inches.

      s is penetration of pile per drop of hammer in inches.

      n is efficiency of blow of hammer and

      C is the temporary elastic compression in inches.

This formula is suitable for use in sand, but not applicable in a clayey soil.

HINDERED SETTLING: When the discrete particles are closely spaced in a suspending medium, their velocity fields interfere during settling. Under this condition , there is an appreciable upward displacement of the suspending medium and the settling is hindered.

HINGE : A device to make a pinned connection beteween members of a structure such that one member can rotate or swing freely about the hinged point or pinned end. A hinged joint can take load, but can not take any moment. There are various types of hinges available in market.

HIP : The meeting edge of two roof surfaces.

HIP CAPPING : The strip of roofing felt covering a hip for its protection

HIP HOOK : A metal hook fixed to a hip rafter for holding the lowermost hip tile.

HIP KNOB : A finial to the ridge of a roof.

HIPPED END : The triangular end of a hipped roof.

HIPPED ROOF : A roof having four slopes, the eaves being at the same level as shown.

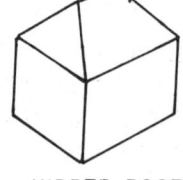

HIP RAFTER : The rafter that forms a hip and on which the jack rafters meet.

HIP ROLL : A rounded timber with a V-cut underneath that covers a hip.

HIPPED ROOF

HIP TILE : The specially made roofing tiles that cover the roofing tiles at the hip.

HOARDING : A high fencing round a building site to safeguard against theft.

HOD : A wooden or light alloy tray having the shape like one piece of a box (measuring 16" x 9" x 9")cut diagonally and fitted to a handle. This is sometimes used by labourers to lift building materials from one level to the other by holding the handle of the tray in one hand and going up a ladder.

HOFFMAN'S KILN : An overground continuous kiln used in burning bricks in a large scale. It is circular in plan and comprises an annular tunnel divided into several compartments. In such a kiln, loading, heating, burning, cooling and unloading can be done simultaneously.

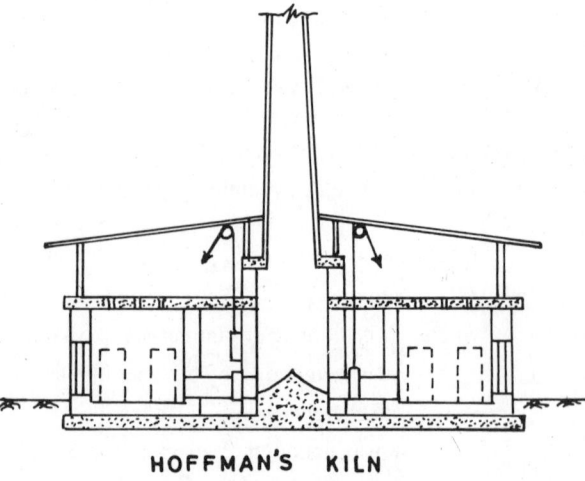

HOFFMAN'S KILN

HOG : A convex surface, called 'Camber'. An upward bending of a beam.

HOGGIN : Well-graded gravel bonded with clay for making a road surface.

HOGGING : Upward bending, opposite to sagging.

HOGGING MOMENT : A negative bending moment (anti-clockwise moment) that causes hogging.

HOGSBACK TILE : A ridge tile.

HOIST : A device to lift or lower a material.

HOLDER BAT : A metal clamp of semi-circular shape fixed to wall for holding a pipe.

HOLDFAST : (i)   A metal spike with an eye is driven into a brick joint such that a bolt can be inserted into the eye for fixing joinery.

(ii)   An anchor for guy ropes.

HOLDING DOWN BOLT : An anchor bolt to hold a structure firmly to masonry or concrete.

HOLDING DOWN CLIP : In flexible metal sheet roofing, the sheet end is folded to form a clip, shaped like a capping for anchorage to the roof boarding.

HOLE CALIPER : An instrument consisting of four arms and an electric resistor motivated by the arms, used for measuring well diameters along a well.

HOLE SAW : Tubular saw or a drill with annular bit to make a ring-shaped groove or to cut out a complete cylinder of metal or wood or any other hard material.

HOLLOW BACKED FLOORING : Flooring with wooden boards which are hollowed out on the underside for the purpose of ventilation.

HOLLOW BED : A bed joint in laying sills with its central portion kept hollow. This prevents their breakage in case of unequal settlement of masonry.

HOLLOW BLOCKS : These are hollow building blocks of burnt clay or concrete, used for making hollow block walls or hollow block floors. Hollow block constructions are light in weight and have thermal insulation. Pre-cast hollow blocks are used for rapid construction.

HOLLOW BRICK : See illustration. This is used in construction of hollow walls or floors.

HOLLOW CHAMFER : Concave chamfer.

HOLLOW CLAY TILE : Similar as hollow block.

HOLLOW BRICK

HOLLOW CYLINDRICAL FENDER: Commonly installed on the quay wall supported by brackets and bars or by frames suspended in chains. When the contact pressure

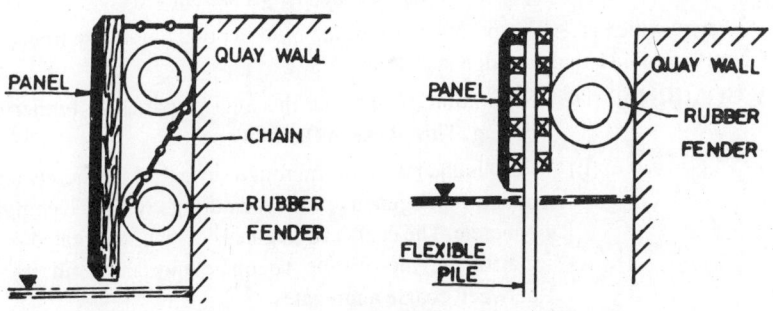

HOLLOW CYLINDRICAL FENDERS

between the rubber fender and the vessel is not wanted, panels hung from a quay wall or supported on piles are used.

HOLLOW CORE DOOR : A flush door made by glueing hard board or plywood onto a skeleton framework. Thus, the shutter becomes light in weight and it looks like a shutter of solid wood.

HOLLOW DAM: A masonry or reinforced concrete dam as shown in illustation is built with buttresses at regular interval supporting the slanting deck slab which takes the thrust of water.

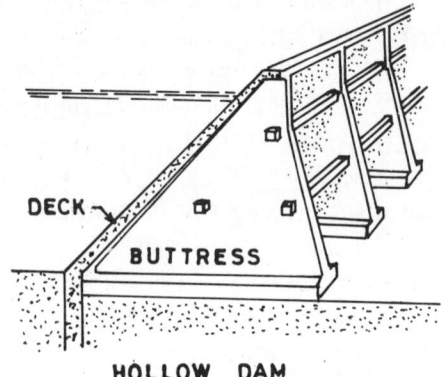

HOLLOW DAM

HOLLOW GLASS BLOCK : Same as building blocks made of glass .

HOLLOW PARTITION : Partition walls made of hollow blocks or hollow bricks .

HOLLOW QUOIN : A masonry having recess to carry the heel post of a lock gate.

HOLLOW ROLL : A method of jointing adjacent flexible metal sheets laid together in a sloping roof by drawing their edges and bending them round to form a cylindrical roll without any wood roll .

HOLLOW SECTION : Tubular section .

HOLLOW TILE : Burnt clay tiles similar to hollow blocks.

HOLLOW TILE FLOOR : A floor made of hollow tiles which is similar to hollow block flooring.

HOLLOW WALL : A cavity wall made of precast hollow blocks or ordinary bricks with a space in between two leaves of the wall.

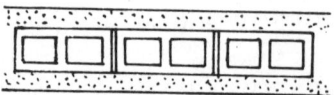

HOLLOW TILE FLOORING

HOLLOW WEB GIRDER : A box girder i.e. a built-up section.

HOLLOW WOOD CONSTRUCTION : Construction with wooden frame covered by plywood or hardboard which is similar to a flash door shutter.

HONE : oilstone which is a very smooth quartz stone used in polishing the sharpened edge of a cutting tool so that the edge lasts long.

HONEY COMBING : (i)   Separation of fibres in the interior part of a timber due to drying. This weakens a timber section.

(ii)   Porous and non-uniform rough surface of a concrete face due to non-homogeneity produced during laying and compacting concrete. The coarse aggregates become segregated from the matrix i.e. there is no adequate mortar to fill the voids between coarse aggregates.

HONEYCOMB SLATING : Diagonal slating with slates, the three corners of each slate being cut-off.

HONEYCOMB STRUCTURE : A structure which is made like a honeycomb to produce architectural effect. This may be made of R.C.C. or metal.

HONEYCOMB WALL : A wall of half-brick thickness made of stretchers only by keeping gaps between them i.e., the bricks are held by bed joints at their ends.

HONING GAUGE : A clamp to hold a chisel at a desired angle for rubbing the cutting edge on a stone.

HOOD : A canopy or shed built over a window opening to throw off rain water.

HOOK : A nail or wedge of tapered flat or round section, whose pointed end is driven into a wall so that articles can be hung from the projected part. There are various forms of hook.

HOOK AND EYE : A door or window fastening which consists of a cabin hook fixed on the frame and it fits into a screwed eye on the door.

HOOKE'S LAW : The stress is proportional to strain within the elastic limit of a material.

HOOK BOLT : A U-shaped bolt whose one end is provided with a head and the other end is threaded. It is chiefly used for fixing A.C. roof sheets to the purlins.

HOOK GAUGE : This is a simple instrument used for measuring water level with accuracy in laboratory tests. It consists of a pointed hook fitted to a vernier that slides along a graduated staff.The hook is lowered and measurement is taken only when the needle point just penetrates the water surface.

HOOK JOINT : A joint between the meeting edges of a door. The S- shaped rebate on one stile fits into the groove of similar shape and size on the other stile.

HOOK **BOLT**

HOOK REBATE : The rebate of 'S' shape cut on one door stile for making a hook joint.

HOOK STRIP : A wooden cleat fixed to a wall to which hooks are screwed for hanging clothes.

HOOPING : Reinforcement in a curve such as in prestressed R.C.C. pipes, R.C.C. circular tanks, etc.

HOOP IRON : Thin strips of steel used as reinforcement in a bed joint of masonry.

HOOP STRESS : The stress developed in the longitudinal section of a cylinder that resists the bursting force. It acts tangentially to the surface of the cylinder.

HOOP TENSION : The tension developed due to the hoop stress produced.

HOPPER : (i)   A triangular deadlight at the side of a hopper light to prevent draught.

(ii)   A feeding trough specially designed to feed materials into a surface or a mixing machine or into a silo for storage.

HOPPER BARGE : A dumb barge for storage of excavated materials.

HOPPER DREDGER : A dredger which acts as a hopper for transporting dredged materials.

HOPPER HEAD: A rainwater pipe head to receive rainwater from a roof area.

HOPPER LIGHT: A light hinged at the bottom with hoppers at each side to prevent draught.

HOPPER WINDOW : A window having been formed by placing hopper lights one above the other.

HORIZON : A plane at right angles to a plumb line.

HORIZON GLASS : A glass half-silvered and half-transparent used in a box sextant for survey work.

HORIZONTAL CIRCLE : A circular graduated plate provided at the base of a theodolite telescope required for measuring horizontal angles with accuracy in a surveywork.

HORIZONTAL CURVE : A curve in a horizontal plane .

HORIZONTAL SHORE : A flying shore to support two adjacent structures.

HORN : An extension in a door or window frame which is embedded in wall for fixing up the frame.

HORNBLENDE : A very heavy prismatic crystal of black or dark green colour, which is a silicate of calcium and magnesium. It is an essential component of many igneous rocks and is abundant in greenstone and syenite.

HORSE : (i)   A temporary framed support made of timber.

      (ii)   A stair string made of timber that carries the treads and risers.

      (iii)   In plumbing, it is a timber finial that is covered with lead.

      (iv)   An equipment used in tile moulding.

HORSED MOULD : Used in plaster moulding. A wooden stock carrying a metal plate cut to the profile of the required moulding is firmly housed into a short board (called horse) which is pushed along the meeting line of the wall and the ceiling and held in position by a running rule. The plaster is then fed onto the moulding.

HORSE SHOE ARCH : Also, known as 'moorish arch'. see illustration.

HORSE-SHOE SEWER : See ilustration.

HORSE POWER : It is a unit of power, abbreviated as 'h.p. one hp means 550 ft-lb of work done per sec. or 33,000 ft-lb/minute. In metric unit, it is 75 kg meter/second. It may be converted into electrical power. 1 hp = 746 watts.

HORSE SHOE
OR
MOORISH ARCH

HORSE POWER HOUR : The work done by spending one h.p. for one hour which is equal to 0.746 killowatt hour. In other words, 1 kwh =1.34 hp hour.

HORSING UP : Forming a 'horsed mould' for running a plaster.

HOSE COCK : A bib cock provided with a fitting for hose at sill height outside a building. This is required for watering the lawn or garden attached to a building.

HORSE SHOE
SEWER

HOSE COUPLNG : A coupling used for jointing two hose pipes.

HOSPITAL DOOR : A single leaf flush door.

HOSPITAL WINDOW : A hopper window.

HOT-AIR HEATER : An air-heater which makes a room warm by injecting hot air into the room.

HOT-AIR SEASONING : Seasoning of timber in a closed chamber by passing hot air.

HOUR ANGLE : A spherical angle subtended at the pole between the declination circle of a star and meridian of the observer. It is the measure of the arc of the Equator intercepted betweeen these two planes.

HOURDIE BRICK : A type of hollow brick having holes in it as shown in illustration.

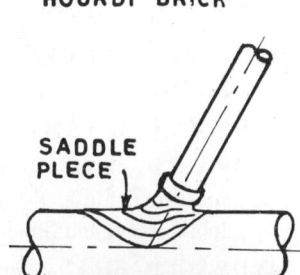

HOURDI BRICK

HOUSE DRAIN CONNECTION : See illustration how a house drain is connected to the street sewer with a saddle piece. The house drain brings all sorts of waste water from a house.

HOUSE TRAP : See illustration. It facilitates easy inspection and cleaning the line, when there is any chokage.

HOUSED JOINT : A joint used in timber work, in which one piece of timber is housed into the notch of another timber piece as shown in illustration.

HOUSED STRING : A close string into which steps are housed.

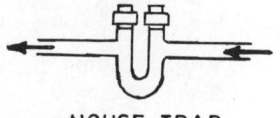

HOUSE DRAIN CONNECTION

HOVELLING : Extending chimney stack walls above the roof level with openings all round the wall for improvement in draught.

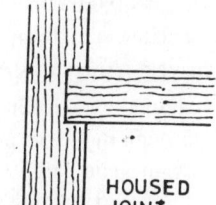

HOUSE TRAP

HOWE TRUSS : A roof truss usually made of steel as shown in illustration, suitable for spans upto 24 mtrs. For small spans, the verticals are made of steel, while the horizontal and inclined members are made of timber.

HUB : The enlarged portion i.e., the bell or socket end of a pipe.

HUMUS : Dark brown or blackish fertile material obtained from decomposition of organic wastes, which may be used as a good soil-conditioner.

HUMUS TANK : The final settling or consolidation tank in a sewage treatment plant from which humus is obtained.

HURDLE WORK: A type of river bank protection by interlacing laths with vertcal poles which forms a low-height fenciing along the river bank. This facilitates in silting.

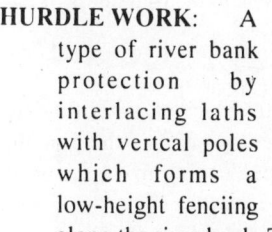

HOUSED JOINT

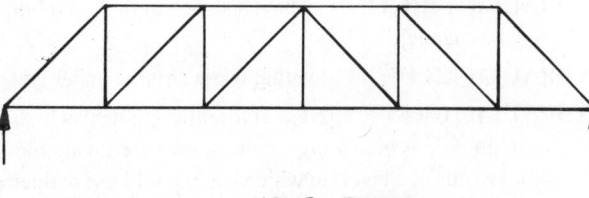

HOWE TRUSS

HYDRALIME : Hydrated lime i.e. slaked lime formed by adding water to 'quick lime'.

$$CaO + H_2O = Ca(OH)_2$$

HYDRANT : A connection to a street water main from which water is drawn during street washing and fire fighting.

HYDRATRED LIME : See 'hydralime'.

HYDRAULIC CEMENT : A type of portland cement that sets and hardens under water.

HYDRAULIC DREDGER : A dredger of suction type for dredging soft materials under water.

HYDRAULIC EJECTOR : A device in which water jet is injected through a pipe at the bottom of a pneumatic caisson and silt, mud and small size loose gravel are drawn out by suction. See illustration. Also known as 'silt ejector' or 'elephant's trunk'.

HYDRAULIC EJECTOR

HYDRAULIC ELEMENTS : The depth and velocity of flow, cross- sectional area, wetted perimeter etc. for discharge of water through a pipe or channel.

HYDRAULIC ELEVATOR : Hydraulic ejector.

HYDRAULIC EXCAVATION : Excavation by injecting water jets at a very high pressure to loosen the materials to be taken out. Thus water carries mud, stiff sand and loose gravel and flows in a channel.

HYDRAULIC FILL : A method of building an embankment or a dam by a slurry of mud, silt, sand and gravel carried in flumes or pipes.

HYDRAULIC FILL DAM : See illustration. A dam having a core of hydraulic

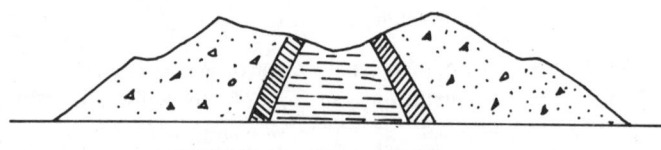

HYDRAULIC FILL DAM

fill of silt, sand, clay and gravel which forms a cut-off wall. Gravels bonded with silt, sand and clay form an impervious wall which prevents the flow of seepage water.

HYDRAULIC FILLING : Filling the abandoned coal mines or any other underground cavities by injecting waterborne materials like silt clay, sand, rubbish, ashes, etc. through flumes or pipes. This produces a compact fill assuring less subsidence of the ground compared to any other fill and at a cheaper cost. This is also known as 'hydraulic flushing'.

HYDRAULIC FRICTION : The resistance to flow of a liquid due to the roughness of the pipe or channel.

HYDRAULIC GLUE : A glue that is not soluble under water.

HYDRAULIC GRADE LINE : Hydraulic gradient which is the slope of water flowing through a pipe indicating the head of water available at every point of the pipe. In other words, the levels to which water could rise in open pipes leading up when tapped from the pipeline.

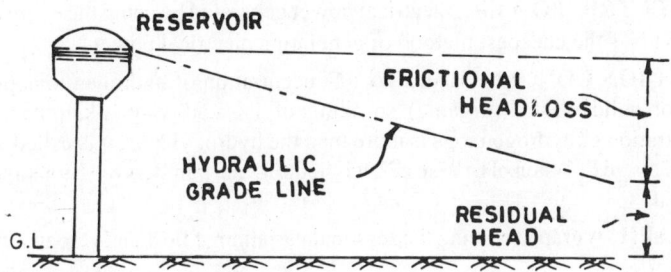

HYDRAULIC JACK : It works on the principle of hydraulic ram i.e. hydrostatic press . In civil engineering, it is used for loading up and testing piles and other structures. It is oil-filled and loading is given by operating a hand pump.

HYDRAULIC JUMP : This is caused when there is an obstruction across the flow of a stream or, channel. See illustration .

HYDRAULICITY : The property of setting of a mortar in absence of excess water.

HYDRAULIC LIFT : A lifting device working on the principle of hydrostatic pressure.

HYDRAULIC LIME : Also known as 'water lime'. It possesses good hydraulic property which is comparable with cement. It is insoluble in water, but makes a thin paste. It sets in a day and becomes hard in four days.

HYDRAULIC MAIN : A water main supplying water under pressure.

HYDRAULIC MEAN DEPTH : Also known as 'hydraulic radius'. It is obtained by dividing the cross-sectional area of water flowing through a channel or a pipe by the wetted peremeter. For a particular slope of a channel, the greatest hydraulic radius gives the largest flow. For a pipe flowing full C. S. area = $\pi d^2/4$ and wetted peremeter is πd. The hydraulic mean depth is $d/4$. where d is the diameter of the pipe.

HYDRAULIC PILE DRIVING : A method of driving a group of sheet piles silently by clamping a driving head onto the piles and forcing them down by hydraulic jack.

HYDRAULIC RADIUS : See 'hydraulic mean depth'. It is usually denoted by 'R'.

HYDRAULIC RAM : Hydraulic press provided with a ram or plunger.

HYDRAULIC RIVETER : A hydraulic press used in rivetting.

HYDRAULICS : The subject dealing with flow of fluids.

HYDRAULIC SILTING : See 'hydraulic Fill'.

HYDRAULIC STOWING : Hydraulic filling.

HYDRAULIC TEST : (i)   A test carried out to find out the pressure that can be sustained by a pressure pipe, water vessel, boiler etc.

(ii)   The water test for newly laid drains and sewers.

HYDRO-DYNAMICS : A disciplene of hydraulics dealing with flow of water through openings, notches, pipes, channels and over weirs.

HYDRO-ELECTRIC POWER : Electrical power generated by using the energy of falling water. This is the cheapest method of generating electrical power.

HYDROGEN-ION CONCENTRATION : Concentration of hydrogen ions present in a liquid or solid wastes (organic) speaking of its acidity or alkalinity. When the concentration of hydrogen ions is more than the hydroxyl ions, it is called acidic. It is denoted by 'pH'. When pH-value is 7, it is neutral, i.e., numbers of H-ions and OH-ions are equal.

HYDROGRAPH : A graph showing the seasonal variation of flow and velocity in a channel or stream.

HYDROGRAPHY : Surveying and mapping for navigability or irrigability of a river, stream or channel.

HYDROLOGIC CYCLE : A water-cycle which is a natural phenomenon as shown in illustration . It is earth's water circulatory system.

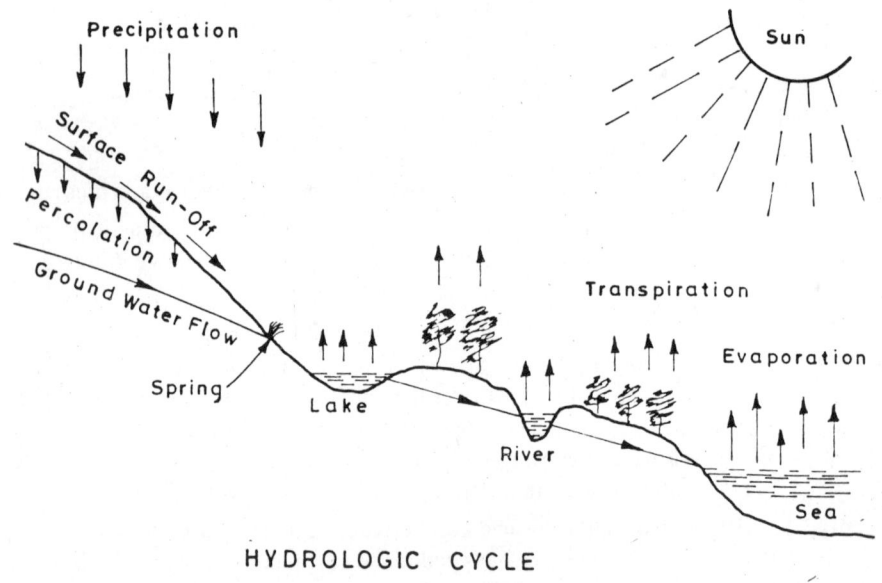

HYDROLOGIC CYCLE

HYDROLOGY : The study of water (underground and surface water).

HYDROMETER : An instrument which when floated in water or any other liquid, indicates the specific gravity of water or liquid by the graduated level of the stem of the instrument submerged in water any other or liquid.

HYDROMETRY : Meaurement of water flow or specific gravity of water.

HYDROSTATIC CATENARY : The curved shape taken by a flume in bridging a gap. It is a peculiar curve of varying radius formed by an inextensible cord or rope when pulled by a force proportion to the length of the cord or rope.

HYDROSTATIC JOINT : A spigot and socket joint in large pipes.

HYDROSTATIC PRESS : Hydraulic press actuated by rams . A small ram is moved up and down to increase the pressure of liquid on the large ram on the other side.

HYDROSTATIC PRESSURE : The pressure exerted by a liquid (at rest) at any point throughout its depth. The magnitude of pressure at any point is given by the density of liquid multiplied by the depth of the point.

HYDROSTATIC PRESSURE RATIO : Rankine's co-efficient of active earth pressure. It is the ratio between the pressure on a vertical plane exerted by a soil and that by a liquid of same density as the soil.

HYDROSTATICS : A part of hydraulics dealing with the pressure of fluids at rest.

HYDROSTATIC TEST : See 'Hydraulic Test'.

HYGI-AERATION SEWAGE TREATEMENT : A package treatment plant for sewage by employing extended activated sludge precess. See illustration.

HYGROMETER : A measuring device to determine the relative humidity of air. The most commonly used hygrometer is dry and wet bulb thermometers. From the readings of dry and wet bulb temperature the humidity of air is determined.

HYGROMETRY : The subject dealing with the mesurement of absolute and relative humidity of air.

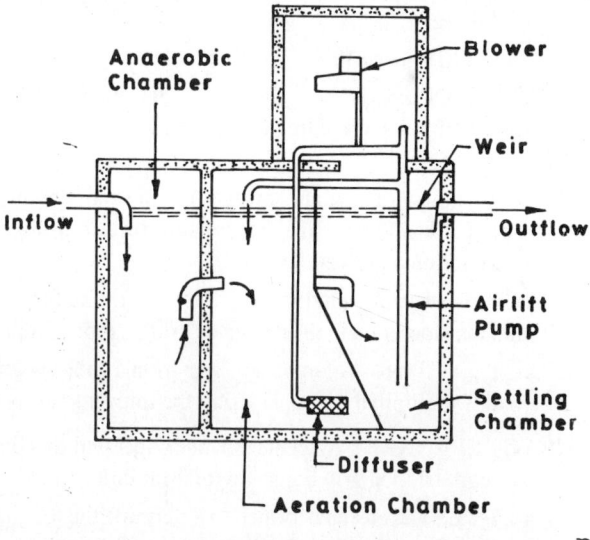

HYGI—AERATION SEWAGE TREATMENT

HYGROSCOPIC CO-EFFICEINT : The percentage (by dry weight) of moisture absorbed by a dry soil mass in saturated air at a particular temperature.

HYGROSCOPIC MOISTURE : Moisture contained in a air-dried soil, which can be driven out, if the soil is dried in an oven at 105°C.

HYGROSCOPIC WATER : Water adsorbed from air that forms a thin moisture film of the surface of soil particles. This water is not available to plants due to great adhesive forces.

HYPERBOLIC PARABOLOID ROOF : A shell roof having the shape of hyperbolic paraboloid, looking like a butterfly in elevation. This produces an architectural effect.

HYPERSTATIC FRAME : Statically-indeterminate frame .

HYPOCHLORITE : chloride of lime and calcium hypochlorite $Ca(OCl)_2$ are used in disinfection of water in a water treatment plant.

HYPSOMETER : An instrument in which the temperature of boiling water is measured to determine the air pressure for indication of altitude or for calibration of thermometer.

HYSTERISIS LOOP : A loop formed in a stress-strain curve beyond the elastic limit of the sample (test piece) and in alternating cylces of tension and compression.

I

I : Notation for moment of inertia or second moment of area of a section.

I-Beam : Usually a rolled steel joist. A flanged beam with a web.

ICE APRON : A ramp on the upstream side of a bridge abutment which slopes up. This helps in lifting floating ice in the stream water and breaking it safely.

ICOS : A method of making diaphragm wall in underground works.

IGNEOUS ROCKS : Rocks formed by solidification of molten mass called 'Magma' coming out on the surface of the earth during volcanic eruption.

IMBREX : The semi-circular overtile which fits over the under tile, called 'Tegula' in Italian roof tiling.

IMBRICATED : A surface made like a tiled roof.

IMHOFF CONE : Used for volumetric determination of settleable sewage solids required for control of sewage treatment plant.

IMHOFF TANK : A deep two-storeyed hopper-bottomed tank for treatment of sanitary sewage from a small community. The foul matters are converted into humus after liquefaction and gassification.

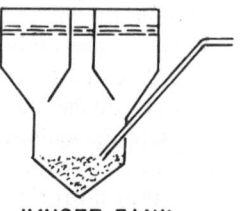

IMHOFF TANK

IMPACT : In structural engineering, it is the striking force taken into consideration in the design of bridge, roof, girder, etc.,

IMPACT FACTOR : A factor ranging from 1 to 2 by which a load is multiplied for considering the impact load.

IMPACT SPANNER : A special spanner operated by compressed air for tightening high strength friction grip bolts and torshear bolts.

IMPACT TEST : A test performed to determine the toughness of a metal, that is, ability to resist any rupture or deformation due to the application of sudden impact load.

IMPELLER : The curved blades of a centrifugal pump or blower, which rotate.

IMPERFECT FRAME : A structural frame which has less number of members than that are required for its stability.

IMPERMEABILITY : Resistance offered by a substance to the flow of a fluid through it.

IMPERMEABILITY FACTOR : Co-efficient of imperviousness or Run-off coefficient which is given by the ratio of rainwater that runs off a surface to the rainfall on it. This is required in calculating run-off from an area.

IMPERVIOUS : The surface that does not allow any flow through it.

IMPERVIOUSNESS : A measure of degree of permeability.

IMPOUNDING RESERVOIR : Artifically built large size water storage reservoir from which water can be available throughout the year.

IMPREGNATED FELT : Roofing felt made of felt, flax or hair impregnated with bitumen, wood tar or coal tar.

IMPREGNATION : The soaking of a material with water-proofing compound or preservative.

INCINERATOR : A special furnace for burning combustible solid wastes. See illustration.

INCISE : TO carve a stone or wood for ornamental work.

INCLINE : A railway track laid with an uniform slope.

INCLINED CABLEWAY : A monocable ropeway having the sloped track cable to facilitate the carrier to run down the slope under its own weight.

INCLINED GAUGE : A graduated sloping staff to read vertical heights.

INCLINED SHORE : An inclined prop or a raking shore to support a structure.

INCLINOMETER : A clinometer or dip needle used in surveying.

INCRUSTATION : Deposition of $CaCO_3$ in the internal surface of a pipe.

INDENT : A gap left in a course of brick work or stone masonry for bonding with the work to be done in future.

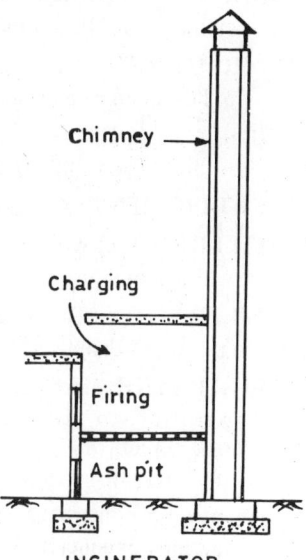

INCINERATOR

INDENTED BOLT : An anchor bolt with indentations on its shaft to increase its grip in concrete.

INDENTED JOINT : A wooden joint in which the timbers to be joined together and the wooden fish plates are cut with matching notches for introduction of hard wood wedges.

INDENTING : Toothing in Timber, steel or masonry works.

INDENTING ROLLER : A roller with a patterned surface is used to make a non-skid texture by rolling it over a green concrete or soft surface.

INDEX : (i)   Alidade used for plane table surveying

(ii)   A number to denote the difference between two limits.

(iii)   A number to denote a ratio

(iv)   A factor

(v)   A power to a number.

INDEX GLASS : The movable reflecting glass of a sextant used in surveying.

INDEX OF LIQUIDITY : The 'liquidity index' which is given as the difference between the water content of a sample and the water content at its 'plastic limit' divided by the 'plasticity index' of the sample. This is reverse of the 'consistency index' of the sample.

INDEX OF PLASTICITY : The 'plasticity index' which is the difference between the liquid limit and the plastic limit of a sample. The higher is the value of plasticity index, more plastic is the sample.

INDEX PROPERTIES : Properties used to distinguish soil characteristics. These are grain size, composition, dry density, moisture content, limits of consistency, etc.

INDICATING BOLT : A door bolt used on a toilet door to indicate its occupancy.

INDIRECT CYLINDER: In plumbing a central heating system for hot water supply, the cylinder must be indirect.

INDIRECT HEATING : Heating of a room by circulation of hot air or steam, the source of heat being at a distant place.

INDIRECT LIGHTING : Lighting a room by means of hidden lamps. Usually, the light is thrown down by the ceiling.

INERT PIGMENT : A pigment that does not react chemically in a paint.

INFILLING : Brick work, wooden boards, glass blocks or insulating boards used as infilling within a building frame.

INFILTRATION : The entrance of water into the ground due to the hydraulic permeability of soil and then its vertical movement down to the ground water table.

INFILTRATION GALLERY : Constructed at right angles to the underflow of valleys or parallel to streams toward which upland flow travels. A gallery may be constructed of masonry or concrete with numerous openings. They are built either in open trenches or by tunneling method. The pipes are laid with open joints surrounded with gravel.

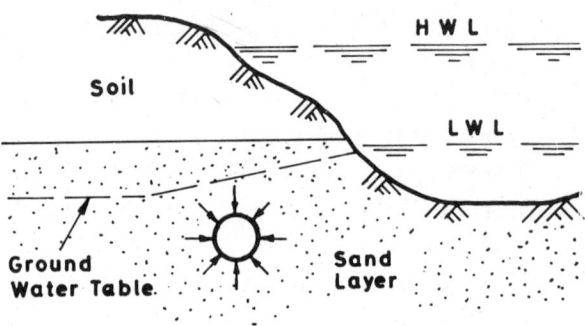

INFILTRATION GALLERY

INFLECTION : Contraflexure.

INFLECTION LINE : A graph in form of a curve extending over the entire span of a beam, used to indicate the effects of loads on a beam.

INGLIS'S FORMULA : A British flood formula devised for India by sir claude inglis by covering almost all catastrophic floods recorded in the country.

$$Q = 7000\, A \,/(A+4)^{0.5} \;;$$

where Q is flood in cusecs and A is catchment area in sq. miles.

INHERENT SETTLEMENT : The sinking of a foundation due to the load it transfers to the soil. This settlement is not affected in any way due to the load conditions on nearby foundations.

INHIBITING PIGMENT : Red lead, zinc chromate, aluminium or graphite powders are the inhibitive pigments added to paints for priming coat with a view to preventing corrosion of a metal surface.

INHIBITOR : (i)   An anti-oxidant material used to reduce skinning the paint in a tank or container.

(ii) Arsenic or antimony compound used to delay the chemical action in pickling acids.

INITIAL SETTING TIME : Time required by a cement paste, mortar or matrix for initial setting. This is measured by a vicat needle apparatus.

INLET TIME : Time required by overland flow to reach the inlet point of a sewer system.

INNINGS : Land reclamation from a marsh, river or sea.

INLAID PARQUET : Parquet flooring glued on wood blocks and fixed to floor boards.

IN-SITU : Fabricated or cast at site.

IN-SITU CONCRETE : Concrete mix prepared and cast at site.

IN-SITU SOIL TEST . Soil tests conducted in field such as load test, vane test, dynamic penetration test, permeability test, etc.,

INSPECTION CHAMBER : A chamber with a manhole at the junction of small diameter sewers or at the point of change of direction of such sewers with a view to having ready access for inspection and facilitating rodding operation for removal of chokage in the lines.

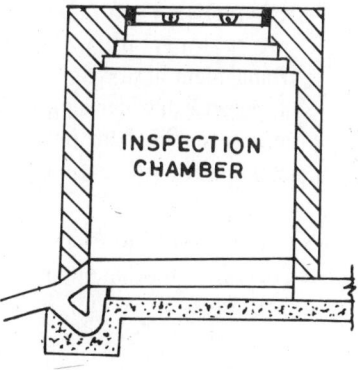

INSPECTION EYE : An access eye with a door provided on the pipe fitting for inspection by opening the door.

INSPECTION JUNCTION : A special pipe fitting with a short pipe provided at the junction of drains, through which the flow-through drains can be inspected.

INSTRUMENT : An appliance used for measurement of quantity or quality of any material.

INSULATED METAL ROOFING : Roofing with light-gauge flexible metal panels, the ceiling being made of insulating fibre boards.

INSULATION JOINT : A joint made in metallic pipe to introduce resistance to the flow of stray electric current along the pipeline.

INSULATING BOARD : Light-weight boards having sound-insulating property used for partition walls, wall facings and ceilings. These are usually fibre boards and wood wool or straw boards.

INSULATING MATERIALS : Materials chiefly used for room insulation against heat, cold and sound are fibre board, wood wool slab, asbestos sheet, asbestos-diatomite sheet, plasterboard, cork board, compressed straw and saw dust board, etc. For sound insulation, sometimes glass blocks or double glazing are used.

INTACT CLAY : A clay having no fissures visible.

INTAGLIO TILES : Tiles with decorative patterns intagliated or engraved onto their surfaces.

INTAKE BELT COURSE : A projecting string course is provided where the wall thickness is reduced.

INTAKE CRIB : A structure usually made of timber, submerged at a depth under water avoiding the interference with navigation. With low entrance velocities they require no regular attendance for maintenance. Square or octagonal wooden cribs are protected

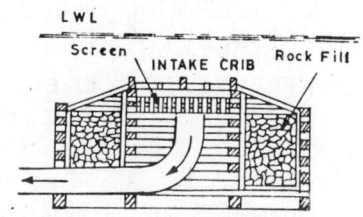

by rip rap. This is required for drawal of water from a river or stream for water supply.

INTAKE HEADING : A term for 'head works' used in USA.

INTAKE WELL : A well built on river bed close to the bank of the river with a view to drawing water from the river for the purpose of water supply to a community, after treatment.

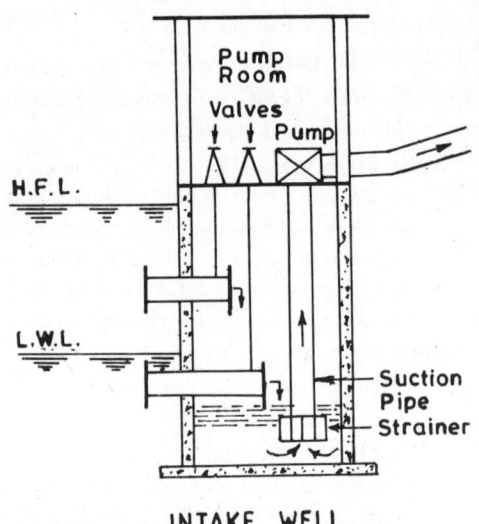

INTAKE WORKS : Structures built into a body of water with arrangement of drawing water for the purpose of water supply. These may be intake crib, pipe intake, intake well, infiltration gallery, etc.

INTEGRAL WATER PROOFING : The water proofing of concrete by mixing some water-repellant materials as admixtures in the mortar or matrix.

INTAKE WELL

INTEGRALLY STIFFENED PLATE : Extruded aluminium sheet having shape like reverse angle section used for decking in construction.

INTEGRATING METER : A meter that records the total quantity of fluid flow by integrating.

INTENSITY OF RAINFALL : The quantum of rainfall in unit time. Usually, average intensity of rainfall per hour is taken into consideration.

INTENSITY OF STRESS : A term for unital stress i.e., load per unit area or stress developed per unit area due to external load applied.

INTERCEPT : In surveying, it is the length of the staff seen between the two stadia hairs of a telescope.

INTERCEPTING DRAIN : A major drain which intercepts the trunk drains.

INTERCEPTING SEWER : A trunk sewer intercepting the main sewers, receives flow from each.

INTERCEPTING TRAP : Sand, grease, debris, etc. are kept out of a drainage system by providing this type of trap which is separator and also called intercepter.

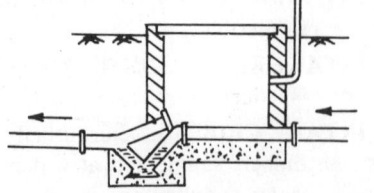

INTERGROWN KNOT : A live knot.

INTERFERENCE BODY BOLT : High strength friction-grip bolt. This bolt can be used in a clearance hole with full grip. The corrugated

INTERCEPTING TRAP

shank near the bolt head gets deformed during driving and it is fully gripped by the corrugations pressing against the hole wall. Thus, the bolt has high bearing and shearing strengths.

INTERFERENCE SETTLEMENT : The sinking of a foundation due to loads on nearby foundations. The inherent settlement of a foundation thus gets interfered.

INTERGRANULAR PRESSURE : Effective pressure of soil.

INTERIOR SPAN : Neighbouring spans of a continuous beam.

INTERLACED FENCING : Interwoven fencing made by weaving together very thin straight boards keeping no space in between.

INTERLOCK : A clutch in steel-sheet piles for interlocking.

INTERLOCKED GRAIN : Twisted Fibres in wood. Spiral grain and ribbon grain are examples. A grain in which the fibres slope one way in one growth ring and reverse the slope slowly in the next growth ring. This is caused when the fibres of a young tree get twisted due to violent winds. It is difficult to work with a wood having interlocked grain.

INTERLOCKING JOINT : A joggled or dowelled or pinned joint made in ashlar work to prevent the movement of stones.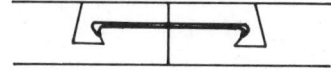

INTERLOCKING PILE : Sheet piles made with projecting tongues to hold each other firmly by interlocking when driven.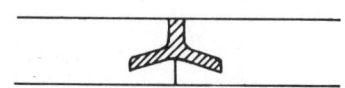

INTERLOCKING TILE : Tiles having grooves at one end and projections at the other with a view to interlocking them during laying so that the movement of tiles is prevented.

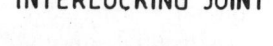

INTERLOCKING JOINT

INTERMEDIATE RAFFER : A common rafter in a wooden truss.

INTERLOCKING   PILE

INTERMEDIATE SIGHT : A staff reading in between back sight and fore sight in a level survey.

INTERMITTENT FILTRATION : Land irrigation with sewage effluent. The filtration of effuent through soil is intermittent due to periodical hydraulic loading.

INTERNAL DORMER : A vertical window in a sloping roof.

INTERNAL FRICTION : Friction between soil or sand grains.

INTERNALLY FOCUSSING TELESCOPE : Anallatic telescope.

INTERNAL GLAZING : Glazing on internal walls.

INTERNAL VIBRATOR : Also known as 'Poker Vibrator' or 'Immersion Vibrator'. A 25 to 75 mm diameter cylinder with vibrating mechanism is used in compacting green concrete by inserting it into the concrete and drawing up.

INTERPOLATION : Defining a point between two known points assuming a smooth linear variation i.e., a straight-line variation.

INTERSECTION : Crossing or meeting of two lines.

INTERSECTION ANGLE : Deflection angle.

INTERSECTION POINT : The point at which two straight lines meet or cut each other.

INTER TIE : An intermediate horizontal member used in a framed wooden partition for stiffening it at door head level.

INTRADOS: The soffit or under surface of an arch.

INTRUSIVE ROCK : Molten mass of rock forced out during volcanic eruption and deposited in thick masses covered by other rocks.

INVAR : An alloy of steel and nickel with other elements, which has low co-efficient of expansion. This is used in making measuring tapes.

INVERT : The lowest point of the internal surface of a drain, sewer, channel, culvert etc.

INVERTED ARCH FOUNDATION :
Suitable, when the bearing power of soil is very low and the foundation depth is less. Inverted arches are constructed between walls at the base as shown. The load is transmitted to a larger area through these arches.

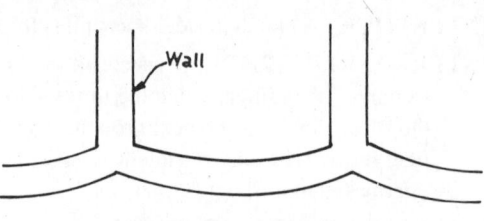

INVERTED ARCH FOUNDATION

INVERTED SIPHON : A device to take a sewerline down below the gradeline, to pass under a stream or a railway track and to raise it again to the regular gradeline beyond. The siphon should consist of two parallel pipes laid straight in grade and alignment and at either end,

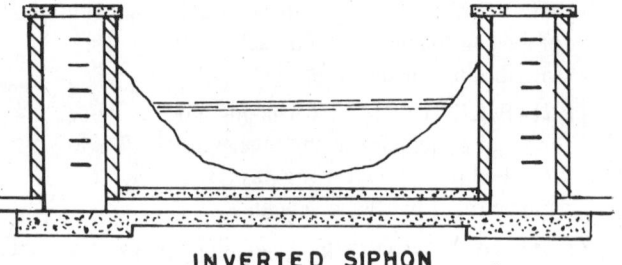

INVERTED SIPHON

manholes should be placed. Self- cleansing velocity must always be maintained in the inverted siphon.

INVERTING EYE PIECE : Astronomical eye piece.

INVERT LEVEL : The level of invert i.e., the lowest point of the inner surface of a drain, sewer, culvert, etc. with reference to a datum.

IRISH BRIDGE : A water splash, a paved ford.

IRISH MOSS : A sea-weed of the atlantic, purple to green in colour, used by painters to make size.

IRON : The most important and valuable metal used in civil engineering constructions.

IRON CEMENT : A mixture of iron turnings, sal ammoniac and flours of sulphur used for mending cracked cast iron parts or for joining cast iron pipes. It is strong enough to withstand the action of water.

IRON CORE : A steel bar covered by a wooden handrail.

IRON FIGHTER : A bar bender.

IRONMONGERY : Hardware, cast iron or wrought iron.

IRON OXIDE : It is used as pigment in a paint.

IRON PAN : Hard pan cemented by iron oxides.

IRON PAVING : A non-skid surface made of cast iron blocks.

IRON SAND : Fine chilled shot fed with water into a cut, during sawing operation of a hard stone.

IRON WORK : Decorative work with cast iron and wrought iron.

IRRIGABLE AREA : The area which is low enough to be irrigated.

IRRIGATING HEAD : The flow required for irrigation of a particular land area.

IRRIGATION : Method of watering land for agriculture. It involves excavating canals and distributories and construction of civil structures.

IRRIGATION REQUIREMENT : The quantity of water required for irrigation for production of crop. The quantum includes water loss but excludes precipitation.

I-SECTION : A beam section, usually a rolled steel joist which has two flanges connected by a web.

ISLAND PLATFORM : A Railway platform having tracks on both sides i.e., a platform located between a main line and a loop line.

ISOCHROMATIC LINES : Coloured streaks in photo-elastic stress analysis, represent lines of same difference of principal stress.

ISOCLINIC LINES : In photo elasticity, dark lines joining all points, where the principal stresses are parallel to the planes of polarisation.

I- Section

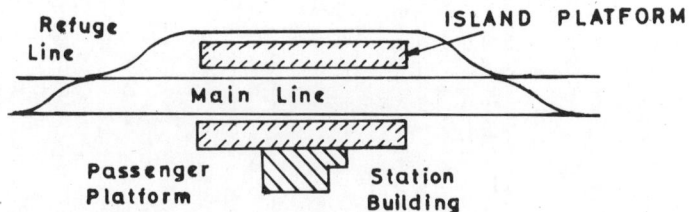

Refuge Line

ISLAND PLATFORM

Main Line

Passenger Platform

Station Building

ISOHYET : A line obtained by joining points of equal rainfall.

ISOLATING MEMBRANE : An underlay.

ISOLATING STRIP : An expansion strip.

ISOMETRIC PROJECTION : It is a form of axonometric projection required in preparing a three-dimensional drawing. Three faces make equal angle with the plane of projection, the line of sight being perpendicular to the plane of projection.

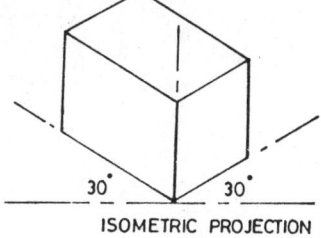

30°   30°

ISOMETRIC PROJECTION

ISOMETRIC SCALE : In an isometric drawing, if the object is drawn to a natural scale it would appear larger than the actual size and it would be in the ratio 3/2. To avoid this, isometric scale should be used. To prepare this scale, the natural scale is placed at 45° to the base line and graduations are marked on the isometric scale by placing it at 30° to the base line as shown.

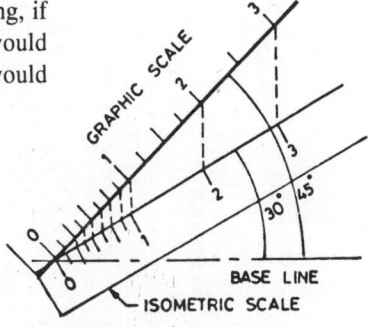

GRAPHIC SCALE

30°   45°

BASE LINE

ISOMETRIC SCALE

ISOTHERM: A line joining places of same
temperature.

ISOTROPIC : Means same physical
properties in all directions.

ITALIAN TILE : Pan and roll tiles; the
curved over-tile is called imbrex and the tray-shaped under-tile
is known as 'tegula'. These are used for roof tiling.

ITALIAN TILE

IZOD TEST : An impact test in which a notched bar is broken by the blow of a pendulum.

J

JACK ARCH : A welsh arch or a flat arch made of brick masonry or concrete over a short span of 1 m. only springing from the bottom flange of a joist. This is used for short span bridge decks or heavy floors and roofs.

JACK ARCH ROOF : This type of roofing was much in use prior to introduction of reinforced cement concrete. See illustration.

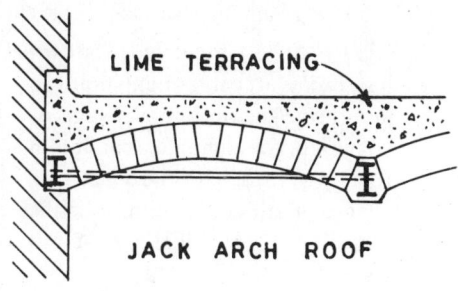

JACK ARCH ROOF

JACK BLOCK METHOD : A method developed from lift slab construction adopted in building a multi-storied house. The floor slabs are cast, matured and stressed up by prestressing cables at ground floor and are jacked up one storey along with the simultaneous construction of supporting walls. The precast concrete blocks of height equal to one stroke of the jacking rams (usually 150 to 200 mm) are laid dry for the total height of the building which are covered by vertical strips of high-alumina cement concrete. Thus, the jack blocks themselves form the supporting walls. The jack block method of construction is very fast and efficient.

JACK PLANE : A bench plane used by carpenters for cleaning timber after sawing.

JACK RAFTER : A short rafter placed in between the ridge and valley or between hip rafter and eave.

JACK RIB : Also known as 'Jack Timber' or 'Cripple'. This is a curved 'Jack Rafter' used in a small dome-roof construction.

JACKED PILE : A pile forced into the ground by jacking against the wall above it as required in underpinning.

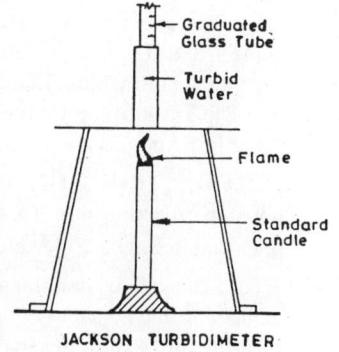

JACKSON TURBIDIMETER

JACK ROLL : A windlass used for hand-hoisting from a pit. Buckets are hung from either end of a rope and an empty bucket goes down while a full bucket comes up.

JACKSON TURBIDIMETER : This instrument is capable of measuring turbidity above 100 ppm. The turbid water is gradually poured in the glass tube until the image of the candle flame ceases to be seen from top. The corresponding graduation in the glass tube indicates the turbidity.

JAMB : The vertical face of a door or window opening, to the full thickness of the wall.

JAMB BRICK : A special brick sometimes used in forming a jamb.

JAMB LINING : A timber lining used for covering a jamb.

JAMB POST : A post made of stone, timber or concrete to form a

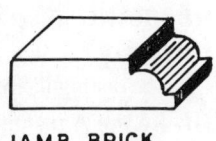

JAMB BRICK

door jamb.

JAPAN : The name used for 'Black Japan' paint.

JAPANESE LACQUER : It is the most durable glossy varnish prepared from the sap of the Japanese Varnish tree 'Rhus Vernicifera'.

JAPANNING : Painting a surface with a stoved 'Black Japan'.

JAR TEST : This test is required to be carried out in a water treatment plant to determine the optimum dosage of coagulant for complete reaction in forming flocs with their settling characteristics depending upon the turbidity, pH and alkalinity of raw water.

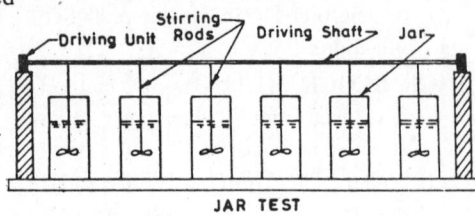

JAR TEST

JENNING'S CLOSET : It is a syphonic closet resembling a 'Wash down closet'. The water in the basin stands at a higher level and a deeper water-seal is formed.

JERKIN-HEAD ROOF : The contrary of a 'gambrel roof' eaves and gabled from there down.

JET PUMP : This pump is free from clogging and it has no valves. It operates by a stream of water entering a venturi-like throat at a high velocity. This causes a partial vacuum which lifts water. The suction head is

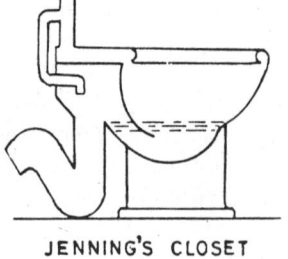

JENNING'S CLOSET

only 2 to 3 m and the pump needs water under pressure for its running. For this, jet pumps are not commonly used.

JETTIED CONSTRUCTION : Construction of timber houses with their upper part jetting out. This type of construction was commonly practised in England in Middle Ages.

JETTING : A method of sinking piles into sands without causing any damage to the adjacent buildings. A concrete pile can be cast by forming hole into the sand by jetting water with or without compressed air.

JETTY : A deck used as a landing stage, supported by piles at the water's edge. A horizontal or sloped structure projecting from a land into the water. The projected structure is supported on piers. Sometimes, jetties are made parallel to the shore and connected to land by a trestle. These can accomodaste one or more vessels at the end.

JET PUMP

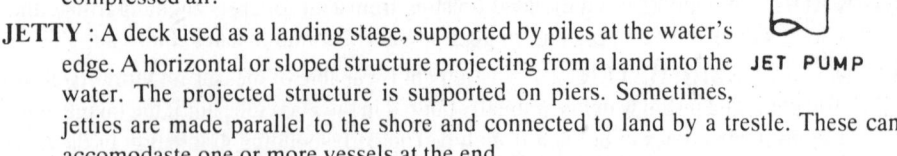

JETTY CYLINDER : A concrete screw pile.

JIB CRANE : A crane with a jib.

JIB DOOR : A door having its face flush with the wall and it is so decorated that it can not be distinguished.

JIG SAW : A reciprocating, power-operated saw used for cutting sharp curves in timber. Also known as 'Scroll Saw'.

JIM CROW : This is a railway tool used in bending rails manually.

JOGGLE : A recess formed in one block which fits a similar projection on another block.

It is obvious that the recess and the projection must be identical. This is mostly used in concrete or stone block work.

JOGGLED : Shaped with indentation or projection.

JOGGLE PIECE : A post to form an abutment for a strut. It is shouldered like the base of a king post.

JOGGLE POST : A term for king post.

JOHNSON COUPLING : A flexible coupling used for jointing C.I. pipes.

JOINERS'GAUGE : A marking gauge used by joiners.

JOINERS' HAMMER : A hammer with a cross peen head. Also known as 'Warrington hammer'.

JOINERY : The term used in making, finishing and fixing timber works like furnitures, doors and windows, skirting, timber linings, etc.

JOINT : A connection between two pieces in masonry, concrete work, timber work, plumbing, pipe laying, steel structure, etc.

JOINT BOLT : A handrail bolt for fixing the handrail.

JOINTER SAW : A sawing machine for cutting stone.

JOINT FASTENER : A corrugated fastener used in Carpentry.

JOINT FILLER : A kind of putty used for filling the space between abutting ends of plaster board or a compressible strip material like bitumen, felt, cork, etc. placed in the space between two concrete units or a strip of lead, aluminium, glass, etc put in between two concrete slabs.

JOINTING COMPOUND : A paste like cementing or binding material inserted into the space between two pieces to make a tight joint.

JOINTING MATERIAL : Gaskets, sheets of rubber, asbestos, bitumen- impregnated paper and washers required in making a joint watertight. These are used in the joints of flanged pipes,pumps, etc.

JOINTING PLANE : A plane used in carpentry.

JOINTING RULE : A long wooden straight edge used by masons in pointing brick walls.

JOINTLESS FLOORING : A type of flooring without close joints, but having very thin joints at about 3 to 6 m. distance apart to allow shrinkage. The materials used are cement-wood, cement- rubber, granolithic screed, asphalt, anhydrite, magnesite, pitch mastic and terrazzo.

JOINT MOULD : A plywood or cardboard template shaped for plaster work.

JOINT RULE : A steel rule with one end cut at 45°, is used by the masons in forming mitres at the junction of cornice mouldings.

JOINT RUNNER : This is asbestos or hemp rope used for making an annular space at the end of a pipe joint which is to be filled with molten lead. This is also known as 'pouring rope'.

JOINT TAPE : Paper or cotton tape fixed over the joints of wallboards.

JOIST : A wooden or steel beam to support a floor. Usually, joists are rolled steel joists (RSJ).

JOIST ANCHOR : Wall anchor to hold a joist.

JOIST HANGER : A steel strap carrying the end of a joist.

JUBILEE WAGON : A tipping wagon which facilitates easy unloading.

JUMBO BRICK : A brick of unusual large size.

JUMPER: (i)   A stretcher stone covering more than one cross joint in a squared rubble massonry.

          (ii)   The mushroom-shaped part of a domestic water tap, the stalk pointing upwards into a hollow guide.

          (iii)   A heavy steel bar with a chisel point commonly used for boring holes in rock or soil by dropping into the hole and twisting it.

JUNCTION: The meeting point of pipes, roads, railway lines, etc.

JUNCTION CHAMBER : It is essentially a 'manhole' consisting of a masonry chamber constructed on sewer lines at junction points to have easy access for inspection and cleaning. See illustration.

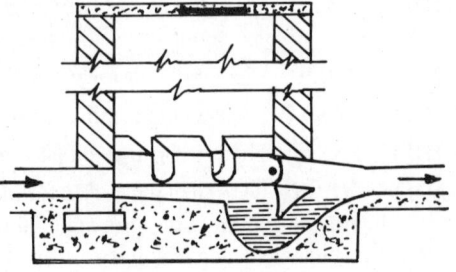

JUNCTION POINT : It is the meeting point of curves, survey lines, pipes, roads, railway tracks, etc.

JUVENILE WATER : New water of magmatic or volcanic origin added to the terrestrial water supply.

JUNCTION CHAMBER

K

KAPLAN TURBINE : A water turbine of propeller type used for generation of hydro-electricity.

KAURI : A fossil copal-resin that was used in early days in hard-drying varnishes.

KEATING'S CEMENT : A white cement made by mixing gypsum with borax and calcining. This cement sets rapidly within 4 to 6 hours, does not effloresce and forms a translucent hard surface that can be painted at once.

KEEL BLOCK : Docking block.

KEENE'S CEMENT : A white cement prepared by mixing gypsum, with a solution of alum and burnt plaster of paris is soaked in this . This cement is suitable for floors, columns, pillasters, skirtings, etc. The superfine quality of this cement receives a brilliant polish. The coarser quality of this cement is of pinkish tinge which produces a very hard surface.

KENNEDY'S CRITICAL VELOCITY : It is the velocity in open channels that will neither deposit nor pick up silt i.e., non-silting and non-scouring velocity in open channel flow.

$$Vc = 0.84 \, d^{0.64} \, ;$$

where d is depth of flow.

KENTLEDGE : Loading by using heavy stone boulders, scrap metals, large concrete blocks, sand bags, and other similar such materials to provide stability to a crane, a reaction over a jack, to test a pile or caisson and to push down a plate into the soil for plate bearing test.

KERATIN : A substance obtained from horns, hoofs, nails and scales or hides of animals which is used as a retarder for plaster of paris.

KERB : A hard stone or precast thick concrete slab used for bordering the footpath of a roadway. Normally, the kerb is 150 to 200 mm high i.e., the footpath is 150 to 200 mm above the road level.

KERB INLET : An opening in the kerb to facilitate entry of silt-laden flow of storm water into the underground sewer via. the street gulley.

KERF : A saw-cut in stone, coal or timber.

KERFED BEAM : A wooden beam with several saw-cuts to facilitate easy bending.

KERFING : Forming saw-cuts on one side of a wooden beam so as to bend it on that side.

KERN : The core of a structural section.

KESSENER TANK : A long tank used in Kessener process to agitate sewage, the flow being from end to end: See illustration.

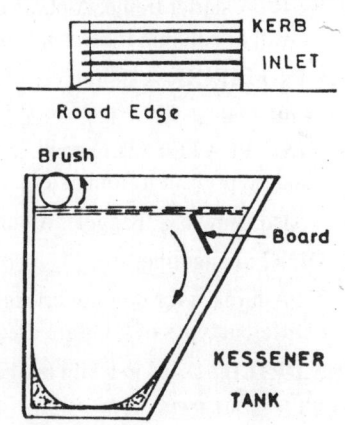

KEY : (i) The frog or recess in a brick or stone required to hold them firmly by mortar in making brickwork or stone masonry.

      (ii) In joinery work, a hardwood wedge let into a joint to strengthen it.

      (iii) A pointing tool used by masons for making a keyed joint.

      (iv) A steel wedge driven between a rail chair and the rail to hold the rail firmly.

KEY BRICK : It is a V-shaped brick used at the centre of an arch to form the crown and to transmit load to both sides by the wedge action.

KEY BLOCK : A key stone similar to a key brick, used in stone arch.

KEY COURSE : A course of stones used at the crown of a vault or wide arch instead of using a keystone.

KEYED BEAM : A wooden beam formed by lap joint into which joggles are cut in each member at the joining portion and hardwood wedges or keys are driven into the holes for strengthening it.

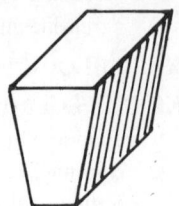

KEY BRICK

KEYED JOINT : A mortar joint with concave pointing.

KEYED MORTISE AND TENON : A Tusk-Tenon joint.

KEY-HOLE SAW : A compass saw tapering to a point.

KEYING IN : Bonding a new brick wall to an existing wall.

KEY MAP : A map showing the location of an area with reference to other landmarks.

KEY PLAN : A small plan showing the locational positions of different units in a scheme.

KEY PLATE : In joinery works it is an escutcheon.

KEY STONE : A key block or a central wedge-shaped arch stone placed at the crown of an arch.

KEY WAY : A slot cut in a shaft and the hub to receive a key to prevant relative movement.

KICK : (i) A frog in a brick.

      (ii) The slope difference between patent glazing and the surrounding roof.

KICKER : A starter frame; A plinth 50 to 75 mm high above the floor forming the base of a wall or column. It helps in clamping the concrete wall or column shutters.

KICKING PIECE : A short piece of timber spiked to the waling to receive the thrust from a raker.

KICKING PLATE : Also known as kick plate. A metal plate fixed to the bottom rail of a door to prevent it from external damages.

KID : Also known as 'Faggot'; A bundle of brushwood to serve as a 'fascine'.

KIDDING : Faggoting.

KILN : A furnace for burning bricks, tiles, lime,and limestones in cement manufacturing. Different types of kilns are used for different purposes.

KILN DRIED : Dried in a kiln by the hot blast of air.

KINETIC ENERGY : The energy of a moving mass due to its weight and motion.

$$K.E. = Wv^2 / 2g$$

where W = weight of the body, v = Velocity

and g = acceleration due to gravity.

KINETIC HEAD : It is the velocity head of a moving mass and is given by $v^2/2g$.

KING BOLT : Also known as 'King rod'. It is a vertical steel rod acting as a king post connecting the ridge and the tie beam of a king post truss.

KING CLOSER : A three–quarter brick used as a closer. One corner of a full brick is diagonally cut vertically from the centre of one end to the centre of one side.

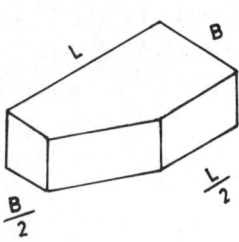

KING PILE : Prior to digging a wide trench, a long pile called the king pile is driven at the strut spacing in the centre of the trench. Thus, the main timbers from each side of the excavation bear on the king pile.

KING POST : Also known as 'Joggle Post'. This is a vertical post in a king post truss connecting the ridge and the centre of the tie beam.

KING CLOSER

KING POST TRUSS : A commonly used wooden roof truss comprising a pair of principal rafters held by a horizontal tie beam and a king post with two struts.

KISS MARKS : Marks where bricks touch in the kiln.

KITE WINDER : When three winders form a right angle by turning, the central one is called the kite winder, since its plan is just like a kite.

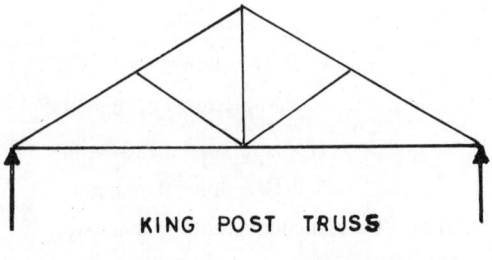

KING POST TRUSS

KNAPEN SYSTEM : A system of drying out damp walls by forming holes into the walls from outside at floor level.

KNAPPED FLINTS : Flints broken across the middle to bring out the dark colour, and shaped for squared face on the wall.

KNAPPING HAMMER : Stone shaping hammer.

KNEE : Also known as 'Elbow'.

   (i)   A naturally-curved short timber.

   (ii)   A convex curve in a handrail.

   (iii) A sharp 90° pipe bend.

KNIFE EDGE LOADING : In design of a bridge, a knife edge load is assumed to be applied along a straight line of zero thickness and per unit length, in addition to a uniformly distributed load.

KNOBBING : Rough dressing of freshly-quarried stones.

KNOCKER : A hinged bar striking on a plate on an entrance door.

KNOCKING UP : Batch mixing and making workable mortar, Concrete or paint.

KNOCKINGS: Small stone chips.

KNOT: The location in a tree trunk wherefrom a branch came out. In a timber, dead knots are harmful.

KNOT BRUSH : A thick brush with its bristles bunched in oval or round shape, commonly used for distempering.

KNOT

KNOT CLUSTER : A group of knots found in a timber where the wood fibres pass round the group.

KNOTTING : Sealing knots in new wood by a quick-drying solution prepared by dissolving shellac in methylated spirit. This is done prior to painting or varnishing a new wood.

KNUCKLE JOINT : (i) A right angled soldered joint between two lead pipes used in plumbing.

(ii) A curb joint in a Mansard roof truss.

KRAFT PAPER : A specially prepared strong brown paper used as a building paper.

K-TRUSS : See illustration.

KUTTER'S FORMULA : This is required to determine chezzy's constant.

$$C = \frac{[1/N + 23 + 0.00155/S]}{[1 + (23 + 0.00155/S)\, N/R^{0.5}]}$$

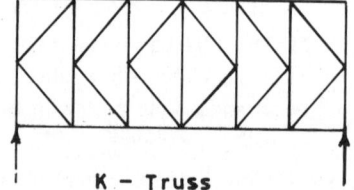

K – Truss

where N = co-effecient of rugosity;

H = Hydraulic mean depth;

S = Bed slope of channel;

K-Value : Thermal conductivity of a material.

KYAN'S PROCESS : A process of preservation of timber by impregnating it with a solution of $HgCl_2$ and $CuSO_4$.

L

LACED COLUMN : A built-up steel column with horizontal and diagonal lacings.

LACED VALLEY : A valley formed by two sloped roofs where a wide board is used in place of a valley gutter and the tiles or slates are laid in such a manner that the slopes do not meet each other.

LACING : (i)   Bracing of light members fixed horizontally or diagonally to angle or channel sections in forming a built-up column or girder or truss.

(ii)   A timber piece nailed to pairs of walings or struts in timbering trenches.

LACING COURSE : A course of stone slabs, tiles or slates laid to strengthen a rubble wall.

LACQUER : Like varnish a glossy finish is produced either for decoration purpose or for coating metal surfaces by using cellulose-based compounds. A thin transparent film is left on the surface by evaporation of the vehicle.

LADDER : A form of stair made of light alloys or timber which is used temporarily at work sites and sometimes permanently for access from one level to the other. There are various types of such ladders.

LADDER BUCKET : A mud-bucket attached to a bucket ladder dredger.

LADDER SCAFFOLD : A scaffold which can be erected quickly on ladders braced together, required for making light jobs at a height.

LADDER TRACK : A number of parallel tracks branched off from a common inclined turnout track which is from a main track as shown in illustration.

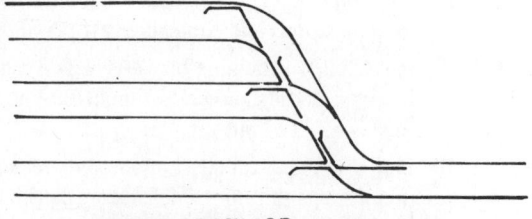

LADDER TRACK OR GATHERING LINE

LADIES URINAL : See illustration. This is a sitting type urinal.

LAG :   (i)   moulded insulation.

(ii)   Heat insulation wrapping material.

LAG BOLT : Coach screw.

LAGGING : (i)   Providing insulation by wrapping hot pipes.

LADIES URINAL

(ii)   Fixing horizontal boards across the centres to support an arch during its construction.

LAGOON : A ditch, pond or low-lying area used for treatment of sewage or sludge.

LAITANCE : Scum or froth appeared on cement concrete surface during laying or on over-trowelled mortar.

LAKE : (i)   A water reservoir on ground surface. It may be natural or artificial.

(ii)   A pigment comprising a dye in an inorganic base.

LALLY COLUMN : A hollow circular steel column (cold rolled).

LAMBERT :   One lumen per square cm. reflected from a polished bright surface.

LAMELLA ROOF: A large-span vault made of timber or metal members connected in a diamond pattern, without any truss. This may also be made of R.C.C. Such a roof facilitates in having a spacious head room and a feeling of volume.

LAMINAR FLOW : A streamline flow which is normally observed in a channel flow.

LAMINAR VELOCITY : The velocity of a liquid in a channel, at which the flow is always laminar and beyond which the flow is turbulent.

LAMINATE : (i)   Built-up sheets glued together with resin.

(ii)   Impregnated sheets of paper, textile, plastics, etc with synthetic resin.

LAMINATED ARCH : A timber arch made of laminated boards.

LAMINATED FIBRE BOARD : Bitumen or resin-bonded fibre boards used for panelling wall and making ceilings.

LAMINATED JOINT : A combed joint used in joinery.

LAMINATED LEAD SHEET : Thin sheets of lead glued to flexible materials which are used to cover walls and sloping roofs to protect against actions of weather.

LAMINATED PLASTIC : Thin sheets of paper, card board, etc impregnated in synthetic resin and bonded together to produce glossy-surfaced stiff boards that can be used to cover walls.

LAMINATED WOOD : Wood veneers and thin plies are glued together under pressure which are used for panelling walls and decorative works.

LAMIN BOARD : Coreboard made of wooden core strips joined together. This is used in making partitions and cubicals in offices and commercial buildings.

LAMP BLACK : A vegetable black used as a pigment by controlled combustion of oily and fatty materials and collection of the shoot. It contains about 90% of carbon and the rest is oily and tarry matter.

LAMP HOLE : A vertical shaft of small diameter constructed over the centre of sewer through which a lamp can be lowered into the sewer for the purpose of inspection of sewer line.

LANCET ARCH : See illustration. This is similar to an equilateral arch, but the centres of arcs lie outside the span on the springing line. This type of arch is used in churches.

LAND ACCRETION : Land reclamation from marshy lands, low-lying areas, rivers and sea by siltation or by dumping waste materials, garbage, rubbish, etc. Sometimes, to encourage siltation, maritime plants or reeds are planted in flood-prone areas.

LAND DRAIN : An agricultural drain for draining water from an agricultural field.

LAND SLIP : Land slide due to slip of a soil mass on a sloping ground. The nature of slide may be rotational slide, shear slide or flow slide.

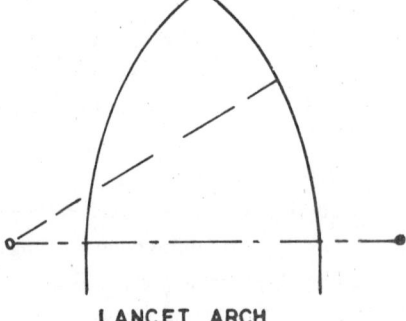

LAMPHOLE

LANCET ARCH

LAND SURVEY : Topographical survey for mappping the land area with physical features.

LAND TIE : A tie-rod used to hold a sheet pile or a retaining wall to a deadman.

LAND TREATMENT : Sewage farming i.e. application of sewage onto a land area (land irrigation) intermittently. The sewage percolates into the ground by intermittent filtration.

LANTERN LIGHT : A raised frame or construction above a flat roof with glazing all round to admit the natural light. This is similar to skylight.

LAP JOINT : A joint commonly used in timberwork, in which two pieces lap over each other and are fastened together either by nailing or by clamping or by bolting with U-straps. The half-lap joint has been developed to make the joint at the same level.

LAPPED TENON : A joint in carpentry in which two tenons entering into a mortise from opposite ends lap over each other within the mortise.

LARCH : A strong, resinous soft wood derived from a deciduous conifer. This is used in piling and joinery works.

LARGE CALORIE : A Kg.–calorie equivalent to 3.97 BTU.

LARRY : (i)   A hoe-shaped tool used for mixing hair with mortar.

(ii)   A mortar in fluid state.

LARRYING : Pouring fluid mortar to fill the vertical joints of bricks that are laid in position.

LASH TERMINAL : Lighter-Aboard-Ship (LASH) terminal does not require deep water port facilities, but only an anchorage area with protectionary measures against extreme wave action .

LATCH : A locking arrangement provided in doors. There are different forms of latch available to serve specific functions.

LATERAL : A small channel taken out from a canal for irrigation purpose.

LATERAL CANAL : A canal made parallel to a river for navigation purpose.

LATERAL FORCE DESIGN : A method of design of buildings in earthquake-prone areas such that a building can safely carry a lateral thrust (horizontal force) in any direction.

LATERAL SEWER : A sewer that receives sewage, sullage or storm water from a small area and discharges into a branch sewer.

LATERAL SUPPORT : Horizontal support given to a wall or column at close interval to prevent buckling or sway.

LATEX EMULSION : Exudation of rubber trees dispersed in water, i.e., latex emulsified in water.

LATH : Any board or metal sheet used as a base for plaster.

LATH HAMMER : A claw-hatchet used by plasterers for nailing laths. Its one end has hammer head and the other end is chisel-shaped with a claw to facilitate withdrawal of nails.

LATITUDE : The angular distance of a point measured from the equatorial plane towards the nearer pole along the meridian of the place on the earths surface. It is $0°$ upon the equator and $90°$ N and $90°$ S respectively on the North and South poles.

LATTICE GIRDER : A open-web girder, the top and bottom booms being connected by vertical and inclined bars.

LATTICE WINDOW : A leaded light in which small glass panes are bedded in lead cames.

LAUNDRY CHUTE: A duct from a bathroom through which dirty clothes are dropped for disposal.

LAUNDRY TRAY : A wide and deep sink fixed to a wall into which clothes are washed.

LAVATORY : A room containing a W.C., a urinal and a wash hand basin.

LAY BAR : A glazing bar fixed horizontally.

LAY-BY : A part of road width out of traffic lanes, where vehicles may be parked.

LAYER BOARD : A board on which the bottommost sheet for box gutter is laid.

LAYERED MAP : A contour map in which the areas bounded by different contour lines are coloured with different colours for identification of areas at a glance.

LAYING TROWEL : (i)   A brick trowel used by bricklayers.

(ii)   A rectangular trowel used by plasterers.

LAY LIGHT : A light fixed to ceiling horizontally.

LAYOUT : A general concept or arrangement for a proposed construction or installation.

LAY PANEL : A panel with its length horizontal.

LEACH : Removal of salts from a soil by passing water through it.

LEACHATE : The water that has come through a medium by leaching action.

LEACHING CESSPOOL : See 'Leaching Pit'.

LEACHING PIT : A cesspool or a pit built underground with honey-combed walls so that the foul water in the pit drains out by leaching action and finds its way into the subsoil.

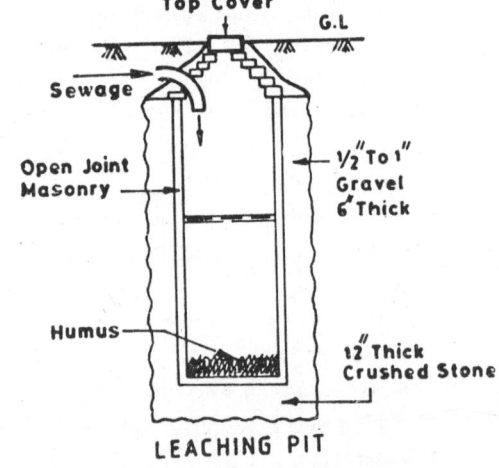

LEACHING PIT

LEAD : A metal used as a building material since the early days. At present it is used mostly in plumbing works. It is feebly lustrous bluish white or greyish metal. It is rapidly tarnished when exposed to air and a dark film is produced on its surface. It is used in jointing C.I. pipes, preparation of types (letters) and blocks for printing and manufacture of paints, important alloys and bullets.

LEAD-CAPPED NAIL : A nail having a lead washer under its head which helps in making a water-tight joint, when driven onto a roof sheet for fixing it to the purlin.

LEAD CESSPOOL : A rain-water head or hopper made of lead, fixed at the end of a parapet gutter to collect and discharge the rain-water into a down pipe.

LEAD CHROMES : Basic lead chromate or a mixture of lead chromate and chrome yellow which is used as a pigment in a paint; It produces yellow to orange colour with high opacity and staining power.

LEAD CHROME GREEN : A pigment composed of lead chrome and prusssian blue. Also known as 'Burnswick Green'.

LEAD DRIERS : Driers that are compounds of lead viz. linoleate, litharge and organic salts of lead, used in paints for quick drying and hardening of drying oils.

LEADED LIGHT : Lead glazing having diamond-shaped glass panes held in lead cames.

LEADER : The man who holds the leading end of a chain or tape in surveying.

LEADER HEAD : A rainwater head.

LEADERS : The guides for the drop hammer in pile driving.

LEAD FLAT : A flat roof covered with lead sheet.

LEAD GLAZING : A leaded light.

LEAD JOINT : A spigot and socket joint in cast iron water mains in which the annular space is filled with molten lead and after that it is caulked with a caulking tool. Sometimes, lead wool is used in the joint instead of molten lead.

LEAD LINE : A string or cord weighted with a hand lead and knotted at intervals, used as a sounding line for taking soundings in hydrography.

LEAD NAILS : Small nails made of copper alloy, used to fix lead sheets for roof covering.

LEAD PAINT : A paint that contains lead pigment, especially white lead. Lead paints are poisonous.

LEAD PLUG : (i)   A small tube of lead inserted into a wall-hole for fixing a screw or bolt tightly.

(ii)   A lead cramp cast into the groove between neighbouring stones laid in a course, to hold them together.

LEAD ROOF : See 'Lead Flat'.

LEAD SAFE : A shallow tray made of lead and fitted with a waste pipe, used to trap the overflow of waste water from a bath tub, sink, basin, etc. and to dispose it off.

LEAD SHEATH : Sheets of lead used to fill the condensation grooves, gaps in steel glazing bars, etc.

LEAD SLATE : A lead flashing which is flexible enough to fit round a pipe with no space or gap left, where it passes through a roof. This is made like a hat with its top cut off.

LEAD SPITTER : A short piece of outlet from a rain water gutter or hopper to a down pipe.

LEAD WOOL : Wool-like light fibres of lead which are required for lead caulking to prevent leakage of water from a spigot and socket joint of C. I. pipes.

LEAF : (i)   A door or window panel .

(ii)   One half of a cavily wall.

LEAFING : The floating property of metallic paints which contain aluminium, bronze or mica powder. The leaf-shaped metal grains protect the paint film and produce a good colour.

LEAN CONCRETE : A concrete made of poor mix like 1 : 4 : 8 or 1 : 6 : 12 (Cement : Sand : Khowa or stone ballasts), usually required for mass concrete work or for making a concrete mat in foundation.

LEAN LIME : A lime which is impure and slakes sluggishly. It expands less than fat lime and sets slowly. The hydraulic property is poor and it is used in inferior constructions.

LEAN MIX: A mix of ingredients for preparation of a lean mortar or lean concrete which has low binding material content i.e., the quantity of cement or lime used is low compared to the quantities of other ingredients like sand and coarse aggregates.

LEAN MORTAR: See 'Lean Mix'. It is harsh.

LEAN-TO-ROOF : A pent roof i.e., a sloping roof from a wall.

LEAPING WEIR : A device to divert the sanitary sewage of a combined sewer into a sanitary sewer through an adjustable crest in the invert of the combined sewer during storm as shown in illustration.

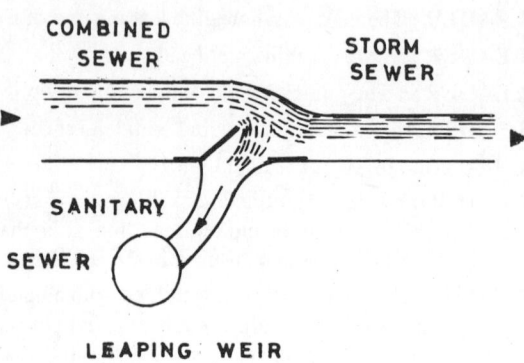

LEAR BOARD : A layer board.

LEAST COUNT : The smallest measurement that can be taken with the help of a vernier or micrometer.

LEAPING WEIR

LECA : Light Expanded Clay Aggregate used in preparation of a light-weight concrete.

LE-CHATELIER TEST : A test to determine the specific gravity of cement and also to determine the expansion of cement by quantitative method. The apparatus consists of a split cylinder with long indicators made of nickel, two glass plates, weights and cold water tray. Le chatelier's cylinder is shown in illustration.

LE CHATELIER APPARATUS

LEDGE : The horizontal timber piece at the back of a batten door.

LEDGED AND BRACED DOOR : A batten door which is braced diagonally between horizontal ledges at the back.

LEDGED DOOR : A batten door with horizontal ledges at its back as shown in illustration.

LEDGEMENT : A string course provided in a building. It is too small a projection all along the walls like a cornice at the level of intermediate floor.

LEDGER : A horizontal pole parallel to a wall in scaffolding, that carries the putlogs.

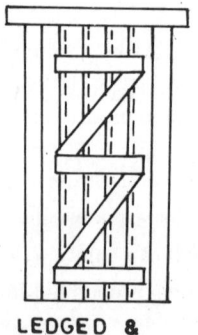

LEDGED & BRACED DOOR

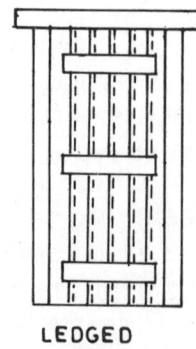

LEDGED DOOR

LEDGER BOARD : A ribbon board used in carpentry work.

LEDGE ROCK : A bed rock.

LEGENDRE'S RULE : The rule employed in ordnance survey. It states that if a triangle is small in comparison with the surface area of the sphere, one-third of the spherical excess from each angle of the spherical triangle should be subtracted prior to application of the usual sine formula.

LEGGATT : A wooden tool used by thatchers to bring the butts of reeds in a line by striking them. This is useful to the economically weaker section of the people to build their thatched houses.

LENTHENING JOINTS : Joints like scarfed, fished, indented, half- lapped, etc which were used earlier to increase the length of a timber member, when long steel sections were not available.

LESBIAN RULE : A straight strip of lead used in plastering and moulding instead of using a 'squeeze' round a mould.

LETTER PLATE : A hinged cover plate (which will shut automatically) fixed to the outside of the front door of a house to receive letters through a slit.

LETTING IN : Housing a timber member into another member.

LEVEE : An embankment constructed to prevent flooding in flood- prone areas which are low-lying lands.

LEVEL : (i)   The elevation or depression of a point with reference to a bench mark.

(ii)   To form a surface horizontal.

(iii)   An instrument with a telescope and a bubble tube required in conducting a level survey with other equipment.

LEVEL BOOK : A field book with vertical lines for noting the staff readings taken with the help of a levelling instrument during a level survey.

LEVEL CROSSING : A crossing between a roadway and a railway. Gates or movable barriers are provided at a level crossing to check the movement of vehicles on the road while a train passes over the crossing. A layout is shown in illustration.

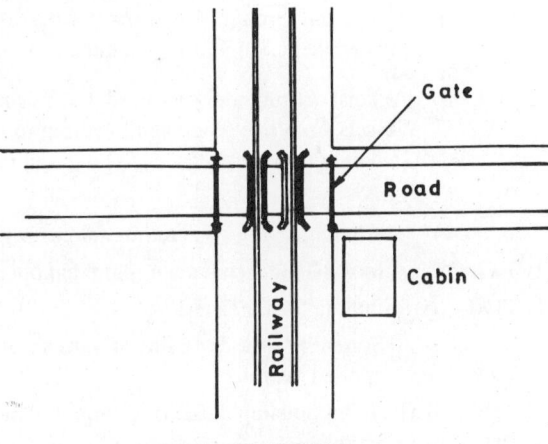

LEVEL CROSSING

LEVELLING RULE : A straight edge with a spirit level for making a surface uniform all throughout.

LEVELLING SCREWS : Foot screws at the base of a theodolite or a levelling instrument required in levelling the instrument at site.

LEVELLING STAFF : A staff made of timber with graduations on it, required in reading the elevation or depression of points during a level survey. The length of a staff is usually 6 ft to 14 ft. The staff readings are observed through the telescope of a levelling instrument stationed at points.

LEVEL OF CONTROL : A measure (standardised from test results) for keeping control over the production process by following the specifications.

LEVEL RECORDER: An instrument (float-operated or pressure operated) required to record continuously the level of water in a channel or canal.

LEVEL TRIER: An instrument with a bubble tube used to measure the slope by counting the number of graduations through which the bubble has moved.

LEVEL TUBE : A graduated glass tube filled with a liquid having an air bubble in it. The bubble remains at the centre of the gradutions when the tube is kept horizontal.

LEVEL ARM : The length of arm from the force applied causing a bending moment.

LEVER BOARDS : Louvres in a door or window, which are adjustable.

LEVER LOCK : A lock having a number of levers which must be moved simultaneously by a key to open or close the lock.

LEVY FACING : A water-tight facing built onto the water face of a dam.

LEWIS : Lifting pins (curved or wedge shaped) used in lifting stone or concrete blocks.

LEWIS BOLT : A form of foundation bolt as shown in illustration. The enlarged serrated conical head of the bolt remains embedded in concrete and a member is anchored by tightening the nut at top.

LEWIS HOLE : A hole cast in a concrete block or cut in a stone to lift it with the help of a lewis.

LIFT : (i) An elevator or an enclosure for lifting man and material from one level to the other in a high–rise building.

LEWIS BOLT

(ii) The height through which a water vessel rises or falls while passing through a lock.

(iii) A hoist (lifting device) used for raising water vessels from one reach and lowering to the other (having different water levels) without the use of a lock.

LEWIS HOLE

(iv) The height shuttered at a time for casting concreste in walls or columns.

LIFT GATE : A lock gate in a channel or dam that opens verticaliy.

LIFTING : (i) Pulling up a material.

(ii) Softening of a dry paint or varnish of first coat, when a second coat is applied over it.

LIFTING BLOCK : A hoisting tackle or differential pulley block.

LIFTING MAGNET : An electromagnet hanged from a crane to lift iron and steel or materials packed in a steel container.

LIFTING PINS : A lewis for lifting concrete or stone blocks.

LIFTING SHUTTERS : Window or door shutters that can be dropped or raised up vertically.

LIFTING TACKLE : A hoisting tackle.

LIFT IRRIGATION : A method of irrigation by lifting up water from one level to the other. There are various devices for lifting water.

LIFT-OFF BUTTS : Loose butt hinges forming a loose hinged joint, which can be separated by lifting one leaf from the other. This device is used in folding shutters.

LIFT SLAB CONSTRUCTION : A technique of rapid construction of building. The columns are erected and slabs for intermediate floors and roof are cast on ground. During the erection and curing of columns, slabs are cast and cured. Thus, the time is saved. When the columns attain strength, the slabs are lifted up one by one and placed at required levels.

LIFT SHAFT : A vertical shaft or well in a building into which a lift or elevator operates.

LIFT STATIONS : With a view to avoiding too deep excavation in laying sewers, pumping stations are installed at suitable locations to lift the sewage up.

LIGHT ALLOYS : Alloys of light metals like aluminium and magnesium.

LIGHT GAUGE RAILWAY : A narrow gauge railway, the gauge length being 2'—6". It is used in hill areas and in other places of less importance.

LIGHT HOUSE : A structure constructed on the sea-rock with arrangement of flash-light for cautioning the ships.

LIGHTNING CONDUCTOR : A copper conductor with lightning arrester built above the roof level of a tall building to protect the building from striking by lightning.

LIGHTNING SHAKE : Failure of timber by compression.

LIGHT–WEIGHT AGGREGATE : Clinker, expanded clay, foamed slag, vermiculite or perlite are used as aggregates to prepare a lightweight concrete.

LIGHT-WEIGHT CONCRETE : A concrete made with light-weight aggregates. Now-a-days,this concrete is used in construction of floor slabs of multi-storied buildings.

LIGHT WELL : A shaft that permits entry of air and light.

LIGNIFIED TIMBER : A high-density plywood which is lignin- impregnated at a high pressure.

LIGNIN : Wood resins that bind the fibres together.

LIMB : The circular base plate (lower most) of a theodolite which pivots on the tribrach. This is graduated from $0°$ to $360°$.

LIME : Burnt $CaCO_3$ which is CaO. This is quicklime. This lime when soaked in water produces hydrated lime, $Ca(OH)_2$.

LIME CONCRETE : A concrete made by mixing lime, surki and khowa or lime, sand and gravel with adequate quantum of water.

LIME MORTAR : A mortar prepared by mixing lime and sand or lime and surkhi with water.

LIME PASTE : Lime putty, which is 'slaked lime' with water.

LIME PLASTER : Plaster work made with lime mortar.

LIME POWDER : The powder obtained by slaking lime in air. It is not used in building construction.

LIME PUTTY : Lime paste.

LIME-SODA PROCESS : A process of softening hard water by removing sulphates of calcium and magnesium. In this process, permanent hardness is removed by addition of lime and soda ash to the water. The amounts of lime and soda ash required, depend upon the quantities of calcium and magnesium sulphates dissolved in water.

LIME STIRRER: A spoon-shaped wooden paddle for stirring lime solution.

LIMESTONES : These are limestone boulders, marbles, white chalk, oolite limestones and calcarious tufa used for manufacturing various types of lime. Hydraulic lime is produced from carboni-ferous limestones and kankars. A very powerful hydraulic lime is produced from magnesian limestones. Fat lime for ordinary use is manufactured from marble, tufa, white chalk and boulders.

LIME-TALLOW WASH : A lime wash mixed with tallow applied on wall surface, which is durable.

LIMEWASH : White wash by using the milk of quicklime with excess water.

LIMIT ANALYSIS : The method of determining the loading, which causes collapse of a structure. In the analysis of statically-indeterminate structures by this method, the application of limit load, limit torque and limit moment is considered.

LIMITING GRADIENT : The ruling gradient i.e., the steepest gradient that can be climbed safely by a motive force in a hill area. This is applied in making hill roads and railways in gradient.

LIMITING SPAN : The longest possible span with economy for a girder, beam or slab. This span varies with the type of structure and materials used.

LIMIT LOAD : The maximum load that can be applied to a structure just before the collapse begins.

LIMIT MOMENT : In fully plastic condition of a loaded beam, the resisting moment that causes stress distribution, is called limit moment.

LIMIT OF LIQUIDITY : The liquid limit i.e., the water content at which the clay is almost in liquid state.

LIMIT OF PLASTICITY : The plastic limit i.e., the smallest water content at which a soil is in plastic state.

LIMIT OF PROPORTIONALITY : The limiting point in a stress-strain curve obtained during tensile test of a specimen, at which the stress is proportional to strain. This point is close to the point showing elastic limit.

LIMIT TORQUE : The torque required to produce an infinite angle of twist for reaching the fully plastic state. This is one-third more than the maximum elastic torque.

LIMPET : Large harbours are usually provided with a limpet, which is a small open caisson that can be lowered into water by a crane by following the profile of the dock wall for repair and other works as and when needed.

LIMPET WASHER : A conical washer used under the nut of a hook bolt to hold a corrugated sheet. This is also used in fixing spokes onto a rim of a cycle.

LINE LEVEL : A small spirit level suspended at the middle of a mason's line in laying bricks.

LINEN TAPE : A measuring tape made of linen cloth used in surveying. This tape does not give accurate measurement, for which metallic tape or steel tape is preferred.

LINE OF COLLIMATION : The imaginary line joining the point of intersection of the diaphragm webs to the optical centre of the object-glass of the theodolite.

LINE OF LEAST RESISTANCE : In blasting operation, it is the shortest distance between the centre of the blast hole and the nearest free face.

LINE OF THRUST : The line along which the resultant thrust acts on a structure.

LINE PINS : Steel pins 75 mm long inserted into the brick work at the ends of the walls to hold the brick layer's line or cord for checking the accuracy of the line of brick work.

LINER : (i) A lining tool used in marking a line of painting.

(ii) A sleeve used in plumbing a line.

(iii) A stretcher board made of timber is spiked onto the opposite members of a timber frame to hold them firmly in their positions.

LINER SPOON : A special device to collect soil samples when it is desired to transport the samples without conducting any field test. Care is taken so that samples are not disturbed.

LINING : (i) Brick, stone or concrete lining on the bed and sloped sides of a canal to prevent scour.

(ii) A gunited or concreted surface given to a shaft or tunnel.

(iii) Acid brick or basic brick lining in a furnace or fire place.

(iv) Wood-wool slab, cork board or fibre board–lining for the purpose of neat and sound insulation.

(v) Plywood or commercial board lining to walls to improve the appearance.

LINING PAPER : Wall paper used for lining walls for decoration.

LINING PLATE : Flexible metal sheets used in roofing, especially in sloping roofs.

LINING TOOL : A tool used for painting lines.

LINK : One hundredth of an Engeneer's chain or Gunter's chain.

LINK DORMER : A 'dormer' connecting one part of a roof to the other having lights at sides.

LINOLEUM : Sheets built-up from hessian canvas and linseed oil used to cover floor. This is washable. The sheets are highly polished with various printed designs.

LINSEED OIL : A valuable oil used in preparation of paints and varnishes. The oil is obtained from the seeds of flax.

LINTEL : A small beam placed over the door or window opening to support the masonry built over it.

LINTEL

LINVILLE TRUSS : Pratt truss.

LIP : A band given on a flush door.

LIP BLOCK : A wooden piece (Lip piece) fixed over a strut, overhangs the waling in timbering a trench.

LIP UNION : A union having a ring-like inner projection used in plumbing work. The ring prevents the penetration of gasket into the pipe and thus partial blockage of the pipeline is avoided.

LIQUEFACTION : (i) A loose sand having void ratio greater than the critical one is subject to liquefaction when there is a flow of water through it. It becomes quick sand and there is a possibility of 'flow slide'.

(ii) Liquefaction of organic wastes during anaerobic decomposition in a septic tank or anaerobic pond or digester.

LIQUIDITY INDEX: The index of liquidity, which is one of the most important index properties of natural clay.

It is expressed as follows :

Liquidity Index = (Water content–plastic limit)/Plasticity Index

LIQUID LIMIT : See 'Limit of Liquidity'. This is determined by Atterberg limit tests.

Liquid limit = Plasticity Index + Plastic Limit.

LITHARGE : A pigment and a drier of pale yellow colour used in preparation of a paint.

LITHOPONE : A white pigment having high opacity which is a mixture of precipitated barium sulphate and zinc sulphide ; This is used in water paints, distempers and in preparation of enamels with synthetic resin.

LITTLE JOINERS : Pieces of wood used to fill holes in timbers.

LITTORAL BARRIER : An obstruction to normal littoral drift along the shore. It may be natural like a tidal inler or man-made e.g., jetties, groynes, and dredger channel. See illustration.

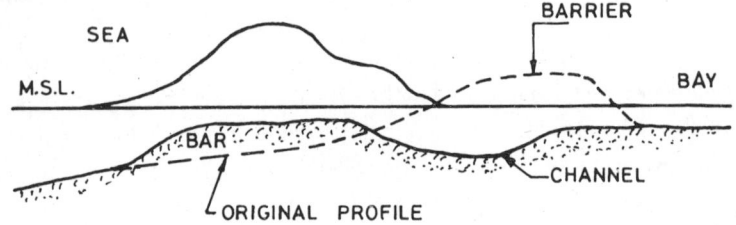

LITTORAL DRIFT

LITTORAL DRIFT : Longshore drift of alluvial material due to longshore flux of energy. It is given by Q=E Sin a, Where Q is quantity of littoral drift, E is long shore energy flux and a is deep water angle of incidence.

LIVE EDGE : An existing painted edge is said to be live edge when it matches with the newly painted adjacent surface without showing any difference in colour and any sign of lap.

LIVE KNOT : A knot found in timbers, whose fibres are intergrown with the surrounding wood. Upto a certain size of the knot, it is permissible in a structural timber.

LIVE LOAD : A load which can be displaced or shifted from one place to the other place on a structure.

LLOYD DAVIS FORMULA : In U.K. a formula was developed to calculate the quantity of storm run-off required in the design of sewers. The formula is

$$Q = 60.5 \text{ C I A}$$

where Q is peak rate of run-off in cfs.,

C is impermeability factor,

I is average intensity of rainfall in inches per hour,

and A is area in acres.

LOAD BEARING WALL : A wall that carries a load in addition to its self-weight and transfers the same to the foundation.

LOADING COAT : A concrete slab laid over ground to resist the upward thrust of water. This acts as a loading slab.

LOADED FILTER : A filter bed of graded gravels is provided at the foot of an earthen dam with a view to stabilising the dam toe by allowing the seepage water to pass through the filter with no accumulation of water under pressure within the dam.

LOAD-EXTENSION CURVE : A curve plotted with the loads applied and corresponding extensions observed in the tensile test of a metal specimen. The nature of load-extension curve is similar to the stress-strain curve.

LOAD FACTOR : The ratio of the load that causes failure of a structure to the design load. Load factor is used in plastic design of structures.

LOAD-INDICATING BOLT : A high-strength, friction-grip bolt which indicates the tension achieved when the bolt is tightened. The projections near the bolt head grip firmly by friction due to compression, when the head is pressed. The tension achieved can be measured by introducing a feeler gauge in the gap under the bolt head.

LOADING BOOM : An overhanging structure which facilitates in loading wagons.

LOADING GAUGE : The limiting dimensions of a loaded vehicle to pass through tunnels, bridges and railway level crossings as shown in illustration. This is provided with a view to avoiding fouling with the overhead construction, live wire, etc.

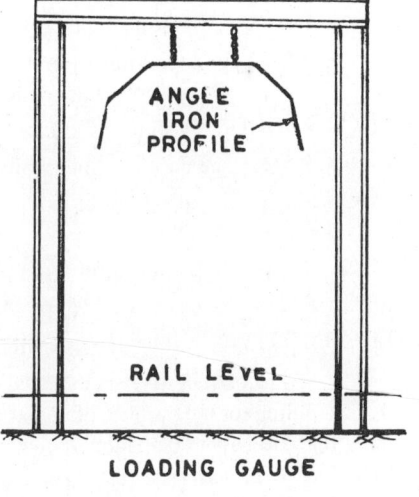

LOADING GAUGE

LOADING SHOVEL : A light-weight, fast-moving mechanical shovel on four rubber-tyred wheels which can be used for digging, loading & unloading, lifting materials and also surface dressing. This can also be used as a fork-lift truck for loading materials and lifting them up to scaffolding.

LOAD TEST : This test is often carried out to determine the bearing capacity and settlement characteristics of a soil at site by applying and increasing loads in stages and noting the stress-strain and consolidation or settlement. See illustration.

LOAM : A mixture of silt, clay and sand roughly in equal proportion.

LOCAL ATTRACTION: Deviation of the magnetic needle of the prismatic compass from the magnetic north at a particular locality due to the magnetic attraction in presence of iron and steel in the locality. Care should be taken against local attraction during a prismatic compass survey.

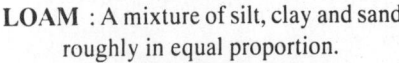

Loading By Sand Bag

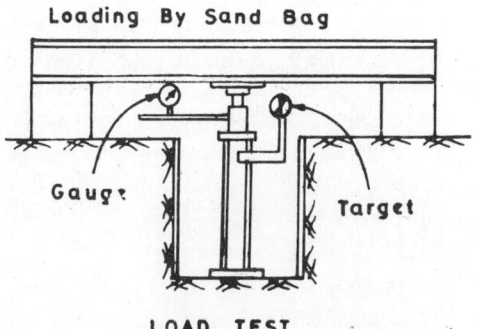

LOAD TEST

LOCATION PLAN: A site plan which shows the dimensions of the location of the proposed construction site.

LOCATION SURVEY : A survey to set out the adopted line on the ground and to obtain required data based on the preliminary survey.

LOCK : (i) A door fastening against theft;

(ii) A chamber separating two reaches of a canal by gates to facilitate passing of vessels from one reach to the other having diferent water levels.

LOCKAGE : The loss of water from one reach to the other during passing of a vessel through a lock.

LOCK BAY : A lock chamber.

LOCK BLOCK : A block of wood to which a lock is fixed in a flush door.

LOCK LEVEL : A hand level.

LOCKING BAR : A hasp and staple arrangement for fastening doors and gates, which may be secured with a padlock.

LOCKING STILE : The lock stile of a door in which a locking device is fitted.

LOCK GATE : A gate that separates water of upper reach from the lower reach.

LOCK JOINT : A joint made in flexible metal sheet roofing.

LOCK NUT : A special nut to be screwed onto a bolt with locking arrangement. This prevents loosening of a nut due to vibration.

LOCK PADDLE : A sluice used for emptying or filling a lock chamber.

LOCK RAIL : The door rail onto which a locking arrangement is fitted.

LOCK SILL : The strip of floor of a lock chamber against which the lock gates rest, when shut.

LOCKSPIT : A narrow V-groove cut on the ground for marking the line of excavation.

LOCK STILE : See 'Locking Stile'.

LOCOMOTIVE : A railway engine which draws the coaches.

LOCOMOTIVE CRANE : A heavy mobile crane which travels on the railway track, required for the purpose of construction and maintenance of railway bridges and also for loading and unloading heavy massive units.

LOCOMOTIVE HAULAGE : Haulage by loco for transport of man and material.

LOESS : Wind-borne silt of size 0.02 to 0.006 mm.

LOFT : (i) A box-like space under a roof which can be used for storage purpose.

(ii) An uppermost floor in a commercial building.

LOFT LADDER : A folding ladder (also known as disappearing ladder) fixed on the top of a trap door to enter into the attic or a hidden space. When the trap door is closed, nothing is visible from below. The trap door is usually hinged to open and hang downwards and the ladder is made of light timber or light alloy which can be folded up.

LOG : A tree trunk after trimming.

LOGARITHMIC GROWTH PHASE : The illustration shows the general growth pattern of bacteria under different phases. During log-growth phase, the cells divide at a rate determined by their generation time and their capability of processing the substrate (food) in the sewage. In this phase, the percentage growth rate is constant.

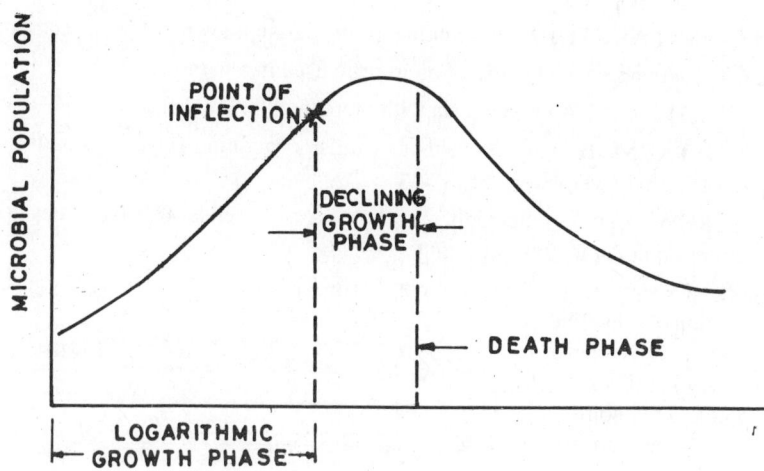

LOG CHUTE : A sloping ramp made of R.C.C. or masonry provided at one end of the weir wall, where wooden logs are to be transported by floating from the upstream to the down stream side of the weir without causing any damage to the structure. A gate is placed at the top of the chute to guide the logs. See illustration.

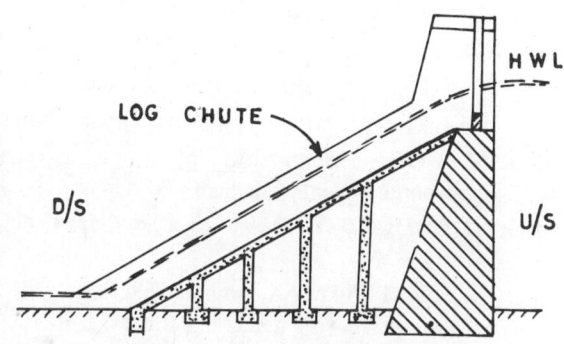

LOG WAY : See 'log chute'.

LONG COLUMN : A column having slenderness ratio (i.e., length of column divided by the minimum radius of gyration) more than 120 or length to diameter ratio more than 30 . A long column is designed to withstand buckling stresses.

LONG DUMMY : A tool used by plumbers for straightening kinks in bent pipes.

LONG FLOAT : A wooden float used by the plasterers and concrete Workers to level a substantial area.

LONG HOPPER CLOSET : This type of closet was in use in early days of sanitary engineering practice. It consists of a long inverted stoneware cone connected with a trap below.

LONGITUDE : The angular distance between the meridian of a place and the standard meridian, measured from 0° to 180° towards East or West .

LONGITUDINAL BOND : A brick bond used in construction of thick walls with courses of bricks laid as stretchers.

LONGITUDINAL PROFILE: A long vertical section through the centre line of a road or sewer to depict the levels.

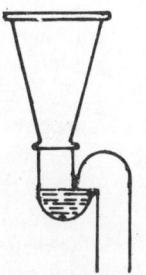

LONG HOPPER CLOSET

LOGITUDINAL SECTION: Longitudinal profile. Also known as 'Long section'.

LONG OIL : An appreciable ratio of oil to resin used in a varnish.

LONG-OIL ALKYD : An alkyd resin with more than 60% of oil .

LONG-OIL VARNISH : An oleo-resinous varnish containing at least 2.5 parts of oil to 1 part of resin by weight.

LONG SCREW : A pipe connector having threads on both ends. This facilitates in opening the pipes quickly for easy cleaning.

LONG SLEEVE JOINT : This is almost a rigid joint with less flexibility in changing direction of a pipeline. It has a long overlap of the socket over the spigot. This joint is suitable in bad soil which is likely to subside.

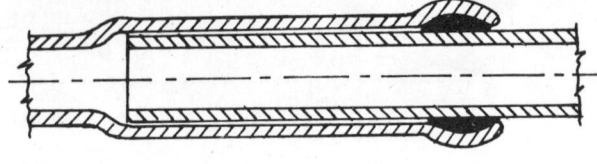

LONG SLEEVE JOINT

LONG SPAN : See 'Limiting span'.

LONG TON : A ton that weighs 2240 lbs.

LOOKOUT : A woden bracket to support the overhanging part of a floor or roof.

LOOP VENT : A ventilation pipe raised up in continuation of a soil pipe.

LOOSE CORE : Segregation of coarse aggregates from a concrete mix during transportation and placement of concrete.

LOOSE GROUND : Granular soil mass which is not compacted well and can be dug out easily.

LOOSE KNOT : A knot in a timber, which is not held firmly by the inter-woven fibres with the wood. This is a defect in a timber.

LOOSE PIN BUTT : A butt hinge with a hinge pin is so fitted that the pin can be withdrawn to remove the hanging shutter without unscrewing the butts.

LOSS OF GROUND : It occurs due to the flow of materials from the adjacent areas of an excavation, when a soft ground containing silt, sand and clay is excavated. So, in such a situation, excavation should be done in sheet-piled cofferdams.

LOSS OF HEAD : The illustration shows the loss of head during filtration of water through a filter bed. The loss is due to the resistance in flow path of water through the bed. In a pipe flow or channel flow, the loss of head occurs due to the frictional re-sistance. Thus, the pressure head gradually decreases with the length of a pipe line of same diameter,

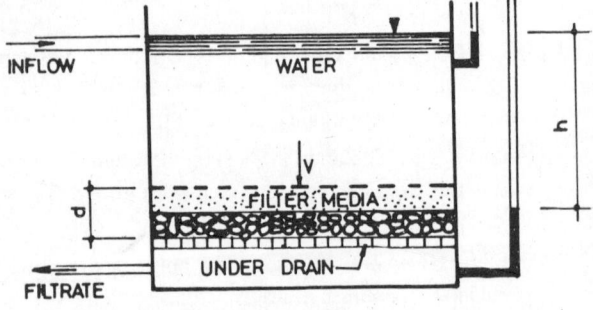

due to the increasing loss of head. Again, the loss of head through a pipe line will be more, if the diameter of the pipe is reduced.

LOST-HEAD NAIL : A wire nail of round section having a very small head which goes partly inside the timber, when hammered.

LOUVRES : Sloping slats are sometimes provided in a window shutter (Venetian shutter) to allow ventilation and to exclude rain water. The louvres may be fixed or movable (hinged to a monkey bar).

LOW-CARBON STEEL : A mild steel that contains a very low percentage (0.05 to 0.25%) of carbon. Mild steel is graded according to its carbon content.

LOW-HEAT CEMENT : A portland cement having low lime content which initially liberates a low heat. Such a cement is used in mass concrete for construction of dams or some massive works.

LOW-WATER VALVE : A valve working on the principle of a fusible plug.

LUFFING CABLEWAY MAST : A cableway tower hinged at its base and held by guy ropes is capable of having an allowable lateral movement to facilitate the movement of a track cable.

LUFFING JIB CRANE : A crane with a jib hinged at its base facilitates working of the hoisting rope at different radii.

LUG : A small projection from a pipe or frame or any other article for fixing it.

LUG SILL : A sill whose ends are built into the jambs.

LUMBER : Square or rectangular timber sawn from logs.

LUMBER CORE : Core board.

LUMP HAMMER : A club hammer.

LUMPING : Renewing a rail track by pulling out the old pair of rails along with sleepers with the help of a crane and fixing a new pair of rails fitted with sleepers under them.

LUMP LIME : Hand-picked lime of good quality.

LUMP SUM CONTRACT : A contract made for execution of a work as per specification at a fixed price (total value).

LURCHING ALLOWANCE : An allowance kept for extra load on the outer girders of a railway bridge that is likely to occur due to sway of a locomative during its movement over the bridge.

LYTAG : A light-weight aggregate obtained by sintering fly ash. This is used in preparation of light-weight concrete.

M

MACADAM : The most common road metal i.e., stones of uniform size which are used in making a road by rolling. The macadam surface may be water-bound or bitumen-coated or cement-bound.

MACADAM SPREADER : A machine to spread macadam uniformly to form a road surface.

MACARTHUR PILE : Uncased or cased cast-in-situ concrete pile with or without pedestal. The steel drive pipe and steel core are driven together. Then, the core is removed and concrete is deposited.

MADE GROUND : A ground formed by land reclamation by filling low- lying areas and marshy land.

MADRAS SCHOOL OF ART TILE : A type of roofing tile as shown in illustration.

MAGNESIAN LIME : A type of quick lime containing 10% to 40% magnesium oxide.

MAGNESITE : Magnesium carbonate, a mineral.

MAGNESITE FLOORING : A type of jointless hard floor that is made by mixing magnesia (MgO) and magnesium chloride (Mgcl$_2$.6H$_2$O) with sand, saw dust or iron dust.

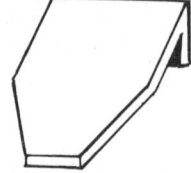

MADRAS SCHOOL
OF ART TILE

MAGNESIUM HYDROXIDE : Mg((OH)$_2$ produced by adding water to Magnesia (MgO).

MAGNETIC BEARING : The horizontal angle of a chain line or a point in surveying from the magnetic north.

MAGNETIC COMPASS : A small handy compact instrument with a magnetic needle used to measure bearing in surveying.

MAGNETIC DECLINATION : See declination.

MAGNETIC NORTH : Centres of magnetic attraction in the north of Canada and Antarctica, which are continually changing, due to which the declination varies throughout the world.

MAGNETIC VARIATION : Variation in declination occuring daily. The variation may be classified as : secular, annual, diurnal and irregular.

MAGNETO GENERATOR : An electric generator having a permanent magnetic field.

MAGNETO METER : An instrument used for measuring the intensity of a magnetic field. This is chiefly employed in geophysical survey to detect the location of magnetic rocks.

MAHOGANY : A kind of durable hard wood which is used in making valuable furnitures.

MAIN : Electric main, water main, etc.

MAIN BEAM : A beam that rests directly over walls or columns and support the subsidiary beams or rafters.

MAIN CANAL : A canal that delivers water to its branches for the purpose of irrigation.

MAIN COUPLE : A timber truss made of principal rafters.

MAIN DRAIN : A major drain that receives waste water from its branches and laterals and leads to the outfall.

MAIN GATE : In mining, it is the haulage road upto a coal face.

MAIN HOLES : Relief holes used in mining.

MAIN SEWER : A major sewer which receives sewage, sullage and/or storm water from its branches and laterals and leads to the disposal site.

MAIN TIE : The principal tension member joining the two ends of a truss at the base level (springing level).

MAISONETTE : An apartment which is self contained, but at two levels having an internal stair. Also known as 'Duplex Apartment'.

MALL : A heavy mallet.

MALLEABLE CAST IRON : A heat-treated and oxidized cast iton which has tensile strength double that of ordinary cast iron. Actually this cast iron is not malleable, but it shows remarkable ductility in tensile test, due to its improved toughness and strength.

MALLET : A wooden hammer, the striking head being made of hard wood.

MALLET-HEADED CHISEL : A type of chisel used by masons which has a round steel head that can be struck with mallet.

MAMMOTH PUMP : An air-lift pump used within the hollow boring rods in mining.

MAN-DAY : The quantum of work done by a man in a day. It varies from man to man depending upon one's expertise.

MANDREL : A cylindrical piece of timber inserted into a distorted lead pipe by giving pressure, so that the distortions are made good.
This is also used to enlarge the pipe diameter.

MANGALORE TILE : A special form of roofing tile.

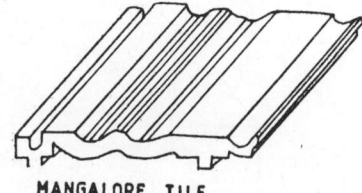

MANGALORE TILE

MANGANESE DRIER : A drier consisting of manganese di-oxide and salts of manganese, used in paint.

MANGANESE STEEL : An alloy steel containing 1% to 10% of manganese, which makes the steel tough and hard.

MANHOLE : An access hole to a tank or a sewer junction chamber, so that a man can enter into the tank or chamber through this hole for inspection and repairwork.

MANHOLE COVER : A cast iron or concrete cover used to close a manhole, which can be taken out or opened as and when needed for entry of a man.

MAN-HOUR : The quantum of work done by a man in one hour. The cost of manpower is assessed on the basis of manhour required to be employed in a job.

MANIFOLD : A major pipe (leader) from which several small pipes branch off.

MANILA ROPE : A rope made of fibrous material like jute or hemp.

MANIPULATIVE JOINT : A compression joint made by flaring (widening) the ends of a copper tube. No gland is required in making such a joint.

MAN-LOCK: An air-lock chameber through which workmen enters into a caisson, shaft or tunnel.

MANNING'S FORMULA: The most commonly used formula in design of sewers.

$$V = (1.486/n)\, R^{2/3}\, S^{1/2} \text{ .... in Imperial Units}$$
$$= (1/n)\, R^{2/3}\, S^{1/2} \text{ .... in Metric Units}$$

where V is velocity of flow in open channel flow condition;

         n is co-efficient of roughness;

         R is hydraulic radius;

and S is slope for chezy's equation.

MANOMETER : A gauge in form of a U-tube filled with mercury or any other liquid to measure the pressure difference. When it is required to measure the pressure difference between two points of a fluid carrying pipe, the two ends of the manometer are connected to the two points of the pipe.

MANSARD ROOF : Similar to a 'gambrel roof'. See illustration. On each side of the roof the top slope is relatively flat compared to the slope of the lower part of the sloped roof.

MANSARD ROOF TRUSS : A timber roof truss as shown in illustration is required to form a mansard roof. It is used for a large span.

MAP : The presentation of topography, physical features, landuse, road network and other utillity services of an area in form of a drawing.

MANSARD ROOF

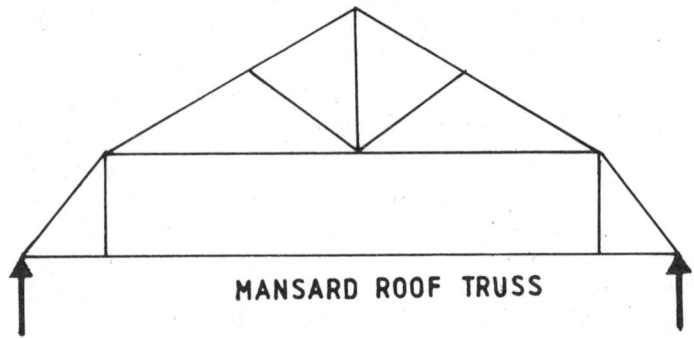

MANSARD ROOF TRUSS

MARBLE : A type of limestone which is valuable for its beauty and durability.

MARBLE FACING : Brick masonry faced with marble stone slabs for better appearance.

MARBLING : Painting a surface to bring the tone of marble shade.

MAREZZO MARBLE : A kind of marble made artificially.

MARGIN : (i)   A border.

        (ii)   The exposed flat surface of a door or window stile or rail.

       (iii)   The projections of a close string of a stair.

MARGIN LIGHT : A glass pane of restricted width at the edge of a sash.

MARGIN TEMPLET : A board with an edge strip to form the width of the margin.

MARGIN TROWEL : Also known as 'Angle Trowel'. A U-shaped steel trowel with square-ended two parallel sides, which is used to form margins and to work internal angles.

MARIGRAPH : A gauge to record fluctuations in the levels of tides in a tidal river.

MARINAS : A small-craft harbour, where a water area of about 200 sq. m is provided for each boat and the land area is approximately 80% of the water area. It has shallow water depth of about 2.5 to 3.5 m. With a view to keeping wave action down, the entrances are made narrow with a front basin.

MARINE BORER : Molluscs and crustaceans belong to the marine borer group, which live in warm water and destroy the timber of marine vessels by eating it. The ship-worm can penetrate any type of wood in water under favourable condition.

MARINE GLUE : A glue composed of 1 part of rubber, 2 parts of shellac and 3 parts of pitch was used earlier for glueing marine vessels prior to the discovery of synthetic resin.

MARINE SOIL : This is a form of transported soil which gets deposited in sea water.

MARINE SURVEYING : Hydrographic surveying to estimate water currents, scouring and silting effects and mapping the nature of sea bed.

MARITIME PLANTS : Plants like rice grass, marram grass and shrubby seablite that grow on foreshores and help in preventing scour.

MARKING GAUGE : A hardwood block with a bar having a steel pin projected at right angles from it. This is used by carpenters for marking lines parallel to the planed surface.

MARKING KNIFE : A steel bar with a point at one end and a cutting edge at the other, used by carpenters for marking timber prior to cutting.

MARL : A soil or rock containing lime.

MAROUFLAGE : Gluing a canvas to a wall by rolling which forms a matt surface to cover it by a mural painting.

MARSHALLING YARD: The yard where trains and wagons are received, sorted out stationwise for respective destinations, new trains are formed and despatched onwards. The essential functions are :

    (i)   Reception of wagons, loaded or empty.

    (ii)   Sorting of wagons including sick siding.

    (iii)   Despatch of trains from departure siding. The yard is provided at important railway junctions. See illustration.

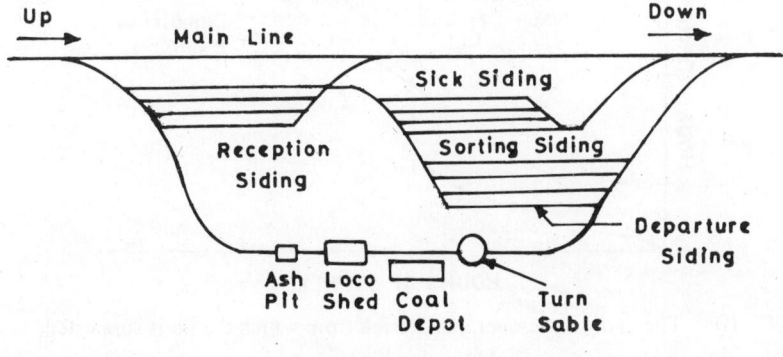

MARSHALLING YARD

MARQUETRY: Decorating a wooden surface by cutting out hollows and then infilling or inlaying into these hollows, ivory, metal, woods of different colours or mother of pearl.

MASH HAMMER : Sledge hammer.

MASKING : The protection of edges of a painted surface during painting by holding a paper or card board mask over them or by using sticking tape or plaster or paper over the edges.

MASON : A brick layer or stone setter.

MASONRY : Brick work or stone work in making walls.

MASONRY CEMENT : A cementing material consisting of portland cement, lime, clay and other plasticizers, which is to be mixed with sand and water for masonry work.

MASONRY FIXINGS : Anchor bolts, cramps, dowels, rag bolt, etc.

MASONRY NAILS : Cadmium-coated hard screws or nails that can be driven into bricks by hammering.

MASON'S JOINT : A mortar joint of triangular projection.

MASON'S PUTTY : Lime putty mixed with stone dust and cement in the ratio of 5 : 7 : 2 used in ashlar work for jointing stones.

MASON'S SCAFFOLD : A scaffold supported on two rows of standards and well-braced baulks.

MASON'S STOP : Mason's mitre, which is an angle joint made by cutting and shaping the corners of regular stones.

MASS CENTRE : Centre of gravity.

MASS CONCRETE : Massive plain concrete.

MASS DIAGRAM : A graph representing the integration of a time- flow curve.

MASS-HAUL CURVE : A curve which shows the amount of excavation and fill in a length. A curve plotted with hourly demand of water vs. hours of supply as shown in illustration.

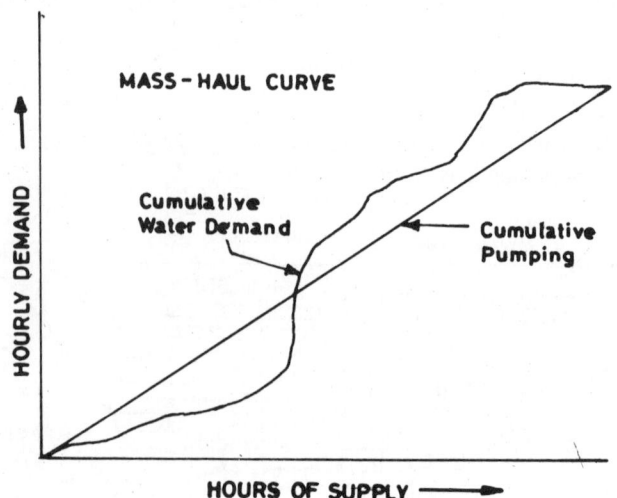

MAST : (i) The vertical member in a derrick from which the jib is supported.

(ii) A slender tower held by guy ropes.

MASTIC : (i)   The exudation from mastic tree i.e. resin, which is used in varnishing by dissolving it in alcohol.

(ii)   A plastic water-proof compound which forms a hard surface. This is used as a sealant for sealing joints and gaps in prefabricated buildings and also jointing pipes, gutters, etc.,

(iii)   Glaziers' putty.

(iv)   A quick-setting water-proof plastering material which can also be used in pointing.

MASTIC ASPHALT : A wearing course to a road or roof, consisting of liquid asphalt laid hot.

MAT : A concrete footing or raft under a wall or column.

MATCH BOARD : Boards used for flooring and wall lining by placing them side by side and joining by tongued and grooved joint.

MATCHED, BEADED AND VEE : A type of joint used in floor boards or ceiling boards. See illustration.

MATCHED, BEADED & VEE

MATCH PLANES : A pair of planes that cut tongue and groove in matchboards.

MAT SINKING : A depression on floor at the inner side of a door into which a door mat is placed.

MATERIALS LOCK : An air-lock chamber through which construction materials are introduced into a pneumatic caisson and rubbish materials are taken out.

MATRIX : Cement-sand or lime-sand or bitumen-sand mix into which coarse aggregates are embedded.

MATT : A poor gloss in paint.

MATTOCK : A tool having one end like a hoe and the other end like a chisel or pickaxe, which is used in cutting hard ground and taking out tree roots.

MATTRESS : (i)   A mat concrete slab.

(ii)   A course of blinding concrete.

(iii)   A flexible layer of brushwood fascines sunk onto a river bed by the weight of stone or concrete blocks, with a view to preventing scour in river bed.

MATT VARNISH : A matt finish in varnishing by using dammar dissolved in turpentine.

MATURING : (i)   Fattening up in plastering.

(ii)   Curing of concrete.

(iii)   Ageing of varnish in varnishing.

MAXIMUM DRY DESNISTY : The dry density obtained by maximum amount of compaction of a soil at its optimum moisture content.

MCMATH FORMULA: An empirical Run-off formula.

$Q = CRA\ 5(S/A)^{1/5}$ where Q is run-off in cfs;

R is maximum rate of rainfall in inches/hour;

S is ground slope, ft/1000 ft; A is area in acres and C is co- efficient of imperviousness.

MCMATH recommended $C = 0.75$, $R = 2.75$ and $S = 15$

MAUL: A large wooden mallet made of hard wood.

MEAN DEPTH : (i)   The average depth

(ii)   The cross-sectional area of a channel or canal or stream divided by the surface width.

MEAN MID-NIGHT : It occurs at the instant, the mean sun crosses the meridian at lower transit.

MEAN NOON : It occurs at the instant, the mean sun crosses the meridian at upper transit.

MEAN SOLAR DAY : An average of all apparent solar days occuring during a revolution of the earth on its orbit.

MEAN SUN : An imaginary point travelling round the ecliptic in exactly the same time as the real sun. The projection of its movement along the equator is uniform.

MEAN TIME : The time at any point on the earth measured from the lower transit of the meridian of the point of observation.

MEAN VELOCITY : The average value of velocities at the surface and at the bottom of a flowing stream.

MEASUREMENT : The quantity surveying to draw up the bill of quanities.

MEASURING CHAIN : Engineers' chain or Gunter's chain used in surveying for measurement of lengths.

MEASURING FRAME : A batch box for measuring aggregates and cement for preparation of a concrete mix.

MEASURING WEIR : A weir having a trapezoidal or rectangular or triangular notch for measurment of quantum of flow.

MECHANICAL ADVANTAGE : The ratio of load lifted by a machine to the force applied.

MECHANICAL ANALYSIS : Analysis carried out to find out the different particle sizes in samples of soil, sand or similar such materials. This is required in gradation or classification of the sample. The sizes of coarge-grained constituents are determined by means of a set of sieves and for soils containing fine-grained material, a wet method of mechanical analysis should be carried out.

MECHANICAL BOND : Bonding or adhesion obtained by using torque steel or indentated reinforcement bar in concrete. This increases the bond strength.

MECHANICAL CORE : Sanitary core i.e. prefabricated pipes ready for installation with minimum work at site. The pipes are used for water supply, drainage of waste water or running electric cables through them in an industrial building.

MECHANICAL EFFICIENCY : The mechanical advantage divided by the velocity ratio. This is expressed in percentage.

MECHANICAL EQUIVALENT OF HEAT : Joule's equivalent. It is the quantum of mechanical energy required to produce one heat unit.

MECHANICAL RAMMER : A ramming machine that lifts up and drops a weight.

MECHANICAL SHOVEL : A power-driven shovel or excavator.

MECHANICAL SAW : A power-driven circular saw or band saw used in sawing timber.

MEDICAL LOCK : A cylindrical air chamber made of steel with air-tight doors and having compressed air supply required for treatment of caisson disease.

MEDULLARY RAY : The radial lines starting from the pith (at centre of a tree) seen, when a tree is transversely sectioned.

MEETING POST : Timber posts fitted at the outer ends of a pair of lock gates.

MEETING RAILS : The rails of a sash window which touch each other, when the window is shut.

MEETING STILE : The middle stile of a folding door.

MEKOMETER : Once used as a range finder in military. It consists of two instruments operated by two persons simultaneously. The left hand instrument consists of two mirrors fixed at 45° and the right hand instrument has one mirror capable of rotation.

MELAMINE FORMALDEHYDE : A synthetic resin used for gluing veneers or laminated boards and also for their surface treatment.

MELAMINE SURFACED BOARD : Resin-bonded board with melamine formaldehyde surface which produces a smooth and decorative appearance.

MEMBER : A part of a structure or a building.

MEMBRANE : A skin-like thin film or layer.

MEMBRANE ANALOGY : For easy understanding of the lines of stress in a twisted structural member, an analogy is developed between the lines of equal shear stress and the contour lines of a soap bubble of the same shape.

MEMBRANE FILTER : Used to determine the number of coliform organisms present in water. This is done by passing a known volume of water sample through this fitler on which the coliforms are retained. This gives a direct count of coliforms.

MEMBRANE THEORY : A theory used in design of thin shells assuming that only shear stress and direct stress (tension or compression) are developed at any section of the shell. The shell deflects like a bubble and it can not resist any bending.

MENDING PLATE : A steel plate with drilled holes which is screwed to the either side of a broken part of a timber.

MENSURATION : A subject dealing with measurement and calculation of lengths, angles, areas and volumes.

MERCHANTABLE BOLE : The usable length of a tree trunk as timber.

MERIDIAN : A plane passing through a point under consideration in surveying and the earth's axis of rotation.

MERIDIAN PASSAGE : The culmination of a star used in geodetic surveying.

MESH : Wire-woven or fibre-woven screen of different grades (pore sizes or openings) used for sieving materials.

METAL COATING : Applying thin films of non-corrodible metals like nickel, copper, chromium, zinc, cadmium, etc. to envelope iron surface. This is done by sheradizing, plating and galvanizing. This brings a good appearance in addition to a protective coating.

METAL CRAMP : A bent metal bar usually of steel to hold stones together.

METAL LATHING : Use of perforated metal sheets or expanded metal to serve as a base for plastering.

METALLING: Surfacing a road with stone chips and bitumen.

METALLIZING: Applying a thin coat of non-corrodible metal.

METALLURGICAL CEMENT : Super-sulphated cement.

METAL PAINTING : Painting a surface with bronzing fluid as a vehicle of paint.

METAL SHEETING : Using sheet metals in roofing, partition, etc.

METAL SHUTTERING : Use of metal sheets for shuttering in concrete work.

METAL SPRAYING : Appying a thin film of metal coating by means of spraying

METAL TRIM : Sheet metals used in bordering a picture frame, skirting and architraves.

METAL VALLEY : A valley gutter lined with a sheet metal of zinc, lead, copper, etc.

METAMORPHIC ROCK : A rock produced by transformation of some rocks due to internal processes acting on them. The agents of metamorphism are heat, pressure and hydrothermal solutions.

METEORIC WATER : Atmospheric water.

METER : A measuring instrument or device with graduations and indicator for reading and with or without an integrating recorder.

METHANE : Marsh gas (CH4). This gas is produced during anaerobic fermentation of wastes and digestion of sludge.

METHYLATED SPIRIT : Methyl alcohol used in varnishing, polishing, etc.

MICA FLAP VALVE : A mica sheet hinged at a 'fresh air inlet' to allow entry of fresh air into a soil pipe.

MICROBES : Microscopic organisms.

MICROMESH DRUM FILTER : Drum type fine screen having non-ferrous wire mesh screen cloth. The drum remaining half-submerged in waste water rotates at a speed of about 4 r.p.m. The screenings when raised above the liquid level, are back-flushed into receiving troughs by high pressure jets of water.

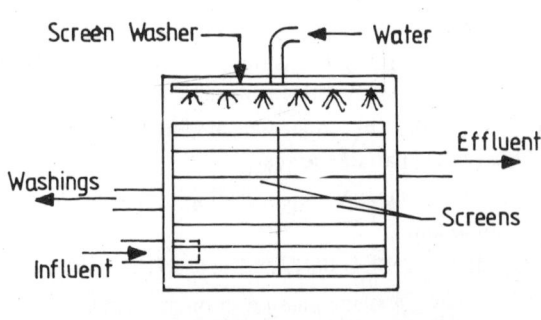

MICROMESH DRUM FILTER

MICROMETER : (i)   A measuring instrument or device to measure one hundredth and even one-thousandth part of an inch.

               (ii) A fitting on the eyepiece of a levelling instrument to read the levelling staff upto one-thousandth part of a foot.

MICROMETER GAUGE : See 'Micrometer'.

MICROMETER THEODOLITE : A theodolite provided with a micrometer in place of a vernier, which enables to read the angles accurately in surveying.

MICRON : One-thousandth of a millimetre i.e. 10^{-3} mm. which is denoted by μ.

MICROSCOPE : An instrument that magnifies a minute object several hundred times. The objects that can not be seen in bare eyes, are observed through a microscope.

MICRO-STRAINER : A rotating drum filter having a fine screen fitted on the periphery of the drum. This is used in removing algae from water prior to treatment and in polishing activated-sludge effluent. The collected solids are back washed by high pressure jets of water.

MIDDLE-THIRD : It is middle one-third of the width of a wall. In design of a masonry or concrete retaining wall or abutment wall, it is aimed to pass the resultant of all the forces acting on the wall including its self weight, through a point anywhere within the middle-third of the base so that no tensile stress is developed and the wall is not overturned.

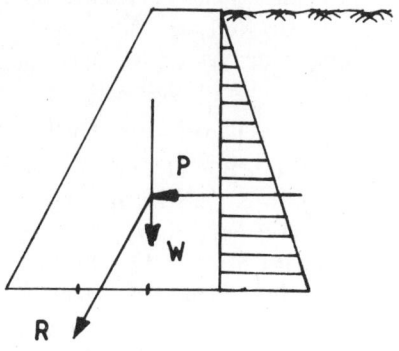

MIDDLE-THIRD RULE

MID-FEATHERED TRAP : See illustration. Now-a-days, this trap is not in use.

MID-ORDINATE : An ordinate in the half-way between two extremes.

MID-POINT : A central point between two extremeties.

MID-RANGE : The average value of the minimum and the maximum i.e. at the middle of two extreme ranges.

MID-SPAN : A point at the middle of a span i.e. halfway between two supports of a beam.

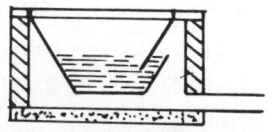

MID FEATHERED TRAP

MIL : One-thousandth of an inch.

MILD STEEL : Basically it is an alloy of iron having low carbon content which makes it ductile.

MILKINESS : The whitish appearance of a varnish film which indicates that the material is defective.

MILK OF LIME : Obtained by slaking lime in water and screening it.

MILL : A machine for puddling, grinding, crushing, shredding and roasting.

MILLED MORTAR : Mortar made in a mixing machine or in a pug mill or country mill.

MILLIMETRE : One-thousandth part of a metre.

MILLIMICRON : One-millionth of a millimetre.

MILL SCALE : A scale of black iron oxide is formed during hot rolling of steel sections. The bars coming out from this process are called 'black bars' owing to their black colour due to formation of scale.

MILL TAIL : A 'tail race'.

MILLWORK : Prefabricated parts of a door, window, partition panel, wooden stair at mills which are assembled at site.

MINERAL BLACK : Black pigment of mineral origin, e.g. graphite, carbonaceous clay, etc.

MINERAL STREAK : Brownish or greenish streaks found in some hardwoods.

MINERAL-SURFACED BITUMEN FELT: A bitumen felt whose top surface being dressed with slate chippings. This is used in covering sloping roofs.

MINERAL WOOL: A flexible, resilient material made from mineral fibres used in making waterproof blankets.

MIRROR CLINOMETER : A handy and compact instrument required for measuring vertical angles and is useful for determining the slope of a ground when chaining in a survey work.

MIST COAT : (i) A very thin film of latex-emulsion paint applied over a distempered wall surface.

(ii) A very thin film of cellulose lacquer applied by spraying over a surface.

MIT CLASSIFICATION : A system of soil classification by the grain size distribution, which is preferred for engineering purposes. According to M I T classification,
Gravel–larger than 2 mm. size
Coarse Sand – 2 mm. to 0.6 mm.
Medium Sand – 0.6 mm. to 0.2 mm.
Fine Sand – 0.2 mm to 0.006 mm.

MITRE : An angle joint or bevelled joint. For a 90° angle joint, the ends of the pieces to be joined are cut at 45° angle.

MITRE BLOCK : A block of wood with saw cuts in it at 45°, which is used for cutting the timber ends exactly at 45° following the profile of the cut.

MITRE BRAD : A corrugated iron fastener used in timber joints.

MITRED AND REBATED : See illustration.

MITRED CAP : A cap provided on a newel post into which the handrail is mitred.

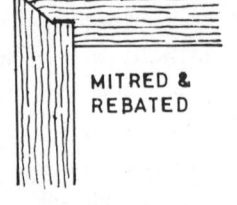

MITRED & REBATED

MITRED CLOSER : A bevelled closer i.e, a brick closer cut at an angle.

MITRED JOINT : See illustration.

MITRED KNEE : A mitred joint which forms a bent like a knee at the point of intersection between a horizontal handrail and a sloping or falling handrail in a stair.

MITRE DOVETAIL : A secret dovetail joint in which the pins are not visible excepting the mitre line.

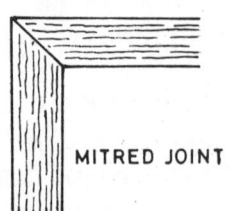

MITRED JOINT

MITRED VALLEY : A hip valley or a close-cut valley forming a mitre.

MITRE JOINT : See 'mitre'.

MITRE MACHINE : A machine used by wood workers for trimming timbers.

MITRE SAW : A tenon saw used in joinery works in timber.

MITRE SILL : The sill of a canal lock.

MITRE SQUARE : A level square.

MITRE TEMPLET : A rebated frame used to guide the chisel during mitring a moulding.

MIX : The mixture of ingredients (dry or wet) to form a paste or concrete. The term is normally used for making a mortar or concrete.

MIXED-FLOW TURBINE : A turbine of inward-flow reaction type, in which the flow acts both radially and axially on the runners.

MIXED GLUE : A glue in which an accelerator is mixed with synthetic resin glue.

MIXER : A mixing machine for making mortar or concrete or a puddling machine for clay.

MIX-IN-PLACE : Stabilization of soil by injecting travel mix, a chemical, without taking out or disturbing the soil.

MNEMONIC : A memory–aid to remember engineering formulae. $H=0$; $M=0$; $V=0$ for equilibrium of a strcucture i.e, the conditions of equilibrium which can be remembered as ' His Master's Voice' Another example like this is

Q = AIR; where Q=Peak rate of run-off, A= Catchment area

I = Average intensity of rainfall per hour, and R= Run-off co-efficient.

This may be remembered as Q = All India Radio

MLSS : Mixed liquor suspended solids in a sewage or waste water.

MOBILE CRANE : A crane mounted on crawler tracks to facilitate its movement from one place to the other.

MOBILE HOIST : A platform hoist supported on a pair of wheels for towing it from one location to the other.

MODEL ANALYSIS : Analysis carried on models of structures, harbours, pipe networks, etc, simulating field conditions, with a view to arriving at a result that can be implemented in large scale in field.

MODULAR MASONRY UNIT : A modular brick or stone.

MODULAR RATIO : The ratio of the modulus of elasticity of reinforcement steel to that of concrete in a reinforced concrete structure. It is denoted by 'm' and expressed as, $m = E_s/E_c = f_s/f_c$.

MODULAR SYSTEM : The system planning of building units or buildings or structures to match with the planning grid of a module.

MODULATED CONTROL : Automatic control of an air-conditioning unit by means of thermostat switch. In a modulated control, the switch on or switch off does not take place rapidly, but gradually by increasing or decreasing the flow rate of cold air or hot air.

MODULE : (i)   A device to maintain a constant head in fluid flow.

(ii)   A unit length or a dimension used by architects in planning a building .

MODULUS OF ELASTICITY : Also known as 'young's modulus'. It is the ratio of stress to strain of a material within its elastic limit and is usually denoted by 'E'.

MODULUS OF INCOMPRESSIBILITY : The ratio of pressure in a soil sample to its volume change by the pressure.

MODULUS OF RESILIENCE : The partial area of a stress-strain diagram (under tensile test of a member) upto the stress at elastic limit.

MODULUS OF RIGIDITY : Also known as 'shear' modulus. It is the ratio of shear stress to shear strain of a material.

MODULUS OF RUPTURE: The fictitious stress obtained at the point of rupture in a testing machine. This is used to compare the ultimate strengths of a member of various sizes and materials.

MODULUS OF SECTION: Section modulus which is denoted by 'Z'. From bending equation we know M=fz and

$$Z = \frac{second\ moment\ of\ area}{depth\ of\ neutral\ axis} = \frac{I}{Y}$$

For a rectangular beam section, where $I = 1/12\ bd^3$ and $Y = d/2$, $Z = 1/6\ bd^2$

b and d being breadth and depth of the beam respectively.

MODULUS OF TOUGHNESS : The total area of a stress-strain diagram (under tensile test of a member) that represents the energy absorbed per unit volume.

MODULUS OF VOLUME CHANGE : The co-efficient of volume decrease of a soil sample which is given by

$$\frac{co\text{-efficient of compressibility}}{1 + initial\ void\ ratio}$$

MOHR'S CIRCLE OF STRESSES : Two dimensional stresses viz normal and shearing stresses can be given a visual interpretation by constructing a circle. This was developed by Otto Mohr in 1882.

MOISTURE BARRIER : Damp proof course.

MOISTURE CONTENT : It is the quantum of water present in a material. It is expressed in percentage. It is given by

$$\frac{weight\ of\ water\ in\ a\ mass}{weight\ of\ the\ dry\ mass} \times 100$$

MOISTURE GRADIENT : The variation of moisture content between the outer and inner parts of a material.

MOISTURE MOVEMENT : A property by which a material increases in its dimensions when its moisture content increases. The material shrinks when the moisture goes out of the material.

MOLE : (i) A tunnel excavator.

(ii) A break water.

MOLE DRAIN : A narrow drain cut mechanically in a stiff clay by means of a mole cutter, used for land drainage instead of laying agricultural drain tiles.

MOLE PLOUGH : A cutter of special shape attached to a caterpillar tractor which is drawn through a soil for making a drain quickly. By this method, a long ditch may be formed in a short time for laying a pipeline.

MOLER BRICK : A brick made of diatomaceous earth.

MOMENT : (i)   The turning effect produced by two equal and opposite forces acting at a distance apart on a body.

(ii)   The moment of a force about a point is given by multiplying the magnitude of the force by its distance from the point normal to the force i.e., the shortest distance.

MOMENT - AREA METHOD : A method of determining slope and deflection in beams due to various types of loading.

MOMENT DISTRIBUTION : A method of distributing moments in continuous beams, portal frames and framed structures. The method is based on successive approximations with the concept of carry- over moment and stiffness of the member.

MOMENT OF INTERIA : Second moment of area i.e., moment of moment o area. The moment of inertia of an area about a line in plane of the area is given by the product of the area and the sequre of the distance of c.g. of the area from the line. It is denoted by 'I'.

MOMENT OF RESISTANCE : The couple or moment produced by the internal forces developed due to the external loads applied. When a bending moment is produced in a structural member, moment of resistance is developed to resist the B.M. For a balanced structure, the bending moment (B.M.) must not exceed the moment of resistance (M.R.). For an econimical section, B.M. = M.R.

MOMENTUM : The product of the mass of a body and the velocity with which it moves,

MONITOR ROOF : A trussed roof for workshops which is provided with a continuous lantern light for the entry of daylight.

MONK BOND : 'Yorkshire bond' or 'Flying bond' which is essentially a modified 'Flemish bond' showing two stretchers and one header in each layer.

MONKEY : A heavy drop hammer used in pile driving.

MONKEY BAR : A wooden bar to which the louvres of a venetian shutter are hinged so that by moving the bar up and down, the louvres can be opened or shut.

MONKEY TAIL : A scroll at the end of a wooden handrail.

MONKEY TAIL BOLT : An extension bolt used in high doors and windows.

MONOCABLE : A single cable in a ropeway which serves double function i.e, haulage and trackway.

MONOLITH : A large hollow square, rectangular or circular foundation of concrete, stone or brick masonry sunk as a box caisson or open caisson and excavated by a grabbing crane. This may also be built from the base of foundation within a pneumatic caisson or a cofferdam.

MONOLITHIC : A jointless construction as a whole in reinforced concrete structure. For example, a column, a beam and a slab are cast in one operation at a time.

MONORAIL : A single rail track of heavy type with proper supports, used for transporting man, material and machine from one point to the other. This may be built on ground or overhead.

MONOTOWER CRANE : A crane having a single tower.

MONOTUBE PILE : A type of cast-in-situ concrete pile, in which a shell of fluted sheet steel is driven into the ground without mandrel and the shell is filled with concrete.

MONUMENT : (i)   A beacon.

(ii)   A stone set by a surveyor for marking the point of reference or a corner or a boundary line.

MONUMENTAL MASON: A stone-carving mason who has the sense of art and architecture in building a mounment by carving stones, cutting letters on it and polishing.

MOORING DOLPHIN: This picks up the pull from the hawsers. This is required in building a berth jetty. See illustration.

MOORING FORCES : The forces on piers and jetties during berthing of vessels. The forces are measured from the kinetic energy of a berthing vessel.

MOORISH ARCH : Horse-shoe arch i.e., an arch of horse-shoe type

.**MOPBOARD** : A skirting board .

MOPSTICK : A circular handrail having a flat surface underneath.

MORTAR : Usually a mixture of cement and sand or lime and surki with adequate volume of water to form a paste. Sometimes, cement is added to a lime mortar or lime is added to a cement mortar. In rural areas, mud mortar is also used.

MORTAR BOARD : A hawk.

MORTAR CUBE TEST : A test to determine the compressive strength of a mortar, which is measured by crushing a well-cured standard-mortar cube in a compression testing machine. Mortar cubes of different mix are tested, from which the mix ratio suitable for a job is selected.

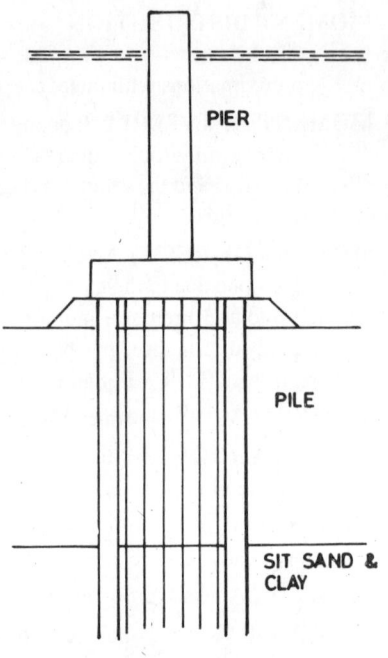

PIER

PILE

SIT SAND & CLAY

MOORING DOLPHIN

MORTAR MILL : A mortar-mixing machine or a pug mill for preparation of mortar. Animal-driven country mill is also in practice in making mortar in rural areas.

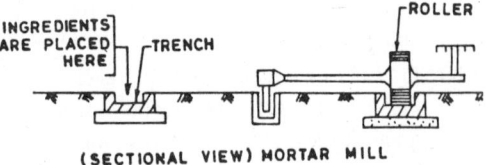

INGREDIENTS ARE PLACED HERE

TRENCH

ROLLER

(SECTIONAL VIEW) MORTAR MILL

MORTISE : A slot made in a member into which the tenon part of another member is fitted for joining the two members.

MORTISE & TENON JOINT : A joint made between two members meeting at right angles. The tenon of one part fits into the mortise of the other part.

MORTISE CHISEL : A stiff chisel used by carpenters to cut mortises in timber members.

MORTISE GAUGE : A two-pin marking gauge used for marking the mortise and tenon prior to cutting.

MORTISE LOCK : A lock fitted in the mortise cut in the door. This makes the lock hidden.

MOSAIC : (i)   A course or layer made of coloured stone chips of small size laid in cement. This is applied on walls and floors. On setting and curing, the surface is smoothened by rubbing and polishing it with carborundum or pumice stone. Different designs are also made in mosaic flooring and walling.

(ii)   A scaled map made from air survey by fitting together the large size vertical photographs of a land area. This is a quick method of preparing a map of a land area depicting the features on land.

MOSAIC CUTTING : The cutting of top surface of a mosaic carpet by rubbing it with pumice stone.

MOTEL : A roadside small hotel of temporary nature for motorists.

MOTORWAY : A pavement or roadway suitable for power-driven vehicles.

MOTTLING : A defect, usually uniform, rounded marks, found on a spray-painted surface.

MOULD : (i) A pattern cut to the profile of a moulding either on a zinc sheet or on a piece of timber. Moulds are also made of cast iron and other metals,

(ii) The wooden or steel formwork for casting concrete This is also known as 'Shuttering'.

MOULD BOARD : (i) The flat blade of a bull-dozer that pushes the earth.

(ii) The steel blade of a blade of a farmer's plough which turns over the chunks of earth.

MOULDED BRICKS : (i) Ordinary bricks moulded by sand moulding or slop moulding manually.

(ii) Special bricks/ornamental bricks prepared by moulding manually.

MOULDED INSULATION : Insulating materials moulded accordingly as per requirement. Sometimes, the insulation is provided to articles by moulding on them.

MOULDED PLYWOOD : Plywood bent by pressure moulding during gluing with thermo-setting glue. Also known as 'Ply plastics'.

MOULDING : (i) A process of casting a material in a mould.

(ii) Forming an article or object by carving on wood or stone manually or in a moulding machine.

(iii) An ornamental work which is nailed or glued to wood panel, furniture and wooden frame.

(iv) An ornamental projection made in a wall or at the head and base of a column.

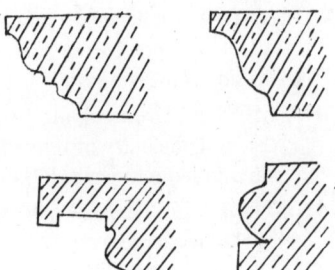

MOULDINGS

MOULDING MACHINE : A machine in which mouldings are cut on wood or stone.

MOULDING PLANE : A carpenter's plane used for cutting mouldings.

MOULD OIL : Oil or oily substances in liquid form are applied on the inner surface of a mould or formwork so that the moulded material does not stick to the mould.

MOUND BREAKWATER : A break water made of rubble mound.

MOUNTAIN RAILWAY : A railway in a hill area, which is of steep gradient and serpentine nature. Sometimes, a rack track is provided in a very steep gradient.

MOVABLE BRIDGE: A draw bridge or bascule bridge or traversing bridge which can be moved horizontally or vertically up, to allow a water vessel or ship to pass.

MOVABLE DAM: A dam having removable part to allow the flood water to flow through it.

MOVING FORMS : A steel formwork having arrangement of sliding, by which it can be moved up with the progress of work. It is also known as 'climbing forms' or 'travelling forms'.

MOVING LOAD : A live load which is movable.

MOVING STAIR : An escalator (power-driven stair).

M P N : Abbreviation for ' Most Probable Number', used in determining the modal number of coliform bacteria in water. It can be expressed as the density most likely to produce a particular analytical result.

MUCK : Waste sludge, rubbish, spoils, etc.

MUCK SHIFTING PLANT : Removal of spoils, mucks, etc. by mechanical means.

MUD BALLS : These are formed due to cementation of flocs together with sand grains on the top of a filter bed, when the bed is not properly washed or when there is too much of turbidity in water to be filtered.

MUD JACKING : A method of raising a concrete platform or road slab by making bores under it and forcing soil-cement slurry through the bores under the slab or platform. The soil-cement slurry is kept on a lorry (mud jack) and is pumped for introducing it under the slab.

MUD SILL : A sole plate kept on ground, over which carpenters work.

MUD VALVE : A type of valve as shown in illustration is used for removing sludge or mud along with water from the bottom of a reservoir or basin.

MULLET : A wooden piece having a groove to the required, depth which is slipped along the edge of a wood panel to examine the correctness of thickness all along the panel.

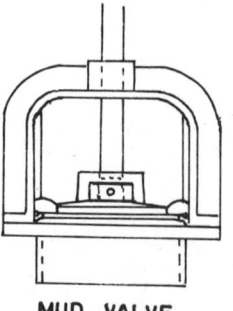

MUD VALVE

MULLION : A vertical member in a door or window frame dividing the entry of light. The divided parts may further be subdivided into panes. Also known as 'Munnion'.

MULTI-BUCKET EXCAVATOR : A type of excavator like a bucket ladder excavator having a number of buckets tied to a chain. This is used for digging sand, gravel or ballasts in road or railway cuttings or in excavation of a canal. The bucket chain is kept in the same slope as that of angle of repose of the material.

MULTIBUOY TERMINAL :   See illustration. The oldest type of offshore mooring, in which the vessel is tied up at a fixed position with the help of multiple chain-anchor moorings, such that the vessel maintains a fixed position.

MULTIPLE ARCH DAM : A light-weight dam built of a number of arches carried on parallel buttress walls, the axes of the arches being sloped at an angle of $45°$ with the horizontal.

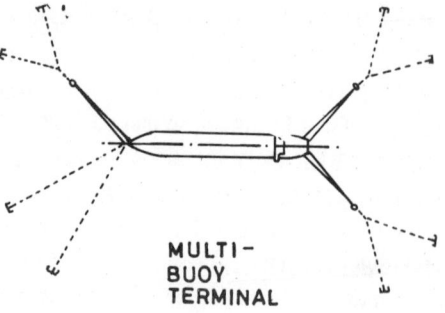

MULTI-
BUOY
TERMINAL

MULTIPLE-DOME DAM : A modified form of multiple arch dam.

MULTIPLE WEDGE : Plug and feather used in blasting of stones.

MULTI-PLY : A ply board having more than three plies.

MULTIPLYING CONSTANT : A multiplier (Constant) used in transforming a value into the result.

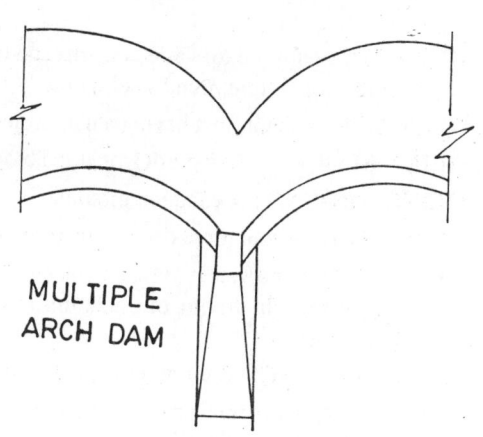

MULTIPLE ARCH DAM

MULTI-UNIT WALL : A wall made of two or more half-brick thicknesses.

MULTI-WHEEL ROLLER : A roller having a number of pneumatic-tyred wheels.

MUNICIPAL ENGINEERING : It essentially covers the utility services like water and sanitation engineering with other allied utility services required in a municipal area or town.

MUNNION : See 'Mullion'.

MUNTIN : (i) A glazing bar or munnion.

        (ii) A vertical member framed into the rails, separating the panels in a door.

MUSHROOM CONSTRUCTION : A slab carried by columns, the column heads being flared i.e, the slab is thickened round the column head, which looks like a mushroom.

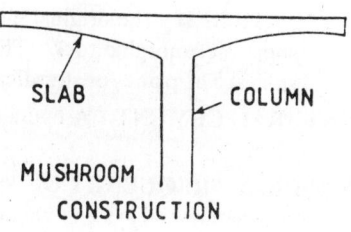

SLAB   COLUMN

MUSHROOM
CONSTRUCTION

MUSHROOM PIER SHAFT : A type of bridge pier occassionally used to avoid skew spans in passing over existing rail tracks or highways. See illustration.

MYER'S FORMULA: A formula to identify the flood-flow characteristics of different drainage basins. It is given by $Q = 100$ $p(a)^{0.5}$ where Q is extreme peak flow, P is percentage ratio of Q to the ultimate maximum flood-flow and a is the area of drainage basin.

MUSHROOM
PIER SHAFT

N

NAIL : Nails are required for various works and they are of various shapes and sizes. Nails are cold-forged from round steel wires.

NAIL FLOAT : A devil float having a nail projected from it.

NAILING BLOCK : A fixing brick or stone block.

NAILING GROUND : A common ground.

NAIL PULLER : A tool more delicate than a claw hammer for drawing out nails.

NAIL PUNCH : A short steel rod with one end tapered to the diameter of a small nail. For quick and straight driving of a nail, this tool is placed over the nail head and is struck by a hammer.

NAKED FLOORING : The frame work of a wooden floor prior to fixing floor boards.

NAPHTHA : A petroleum product obtained by distillation of petroleum or a organic material containing paraffins and olefins at temperatures ranging from 160^0 C to 270^0 C. This is used as a thinner in a paint.

NAPPE : A water sheet flowing over the crest of a dam or weir.

NARROW GAUGE : This is a railway gauge (2'- 6") which is narrower than the standard gauge of 4'–8.5".

NARROW-RINGED TIMBER : A close-grained timber obtained from a tree which has grown slowly. This timber is stronger than wide- ringed timber.

NATURAL ASPHALT : Asphalt obtained naturally from the voids in limestone or sandstone.

NATURAL BED : The original plane of stratification found in sedimentary rocks and in some metamorphic rocks. The flattened surfaces are parallel to the natural bed of stones. The planes of stratification have a reduced cohesion.

NATURAL CEMENT : A hydraulic cement obtained by burning limestone containing clay.

NATURAL FREQUENCY OF A FOUNDATION : It is frequency of free vibration of a foundation-soil system which must be different from that of any machine carried by the foundation with a view to avoiding resonance.

NATURAL GAS : Usually methane gas obtained naturally from underground sources. This is used as a fuel.

NATURAL HARBOUR : A harbour made by following the shape of the coastline.

NATURAL ROCK ASPHALT : Natural asphalt obtained from rocks of limestone or sandstone.

NATURAL SCALE : An object is drawn to natural scale means same scale has been used to draw both vertical and horizontal lines.

NATURAL SEASONING : The driving out of sap from a freshly cut timber by keeping it under a shade, but exposed to atmospheric air and temperature.

NATURAL STONE : A stone that is quarried from a rock, cut and shaped.

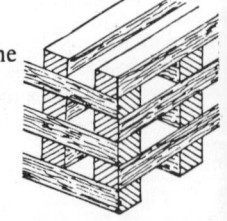

NATURAL SEASONING

NAUTICAL ALMANAC : An astronomical calender for the use of astronomers, navigators and surveyors.

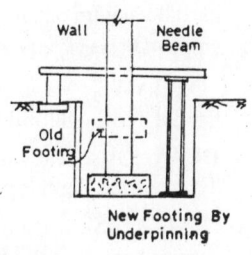

Wall / Needle Beam

Old Footing

New Footing By Underpinning

NAVIERS HYPOTHESIS : An assumption which is nearly true, is used by civil engineers in simplifying calculations for design of beams.

The stress at any point in a beam due to bending is proportional to its distance from the neutral axis.

NAVIGATION : A river canalized for water transport. The waterways must have a minimum draught for the water vessels.

NAVVY : A labourer specialized in digging foundation trenches, drains and in formation of roads and railways including tunnelling.

NEAT LINES : Lines that define the sides of excavation with a view to avoiding overbreak.

NEAT SIZE : The size of a wooden piece after sawing and planing.

NEAT WORK : The visible brickwork above the footings.

NEEDLE BATH : A shower bath with jets of water striking the bather horizontally from all directions.

NEEDLE BEAM : A short beam passing through a shored wall, carries the wall and transfers the load to dead shores.

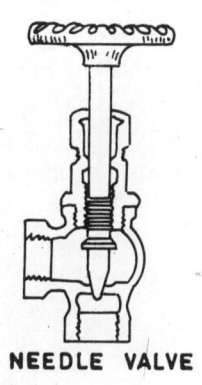

NEEDLE INSTRUMENT : A surveying instrument controlled by the magnetic needle of a compass.

NEEDLE SCAFFOLD : A scaffold hung on needles (needle beams) driven into the wall.

NEEDLE TRAVERSE : A compass traverse used in surveying.

NEEDLE VALVE : A cone-shaped valve used for controlling flow of water to turbines.

NEEDLE VALVE

NEEDLE WEIR : A framed weir carrying heavy vertical timber needles. The needles can be withdrawn when it is required to lower the water level.

NEEDLING : Insertion of a needle into a wall.

NEGATIVE HEAD: During filtration of water, as clogging occurs, the friction loss in top layer of sand increases greatly and when this loss becomes greater than the head of water above the sand surface, the water column below acts as a draft tube and a

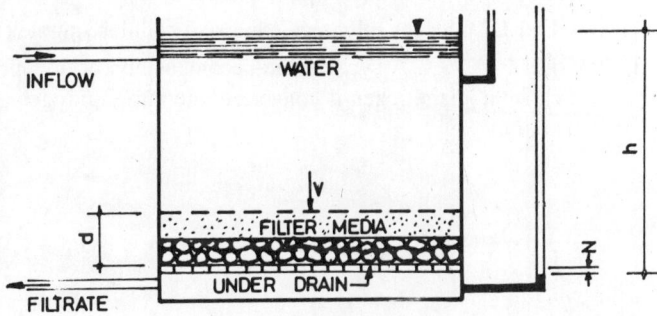

INFLOW

WATER

FILTER MEDIA

UNDER DRAIN

FILTRATE

partial vacuum occurs. This condition is called 'negative head'. See illustration.

NEGATIVE MOMENT: Anticlockwise moment; Hogging moment.

NEST OF SAWS : Various saw blades of different grades that can be fitted in the same handle and used for different purposes as and when required.

NET DUTY : Also known as 'Farm Duty'. The quantum of water required by a farm expressed as depth of water over the area irrigated.

NEUTRAL AXIS : It is the axis or plane in a beam section where the stress is zero. In a simple beam of symmetrical section and homogeneous material, the neutral axis passes through the centroidal axis of the section, above which the fibres are subjected to compression and below which the fibres are in tension.

NEUTRAL PRESSURE : The hydrostatic pressure in the pore water of soil.

NEWEL CAP : A wooden top of a newel post.

NEWEL DROP : A decorative projection of a newel post.

NEWEL JOINT : A joint between the newel and the handrail.

NEWEL POST : A post provided in a flight of stairs. It carries the ends of outer string and the handrail and supports them at a corner.

NIB : A downward projection at upper end of a roof tile for hooking over the batten.

NIGGED ASHLAR : Roughly dressed granite.

NIGGERHEAD : A capstan used for sinking bored piles; a projected end of a driven shaft round which a rope is turned and pulled tight .

NIGHT VENT : An opening light at the top of a casement window.

NIPPER : Used for lifting heavy stone blocks. Two curved levers are hinged together near the middle (like a scissor) and their upper ends are tied to a crane. The jaws (lower ends) of the nipper hold firmly the stone block and it is lifted by the crane.

NIPPLE : (i) A-short piece of pipe threaded outside at both ends, required for joining two internally threaded pipes or couplings.

(ii) A small valve at high points of a hot water pipeline system with a view to releasing air.

NISSEN HUT : A hut of arch shape, built of corrugated sheet. The construction is very fast (several hours) and cost is low.

NITRIDING : Case hardening of steel by keeping it at about $500^{\circ}C$ in ammonia gas for about 2 to 3 days such that nitrogen is absorbed into the surface of the steel. In this process, no further heat treatment is needed.

NITROCELLULOSE : A solid high explosive-cellulose nitrate or guncotton.

NITROGEN CYCLE : A cycle in the decomposition of organic nitrogenous matters in which ammonia nitrogen is converted into nitrite nitrogen and nitrate nitrogen. See illustration.

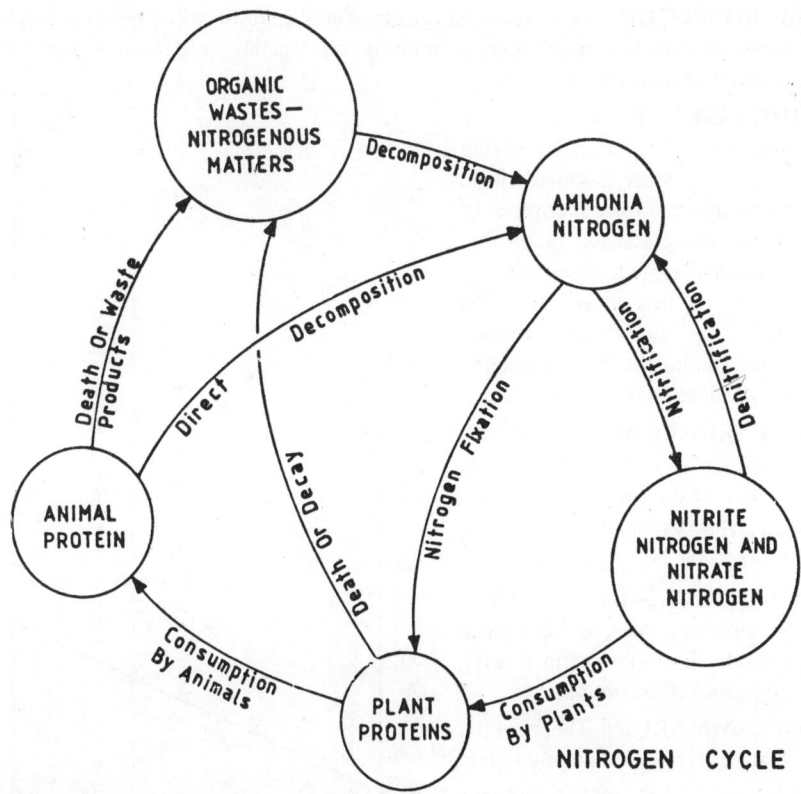

NITROGEN CYCLE

NITROGEN HARDENING : Nitriding.

NITROGLYCERINE : A liquid, highly explosive. When detonated, it evolves CO_2, N_2, O_2 and H_2O. It forms the major part of dynamite.

NODE : A junction point of two or more members in a structural frame or a junction of pipes in a pipeline network.

NODAL DEMAND : Demand of water at the nodes of a pipeline system.

NO-FINES CONCRETE : Concrete made without using sand. It consists of 2 parts of cement with 1 part of dry slaked lime, 2 fine slag and 2 coarse slag. This concrete is not reinforced and it contains large voids in it providing no capillary passage for water. No-fines concrete should not be rammed or vibrated. The external surfaces of this concrete must be rendered well for srengthening and protecting it from atmospheric actions.

NOG : A fixing brick or stone.

NOGGING PIECE : Horizontal short timbers required to stiffen the verticals of a framed wooden partition.

NOISE ABSORPTION : Noise level in a room can be reduced by using noise-absorbent material either during construction or afterwards.

NOISE INSULATION: The transmission of noise can be minimised by using hollow block construction and noise-insulation boards.

NOISE POLLUTION: The pollutional effects of noise are impaired hearing, headache, nervous strain, loss of efficiency, irritability, etc. The threshold between harmless and harmful noise is about 100 db.

NOMOGRAM : Nomograph or alignment chart used for saving time and energy in calculations. In its simplest form it consists of three lines graduated in suitable scales and are placed such that by joining the two known values on two lines, the third unknown value can be obtained by reading the graduation on the third line.

NON-BEARING WALL : A wall that carries its self-weight only and the wind load during a storm.

NON-COHESIVE SOIL : A cohesion-less soil i.e., sand, kankar, etc.

NON-COMBUSTIBLE : The material that does not burn. Such material is important in making fire-proof construction.

NON-FLAMMABLE : The material that burns but without any flame.

NON-HYDRAULIC LIME : Lime reach in calcium. This is fat lime or rich lime, whose setting power is poor.

NON-INFLAMMABLE : The material which does not burn with a flame.

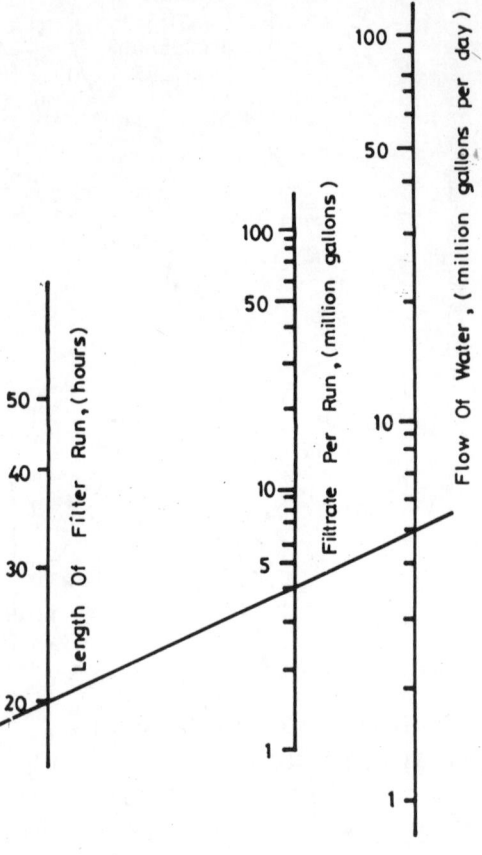

NOMOGRAPH

NON-MANIPULATIVE JOINT : A compression joint used in pipelines.

NON-METALLIC MINERALS : Stones, Asbestos, Asphalt, Lime, Coal, Gems, Petroleum, Fossil gums, Abrasives, Natural gas, Pigments, etc.

NON-RETURN VALVE : One-way valve i.e., a valve that allows flow in one direction only and any return flow is checked.

NON-SIPHONING TRAP : A fixture trap in which resistance to siphonage is created not by venting, but due to tortuous passage as shown.

NON-SLIP FLOOR : A rough concrete floor which is non-skiddy. Sometimes, non-slip or non-skid floor is made by treating a concrete surface while green, with iron filings or carborundum powder.

NON-SIPHONING TRAP

NON-TILTING MIXER : A concrete mixer of drum shape having two openings which rotates horizontally. The mixed concrete comes out from the side baffles when it is received through a chute.

NORIA : A primitive device for lifting water with the help of an endless chain of metal, wooden or bamboo buckets. The system is mechanically efficient. Its modified form is 'Persian Wheel'.

NORMAL STRESS : Direct stress developed in a member due to application of direct load.

NORMAN BRICK : A brick used in USA, measuring 300 x 100 x 70 including mortar thickness.

NORTH-LIGHT ROOF : A sloped roof having one steep slope and one mild slope. The steep slope is glazed and faces north in the northern hemisphere and it is made reverse in the southern hemisphere. Also known as 'Saw-tooth Roof'.

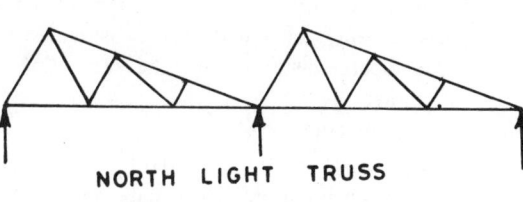

NORTH LIGHT TRUSS

NOSE : (i)   A blunt projection or overhang.

        (ii)   The lower end of the door shutting stile.

NOSING : A half-round projection of the edge of the tread of a stair, window sill or a roof slab.

NOSING LINE : The line touching the edges of the stair nosings.

NOTCH : A groove in a construction.

NOTCH BOARD : A cut string.

NOTCHED WEIR : A measuring weir with a notch.

NOTCH EFFECT : The effect of notching in a member, which increases stress locally. Notching in a highly stressed member should always be avoided.

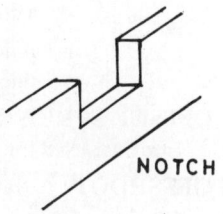

NOTCH

NOTCHING : (i)   For joining two timber pieces, notching (cutting a groove) is done in one of the pieces.

             (ii)   Excavating in a series of horizontal steps for advancement in cutting.

NRC EQUATION : The equation developed by the National Research Council to find out the efficiency of a trickling filter for BOD removal. The efficiency is given by

$$E = 100 / [1 + 0.0085 \, (w/VF)^{0.5}]$$

where,   w = weight of BOD applied, 1bs/day

            V = Volume of the filter, acre ft.

and   F = Number of effective passages of sewage through the filter.

N-TRUSS : Pratt truss. See illustration.

N-VALUE: Rugosity co-efficient or co-efficient of roughness. This value is of importance in the design of sewers, drains and canals.

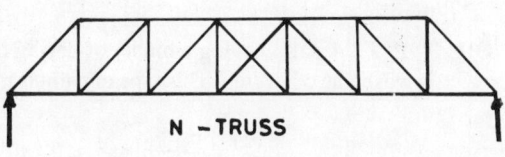

N –TRUSS

O

OAK : A kind of hardwood growing in temperate climates. This is a valuable structural timber owing to its hardness and twisted fibres.

OAKUM : Tarred untwisted hemp used as a caulking between precast concrete blocks.

OAK VARNISH : An oil varnish used for indoor works, with high ratio of oil to resin.

OBLIQUE BUTT JOINT : A butt joint in which the meeting angle is other than 90°.

OBLIQUE GRAIN : The diagonal grain found in certain timbers.

OBLIQUE OFFSET : The oblique distance of a point from a survey line (chain line); This measurement is taken where the measurement of perpendicular distance from the surveyline is interrupted.

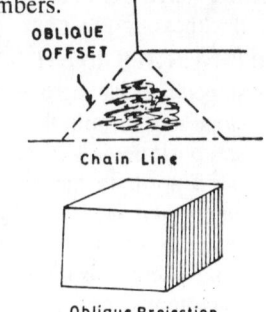

OBLIQUE PROJECTION : The simplest form of drawing a three-dimensional object. This is obtained when the projectors make an angle of 45° in any direction with the picture plane.

Oblique Projection

OBSCURED GLASS : The glass which is made translucent with a view to making it visionproof with entry of light. This glass is prepared either by sand blasting or by moulding it with irregular surface.

OCHRE : Hydrated iron oxide, yellow or yellowish brown in colour, used as a pigment.

OFFSET : (i)   The distance of an object or a point from a chain line (survey line).

(ii) A ledge in a wall due to change in wall thickness

OFFSET SCREW DRIVER : A special type of screw driver used for turning screws at 90° to its length.

OFFSHOOT : A water table.

OFF-WHITE : A colour used in painting which approaches to white colour, but not white.

OGEE ARCH : A special form of arch mostly used in producing architectural effects in earlier days. Two serpentine curvatures starting from the two ends of the springing line meet at the apex of the arch.

OGEE JOINT : A type of spigot and socket joint used in pipe lines.

OIL-BOUND DISTEMPER : A distemper containing drying oil. Also known as oil-bound water paint.

OIL LENGTH : In a varnish medium, it is the ratio of oil to resin.

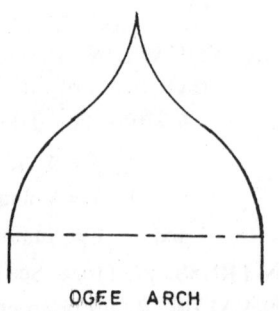

OGEE ARCH

OIL PAINT : A paint having a binder of drying oil, which is washable on drying. This type of paint is most commonly used for all sorts of work.

OIL SLIP : An oilstone slip.

OIL STAIN : A very thin oil paint having a small quantum of pigment, used for staining wood walls and floors.

OIL STONE : A hone.

OIL-WELL CEMENT : A hydraulic cement having a slow initial setting power at a high temperature required in drilling oil-wells.

OLEO RESIN : A resin obtained either from dead wood by wood distillation or by bleeding a living tree. Mostly, it is pine gum.

OLEO-RESINOUS VARNISH : A varnish having vegetable drying oil and natural or synthetic resin.

ONE-PIPE SYSTEM : (i) This is a system used in house plumbing, where a single vertical pipe receives sanitary sewage, sullage and other waste liquids and all the branch pipes for ventilation are connected to one vertical anti-siphonage pipe.

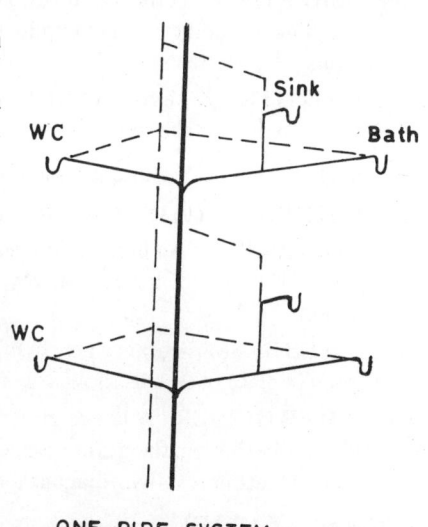

ONE PIPE SYSTEM

(ii) In heating system, when all the flow and return connections to the radiators are made with the same pipe, it is called one-pipe system.

OPACITY : It is the hiding capacity of a paint.

OPEN CAISSON : A caisson, either a cylinder or a drop shaft with both top and bottom being open. See illustration.

OPEN-CELL PROCESS : A process used in creosoting timber.

OPEN CHANNEL : A conduit acting as a channel, where the upper surface of the liquid never comes in contact with the crown of the conduit.

OPEN CUT : An excavation in the open.

OPEN DEFECT : A split, knot hole or wormhole found in timber.

OPEN FRAME GIRDER: A vierendeel girder which has top and bottom booms connected with vertical members at intervals.

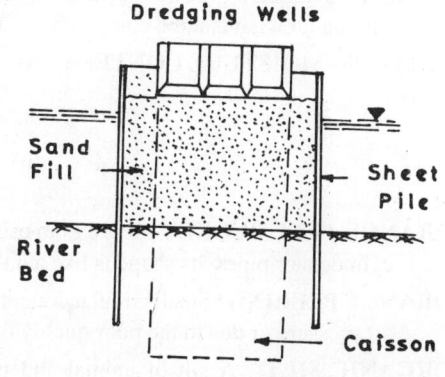

OPEN CAISSON

OPEN GRAINED: A timber with wide annual rings.

OPEN HEARTH PROCESS : Also known as Siemen's-Martin process used in manufacturing steel. In this process, pig iron, scrap steel and iron are melted together by regenerative gases to produce better quality steel with the removal of C, Si and P.

OPEN MORTISE : A mortise open on three edges required for joining table legs.

OPEN NEWEL STAIR : A geometrical stair most commonly used without newels as shown.

OPEN ROOF : A roof having no ceiling i.e., rafters and purlins are visible from floor.

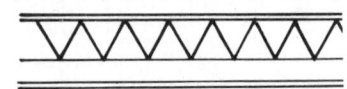

OPEN TRAVERSE

OPEN SHEETING : Vertical poling boards without touching each other are held up by walings and struts.

OPEN SLATING : Laying of slates with gaps between them in the same course.

Open Web Beam

OPEN STAIR : A stair whose sides are open.

OPEN STRING : A cut string used in joinery works.

OPEN TRAVERSE : An unclosed traverse where the last line does not meet the starting point. This type of traverse survey is needed for roads, railway tracks, streams and rivers.

OPEN WEB BEAM : A beam where bracings are used in place of web as shown.

OPEN WEB GIRDER : A lattice girder, where the top and bottom flanges are connected by verticals at intervals with diagonal bracings.

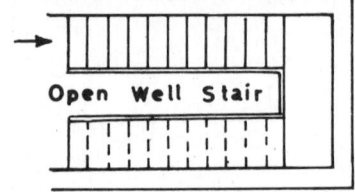

Open Well Stair

OPEN WELL STAIR : A stair with a spacious well in between two flights.

OPERATION WASTE : Water lost from an irrigation system through spillways or other devices, when the flow of water is diverted.

OPTICAL SQUARE : A small instrument extensively used in setting out right angles by holding it on hand at the construction site.

OPTIMUM MOISTURE CONTENT : (i) It is the moisture content at which the compaction produces the highest dry density.

    (ii) The moisture content that produces a complete reaction in least time.

ORANGE-PEEL BUCKET : It is a grab used in taking out sand, silt and clay from small cylinders or pipes. Its shape is like a half orange cut into segments.

ORANGE PEELING : Small circular craters formed in a spray paint finish due to incorrect air pressure or due to the poor quality of solvent used.

ORGANIC SILT : A silt of animal and plant origins, having dark black colour and obnoxious odour. This silt is highly compressible.

ORIEL WINDOW : An overhanging window on corbels in a upper storey.

ORIFICE METER : A meter used to indicate the quantum of flow through a pipe. A plate with a small hole in it, is placed across a pipe through which a liquid is flowing and from the measurement of pressure difference between the two sides of the pipe, the flow quantum is determined.

ORPIMENT : A lemon-yellow mineral pigment which is arsenic sulphide, used in a paint.

ORTHOGONAL : A term used for orthographic; It means 'at right angles'.

ORTHOGRAPHIC PROJEC-TION : This comes under parallel projection system. In engineering drawing, plan, elevation and end view of an object are drawn with the help of orthographic projection. The projection may be thought of as being formed by perpendicular projectors extended from the object to the plane.

ORTHOTROPIC : The physical properties of any material in two or three directions at right angles to each other.

OUTBAND : A strecher stone visible in a reveal.

ORTHOGRAPHIC PROJECTION

OUTFALL : The ultimate point at which a sewer or a drainage channel discharges its flow into a stream or river or sea.

OUTFALL SEWER : The sewer which carries the total flow from a tributary area and discharges at the outfall, the ultimate point. Normally, it leads from a pumping station.

OUTLET : An opening through which a fluid goes out of a system.

OUTRIGGER : A beam projected from a building to carry an utility article or for other purposes.

OUTSIDE GLAZING : Glazing done at the outside of a frame.

OUT TO OUT : Overall dimension.

OVAL WIRE BRAD : A brad nail manufactured from oval wire.

OVERFLOW DAM: This is designed as a hollow gravity dam to allow the flow of water over the section. The section is so designed that no damage is caused to the structure

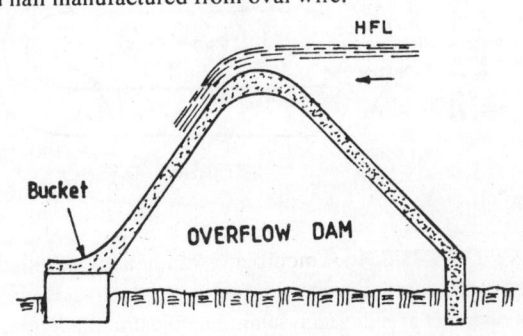

due to the free-falling nappe. The down stream side of the dam is provided with an up-turn bucket to dissipate the energy of falling water.

OVERFLOW STAND: A stand pipe into which water rises and overflows.

OVERFLOW WEIR : It is usually provided in a manhole, where sewage runs in a channel. Any excess sewage in the channel automatically flows out of it and enters into the storm sewer. Normally the channel receives the dry weather flow from the sanitary sewer.

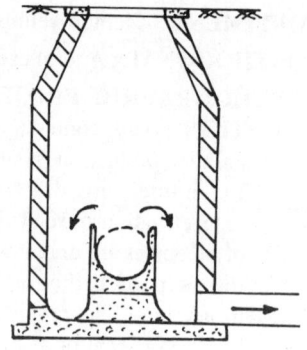

OVERHAND WORK : Laying of facing bricks from inside a building either by standing on a floor or on a scaffold.

OVERHANG : A projection beyond a wall.

OVERHANGING EAVES : Rafters, purlins, tiles and eaves board projecting over the wall plate and clear of the wall.

OVERFLOW WEIR

OVERHEAD DOOR: A door which opens up verticallly and slips into a horizontal position at the door head.

OVERHEAD ROPEWAY : An aerial ropeway.

OVERHEAD TRAVELLING CRANE : Power-operated hoisting crane carried on a gantry girder between rails. The hoisting crab can travel from end to end of the girder covering the entire workshop area for lifting heavy articles from one place and transfering it to the required position.

OVERSAILING COURSE : A brick or stone string course.

OXIDATION : (i) Absorption of oxygen from atmospheric air to form a durable paint film by hardening of drying oils.

(ii) Oxygenation of unstable organic matters by natural or mechanical aeration.

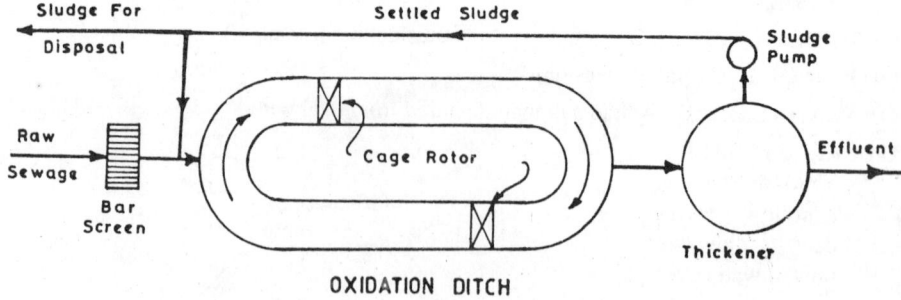

OXIDATION DITCH

OXIDATION DITCH : A modified form of activated sludge process. In essence, it is an extended aeration process, capable of taking shock loadings without upset. The ditch is operated as a closed system. See illustration.

OXIDATION POND : A shallow pond of 1.5 m. depth where sewage gets purified by the action of atmospheric oxygen and sunlight, the driving force being the photosynthesis supported by symbiosis between aerobes and algae.

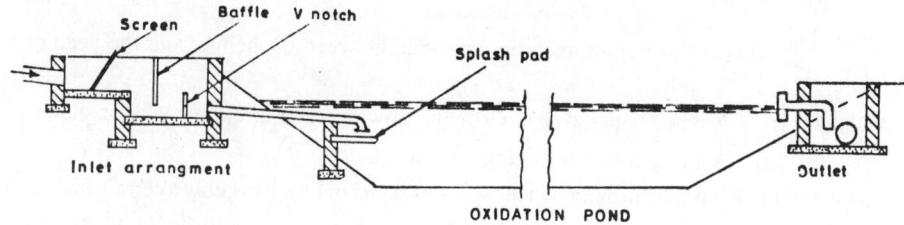

OXIDATION POND

OXYGEN SAG CURVE : Also known as 'dissolved oxygen Sag'. This is a spoon-shaped profile of the dissolved oxygen deficit formed due to the interplay of the deoxygenation of polluted water and its reoxygenation from the atmospheric air. See illustration.

OZONE: It is a strong chemical oxidant used for disinfection or deodorizing water.

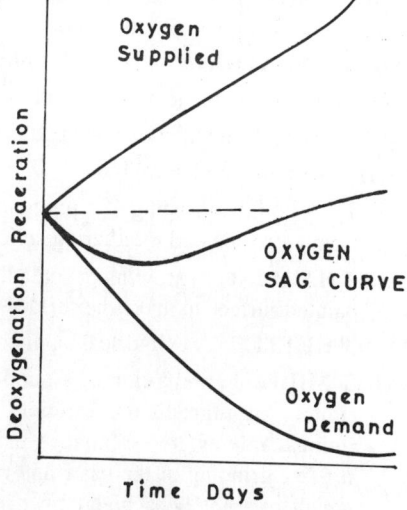

P

PACKING : (i)   Materials used in filling the gap in a construction joint.

      (ii)   Hemp or asbestos packed in the space left between gland and stuffing box of a valve or pump.

      (iii)   Cushioning material provided between the helmet and the head of a pile.

PADDLE : (i)   A wood panel used to close the flow of water through a lock or sluice.

      (ii)   A stirrer used for mixing chemicals.

PADDLE HOLE : A hole provided in the wooden panel of a lock to allow a certain quantum of flow through it.

PADDLE MIXER : A mortar mixer consisting of two horizontal shafts with paddles rotating in opposite directions.

PAD FOUNDATION : An isolated column foundation.

PAD SAW : A saw blade of a tool pad.

PADSTONE : A concrete or stone pad in a wall to support a beam or truss.

PAGE : A small wooden wedge.

PAINT : A liquid consisting of a pigment, a vehicle and a drier, used to protect a surface from corrosion and weathering actions.

PAINT HARLING : Throwing or spreading paint-coated small chips of stone onto a sticky painted surface to make the surface rough.

PAINT KETTLE : A cylindrical paint container used during painting a surface by brush.

PAINT MILL : In prepartion of a good paint, proportioning of ingredients, efficient grinding and intimate mixing are essential. A paint mill is used for fine grinding of the paste and through mixing which can not be achived by hand and paddle method.

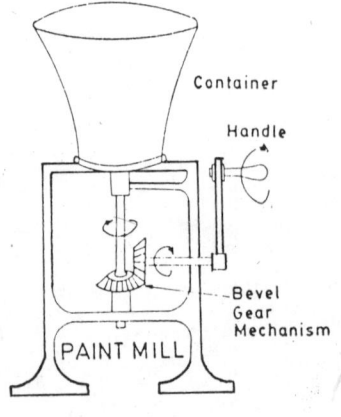

PAINT REMOVER : Organic solvents or soda-based liquids used in dissolving old paints and brushing off.

PAINT SYSTEM : A system comprising application of different coats of paints in sequence. The first coat is the priming coat or sealing coat, the second coat is the under coat with opacity and the third coat is finishing coat to bring gloss in the painted surface.

PALE BOILED OIL : Boiled linseed oil with a small quantity of drier, through which air is bubbled at a temperature of $130^\circ C$ of the mixed oil. This produces a glossy paint.

PALISADE : An enclosure made by a fencing of pointed wooden or metal bars known as pales.

PALLET : (i)   A fillet for fixing.

      (ii)   A tray used for lifting materials with the help of a fork lift.

PALLET BRICK : A brick with a rebate at one edge to receive a fillet for fixing.

PALMER BOWLUS FLUME : A device for measurement of flow in a variety of open channels. The principle is based on venturiflume. Such a flume is sometimes installed in a manhole to measure the flow in a sewer.

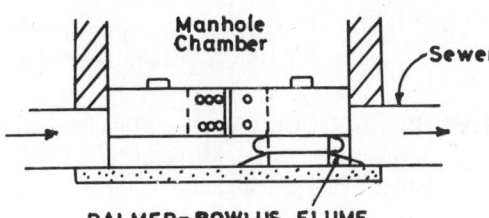

PALMER-BOWLUS FLUME

PAN AND ROLL TILE : Italian tile for roofing.

PAN CLOSET : A type of water closet. See illustration.

PANE : (i)   A panel

(ii)   A square or rectangular glass sheet fixed between glazing bars.

(iii)   The hammer peen.

PANEL : (i)   A single span of a continuous slab.

PAN CLOSET

(ii)   An infilling of timber board or sheet glass let into grooves of a panelled frame for a door or window shutter or a partition.

(iii)   The brickwork infilling in a framed structure of steel or concrete.

(iv)   The space between adjoining vertical members in a lattice girder.

PANELLED DOOR : A door shutter made of panels set in a frame.

PANELLED FRAME : A frame with panels used for partition walls.

PANELLED WINDOW : Same as a panelled door.

PANEL MOULD : A mould for casting panels of plaster.

PANEL PIN : A wire nail with a very small head used in panel work.

PANEL POINT : A node or point in a truss, where a vertical meets a chord.

PANEL SAW : A cross-cut saw.

PANEL WALL : Brickwork infilling in a skeleton frame of a building, which serves as a wall. Usually, this is not a load-bearing wall.

PAN HEAD : A form of head used in screws, bolts and rivets.

PANIC BOLT : A bolt used at exit doors of auditoriums, which is opened by applying pressure onto a horizontal bar in the door at waist height.

PAN MIXER : A mixing machine for preparation of mortar.

PAN TILE : Italian roofing tile. See illustration.

PANTAGRAPH: (i)   A simple instrument used in a drawing office for making reduced copies of drawings. It consists of a framework of tubular sections freely-hinged at joints forming a parallelogram.

PAN TILE

(ii)   A diamond-shaped frame of steel rods hinged at ends, which is placed over an electric locomotive for drawing electical power from an overhead wire.

PAP : The downward vertical outlet from an eaves gutter for draining out rain water.

PARABOLA : The shape obtained by cutting a cone parallel to its one edge.

PARABOLIC DRAIN : A surface drain having cross-section like a parabola as shown in illustration.

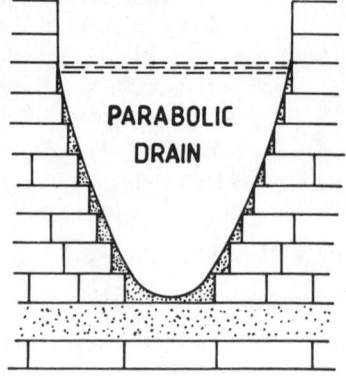

PARABOLIC DRAIN

PARABOLIC RULE : Simpson's rule for calculation of areas of irregular shape of land. The curved boundaries are assumed as portions of parabolic areas. See 'Simpson's rule'.

PARALLAX : A condition when the image formed by an object is not situated in the plane of the cross hairs. Due to this, the line of sight may apparently be made to intersect different points according to the position of eye. Thus, accurate sighting is impossible unless parallax is eliminated.

PARALLEL COPING : A coping of uniform thickness provided over a sloping wall. This is not weathered.

PARALLEL GUTTER : A box gutter which is a parapet gutter.

PARAPET : A low-height wall provided round the edge of a roof, bridge, culvert, balcony, etc. for safety.

PARAPET GUTTER : A box gutter as shown in illustration.

PARAPET GUTTER

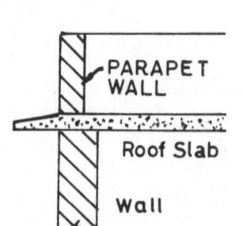

Common rafter

PARAPET WALL : A parapet. See illustration.

PARGE : A mixture of coarse mortar, cowdung and hair.

PARGING : (i) Pargetting i.e., flue lining with parge.

(ii) Decorative plastering on outer walls of a building.

PARIAN CEMENT : Gypsum plaster.

PARING CHISEL : A long, sharp, bevel-edged thin chisel which is not to be struck with a mallet. This is used by controlling the movement and pressure of hands.

PARAPET WALL

Roof Slab

Wall

PARING GOUGE : A thin, long gouge with inside sharpness.

PARKERIZING : A process of phosphating steel.

PARKWAY : A pathway passing through a park.

PARLIAMENT HINGE : H-hinge used in fixing door and window shutters which keeps the shutter lying over the wall surface, when opened.

PARQUET FLOOR : A wooden floor made by gluing hardwood blocks in geometrical pattern over wooden floor boards and polishing the surface. It produces a beautiful appearance. Now-a-days, ready-made parquet boards fixed on plywood are available to make a large floor quickly.

PARQUETRY : The inlaid geometrical patterns of a parquet floor.

PARQUET STRIP : A wooden floor consisting of strips of hardwood boards set together by tongued and grooved joints.

PARSHALL FLUME : A modified form of venturi-flume used in measuring flow through an open channel. See illustration.

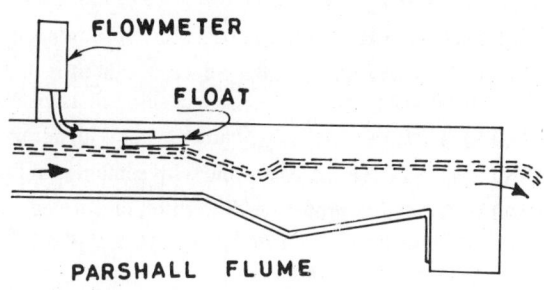

PARSHALL FLUME

PARTIALLY FIXED : A beam end is said to be partially fixed, when the full fixing moment is not developed at the support.

PARTIALLY-SEPARATE SYSTEM : A system of sewerage in which the sanitary sewage with sullage from kitchen and bathroom flows through one conduit and the storm water flows through another conduit. This system helps in maintaining the non-silting velocity in sanitary sewers.

PARTICLE BOARD : A type of chipboard made of wood chippings and saw dust.

PARTICLE SIZE ANALYSIS : A mechanical analysis to obtain gradation of particles present in proportions by weight in a soil or sand and thus a grading curve is drawn to reveal the grain size distribution in the soil or sand sample.

PARTING AGENT : A parting compound or release agent like mould oil, sealant or grease, laid over the forms or inside a mould prior to casting so that the forms or the mould can be parted easily leaving a good surface of the cast material.

PARTING BEAD : A narrow strip of timber fixed vertically to the pulley stiles of a sash window.

PARTING SLIP : A long narrow slip of timber hanging vertically from the pulley to the bottom of a sash window frame.

PARTING TOOL : A V-shaped gouge used in carving and turning wood.

PARTING WALL : A party wall separating two apartments or properties and can be used by both parties as a commom wall. A party wall must have a foundation.

PARTITION : A wall of temporary or permanent nature separating two adjacent rooms. A partition wall may not be of full height as that of the rooms and it may not have a foundation. This is not a load-bearing wall.

PARTY WALL : See 'Parting Wall'.

PASCAL'S LAW : At any point in a fluid the intensity of pressure acts equally in all directions. If a pressure is applied at any point in the fluid, it will be transmitted equally to all points.

PASSINGS : The length by which one sheet overlaps the other in roof flashings, ridge coverings, etc.

PASSING PLACE : A local widening of a narrow street or lane to facilitate passing of vehicles.

PASSIVE EARTH PRESSURE : The passive earth force or resistance offered by a vertical face of earth, when subjected to deformation by a horizontal force due to active earth pressure.

PASSOMETER: A distance measuring instrument in watch form, used in route surveys, which counts the distance when attached to a person who is pacing a distance on walking. The number of steps are recorded automatically.

PASTE DRIER: A stiff paste formed by mixing a drier with a drying oil and an extender.

PASTE FILLER : A filler in paste form used in paints.

PATENT GLAZING : Glazing a window shutter or a fixed panel without use of putty. The glass sheets are bedded on a cushion of asbestos.

PATENT PLASTER : A gypsum plaster with admixtures.

PATENT STONE : A cast stone with cement, sand and coloured stone chips.

PATINA : A thin protective film of oxide formed on the surface of certain metals, when kept exposed for some days. This durable film protects the metal from further atmospheric actions.

PATTERN STAINING : Discolouring of plasterwork at particular places due to different conductances of backings. This occurs due to electro-chemical reactions.

PAVEMENT : A hard floor or platform or footway made of wood blocks, stone sets, bricks, tiles or concrete.

PAVEMENT LIGHT : A 'Vault light' made of glass built into concrete or iron frame over a basement to permit the entry of daylight.

PAVEMENT PRISM : A solid glass block that acts as a prism for the entry of daylight.

PAVILION ROOF : A polygonal or pyramidal roof with equal hips all round.

PAVING : Surfacing over a ground by laying tiles, bricks or concrete slabs.

PAVING BRICK : A hard brick that can be used for paving.

PAVING FLAG : A flagstone used for paving footways.

PEARL ESSENCE : A paint called 'mother of pearl' made by dissolving scales of fish in a lacquer.

PEAT : Gelatinous dead vegetable matters of dark brown colour preserved by humic acid in ground. This can be used as a fuel, when taken out from ground and dried.

PEBBLE DASH : Making an external plaster rough by throwing small stone chips onto the plastered surface while plastic.

PEBBLE WALLING : A wall made of rounded pebbles, which looks like a beautiful flint walling.

PECKY TIMBER : A timber showing decorative grains and some spots of design.

PEDESTAL PILE : Mac Arthur compressed, uncased or cased piles and Western Button piles having pedestal of concrete at the bottom.

PEDOLOGY : A branch of geology dealing with the outer 1.5 m of the earth's crust, which is formed of disintegrated rock materials by weathering action.

PEDOMETER : A measuring instrument which reads distances directly instead of counting number of paces. It is attached to a wheel and when a route is wheeled, it counts the number of revolutions.

PEELER LOG : A log which is suitable for rotary cutting.

PEELING : (i)   Dislodgement of plaster from the backing.

(ii)   Slicing of veneers by rotary cutting.

PEEN HAMMER : A hammer having cutting peens on both sides.

PEG : A small wood dowel or metal pin.

PEGGIES : Small slates of assorted size.

PEGMATITE : A class of igneous rocks containing feldspar, quartz and other minerals.

PEG STAY : A stay for casement windows.

PEG TEST : A test carried out in surveying to adjust the line of collimation and the level. See 'Two-Peg Test'.

PEG TOP PAVING : Paving with small setts.

PELTON WHEEL : A type of impulse turbine which is a wheel fitted with buckets on its periphery. The wheel rotates when the buckets are struck by jets of water. See illustration.

PENCOYD FORMULA : A formula used for finding out the fractional impact allowance to be added for a given span in the design of railway bridges. The proportional live load to be added due to impact, diminishes with the increase in span length. The fractional impact allowance is determined as
1000/(Span in m+1000)

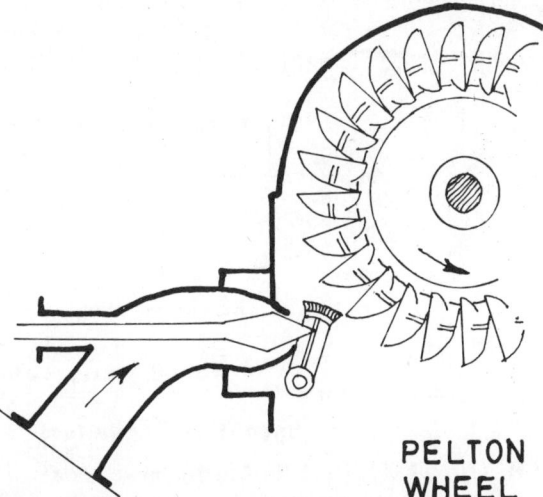

PELTON WHEEL

PENDANT POST : A timber standing against a wall from wall plate down to the corbel to support a hammer-beam roof.

PENDULUM SAW : A circular swinging crosscut saw which is used to cut a log transversely.

PENETRATION TEST : (i) A field test that indicates the load- bearing capacity of a soil. The test may be either static or dynamic.

(ii) An important test to measurse the hardness or consistency of asphalt and bitumen for use in road making.

PENETROMETER : (i)   An instrument used in penetration test. It is forced into a soil at a measured pressure.

(ii)   A blunt-point needle that penetrates a sample of asphalt at 77^0F. , when the needle is loaded with 100 gms. applied for 5 seconds.

PENNING GATE : A rectangular sluice gate that opens vertically upwards.

PENSTOCK : (i)   A pressure main to feed water to a water turbine.

(ii)   A penning gate which is lifted up and lowered manually. This is used at the end of an outfall sewer discharging into a tidal river. This is also used for sludge draw-off from a sedimentation tank.

PENT HOUSE: (i)   A small apartment with a walking space all round, built on the flat roof of a building.

(ii) A projected hood over a door or window for protection against rainwater.

PENT HOUSE ROOF: A lean to or pent roof that has slope in one direction only.

PERAMBULATOR : A Viameter or wheeled pedometer, which is used for automatic recording of number of revolutions for measurement of distance in route surveys.

PERCHED AQUIFER : A special form of unconfined aquifer as shown in illustration.

PERCHED WATER TABLE : The water level of a perched aquifer temporarily or permanently above the standing water level under ground. See illustration.

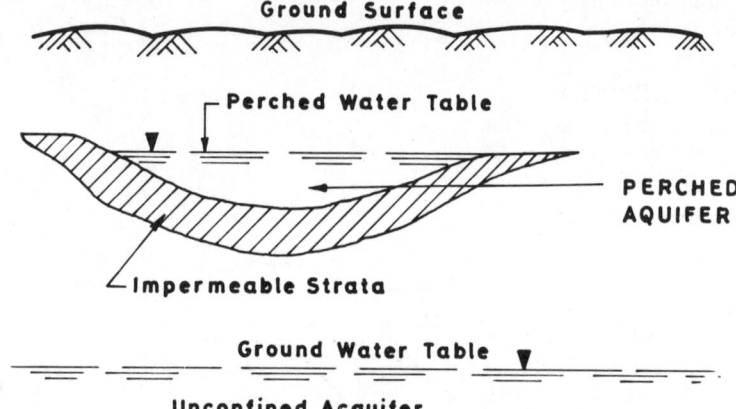

PERCOLATING FILTER : A bed of broken stones or brick bats having slime layers over their surfaces to facilitate biosorption, when a sewage percolates through the bed.

PERCOLATION : The movement of fluid through the pore spaces in soil or any other porous material.

PERCUSSION DRILL : Also known as 'Churn Drill' or 'Cable Drill'. It is a heavy-duty rig for drilling holes in clay or rock for water or oil well .

PERCUSSION TOOLS : Tools used in percussion drilling by striking rapid blows by compressed air.

PERCUSSIVE ROTARY DRILLING : A modern technique of drilling through rocks by applying rotational as well as percussion motion (Vibration) on the bit.

PERFECT FRAME : A structural frame having least number of members neither deficient nor redundant. If 'n' be the number of nodes, then for a perfect frame number of members should be $(2n-3)$.

PERFORATED BRICK : A brick made with perforations, which makes the brick light in weight .

PERFORATED GYPSUM PLASTER BOARD : A gypsum plasterboard having uniform perforations all throughout, which gives high sound-absorption quality.

PERGOLA : A structure having intersecting beams supported by cantilevers and a few posts, which is kept open to sky and made covered with creepers, climbing roses and other plants for beautification at the entrance of a building or above the roof level.

PERIDOTITE : Dark-coloured coarse and medium crystalline igneous rocks such as augite, Hornblende, olivine.

PERIHELION : When the earth is at the nearest point of its orbit (elliptical path) to the sun, it is said to be in perihelion.

PERIMETER DIFFUSER : In room heating with warm air, a diffuser is kept under a window to neutralise its cold effect.

PERLITE : A glass of volcanic origin used as an insulating aggregate in concrete. It expands and forms spherical glassy light particles, when heated.

PERLITE PLASTER : Gypsum plaster containing perlite aggregate instead of sand. It is light in weight and serves as a good insulator.

PERMAFROST : Permanently frozen-ground, where a building foundation should be built below the active depth of the frozen-ground and care should be taken against thawing by providing insulation under the ground floor.

PERMAFROST REGION : Large areas in polar regions of the earth, which are occupied by permanently frozen-ground.

PERMANENT ADJUSTMENT : The permanent adjustments for a theodolite are the adjustments of the

      (i)   Parallel plate bubble tubes

      (ii)   horizontal axis of the telescope

      (iii)   bubble tube on the telescope

      (iv)   line of collimation laterally as well as vertically.

PERMANENT SET : Permanent deformation in a structural member or test specimen, when the loading reaches beyond yield point. The deformation remains, even if the loading is released.

PERMANENT SHUTTERING : A lining of woodwool or insulating board used inside the formwork such that it encases the concrete and remains attached to it permanently.

PERMANENT WAY : A railway track comprising rails, sleepers and ballasts with fittings and fixtures on a stable ground.

PERMANENT WAY

PERMEABILITY : The rate of flow of a fluid under pressure through a soil depending upon the porosity of soil and the pressure head of the fluid. This is measured in field by pumping out water at a constant rate from a well such that the water level in the well remains constant.

PERMEAMETER : An instrument used in laboratory for determining the co-efficient of permeability of a soil. This may be a constant-head permeameter or a falling head permeameter. The former is used for sand and gravel and later for silt and clay.

PERPEND : (i)   The sloping joint between two adjacent tiles.

       (ii)   The corner of a brick wall first built, with correct plumbing so that it can serve as a guide for the walls between, to be constructed subsequently.

PERSPEX: A glass-like transparent acrylic resin (thermo- plastic), lighter and stronger than glass, used in making sheets for roofing and decorative works. The sheets can be joined together by cementing or gluing or by using organic solvent.

PERSONAL ERROR: In instrumental surveying, each surveyor has a difference in reading an observation (value) from that by other surveyors. This is due to the personal approximation, judgment and other factors. For accuracy, the person concerned applies his own sense of adjusting the values accordingly to bring the results close to the truth.

PERVIBRATION : A term for internal vibration of a concrete structure.

PET COCK : A very small drain cock or valve used in a piping system to release air from the upper part of the pipeline.

PETRIFYING LIQUID : A particular thinner used in distempers or in special paints to give a protective coating to a masonry work.

PETROGRAPHIC MICROSCOPE : A special microscope fitted with a polarizer and an analyser prisms required for examining thin sections of rocks to find out the mineral contents in the rock and the particle sizes.

PETROL TRAP : A type of intercepting trap with two or three compartments used in petrol or diesel pumping stations to receive the liquid wastes from the yard of the pumping station as well as car washings. This prevents the floating petrol and other oils flowing into the sewerline.

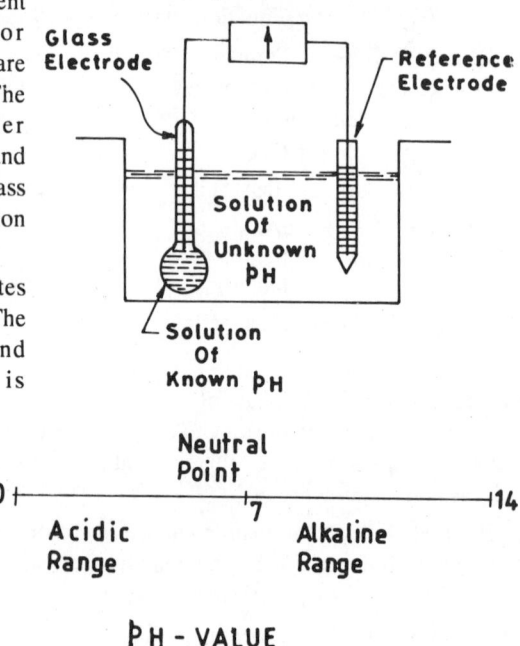

ÞH-METER

pH-METER : A measuring instrument to determine the acidity or alkalinity of a solution. There are various types of pH-meter. The illustration shows a meter connected to a glass electrode and a reference electrode. The glass electrode is filled with a solution of known pH.

pH-VALUE : A value that indicates hydrogen ion concentration. The product of hydrogen ions and hydroxyl ions in water is equivalent to 10^{-14}. When the concentrations of these two ions are equal, each is equivalent to 10^{-7}. It is customary to employ the logarithm of the reciprocal of the concentration. The value thus ranges from 0 to 14. A pH-value between 0 to 7 indicates acidity and that between 7 to 14 indicates alkalinity. See illustration.

PHENOL FORMALDEHYDE RESIN : A synthetic resin which is used as a thermosetting, moisture-resistant glue for joining veneers and plywood in a hot press.

PHON : A unit of noise or loudness of sound. A soft voice is about 40 phons, while a painful sound or noise is about 120 phons.

PHOSPHATING : A pretreatment of a metal surface by applying hot phosphoric acid which should be followed by a finishing coat of lacquer, paint or wax for protection of the surface.

PHOSPHORESCENT PAINT : A paint having sulphide of strontium, Calcium or zinc which emits light for some time after a visible light has fallen on it.

PHOTOELASTICITY : A method of determining the distribution of stresses in models of soil structures under load, in which polarized light is passed through the transparent model of the structure showing isochromatic and isoclinic lines. The magnitude of principal stress at different points is indicated by this technique.

PHOTOGRAMMETRY : Preparation of maps by air photographs i.e, aerial surveying. A contour map can also be prepared from air survey by using a plotting instrument.

PHREATIC WATER : A term occassionally used for ground water.

PHYLLITE : Fine to microscopic crystalline metamorphic rocks having foliated structure, such as slate.

PHYSICO-CHEMICAL TREATMENT : A process of treating waste water (usually industrial effluent) by physical and chemical actions as shown in illustration. Required chemicals are added to the waste water after passing through screen and grit chamber and prior to clarification. The clarified liquid is filtered and then passed through a bed of activated carbon or through a carbon adsorption unit. Lastly, it is kept in a holding pond wherefrom the treated effluent goes out.

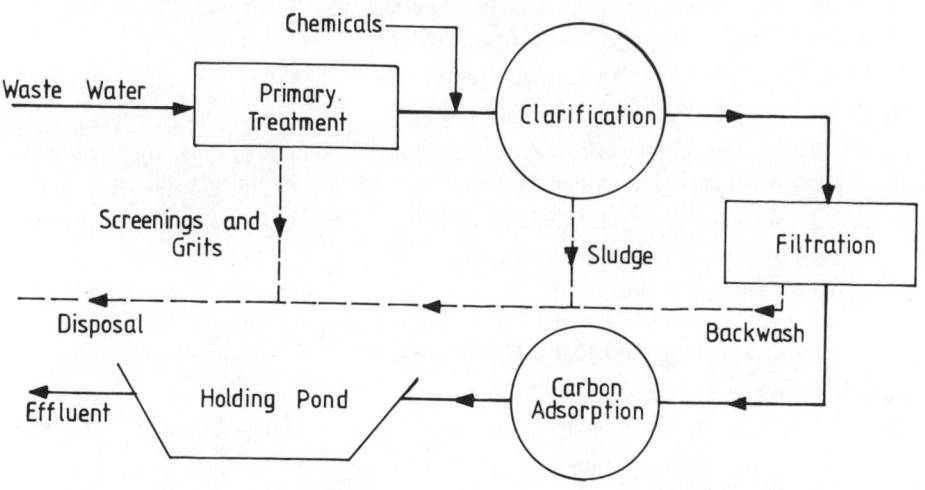

PHYSICO – CHEMICAL   TREATMENT

PICK : A tool with two long sharp prongs used for digging stiff clay or any other material.

PICK AND DIP : A method of laying bricks rapidly by picking up a brick in one hand and buttering it with mortar by another hand

with the help of a trowel.

PICK AXE: (i)   A navy pick.

                 (ii)   A commonly used tool for digging.

PICKET: A ranging rod or pole used in surveying.

PICKING : Dressing a stone surface by striking it with a steel point tool to form numerous small pits.

PICKING UP : (i) Joining live edges or lines during painting a surface.

(ii) Pulling up.

PICKLING :   (i) Bethell's process of creosoting timber under pressure in a chamber.

(ii) Preparation of a metal surface prior to galvanising or painting, by dipping it in hot sulphuric acid, washing it with water and again dipping it in hot phosphoric acid for removal of mill scales, rusts, grease, etc.

PIER :   (i) An intermediate support in a continuous span bridge, usually made of concrete, brickwork or stonework.

(ii) A break water used as a quay wall.

PIER CAP : A thick slab or plate provided at the top of a pier for distribution of load coming over a pier.

PIEZOMETER TUBE : A tube with its top open to measure moderate pressures of a fluid. It is a stand pipe installed at the flow path of a fluid.

PIEZOMETRIC LEVEL : The water level in a piezometer tube due to rise of water through the tube.

PIEZOMETRIC SURFACE : An imaginary surface coinciding with the hydrostatic pressure level of the water in a confined aquifer.

PIGEON-HOLED WALL : A wall with honeycomb.

PIG LUG : A dog ear used in flexible metal roofing.

PIGMENT : Minerals or lakes finely ground to form an insoluble opaque powder, which is used in making paints by mixing it with a vehicle, a drier and an extender.

PILASTER : A pier of rectangular section projecting out from the face of a wall and at right angles, which act as a buttress to the wall.

PILE : A solid shaft or post or column made of timber, steel or concrete which is driven underground in a soft soil to increase the bearing power of soil and to carry the load of a structure coming over it. R.C.C. piles are either precast or cast-in-situ.

PILE BRIDGE : A bridge supported by piles.

PILE CAP : (i) A helmet on the top of a pile to prevent damage of the pile head due to hammering.

(ii) A plate or slab placed on the top of a group of piles for even distribution of load.

PILE CORE : A steel core introduced into a hollow steel cylinder pile for driving the pile by hammering.

PILE DRIVING : See illustration. The pile is driven into the ground by dropping the monkey over it.

PILED FOUNDATION : A foundation built over the piles such that the load is transferred to a hard stratum through piles.

PILE DRAWER : A pile extractor which takes out pile from ground.

PILE DRIVER : A frame with a hoist and a monkey required for pile driving as shown in.

PILE GROUP : Several piles driven close together to form a group to carry a heavy load.

PILE HAMMER : A steam hammer or a drop hammer made automatic (300 blows per minute) or semi-automatic (40 to 60 blows per minute).

PILE HEAD : The top of a pile.

PILE HELMET : A steel cap put on the pile head with a cushioning material in between, to protect the pile head from damage during blowing by a hammer.

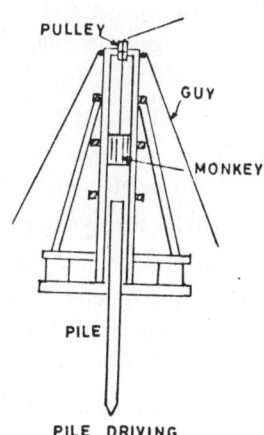

PILE HOOP : A driving band provided at the top of a wooden pile to protect the pile head.

PILE RING : A pile hoop.

PILE SHOE : The conical tip of a pile made of good quality cast iron which helps in penetrating the pile into the ground.

PILING : Thickening of a quick-drying paint and formation of an uneven surface after painting.

PILLAR : A column or post made of timber, cast iron, brick or stone masonry and concrete.

PILLAR QUAY : A quay built of pillars as shown in illustration for berthing of vessles.

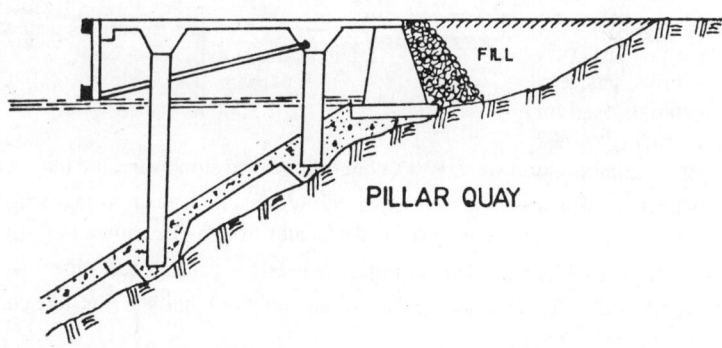

PILLAR TAP : A tap or cock fitted to a wash hand basin and bath tub.

PILOT HOLE : A test hole or a guiding hole of small diameter.

PILOTIS : A reinforced cement concrete column erected and projected up through an open space in ground floor to support a structure above.

PILOT LIGHT : A small flame which is always burning in a burner to ignite the main gas burner as soon as the gas knob is open.

PILOT NAIL : A spike temporarily kept driven to hold the timber pieces in position for nailing with the required nails.

PIN :   (i)   A dowel or nail or wedge.

   (ii)   The tenon that is inserted into a joint.

PINCH BAR: A small form of a crow bar. A bar provided with a bent claw at one end and chisel edge on the other end.

PINCH ROD : A rod used for checking the gap length.

PIN JOINT : A hinged joint in a frame or structure.

PINE OIL : A good solvent used to give the spreading property of a paint. This is obtained either from oleo-resin of pine trees or synthetically.

PINE SHINGLES : Shingles made from pinewood planks.

PIN HOLE : Worm hole found in green timber.

PINHOLING : A defect in spray painting. Tiny holes are formed due to moisture or too low pressure in the spray gun or too thick coating.

PIN KNOT : A small knot less than 10 mm diameter. Also known as 'Cat eye'.

PINK PRIMER : A paint prepared with white and red lead pigments for use as priming coat on a timber.

PINNED : A hinged joint with pins.

PINNINGS : Setting of stones of different colour and texture in rubble masonry such that it produces a chequered effect.

PINOLEUM : Narrow, thin timber slats interoven with canvas strips to make sunblind sheets.

PIN RAIL : A wooden bar with wooden hooks for hanging materials.

PINTLE : The pin used in a door hinge or lock gate hinge.

PIPE : A conduit used to carry fluid like water, sewage, chemical solutions, petrol, steam, milk, gas, etc. The materials used for manufaturing pipes 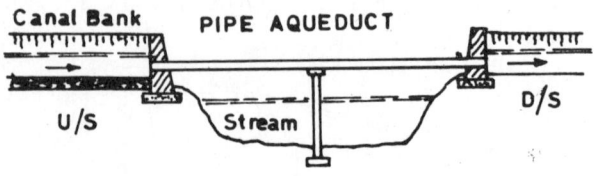 are cast iron, ductile iron steel, PVC, concrete, glazed stoneware and burnt clay.

PIPE AQUEDUCT : When an irrigation canal of low capacity is to cross a natural drainage channel, pipe aqueducts are built instead of water troughs, as shown in illustration.

PIPE CUTTER : A tool having hard cutting discs used for cutting steel pipes.

PIPE HOOK : A hook whose pointed end is driven into a wall and the ring-like curved end holds the pipe.

PIPE JOINTING CLIP : A metal ring or collar enveloping a pipe joint.

PIPE PILE : Open-ended steel pipe is driven underground without any core, by means of air and water jets and then filled with concrete.

PIPING : The movement of a water stream or sand through a dam or any other water-retaining structure due to subsurface boil. Also known as subsurface erosion.

PIPE SLEEVE : A sleeve built into a wall through which a pipe passes.

PIPE STOPPER : A plug to prevent flow through an open end.

PIPE TONGUES : Foot prints.

PIPE WRENCH : A heavy duty wrench with serrated or indented jaws for gripping a pipe firmly for screwing or unscrewing.

PISTOL BRICK : A special brick as shown in illustration.

PISTON PUMP : A pump in which a piston moves to and fro to deliver water intermittently. See illustration.

PIT : An excavation for exploration of the ground or for obtaining soil, sand, etc. or for verification of the underground utility lines.

PIT BOARDS : Well curbing.

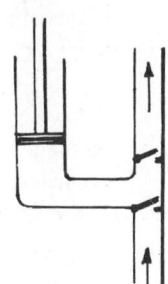

PISTOL BRICK

PITCH : (i)   The centre to centre distance of two rivets in rivetting.

(ii)   The distance (horizontal) from one nosing to the other in a stair.

(iii)   The ratio of height to span of a stair or the angle of stair slab to the horizontal.

PITCH BOARD : A triangular templet or a gauge board for moulding steps in a stair.

PITCHED ROOF : A sloped roof usually made of two slopes meeting at a ridge. A roof is sometimes made of four slopes.

PISTON PUMP

PITCHED WORK : Brick or stone revetment made on the slopes of a river bank or canal or reservoir.

PITCH FIBRE PIPES : Black pipes made of 25% wood or asbestos fibres bonded with 75% coal tar pitch. These pipes are suitable for carrying storm water, sullage and sanitary sewage. These are light in weight and can be laid under ground easily and quickly.

PITCHER PAVING : Making a pavement with stone setts, usually of granite.

PITCHING : (i)   Setting bricks or stone blocks on the sloping bank of a river or canal or tank to protect from scouring action.

(ii)   Penning or soling with stone setts to form a road foundation.

PITCHING FERRULE : A short length of steel pipe cast into a R.C.C. pile to facilitate withdrawal of the pile, when required. The location of introducing the pipe is calculated on the basis of least B.M.

PITCHING TOOL : A hammer-headed chisel used for dressing stones roughly at quarry.

PITCH MASTIC : A composition of coal tar pitch and aggregates spread hot for making a jointless floor.

PITH : The dark-coloured centre of a log.

PITH FLECK : A discolouration like pith, caused by the attack of insects.

PITH KNOT : A knot having a pith hole.

PIT PRIVY : A low-cost privy for economically weaker section of people living in rural areas or semiurban areas. It essentially consists of a small pit either dry type or wet type with a squating slab on the top of the pit as shown.'

PITH RAY: Rays starting from the pith as seen in the transverse section of a log.

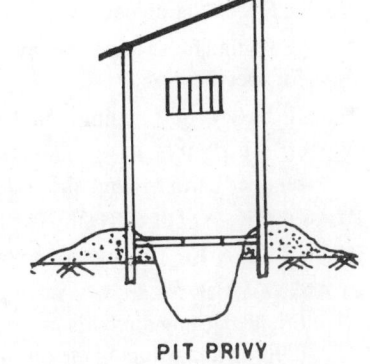

PIT PRIVY

PITOMETER: A meter to discover waste of water due to blown joints in water mains and unauthorised and unmetered connections. It consists of two bent tubes introduced through a corporation cock into the street water main. The orifice of one tube points upstream and that of the other tube points downstream. The pressure difference is measured by means of a manometer.

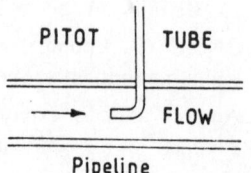

PITOT TUBE : An instrument by which the velocity head of a flowing liquid can be measured . It is simply a tube with its lower end bent at 90°. It is placed vertically in a flowing liquid such that its lower end faces the direction of flow. The liquid flows up the tube and when the level in the tube stands constant, the velocity of flow is estimated by the height of the liquid in the tube.

PIT-RUN GRAVEL : Gravel with impurities dug out from a pit.

PIT SAWING : The primitive method of ripping a log by sawing manually with the help of a long saw having handles each end. In this method, the log is held slightly inclined by means of stakes amd wedges. A man stands in a pit and another man at the top of the pit. Sawing is done by to and fro motion of the saw (by pull and push given by the two men at the two ends). In present days, this is found to be practised in rural areas of undeveloped countries.

PITTING : Blowing of plaster which forms an uneven surface.

PIVOTAL ERROR : The error although small, is likely to occur at the pivot of a theodolite. It is difficult to prevent its occurrence. Sometimes, due to too much of wear and tear of the pivots the error may be appreciable.

PIVOT BRIDGE : A swing bridge which swings about its pivot to allow passing of vehicles.

PLAIN CONCRETE : A concrete without any reinforcement steel. It may be cement concrete, lime concrte or mud concrete.

PLAIN TILES : Commonly used clay tiles for roofing which are of flat type with a slight convexity upwards.

PLAN : (i)   The top view of an object in an orthographic projection.

       (ii)   The layout plan of a building showing arrangement of rooms and other spaces.

PLANCEER PIECE : A soffit board to an eave.

PLANCON : A hardwood log which is roughly octagonal.

PLANE : A tool for shaping and smoothing timber. There are various types of plane required for specific jobs.

PLANE IRON : The cutting iron or blade of a plane supported by a back iron.

PLANE OF RUPTURE : The plane along which a failure takes place, e.g. failure of a retained earth by land-slide along a plane.

PLANE OF SATURATION : The level to which a material is saturated with a liquid.

PLANE STOCK : The body of a plane.

PLANE TABLE : A drawing table pivoted on a tripod with a swivel joint so that it can be oriented in any direction in the horizontal plane. This is used in plane table survey for plotting and mapping the physical features in field.

PLANE TABLING : A method of land survey in which both plotting and mapping of physical features are done quickly and simutaneously with the progress of the survey-work. The major appliances and instruments required are plane table, alidade, spirit level, compass, tape and plumb bob.

PLANIMETER : A very handy instrument used for measuring areas accurately from a map or plan. The principle is based on the motion of a rod on a plane surface. It consists of two bars hinged at one end. The other end of one of these bars is provided with a needle point and a weight for fixation. The free end of the other bar is provided with a tracing point which is moved over the boundary line of the area. The instrument has horizontal and vertical dials graduated and provided with vernier scales. On complete revolution of the tracer roller following the boundary line, the reading obtained from the dials is multiplied by the instrument constant to determine the area.

PLANK : Timber board of 20 mm to 80 mm thickness obtained by sawing a squared log.

PLANNING : In civil engineering practice, it is the formulation of a concept plan and layout based on norms and design criteria and the strategy for making it technically feasible and economically viable.

PLANNING ENGINEER : A planner who is basically an engineer. He may be a landuse planner, traffic & transportation planner, water & sanitation planner, architect planner or town & country planner.

PLANNING GRID : A network of lines showing modules which is used by architects to arrive at a building layout.

PLANOMETRIC PROJECTION : A pictorial view of an object develped on its plan by drawing oblique parallel lines from the corners.

PLANT : In civil engineering, it relates to production of certain building components, mixing of ingredients and processing certain materials A few examples of these are: board manufacturing plant, fabrication plant, asphaltum plant, concrete mixing plant, water treatment plant,sewage treatment plant, etc.

PLANTING : Fixing certain ornamental works (which are prepared separately) to a building or any other structure by buttering with mortar or by nailing or gluing.

PLANT MIX : A mortar or concrete is called 'plant mix', when the mixing operation is done at a central mixing plant.

PLASHING : Weaving living sticks horizontally in a hedge.

PLASTER : A mortar that hardens on setting, when applied to a wall or ceiling. It may be mud, lime, gypsum or cement plaster.

PLASTER BASE : The surface or ground for plaster. It may be brickwork, wood, metal or gypsum lath.

PLASTER BEAD : An angle bead in plastering.

PLASTER BOARD : Gypsum plaster board.

PLASTER DAB : A lump of gypsum plaster sticking to a wall, which is used as a fixing material for wall tiles.

PLASTERER'S PUTTY : Lime putty.

PLASTER OF PARIS: Hemihydrate plaster i.e., gypsum converted to $CaSO_4.^{0.5} H_2O$, by driving out some of its moisture content by heating. It is an excellent plaster for casting. It sets within 10 minutes time after casting.

PLASTER SLAB: Precast slab of gypsum plaster and sand used in making false ceiling, partition, etc. ,

PLASTICS : An organic substance made from resins, plasticizers, pigments and fillers. It can easily be moulded into desired shape and size under heat and pressure. Thermoplastic variety softens on heating and becomes hard on cooling. This class of plastics can be reused. The thermo-setting plastic requires great pressure and momentary heat during moulding. This can not be remoulded once it is set.

PLASTIC DEFORMATION : The deformation of a ductile material when it has exceeded its yield point under tensile test. The deformation (plastic flow) takes place during this stage without application of any extra load.

PLASTIC DESIGN : The design of framed structures assuming that plastic hinges form at points of maximum bending moment in individual member of the frame.

PLASTIC EMULSION : A paint that produces a highly lustrous, smooth, washable surface.

PLASTIC FLOW : Plastic deformation.

PLASTIC GLUE : Synthetic resin glue.

PLASTIC HINGE : A point of maximum bending moment as assumed in plastic design of a framed structure.

PLASTICITY : In soil science, it is a characteristic of clays which can be determined in field. At a certain moisture content a clayey soil becomes plastic and it can be rolled into a thread. When the non-plastic condition reaches due to loss of moisture, the thread crumbles during rolling.

PLASTICITY INDEX : The index of plasticity. It is the numerical difference between the liquid limit and plastic limit. The value is obtained by carrying out Atterberg's limit tests on a soil sample.

PLASTICIZER : An admixture used in a mortar or concrete to increase the plasticity or workability of the mix.

PLASTIC LIMIT : It may be defined as the minimum moisture content at which a soil can be rolled into a thread of 3 mm. diameter without crumbling. It is the minimum water content at which a soil becomes plastic.

PLASTIC MODULUS : The modulus of section used in plastic design of structures.

PLASTIC STEEL : A synthetic resin with iron filings used for repairwork in different materials.

PLASTIC WOOD : A mixture of nitrocellulose, resin, wood flour and plasticizers is made into a paste with suitable solvents and is used in filling cracks and holes in wood.

PLASTIC WELDING : Forge or pressure welding when iron or steel is in plastic stage by heating.

PLASTIC YIELD : Plastic deformation.

PLAT : A map or plan showing a plot of land with subdivisions, if any and the land ownership.

PLATE : A metal sheet or thin board.

PLATE BEARING TEST : A test to determine the bearing power of soil by digging a trench of required depth, placing a stiff steel plate of 12" x 12" at the bed of the trench and gradually loading it until it fails by sinking. This gives a measure of bearing capacity of soil.

PLATE CUT : A foot cut in woodwork.

PLATE FLOOR : A thin R.C.C. floor with a flat soffit.

PLATE GIRDER : A built-up girder by welding or rivetting plates and angle sections with stiffeners at intervals. The sectional area of flanges is provided on the basis of the shearing force and bending moment to be taken up. See illustration.

Stiffener

PLATE GLASS : A glass cast in form of sheets with smooth surface.

PLATED BEAM : A beam made of plates accordingly as required to carry the load.

PLATEN : (i) The smooth steel plates which press the concrete cube during compression testing in a machine.

PLATE GIRDER

(ii) The hot steel plates used in hot pressing for making plywood by using thermosetting glue.

PLATE SCREWS : Foot screws of a thedolite.

PLATE VIBRATOR : A vibrator with a base plate used for compacting a fill.

PLATFORM FRAME : The floor platform made of a frame resting on the wall frames properly braced in a timber-frame house.

PLATFORM GANTRY : A gantry to bear a portal crane.

PLATFORM HOIST : A hoist that lifts a platform carrying materials to the required height during construction.

PLATFORM ROOF : A flat roof.

PLENUM CHAMBER : A chamber containing air at pressure.

PLENUM SYSTEM : A system of air conditioning a large building by blowing cold clean air into rooms above the door level and withdrawing out hot foul air at floor level.

PLIERS : A scissor-like tool with cutting blades built into its jaws. It is a gripping and cutting tool.

PLIES : Sheets of veneer.

PLINTH : Pedestal of a wall or column. The widened wall below ground floor level.

PLINTH BRICK : A brick used in making a plinth.

PLINTH COURSE : The top course of a plinth.

PLINTH HEIGHT : The height of wall from ground level to plinth level.

PLINTH BRICK

PLINTH LEVEL : The ground floor level.

PLOUGH: (i) A plane used for making grooves at a distance from the edge of wood.

(ii) An agricultural implement which penetrates the top soil and moves forward.

PLOUGH STEEL: A steel from which wires are drawn for making ropes.

PLOUGHED AND TONGUED : A feather joint used in woodwork as shown in illustration.

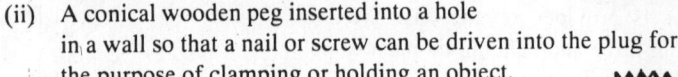

PLOUGHED & TONGUED

PLUG : (i) A pipe fitting to close an open end of a pipe by screwing as shown.

(ii) A conical wooden peg inserted into a hole in a wall so that a nail or screw can be driven into the plug for the purpose of clamping or holding an object.

(iii) A socket outlet in electrical connection.

(iv) A screw plug.

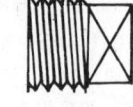

PLUG

PLUG AND FEATHER : A conical wedge (plug) is placed between two flat wedges (feathers) in a hole made in a rock for dislodging. When all the plugs inserted in a series of holes made in a rock are hammered simultaneously, the rock cracks along the line of holes due to pressure.

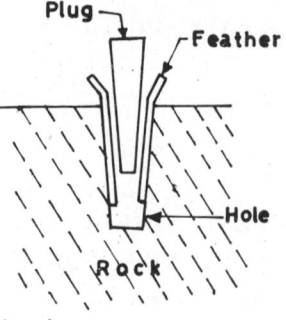

PLUG CENTRE BIT : A tool bit used for widening a drilled hole.

PLUG COCK : A cock having a tapered plug. In open condition, a fluid passes through the hole in the plug. The cock is closed by turning the plug through 90°.

PLUGGING : Closing a hole in masonry by inserting a plug into it.

PLUG TENON : A short tenon which projects out from the foot or head of a post for fixing in a truss or frame.

PLUMB : Vertical .

PLUMBAGO : A quality of graphite used as a refractory material .

PLUMB BOB : A conical weight suspended from a string to indicate the vertical line.

PLUMB CUT : The vertical cut at the foot or ridge point of a rafter.

PLUMBER : A person skilled in laying pipelines and connecting fittings and fixtures.

PLUMBER'S SOLDER : An alloy of lead and tin used for making joints in lead pipes.

PLUMBING : The pipework for water supply and sanitation in a building or elsewhere.

PLUMBING FORK : A U-fork like hair pin bend, used for setting a survey station in a plane table survey.

PLUMBING FORK

PLUMBING SYSTEM : The plumbing system for disposal of foul water from various units of a building may be either one pipe installation or two-pipe installation. The illustration shows a schematic presentation of a plumbing system for water supply to different floors in a building having no water reservoir at roof level i.e., the water is directly fed from a street water main. The flow and pressure are controlled by reducing diameters of pipes, using nipples, short pieces and control valves.

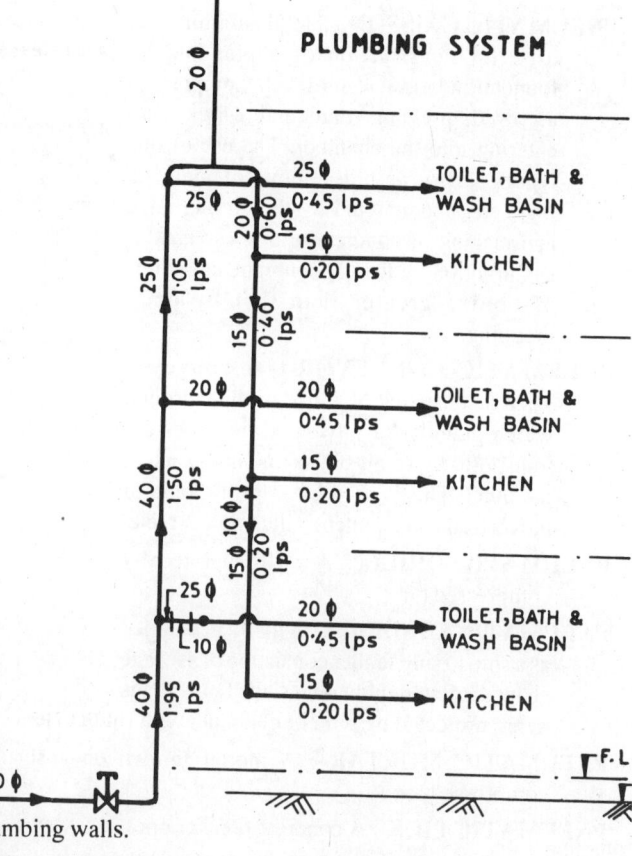

PLUMB LEVEL : A spirit level with a small bubble set at right angles to the main bubble to indicate the vertical or plumb direction. When the main bubble is held vertical, the small bubble must come in central position.

PLUMB LINE : A weight hung from a string to stretch it in a vertical direction which is the plumb line.

PLUMB RULE : A straight edge from which a plumb bob is hung with the help of a thread and it is used for plumbing walls.

PLUMMET : A plumb bole.

PLUNGER PUMP : A pump actuated by a plunger to deliver water intermittently. See illustration.

PLUS SIGHT : A back sight in surveying.

PLY : A veneer which is a very thin sheet of wood used in making plywood by gluing several plies together.

PLY GLASS : Glass fibres or plastics sandwiched between two glass sheets which is used for diffusing light, where direct sunlight is to be avoided.

PLYMETAL : A plywood faced with G.I. or aluminium sheet. This is practised, where the plywood is likely to be moistened as in bathroom or toilet doors.

PLYWOOD : Sheets of veneers or wood plies are glued together by heat and pressure to obtain a thick board, which is called plywood.

PLYWOOD CLIPS: Special clips used for fixing plywood to a wooden or metal frame. These are made of plywood which can easilly be glued to a frame.

PLYWOOD PARQUET: A form of parquet made by fixing plywood tiles to a subfloor (wooden base).

PNEUMATIC CAISSON : See illustration. It is used for construction of a pier and its foundation below water level. Compressed air used, prevents mud and water from entering into the chamber. The use of this type of caisson permits removal of boulders, logs, etc. encountered by the cutting edge and placing of concrete in dry. the cost of construction by using pneumatic caisson is of course greater than that by open dredging.

PNEUMATIC CAISSON

PNEUMATIC CONVEYOR : A conveyor system consisting of a tube or pipe through which granular or powdered materials are transportd by an air blast from one place to the other. This is used for transportation of pulverized coal, cement, silica powder, etc.

PNEUMATIC DRILL : A drill operated by compressed air.

PNEUMATIC FENDER : Originally developed as a ship to ship fender consisting of sausage shaped inflated rubber bags, the fender bags being protected by wire or chain net with rubber sleeves or tyres.

PNEUMATIC MORTAR : A mortar thrown on a surface by guniting operated by compressed air.

PNEUMATIC PICK : A concrete breaker operated pneumatically.

PNEUMATIC RIVETER : A rivetting machine having a percussive compressed air tool provided with a snap for making the rivet head.

PNEUMATIC SEWAGE EJECTOR : It is used for lifting a sewage volume from a lower level to a higher level of a sewer line under moderate head. First, the sewage fills a chamber through a non-return valve. At a certain level of sewage in the chamber the float opens the air valve and compressed air is admitted into the chamber which forces the sewage to flow up.

PNEUMATIC SHAFT SINKING : Shaft sinking by means of a drop shaft fitted with an air-tight working chamber having an air-lock for accessibility or workers and to lower materials. The air-lock door opens downward and gets closed by upward pressure of air. The arrangement is similar to that of a pneumatic caisson.

PNEUMATIC TYRED ROLLER : A multiwheel roller used in compacting massive earth fills and earthen dams. It is essentially a towed roller having several wheels of rubber tyres loaded with water tanks or ballast containers. It compacts soil in 100 to 150 mm. thick layers.

PNEUMATIC WATER SUPPLY : A system of water supply in which the water of a closed tank is forced up into a building by compressed air.

POCKET : (i) A hole in the pulley stile of a window frame.

(ii) A hole in a wall into which a beam is to be inserted.

POCKET CHISEL : A small chisel used for cutting a hole in the pulley stile of a window frame.

POCKET ROT : A decay of timber in form of a small patch which forms a hole in the long run.

POCK MARKING : Local depressions formed on drying of a painted surface.

POD AUGER : A primitive form of auger, in the upper part of which there is a pod to receive the wood chippings.

PODGER : An open-jaw spanner having a pointed handle used in construction of railways and steel framework. The pointed end is used to bring two or more holes in alignment for use of bolt-nuts to make a joint.

PODZOL : The top soil which is more or less acidic in nature, from which soluble materials move into underlying layers by leaching action.

POINT BEARING PILE : An end bearing pile that supports load by tranferring it to a hard stratum through its tip.

POINT GAUGE : A measuring device, in which a pointer is made to slide along a graduated scale to indicate the level of a liquid by touching the liquid surface.

POINT INDICATOR : This is a signal interconnected with the rail track points. When the points are set for a track, this is operated simustaneously with the help of levers. It consists of a box with circular discs. Two white and two green glasses are fixed on the sides of the box with a lamp inside. The box can be roatated about its vertical axis.

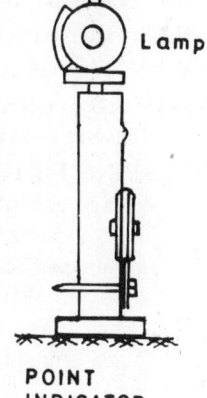

POINT LOAD : A concentrated load which acts at a point.

POINT OF INFLECTION : (i)   The point at which log growth phase of microbes ceases and the declilning growth phase starts as shown in illustration.

(ii)   The point of contraflexure where the bending moment is zero due to change from positive to negative bending moment in a structural member. It never occurs in a simple beam or in a cantilever.

POINTS: Tapered and pointed rails are so arranged and hinged at a railway crossing or junction that trains can be directed from one track onto the other. The movable rail is hinged at the rail

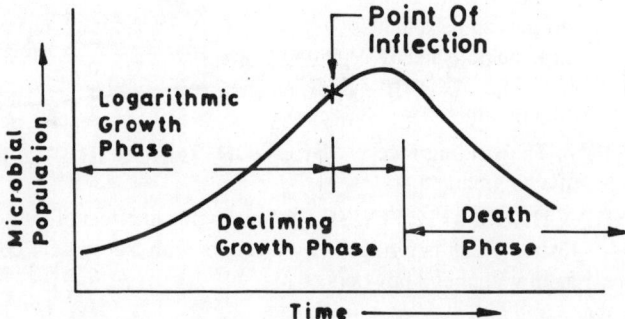

heel and the toe gets locked against the stock rail.

POINTING: The mortar joints are raked out from the surface of a masonrywork and these are finished with a strong mortar to resist weathering action as well as to improve the appearance of the surface.

POISSON'S RATIO : The ratio of lateral strain to longitudinal strain. It is usually denoted by μ.

POKER VIBRATOR : An internal vibrator which is introduced into green concrete to bring the homogeneity and compaction in concrete.

POLARISCOPE : An apparatus consisting of two nicol prisms–a polarizer and an analyzer, used in petrographic microscope for model analysis.

POLARIZED LIGHT : A light having vibrations in one plane. This is obtained in a petrographic microscope with the help of a Nicol prism which filters all light waves excepting those vibrating in one plane.

POLAR MOMENT OF INERTIA : The second moment of area about an axis perpendicular to its plane i.e, about the polar axis.

POLDER : Reclaimed land area from the sea by constructing dykes and filling with boulders.

POLE : A rod or bar or post.

POLE PLATE : A horizontal timber supporting the end of a tie beam or principal rafter at its base.

POLING BACK : Digging earth behind timbering or sheet piling.

POLING BOARDS : Vertical boards which support the trench sides.

POLING FRAME : A tucking frame.

POLISHED WORK : A work to produce a glass like smooth surface by rubbing with abrasives and buffing.

POLISHING FILTER : A filter bed of gravel and sand, through which effluent from a sewage treatment plant is passed for further clarification removing B.O.D. and suspended solids. See illustration.

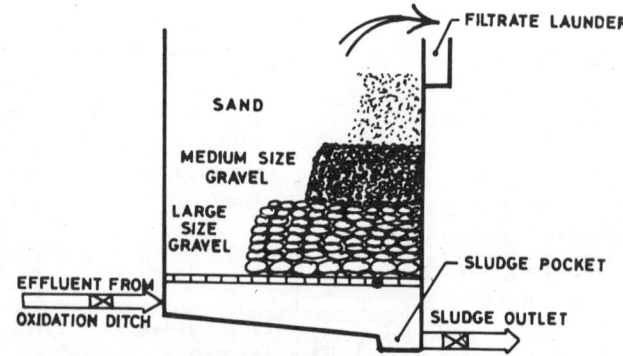

POLISHING VARNISH : A rubbing varnish that sets so hard that it can be polished by rubbing with mineral oil.

POLL : The striking face of a hammer.

SECTION THROUGH POLISHING FILTER

POLYCHROMATIC FINISH : A finish with an effect of multi-colours or with a metallic lustre, which is produced by lacquers with transparent colouring matters or enamel paint with metal powders.

POLYESTER RESIN : Glass-fibre-reinforced synthetic resin.

POLYGONAL RUBBLE : Rubble wall faced with multi-sided stones as shown in illustration.

POLYMERS : Usually solid organic compounds having a long chain of molecules. Synthetic resin, casein and polyvinyl chloride are examples of polymers.

POLYMERIZATION : Condensation or hardening of synthetic resin.

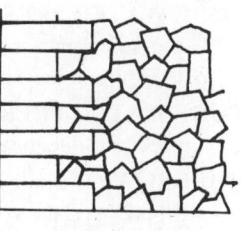

POLYGONAL WALLING

POLYSTYRENE : A thermoplastic material used for making glass like blocks for transparent partitions and wall tiles for beautification of walls. Its white colour changes to light yellow on exsposure to ultra-violet light.

POLYSULPHIDE SEALANT : A sealant having permanently high resilience, which is stable over a wide range of temperature.

POLY TETRA FLUORETHYLENE : A class of plastics used as a jointing compound.

POLYTHENE : Acid–and alkali–resistant material, which is used in making insulators, sheets, pipes and pipe fittings.

POLYVINYL CHLORIDE : A vinyl resin which is a polymer. It is an inert rubbery incombustible material with which waterproof sheets, acid, alkali–and oil–containers, flooring, walling and ceiling tiles, pipes and pipe fittings are manufactured. This is chemically resistant to corrosive fluids and stronger than thermoplastics. The maximum service temperature is between 130°F and 150°F.

POMMEL : (i)   A finial to a roof

(ii)   A light-weight, hand-operated rammer with a iron foot.

POND : (i)   A water reservoir made by excavation of earth.

(ii)   A reach or stretch between two canal locks.

PONTOON : A flat bottomed vessel which is sometimes used to carry a crane or to support the end of a floating bridge or to carry plants and materials.

PONTOON BRIDGE : A floating bridge of temporary or permanent nature which is supported by a R.C.C. pontoon moored to the river bed.

POOLE'S TILES : Flat, clay tiles of interlocking type used as roofing tiles. Each tile has two waterways at the two sides, a central ridge extended half-way up and two holes for nailing with no nib.

POOR LIME : An impure lime containig earthy matters which is used for inferior quality of work of temporary nature.

POPPING : Blowing off thin plaster.

POPULATION DENSITY : The number of persons settled in an unit area. It is commonly expressed as 'ppa' (persons per acre).

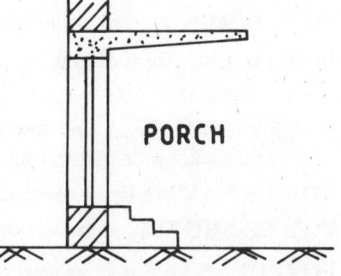

PORCH

PORCH : A cantilever hood without any prop at the free end, provided at the entrance to a building as shown in illustration.

PORES: Inter-granular spaces or small cavities in a granular material.

PORE WATER PRESSURE : The pressure exhibited by water in a saturated soil. The pressure is measured by inserting a tube into a saturated soil with its open end leading to Bourdon pressure gauge.

PORE WATER-PRESSURE CELLS : A sensitive instrument to measure the pore water pressures due to variation in loading.

POROSITY : The ratio of volume of voids to the total volume of solids. It is given by $e/1+e$, where 'e' is the void ratio.

PORTABLE CRANE : A small crane having a power-driven hoist is capable of moving about on wheels. ·

PORTAL : See 'Portal frame'.

PORTAL CRANE : A jib crane or gantry crane carried on four-legged portal frame.

PORTAL FRAME : A frame consisting of two vertical poles rigidly connected at top by a horizontal member as shown in illustration.

PORTICO : It is similar to porch having supports at the extended ends as shown in illustration. It is made larger than a porch.

PORTLAND CEMENT : It is a cementing material which resembles portland rock on setting. It is extensively used in civil ingineering constructions. Its chief constituents are lime, silica, alumina and ferric oxide. Lime, silica and ferric oxide give strength, while alumina imparts setting property.

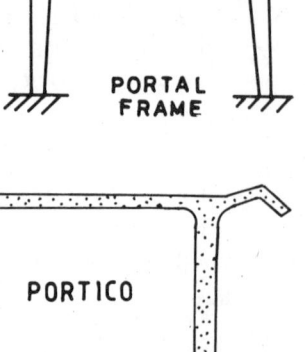

PORTAL FRAME

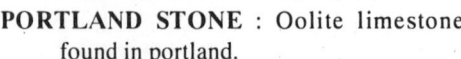

PORTICO

PORTLAND STONE : Oolite limestone found in portland.

POSITION HEAD : The elevation head of a liquid due to its position.

POST : A vertical member in a construction.

POST CHLORINATION : The chlorination of water after filtration and prior to supply to the water consumers. It is essential to make the water free from disease-causing bacteria.

POST HOLE AUGER : A tool for boring holes in soil.

POST TENSIONING : Prestressing concrete members in a structure by pulling stranded rods or cables through the holes kept in the members during casting.

POT FLOOR : A floor made of hollow tiles.

POT TILE : A form of clay tile, as shown in illustration, used in covering a sloping roof.

POURING ROPE : A joint runner used to form a ring hole around a spigot and socket joint for sealing it with molten lead.

POWER RAMMER : A compacting machine.

POWER SHOVEL : A power-driven excavator.

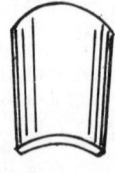

POT TILE

POZZOLANA : Finely ground volcanic ash and other minerals which is used as a cement.

PRATT TRUSS : See 'N-truss' and 'Whipple-Murphy truss.'

PRECAST CONCRETE : Concrete blocks, slabs, posts, lintels and parts of a concrete frame, which are precast and cured in a factory. The castings are transported to the work site. This facilitates in rapid construction.

PRE-CHLORINATION : When too much of organics and microbes are present in a raw water and it has colour, taste and odour, prechlorination is a must prior to its treatment in a plant.

PRECIPITATION : (i)   Ranfall or downpour.

(ii)   Settlement of suspended matters in a fluid.

PRECISE LEVELLILNG : Levelling with a precision levelling instrument such that the difference in level between two points at 'L' distance apart in two successive observations does not differ more than $0.02L^{0.5}$.

PRE-CONSOLIDATION LOAD : From the consolidation test results a consolidation curve can be plotted, from which the maximum load to which the soil was subjected in the past, can be estimated.

PREFABRICATED STRUCTURE : A structure whose components are prefabricated in a factory and are assembled at construction site.

PREMIXED CARPET : A carpet of stone chips premixed with bitumen is laid hot on a damaged road surface and rolled.

PREMIXED PLASTER : Dry mixes of plasters containing vermiculite or perlite are brought to site in bags, which are made in form of a paste by adding water to the mix.

PRESERVATIVES : Materials to be used as preservatives vary in a wide range depending on the materials to be preserved. Zinc chloride, zinc sulphate, copper sulphate, zinc tannin, sodium fluoride and creosote oil are normally used for preservation of timber. The preservatives for cast iron and steel are tarring, painting, phosphating, metal coating, etc.

PRESS : A manual or mechanical device to press glued veneers to form plywood, impregnated fibres to manuafacture fibre boards, and other materaials between two steel plates for making sheets, boards, slabs, etc. A press is also used in punching holes, shearing and notching plates, etc.

PRESSED BRICKS : Bricks moulded by giving pressure to make their edges sharp with a smooth surface.

PRESSED GLASS : Glass pressed to brick shape and blocks for use in pavement lighting and making partition walls.

PRESSED STEEL : Steel sheets pressed hot to make different small frames required in various jobs.

PRESSURE CREOSOTING : Creosoting timbers in a chamber under pressure for preservation of timber.

PRESSURE FILTER· Essentially a rapid gravity filter built within a cylindrical vessel or container placed horizontally or vertically as shown in illustration. Pressure is applied to raise the water. This is suitable for water supply to a small area.

PRESSURE GAUGE : An instrument to measure the pressure of fluids flowing through pipes. There are various forms of pressure gauge.

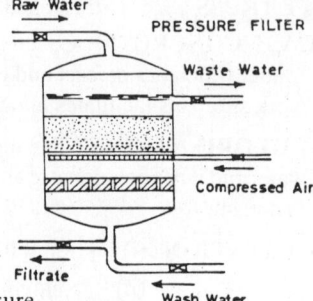

PRESSURE GUN : An appliance similar to a grease gun, used for applying a sealing compound to a joint or for other purposes.

PRESSURE HEAD : The head of liquid available at any point in a pipeline due to its pressure.

PRESSURE RIDGE : A ridge of fresh water formed adjacent to and parallel to the sea-coast, enough high above the sea level to repel salt water of the sea. The pressure ridge should therefore be located inland from the saline front to control intrusion of salt water into the aquifer. The ridge is formed with recharge wells.

PRESSURE TANK : A tank in which a fluid is kept under pressure.

PRESSURE WELDING : Welding by applying pressure.

PRESTRESSED CONCRETE : This is more advantageous compared to ordinary concrete, when long spans, shallow depth and light self loads are desired. Prestressed concrete structures require about 25% less concrete than standard designs of ordinary concrete.

PRESTRESSING : First, a structural member is kept under compression and then steel is placed in tension before the loads are applied to the member. The reinforcement steel is not bonded to the concrete. When the bars are fully stressed their ends are locked in place.

PRETENSIONING : A process in which the reinforcement is placed under tension before the concrete is cast around it. When the concrete is cured and tensioning device is released, the bond on steel surface prevents slippage.

PRICKING-UP COAT : The first coat of plaster on a lath.

PRICK POST : A queen post in a truss or any intermediate post in a frame.

PRICK PUNCH : A small punching tool similar to a nail punch, used to prick holes in sheet metals.

PRIMARY COAT : The base coat on a wooden surface. For iron and steel surfaces, it is a coat with a base of red lead, baryta or iron oxide. Linseed oil without a pigment should not be used for priming coat.

PRIMARY GLUING : Gluing veneers for making plywood.

PRIMER : (i)   A priming coat applied on a surface prior to painting.

    (ii)   A bituminous adhesive applied to a roofing felt.

    (iii) A bituminous spray in soil stablization.

PRIME VERTICAL : The vertical circle passing through east and west points and tracing a plane on the celestial sphere through Zenith and Nadir which is at right angles to the meridian of the place.

PRIMING : (i)   The first filling of a reservoir prior to its functioning.

    (ii)   Filling a siphon or a pump with water for starting its operation.

PRIMING COAT : A first coat on a surface prior to application of the undercoat. The priming coat fills up the pores, hair cracks and depressions and forms the base for painting. For steel surfaces, the priming coat contains inhibiting pigments.

PRINCESS POST : Intermediate post used between queen post and the wall (end support) to strengthen a queen post truss.

PRINCIPAL POST : The end post or the door post in a framed partition.

PRINCIPAL RAFTER : The main rafter on either side of a roof truss to carry the purlins and common rafter.

PRINCIPAL STRESS : The stress across the principal planes of a structural member. The maximum and minimum normal stresses occuring on planes of zero shearing stress are called principal stresses.

PRISMATIC ASTROLABE : An instrument used for determination of latitude and time by observation of stars at a constant altitude of 60°. There is an equilateral prism in front of the object glass of the telescope with its edges horizontal. The direct rays from a star entering normal to the uppper face of the prism are refracted from the lower face and proceed horizontally and normal to the back vertical face of the prism. At least four stars should be observed, the stars being in four quadrants. Its recent development is a 45° prismatic astrolabe.

PRISMATIC COMPASS : A handy compact instrument consisting of a magnetic needle with an attachment of a circular graduated disc placed in a small shallow cylindrical box. At one edge of the box, a frame with a hair-line sight is hinged, while opposite to this is a prism . The instrument is used to read bearings of lines in a survework.

PRISMATIC GLASS : A glass pane having small parallel triangular prisms on one surface. When this glass is placed in a window with triangular prisms horizontally on the outer surface, it admits more light into a room.

PRISMATIC TELESCOPE : A telescope with its eye piece having a prism to reflect at 90°. This facilitates in taking steep sights without the help of an auxiliary telescope.

PRISMOIDAL FORMULA : A formula used to compute the volume of earthwork in excavation. When A_1 and A_2 are the two end areas in a length L and A_m is the area of the mid-section, then volume = $L/6 \times (A_1 + A_2 + 4A_m)$.

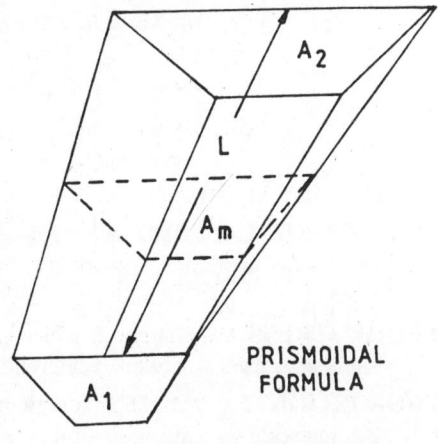

PRISMOIDAL FORMULA

PRISM SQUARE : An optical square having two prisms set in a small cylindrical box, used to set out right angles in field when the offsets to be taken are too long. The included angle between two reflecting surfaces of each prism is 135°.

PROBABILITY CURVE: A curve representing the probability of occurrence of an item or an error in surveying or computations. The ordinate of the curve at any point represents the relative frequency or probability of the occurence. In fact, it is a Gaussian Curve.

PROBABLE ERROR: It is about two-thirds of the standard error or standard deviation or mean square error. In surveying, this term is commonly used. It may be said that errors are equally likely to be numerically greater or less than the probable error.

PROCTOR COMPACTION TEST : A laboratory test conducted to determine the optimum moisture content of a soil sample for maximum compaction. The required apparatus for this test is shown in illustration. The test is carried out by varying the moisture content of the soil sample.

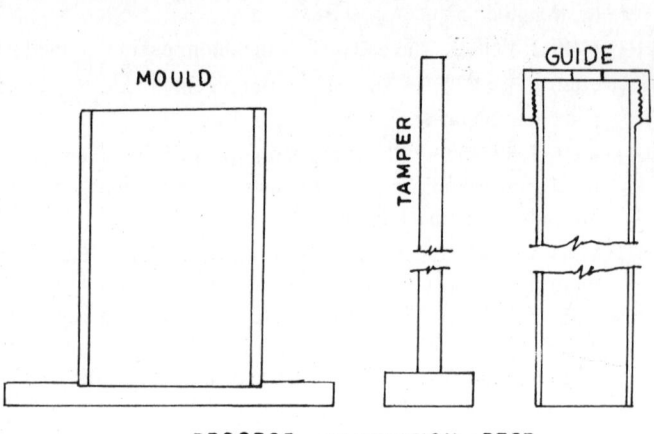

PROCTOR CAMPACTION TEST

PROCTOR PLASTICITY NEEDLE : An instrument having a penetration needle, a loading device and a spring balance, used for measuring the resistance offered by a soil to penetration. A standard needle is pushed into soil at the rate of 12 mm. per second and the force required to push is indicated by the spring balance.

PROGRESSION : Traversing by setting the plane table at each station by rotation, measuring bearings by a prismatic compass, aligning foresight and back sight by using alidade and measuring distances between stations with the help of a chain or tape. By this method, traversing as well as mapping in field can be carried out.

PROGRESS CHART : A bar chart showing the itemwise work progress of a project.

PROJECTION : (i) Extension of a structural member or fixing a moulding projected from a surface or overhanging part of a structure.

(ii) Prediction of future population by following the past trend of population growth.

(iii) Drawing parallel lines from each point of an object in making plan, elevation and end views of the object in orthographic projection.

PROJECTIONAL DRAWING : A drawing prepared by orthographic projection.

PROJECTING SCAFFOLD : A hanging or bracket scaffold made at a level above ground to facilitate workmen to work standing on the working platform supported by the bracket.

PROOF STRESS : The stress at which a permanent set is caused in a metal. This is sometimes used as a basis to determine the safe working stress of a metal.

PROPORTIONAL COMPASS : A drawing instrument used to copy a plan or map to any desired scale by adjustment of ratio of the distances by sliding the arms against the graduations marked on the face of the instrument.

PROPORTIONAL WEIR : A weir as shown in illustration is installed at the outlet point of a rectangular grit channel to maintain a cons- tant velocity with varying depth of flow in the channel.

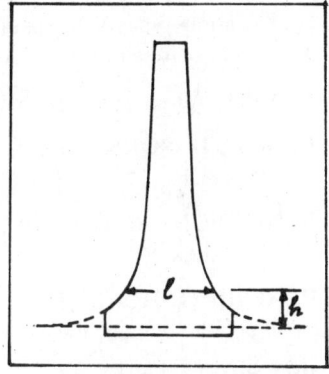

PROTECTIVE METAL FINISH : A thin protective coat of metal applied to iron and steel surfaces to make them rust-proof. The protective metal finishes are done by electroplating, galvanizing, sheradizing, phosphating and plating with cadmium, nickel, chromium copper.

PROTRACTOR : An instrument used for plotting on a plan the values of angles obtained in field during surveying. For accuracy, sometimes it is provided with a vernier scale.

PROPORTIONAL WEIR

PROVING RING : A standardized steel ring (which is precisely calibrated by measuring its diametral deformation for different loads) is used with a dial gauge to measure the load applied to a structure during testing.

PRUSSIAN BLUE : A blue pigment obtained synthetically consisting mainly of $Fe(CN)_6$ and potassium. When used in a paint, it producess a metallic lustre. Also known as chinese blue.

PRY BAR : A nail puller which withdraws nails.

PUDDLED CLAY : A clay which is well-mixed.

P-TRAP : A trap commonly fitted with the W.C. pan to form the water seal. See illustration.

PUFF PIPE : An anti-siphonage pipe.

P -TRAP

PUG MILL : A simple machine as shown in illustration is used for mixing brick earth thoroughly for preparation of bricks. It is a conical (truncated) iron tub with a centrally placed vertical iron shaft carrying horizontal arms with knives. The iron shaft is rotated by revolving the long arm attached to it.

PULSOMETER PUMP : A displacement pump used for pumping muddy water.

PULVERIZED COAL : Coal pounded to ingredients of less than 0.01 inch diameter which burn very rapidly. This is fed to the boiler furnace by transporting it from the grinding mill by a current of air.

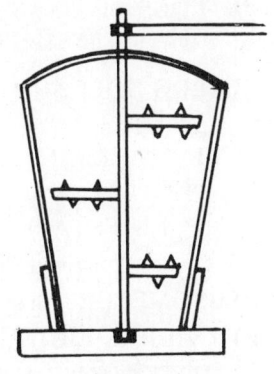

PULVERIZING MIXER : A mixer which is used in mixing stabilisers with the soil for soil stabilization. It pulverises

PUG MILL

the soil with its revolving tines and passes over with the stabilser spread over the soil for intimate mixing.

PUMICE : A sponzy rock used as as an abrasive and also as aggregate in light-weight concrete.

PUMP : A machine used to raise water and other liquids from a lower level to a higher level. Various types of pumps are used to serve specific purposes. Different types of common pumps are air lift, displacement, reciprocating, rotary, diaphragm and centrifugal.

PUNCH : A tool or device to make holes in sheet metals.

PUNCHED WORK : Broached work, a type of rough dressing of stones by striking with a punch.

PUNCHEON : A short post used in the middle of a queen post roof truss or a trussed partition.

PUNCHING SHEAR : The purest form of shear. For example, failure of a heavily-loaded column by punching through its base slab.

PUNNER : A hand rammer consisting of a flat-bottomed heavy block of iron with a long wooden handle.

PURLIN : A horizontal timber placed in between a principal rafter and a common rafter.

PURLIN ROOF : A small roof formed by placing purlins on cross walls instead of using any roof truss.

PURPOSE-MADE BRICK : A special brick made to serve a specific function.

PUSHER TRACTOR : A crawler tractor used to push materials to feed a large bowl scraper.

PUSH PLATE : A polished metal plate fixed on the lock stile of a door at the level of waist or hand to protect the door surface from disfiguration or discolouring by hands.

PUSH-PULL RULE : A steel tape that can be rolled by pushing into a small box and can be taken out by pulling.

PUTLOGS : Small horizontal bars to carry scaffold boards in a scaffold. One end of a putlog rests in a hole kept in brckwork.

PUTLOG HOLE : Hole kept in a brickwork during construction to support the end of a putlog.

PUTTY : A material used to fill cracks and crevices, holes, depressions and gaps. These are white lead putty, mason's putty, lime putty, glaziers'putty, etc.

PUTTY KNIFE : A knife used for applying putty.

P V C : See 'Polyvinyl Chloride'.

PYCNOMETER : An instrument or a device to measure the density of a soil.

PYRAMIDAL LIGHT : A roof light is so made that the sloping glass shets meet at a point, its base being a regular polygon.

PYRAMID ROOF : A pavilion roof of pyramidal shape.

PYROCONE : A type of small incinerator installed at the base of a solid waste chute in a multistoried building, hospital and public buildings. See illustration.

PYROLYSIS : A process to convert organic fraction of solid waste into gas, liquid or char of a substantial energy value.

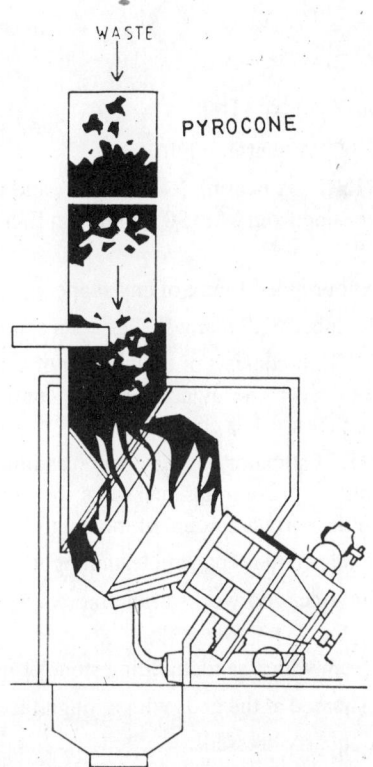

PYROLYSIS REACTOR : See illustration. A fluidized-bed reactor as shown in illustration where the heat transfer, rate is rapid. It gives low liquid yields and high gas yields from municipal solid wastes. The hot gas can be used to pyrolyse the feed in the bed. The char produced should be removed from the bed so that pyrolysis gas does not react with oxygen.

PYROMETER: An instrument having a thermocouple to measure the temperature of a furnace.

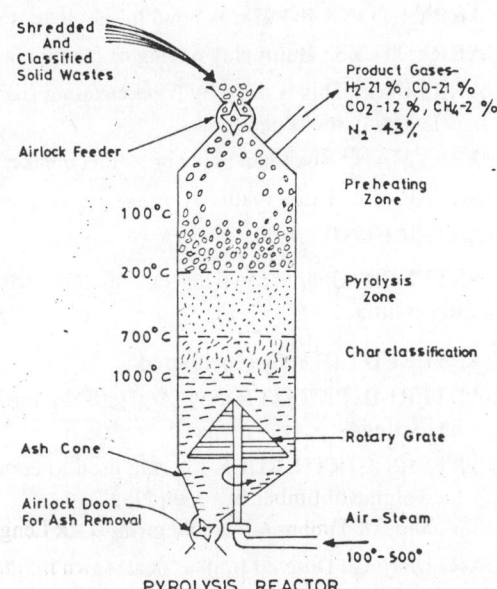

PYROLYSIS REACTOR

Q

Q-TRAP : See illustration. Also see 'Trap'.

QUADRANT : An arc of $90°$ A quarter circle.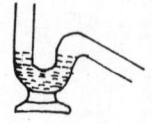

QUADRANTAL BEARING : A bearing less than $90°$ and numbered in four quadrants, increasing from $0°$ to $90°$ from N to E, S to E, S to W and N to W.

QUADRILATERAL : A four-sided figure of any shape.

QUAGGY TIMBER : A timber with many shakes at its centre.

QUANTITY SURVEYING : It consists of drawing up of estimate of quantities or bill of quantities, contract between the owner and the contractor, and the arbitration of disputes, if any, when the work is completed.

QUANTITY SURVEYOR : Estimator, A person who estimates the quantities, looks after the technical accountancy of contracts, measures the work done by the contractor, and advises the client on the correct amount of money to be paid to the contractor.

QUARREL : A diamond shaped glass pane in leaded lights.

QUARRIER : Quarryman; one who works at the face of a quarry, drilling and breaking rock.

QUARRY : An open pit from which sand, building stone or mineral is taken out.

QUARRY DRESSED : Squared at the ends with a rough face.

QUARRY FACED : See 'quarry dressed'.

QUARRY PITCHED : See 'quarry dressed'.

QUARRY SAP : The moisture content in a freshly cut stone from a quarry.

QUARRY STONE BOND : A bond in rubble masonry.

QUARRY TILES : Burnt clay paving or facing tiles.

QUARRYING : This is done by (i) excavating (ii) heating, (iii) wedging (iv) drilling and (v) blasting rocks and mines.

QUARRYMAN : Rockman ; Hewer ; stone breaker ; rockgetter. See 'quarrier'.

QUARTER : Flank of a road.

QUARTER BEND : A $90°$ bend.

QUARTERED : Quarter sawn i.e., cut into four quarters by quarter sawing. se illustration.

QUARTERED LOG : See 'quartered'.

QUARTERED PARTITION : A partition built of quarterings.

QUARTER - GIRTH RULE : A rule used to compute the volume of timber in a round log.
Volume of Timber = [Middle girth/4]^2 x Length

QUARTER SAWN

QUARTERING : Derived from a 'deal' sawn into four equal pieces, cross section of each being 2" x 3" roughly; used as 'studs' in a framed partition.

QUARTERING WAY : Roughly perpendicular joints and cleavage planes in most rocks. It is the easiest plane of splitting.

QUARTER LANDING : See illustration.

QUARTER PEG : A peg placed at the quarter width of a road which helps to define the road surface.

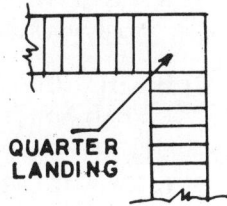

QUARTER
LANDING

QUARTER ROUND : See 'quadrant'.

QUARTER SAWING : The log is first cut into quarters and then each quarter is sawn radially as far as practicable, with no growth ring at an angle of less than 80° to the surface of the quartered timber. It is also called 'rift sawing'.

QUARTER-SAWN TIMBER : Timber obtained by quarter sawing.

QUARTER-SPACE LANDING : A square landing of the width as length of a tread.

QUARTER STUFF : A board of 1/4 in thickness.

QUARTZ : Crystalline silica (SiO_2); The main part of sandstone or gravel and the transparent part of granite.

QUARIZITE : Sandstone cemented by 'Quartz' (98% silica).

QUAY : A site on the sea-cost for loading and unloading of vessels.

QUAY WALL : A wall constructed for one or more berths continuously bordering on and in contact with the land in a dock. This may be built as a block wall or caisson or bulk head wall or sheet pile wall.

QUEBEC BRIGE : One of the largest cantilever bridges in the world (1800 ft. central span), built in 1917.

QUEEN : A 36" x 24" slate.

QUEEN ANNE ARCH : A combination of a central semi-circular arch with camber arches on both sides on the same springing line.

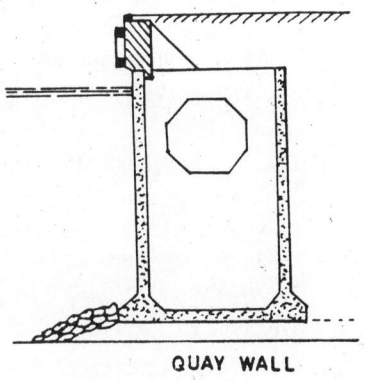

QUAY WALL

QUEEN-BOLT : In Queen post trusses, sometimes steel bolts are used in place of Queen posts. These are 'Queen bolts'.

QUEEN-CLOSER : A brick cut in half along its length. In India, it measures 9.75" x 2 .375" x 2.75". In England, It is 9" x 2.25" x 2.625". See illustration.

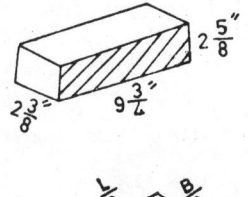

QUEEN-CLOSER QUARTER : See illustration.

QUEEN-POST ROOF TRUSS : This is used up to a span of 45'. It consists of two queen posts, two principal rafters, two common rafters, struts, tie beam, straining beam, purlins, etc.

QUETTA BOND: It is like a 'rat trap bond' in which bricks are laid in bed, each course is made in 'Flemish' bond and gaps are left in the middle of the

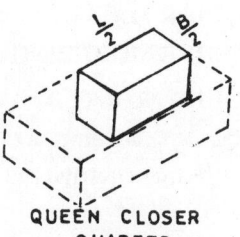

QUEEN CLOSER
QUARTER

wall so that reinforcement steel may be introduced vertically and the gaps are filled with grout. The usual wall thickness is 15".

QUICK GOUGE : A gouge whose cutting edge has a small radius of curvature.

QUICK-HARDENING CEMENT : Rapid hardening cement.

QUICK-HARDENING LIME : Hydraulic lime.

QUICK-LEVELLING HEAD : A ball and socket fitting under a dumply level instead of three levelling screws. This facilitates levelling up, but this is not so common in theodolites as in levels.

QUICK LIME : Caustic lime (cao), obtained by heating limestones in a rotary kiln to a temperature of 1700° F. $CaCO_3 = CaO + CO_2$.

QUICK SAND : It is the sand held in suspension by water. This happens due to the fast upward movement of water through the sand. Quick sand has therefore no bearing power.

QUICK-SET LEVEL : A level provided with a quick-levelling head.

QUICK-SETTING CEMENT : It is ground finer than rapid hardening cement and contains a little amount of gypsum. This cement sets in 3 minutes and final setting takes place within 30 minutes. This is chiefly employed for subaquous works.

QUICK SWEEP : A circular work having a small radius.

QUICK TEST : Undrained shear test, box shear test, or triaxial compression test of a sample of soil.

QUIESCENT FLOW AND QUIESCENT SEDIMENTATION : Fill and draw principle. Water is run into a tank and when the tank is full, the water is allowed to stand for 2 to 3 hours. Thus, all suspended solids settle down. The water is then drawn off and the tank is refilled. Quiescent sedimentation requires a large number of tanks.

QUIRK : A narrow groove alongside a bead.

QUIRK FLOAT : A plasterer's trowel for finishing mouldings.

QUIRK MOULDING : A moulding having a small groove in it.

QUIRK ROUTER : A form of plane for shaping quirks.

QUOIN : A corner or the external angle of a wall.

QUOIN BONDING : The mode of arranging bricks in making a return wall.

QUOIN HEADER : A header placed at the corner of a wall which is therefore a stretcher in the side wall.

QUOIN POST : A heel post.

R

RABBET : A step-shaped rectangular recess cut in the edge of a timber. Also, known as 'rebate'.

RABBET PLANE : A rebate plane used to cut rectangular recess.

RACE : A channel to or from a water-wheel.

RACK : (i)   A toothed bar as used in a rack-railway track, over which a cogged-wheel (toothed) moves.

      (ii) A trash rack as used in a sewage pumping station to arrest and remove screenings from the sewage.

RACKED TIMBERING : Diagonally braced timbering.

RACKING : Tendency to distort a frame.

RACKING BACK : In building a brickwall from middle part to the corners, the gradual increase of height to the corner is known as racking back.

RACKING COURSE : A layer of graded stones uniformly placed over stone-pitching to fill the voids.

RACK RAILWAY : A railway track with a toothed bar between rails, used in a mountain railway having steep gradient.

RADER : In aerial survey for taking vertical photographs, the position and altitude of the aircraft is ascertained by rader.

RADIAL BRICK : Radius brick or compass brick used in construction of arches and circular brick columns.

RADIAL GATE : A tainter gate or segmental sluice gate with a curved water face.

RADIAL ROAD : A road which comes out radially from the centre of a city or town.

RADIAL SETT PAVING : Paving of small tiles or slates or slabs to form fan shape.

RADIAL SHAKE : This is one of the defects found in timber. Splits are formed radially.

RADIAL SHRINKAGE : The shrinkage of timber normal to the growth rings during drying. This shrinkage is nearly half the tangential shrinkage.

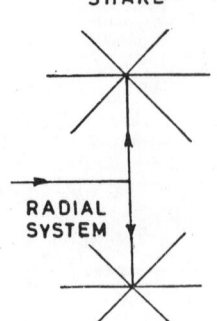

RADIAL SHAKE

RADIAL SYSTEM

RADIAL STEP : A winder in the curved portion of a stair or in a circular stair.

RADIAL SYSTEM : A system of water distribution by laying water lines radially for equidistribution of water at uniform pressure.

RADIANT HEATING : (i)   Stoving a paint or varnish finish by radiation of heat.

      (ii) Coil heating.

RADIATION : A method of plotting features in plane table surveying by radiating lines with the alidade. See illustration.

RADIATOR : A grilled container of water used in central heating system.

RADIO-FREQUENCY HEATING : A device to heat thick plywood rapidly for gluing.

RADIUS OF GYRATION : Usually denoted by K and is
given by $K = (I/A)^{0.5}$; where I =
moment of inertia and A is the cross-
sectional area of the member. This is
required to calculate the slenderness
ratio of a strut.

RADIUS ROD : Also known as 'gig stick'.
A strip of timber 50 mm x 50 mm in
cross-section is pivoted at the centre
of an arch to be moulded. The gig stick
holds the mould firmly.

RADIUS SHOE : A zinc plate screwed to
one side of the radius rod over its
centre point.

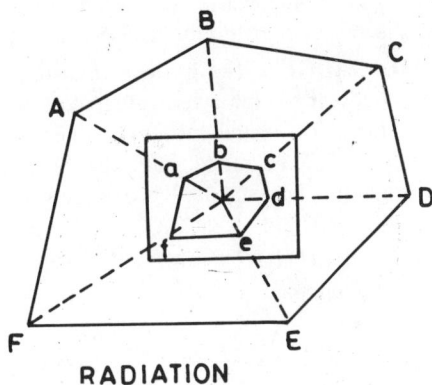

RADIATION

RAFT FOUNDATION : Mat Foundation. A reinforced concrete foundation slab designed
to act as a mat foundation to support
the total structure above it. See
illustration.

RAFTER : In a sloping roof, it is a
timber piece extending from the
eaves board to the ridge board. It is
either a principal rafter or a
common rafter.

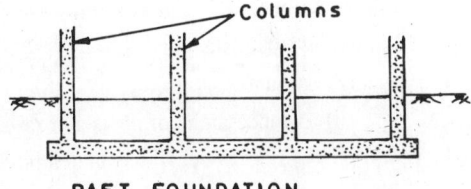

RAFT FOUNDATION

RAFTER FILLING : Brick or concrete block infilling between principal rafter and
common rafter at the wall plate.

RAG BOLT : A foundation bolt used for machine foundation. The conical bolt head is
serrated which remains in concrete.

RAG FELT : Bituminous felt used in sloping roof and for other purposes.

RAGLET : A thin groove often dovetailed in stone or brickwork to receive the lead flashing
end.

RAGSTONE : Coarse-grained sandstone.

RAG WORK : Thin flat stones built into a wall of rubble masonry.

RAIL : (i)   A horizontal timber plate joining vertical stiles on either side of a door or
window shutter.

(ii)   The track in a rail road.

(iii)   A longitudinal horizontal member in fencing.

(iv)   The upper part of a balustrade.

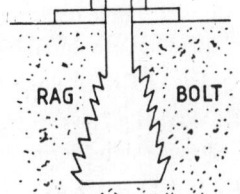

RAIL BENDER : A Jim Crow used for bending rails.

RAIL BOLT: A handrail bolt.

RAIL CHAIR: A cast iron chair to hold a bull-headed rail firmly as shown in illustration.

RAIL FASTENING : Fastenings required to hold a rail are chair, key, wedge, spike, screw, fish plate, anticreep anchor, etc.

RAIL GAUGE : The horizontal distance between the inner vertical faces of two rails. The usual railway gauges are broad gauge, metre gauge and narrow gauge. The standard broad gauge is 4'—8.5"

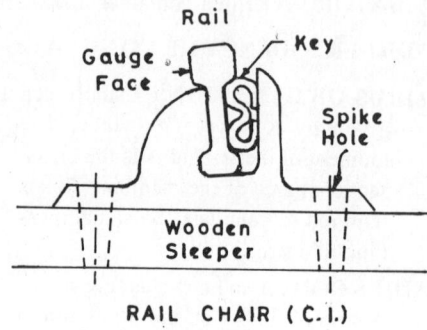

RAIL CHAIR (C. I.)

RAIL GRILLAGE : A grillage formed of old rails.

RAIL KEY : A steel wedge having spring action is required to fix a bull-headed rail into a rail chair.

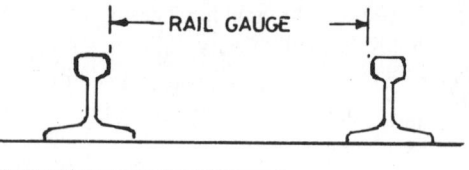

RAIL TEST : A test carried out to determine the brittleness of rails. A weight of one tonne is dropped onto a rail from different heights, the rail supports being at 1m. apart.

RAIL TIE : A mild steel sleeper.

RAILING : (i) A band

(ii) A fence made of posts and rails.

RAILWAY CURVES : A set of templates having different curvatures, used in drawing office.

RAILWAY TRANSIT : A theodolite with no vertical circle.

RAINDROP FIGURE : A mottled figure that appears on timber with ribbon grain.

RAINFALL INTENSITY-DURA-TION CURVE : The rainfall intensity decreases with the increase in storm duration. From a curve as shown in illustration, the rainfall intensity for any storm duration can be determined. First of all, the curve is to be plotted with the observed data.

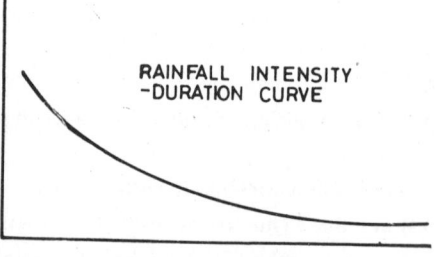

RAIN GAUGE : As shown in illustration, it is a simple device to measure the quantum of rainfall over a period of storm.

RAIN LEADER : A down pipe to carry rain water.

RAIN WASH : The movement of rock or soil i.e., scour of surface down a slope during rainfall.

RAIN WATER HEAD : The head or mouth of a down pipe to receive rain water.

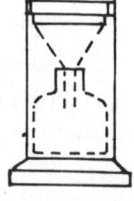

RAIN GAUGE

RAIN WATER HOPPER : Hopper-shaped rain water head for receiving rain water and transferring to down pipe.

RAIN WATER PIPE : The down pipe to carry rain water from a roof.

RAISED PANEL : A door or window panel which is thicker at the middle than the edges.

RAISING PLATE : A wall plate or poll plate.

RAKE : Batter or inclination to the vertical.

RAKED JOINT : A mortar joint in which about 12 mm mortar is taken out from the face to form a grip for the plaster.

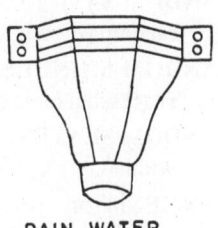

RAIN WATER HEAD

RAKE MOULDING : A battered moulding at the top of a barge board, below the tiles on a gable end of a sloping roof.

RAKING BACK : See 'Racking Back'.

RAKING BOND : Diagonal Bond.

RAKING COPING : A coping on a slope.

RAKING CORNICE : A cornice over a gable.

RAKING FLASHING : A cover flashing as used at the junction of a chimney and a sloping roof.

RAKING OUT : Cleaning out mortar from a joint prior to pointing or plastering.

RAKING PILE : A batter pile i.e., a pile driven inclined to vertical. See illustration.

RAKING RISER : An inclined riser in a step that gives more foot hold i.e., the tread is bigger than its going.

RAKING SHORE : An inclined shore to hold a structure temporarily.

RAM : (i)   The moving cylindrical block in a ram pump.

(ii)   Weight of a drop hammer.

(iii)   The plunger.

RAMMED EARTH CONSTRUCTION : Cob construction.

RAMP : (i)   A Sloping floor or path or road

(ii)   A drain of very steep gradient for a short length.

(iii)   A bend in a handrail with upper side concave.

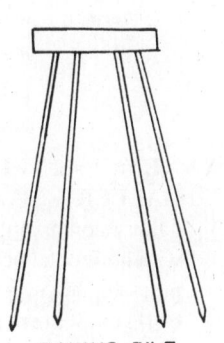

RAKING PILE

RAMSDEN CHAIN : A 100 ft. long steel chain subdivided into forty links, was used earlier for ordnance survey under a constant pull of 28 lbs applied by means of a weight.

RAM'S HORN FIGURE : A ripple.

RANDOM ASHLAR : Courses with stones of varying depths in a masonry work.

RANDOM RUBBLE: Rubble masonry built of stones of irregular shape and size. But, this imparts beauty, if properly placed with selection of their shape and size. See illustration.

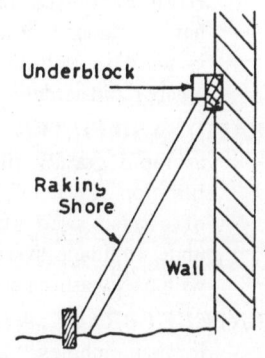

Underblock

Raking Shore

Wall

RANDOM SAMPLE: Samples taken at random without bias.

RANDOM SHINGLES : Shingles of varying shapes, but more or less of uniform length.

RANDOM SLATES : Rustic slates of varying width.

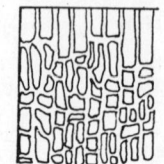

RANDOM RUBBLE MASONRY

RANGE : In surveying, to align points by eye through a telescope or with ranging poles.

RANGE LINE : Used in surveying adjacent township lines.

RANGE MASONRY : Regular coursed ashlar work.

RANGE POLE : Wooden pole used to range points in surveying.

RANGING LINE : The line joining the points of ranging rods and the object observed.

RANGING ROD : Range pole.

RANGING ROD

RANK SET : A set of back iron of a wooden plane that keeps a space between it and the cutting iron.

RANKINE'S THEORY : For each foot depth of earth retained, the pressure on a vertical wall retaining earth with a horizontal surface is

$$w \times ((1 - \sin \theta)/(1 + \sin \theta));$$

where, w = density of earth

and θ = angle of friction of earth.

$(1 - \sin \theta) / (1 + \sin \theta)$ is the co-efficient of active earth pressure.

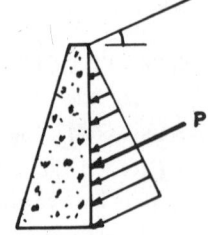

RANKINE'S THEORY

RANNEY   INFILTRATION GALLERY : An infiltration gallery comprising a central well with horizontal perforated branch pipes laid below river bed level, to collect water as shown in illustration.

RAPID HARDENING CEMENT : High early-strength cement having higher lime content. This is used in sub-aquous works where rapid hardening is required.

RAPID SAND FILTER : Also known as rapid gravity filter, used in filtration of water. The filter bed comprises layers of coarse sand and graded gravel with an under-drainage system. Such a filter is provided with backwashing arrangement.

RATCHET BRACE : A ratchet drill used by carpenters for making holes.

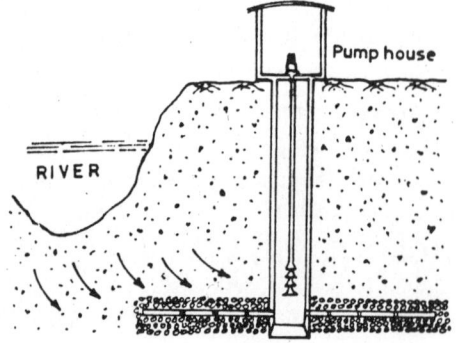

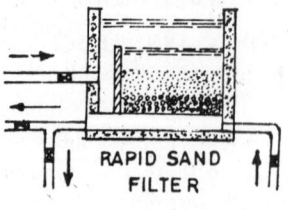

RAPID SAND FILTER

RATCHET AND PAWL MECHANISM : A cog-wheel with which a single tooth (Pawl) is engaged such that it resists movement in opposite direction.

RATING : A relation between water level and discharge of a stream or channel.

RATING CURVE : A duration curve of a stream or channel.

RATING FLUME : A flume having still water into which pitot tubes or current meters are drawn at known velocities for the purpose of calibration.

RATIONAL METHOD : The most widely used Lloyd-Davis formula for calculating rates of storm water run-off for the purpose of sewer design. The run-off is given by

Q = CIA, where C is co-efficient of run-off,

I = Average intensity of rainfall, inches per hour, and

A = Catchment area in acres.

RAT-TAIL FILE : A round file gradually tapering from a larger diameter near handle to a smaller diameter at the tip.

RAT-TRAP BOND : A brick bond in which bricks are laid on edge. For a 9″ thick wall with a cavity of 3″ thickness, the headers and stretchers are laid alternate. The wall built with rat-trap bond is quite cheap as well as strong enough.

RAW BOLT : A bolt screwed into a hole lined with lead plug in masonry for anchoring a machine. The lead plug is gripped by the bolt and pressed against the hole in masonry.

RAW LINSEED OIL : Linseed oil, refined but not boiled.

RAWL PLUG : A drill bit used for making small holes in masonry by hammering and turning the bit manually.

RAYMOND PILE : Cast in place taper, or step-taper piles. During driving, a steel mandrel is inserted in a shell and the mandrel is driven into the ground and removed. Then, the shell is filled with concrete.

RAYMOND STANDARD TEST : A dynamic penetration test carried out to compare the bearing capacity of soils. First, a hole is bored upto the required depth of foundation. A soil sampler is introduced into the hole and pressed 6″ into the soil below the foundation level. Then, the sampler is given a number of blows by means of a 140 lbs hammer dropped from a height of 30″. The number of blows required to drive it upto 12″ into the soil is counted and the relative density of the soil is found out.

REACH : The stretch of water in a channel between two locks.

REACTION : The force in a support or a member to react with the load applied.

REBATE : See 'Rabbet'.

REBATED & FILLETED JOINT : See illusration.

REBATED JOINT : See illustration.

REBATED,TONGUED & GROOVED JOINT : See illustration.

REBATED & FILLETED

REBATED WEATHER BOARDING: A wedge-shaped weather boarding with a rebate along the inner face of its lower edge to drain out the rain water.

REBATED

RECESSED HEAD SCREW: A screw head with a cross-shaped recess into which a cross-shaped screw driver fits in. This screw is used in joinery works.

REBATED, TONGUED & GROOVED

RECESSED POINTING : Pointing with mortar joint set back about 6 mm from the wall face.

RECIPROCAL LEVELLING : The error due to the line of sight not being parallel with the axis of the bubble and error due to curvature and refraction are eliminated by applying corrections obtained from reciprocal levelling. To eliminate instrumental error in levelling between two points, levels are taken on these from two set-ups, one near each point. The true difference is the average of the level difference.

RECIPROCATING PUMPS : These are either lift pumps or force pumps. These are suitable for pumping against varying heads with high efficiency. They may be single-throw, double-throw or three-throw pumps.

RECONNAISSANCE SURVEY : Surveying an area from observations in field and without use of any instrument/apparatus. This is required to have an idea about the area and its physical features for drawing a concept level plan for a project to be taken up.

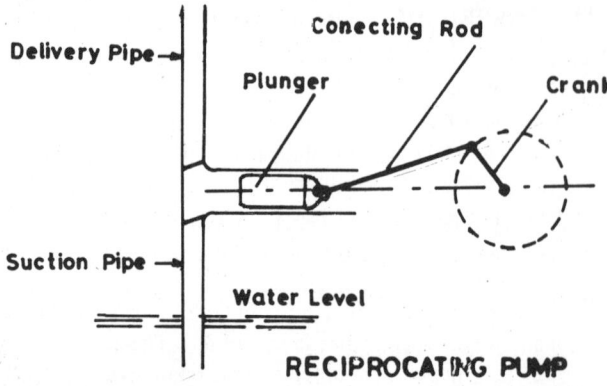

RECIPROCAL LEVELLING

RECIPROCATING PUMP

RECORDING GAUGE : An automatic gauge that records the water level or velocity of flow in a stream, channel or river. There is also automatic recorder for measuring the flow and pressure in a conduit.

RECOVERY PEG : In surveying, it is a peg of reference with relation to a permanent level, distance and direction, which is used as a bench mark. This is re-established at its position when it is disturbed.

RECTANGULAR WEIR : A weir having a rectangular notch for fluid flow.

RED LEAD : An inhibiting red pigment used in lead paints for painting on timber or iron. It is Pb_3O_4.

RED OXIDE : Iron oxide red in colour, used as a pigment which does not inhibit corrosion. This is chiefly used for the primary coat on iron and steel.

REDUCED BEARING : A bearing that is smaller of the two angles which a chain makes with the meridian, so that it is never greater than 90°. It is the angle always measured from the meridian.

REDUCED LEVEL : The level (altitude) of a point with respect to a datum plane assumed. This is determined either by collimation method or by rise and fall method.

REDUCED NATURAL FREQUENCY : The natural frequency of vibration of a foundation at an average ground pressure of unity. This may be written as: Reduced Natural Frequency is equal to natural frequency multiplied by (average ground pressure)$^{0.5}$. This decreases as the area increases.

REDUCER : A reducing pipe fitting from higher diameter to lower diameter.

REDUCING POWER : The strength of a white pigment when mixed with a coloured pigment. The greater the reducing power of the white, more is the paleness of the tint.

REDUCING SOCKET : A reducer used in a pipeline work. See illustration.

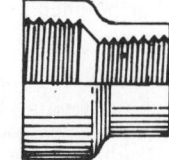

REDUCTION FACTOR : For a given slenderness ratio, the permissible load on a long column is reduced by applying the reduction factor to the allowable load on a short column.

REDUCTION OF LEVEL : Calculation of difference in level between two points from the staff readings in a level survey.

REDUCING SOCKET

REDUNDANT FRAME : A frame that has more numbers of member than is required for it to be a perfect frame.

REED : A plant used in making mattresses, thatched roofs and partition walls of very temporary nature.

REFERENCE MARK : A point of reference used in surveying to take bearings of other points with reference to the point.

REFERENCE PEG : See 'recovery peg'.

REFLUX VALVE : A non-return valve i.e., water can flow in one direction only through the valve. This is usually provided in pumping main near pump house.

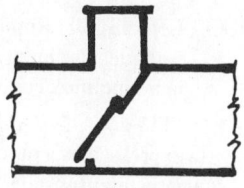

REFRACTORY : Materials that can withstand very high temperatures. These are used in lining furnaces, boilers, converters, crucibles and pyrometer tubes. These are classified as acidic, basic and neutral.

REFLUX VALVE

REFRACTORY MORTAR : A mortar prepared with 3 parts of grog and 1 part of high-alumina cement or 3 parts of grog and 2 parts of fireclay, used in laying refractiory bricks for lining furnaces.

REFRIGERANT : A fluid usually freon or sulphur di-oxide used in a refrigerator to produce cooling effect on its vaporization, which is converted again to liquid by a compressor.

REFUGE : A traffic island to divide the traffic flow and to provide a safe zone for the pedestrians.

REFUSAL: A pile is said to be driven to refusal, when it penetrates only about 12 mm in five blows.

REGIME: A channel or stream or canal is said to be in regime, when its rate of flow is such that the bed is neither scoured nor any material is deposited on the bed. This hardly occurs throughout the entire length of the water course. This however may be found in certain stretches of the watercourse.

REGLETTE : A 12-inch scale used in surveying which is divided into tenths and hundredths of a foot.

REGULAR COURSED RUBBLE : Coursed ashlar work or walling with coursed rubble in courses of uniform height.

REGULATING COURSE : A course of stones laid over an old road surface with a view to giving a new shape prior to surfacing.

REGULUS METAL : Antimonial Lead which is an alloy of 90% lead and 10% antimony, used in wall cladding.

REINFORCED BITUMEN FELT : A light felt made of jute hessian or fibres impregnated in bitumen.

REINFORCED BRICKWORK : Brickwork reinforced with mild steel (m.s.) rods, wire mesh or expanded metal. Brickwork suitably reinforced and bonded is capable of withstanding the shock loads due to earthquake. M.S. rods are usually introduced vertically, during construction of a reinforced brick wall.

REINFORCED CONCRETE : A concrete which is reinfoced with mild steel rods or wire mesh. In a reinforced concrete member, concrete takes compression and reinforcement steel is in tension. Reinforced concrete is extensively used in Civil Engineering constructions.

REINFORCED MASONRY : A brick or stone masonry reinforced with steel.

REINFORCED WOOD WOOL : Building slabs made of wood wool reinforced or stiffened longitudinally by wooden battens or pressed steel.

REINFORCEMENT : Strengthening a material by introducing some foreign material. Mild steel rods, expanded metal and wire mesh are used for reinforcement.

REITERATION : Repitition of a method successively until the correct result is obtained, e.g., measurement of an angle in surveying with a theodolite, balancing head or flow in a pipeline network.

RELATIVE COMPACTION : It is determined by Proctor compaction test of soil and is expressed as a percentage. It is the value obtained by dividing the dry density of a soil in its natural state (in situ) by the maximum dry density of the soil.

RELATIVE DENSITY : A measure of the density of a sand to indicate its compaction. The densities of the sand in its loosest possible dry state and in its densest possible state are measured in the laboratory and the relative density is determined as follows:

Relative density
= (Maximum density – Field density)/(Maximum density – Minimum density)

RELATIVE HUMIDITY : The ratio of the weight of water-vapour in air to the weight of water-vapour in saturated air, the temperature of air remaining constant.

RELATIVE SETTLEMENT : The differential settlement in a structure.

RELEASE AGENT : Parting agent or parting compound like mould oil, sealant, etc. laid over forms for casting concrete which facilitates in easy removal of formwork and ensuring a good finish of the concrete surface.

RELIEF HOLES : The holes that are fired after the cut holes in tunnelling or shaft sinking work.

RELIEF SEWER : A sewer built parallel to the existing sewer to give relief to the existing sewer.

RELIEF WELL : A borehole made at the toe of a massive dam to relieve high pore water pressure.

RELIEVING ARCH : An arch built over a wooden lintel or wooden frame to carry the load of wall above it.

RELIEVING PLATFORM : A deck slab at the end of a retaining wall of a jetty, resting partly on the wall and partly on the raking or bearing piles. It is so constructed that the retaining wall does not get surcharged due to the movement of heavy wagons over the slab.

RELISH : The projected part of a haunch from the point of a tenon.

REMOTE CONTROL : Operation and control of a machine or a mechanism from a distant place by electrical control circuit or by telecontrol.

REMOULDED CLAY : When a sample of undisturbed clay is kneaded with its original water content, it becomes softer. This characteristic may be investigated by remoulding the clay at its initial water content and by carrying out unconfined compression test on the clay. If a remoulded clay with a sensitivity greater than 1 be allowed to stand without any disturbance, it may regain a part of its original strength and stiffness.

REMOULDING INDEX : It is given by the ratio of load per cm. of compression of undisturbed clay to the load per cm. compression of the remoulded clay, the clay constituents remaining the same.

RENDERING : Application of a coarse material to a cement mortar for use in plastering or covering a wall. Surface rendering may be of various types depending on the choice, the situation and the environment.

RENDERING COAT : The first coat of plastering on a wall.

REPRESENTATIVE FRACTION : The scale of a map or drawing is expressed as a representative fraction (R.F.), which is a ratio between a length of one unit on the paper to several similar units on the ground.

REPRESENTATIVE SAMPLE : A sample that represents the whole. To prepare such a sample, proportionate sampling is done from various parts of the whole and thoroughly blended.

RESECTION: In a plane table survey, this is a method to locate a point away from the base line, on one end of which the plane table is fixed. This method can be used for locating station points only, prior to starting the detailing in a plane table survey.

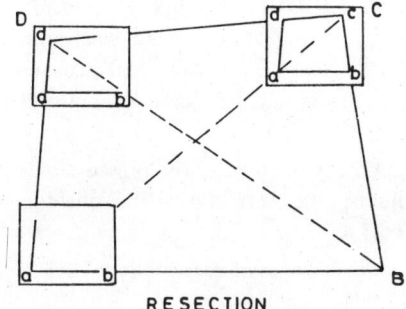

RESECTION

RESERVOIR: A tank or lake used for storage of water. The tank may be underground, overground or overhead.

RESERVOIR VOLUME : The volume of water to be stored in a service reservoir, (elevated ground reservoir or overhead reservoir), the storage volume is usually kept as two times the peak volume. 50% of the reservoir volume is kept for fire fighting and emergencies and 50% to meet the demand during the peak hours. The depletion during average day supply is taken as 25% of the reservoir volume with pumping. See illustration.

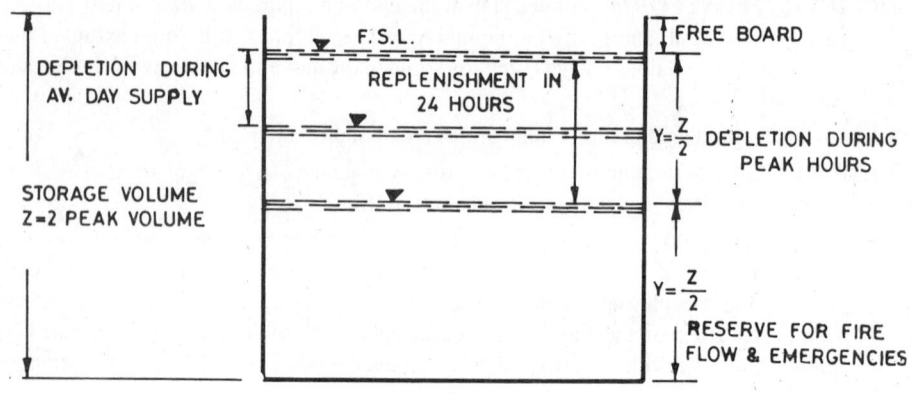

RESERVOIR VOLUME

RESIDUAL CHLORINE : The amount of chlorine remaining in a potable water after disinfection in a water treatment plant. This may be present in form of free chlorine residual or combined chlorine residual. This is normally kept at 0.2 to 0.3 mg./l after 30 minutes contact period. The chlorine residual helps to prevent further contamination during its travel from treatment plant to the consumers.

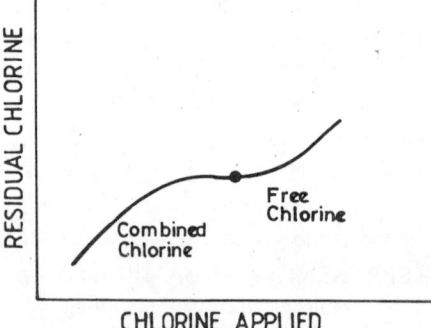

RESIDUAL ERROR : The error left even after an accurate measurement i.e., the error of negligible amount that can not be eliminated. This is also known as apparent error. If x be the assumed true value of a measurement and $x_1, x_2,.........$be the deviations from it, the theory of least squares is applicable. The deviations $x_1, x_2.......$ are the residual error.

RESIDUAL PRESSURE : The pressure of a fluid in a pipeline avaliable at the tapping point i.e., the pressure after frictional losses.

RESIDUAL SOILS : Disintegrated materials of indefinite grain size above the rock crust, in various stages of cementation. Their characteristics differ in a wide range.

RESIDUAL TACK : In painting, this is a fault in a finishing coat which does not harden. This is caused due to the use of dehydrated vegetable oils in a paint.

RESILIENCE : The strain energy stored (per unit volume) in an elastic material. In other words, it is the energy that can be absorbed per unit volume without any deformation of the material.

RESIN : Natural resin is obtained from exudation of certain species of plants. Synthetic resin is an organic compound prepared artifically. Resins of various kinds are used in paints and varnishes.

RESIN-BONDED : A material bonded with synthetic resin.

RESIN-IMPREGNATED WOOD : Wood impregnated in resin such that the wood contains about 50% resin. With this, the hardness, electrical insulation and resistance to moisture increases to a good extent.

RESOILING : Levelling and dressing the ground after completion of a construction work and replacing the topsoil for plantation.

RESONANCE : When the vibration of a running machine synchronises with the natural frequency of its foundation, the machine foundation will have excessive vibration which is not desired. Similar is the case with the foundations of Civil Engineering Structures.

RESORCINOL FORMALDEHYDE RESIN : A synthetic resin.

REST BEND : A bend having a base like duck foot.

RETAINING WALL : A wall to retain earth, water or any other material. A retaining wall may be of gravity type, cantilever or counterfort type. Its base has a toe and a heel. The failure of a retaining wall occurs either by sliding or by overturning.

RETARDER : (i)   Powdered Gypsum is mixed with cement to act as a retarder i.e., to slow down the setting time to avoid shrinkage and to attain strength gradually.

(ii)   Braking bars laid parallel to the running rails in a locoyard are used by the signalman to retard the speed of a moving wagon and to stop it ultimately.

RETICULE : Intersecting fine spider webs on the diaphragm as the optical focus of a telescope used in a surveying instrument to define the line of collimation.

RETIRING EMBANKMENT : These are constructed similar to dykes, but away from the natural bank of a river as shown. They provide a large area for carrying flood water which helps in keeping down the H.F.L.. The river water stands in between the banks for a longer period forming

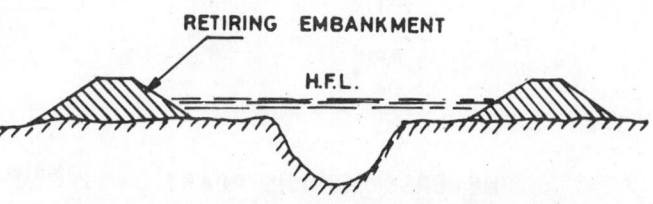

a temporary storage. Although costly initially, it is economical in long run.

RETREADING ROADS: Damaged roads are retreaded by repairing the ruts and pot holes and then by applying premixed chipping carpet over the surface with rolling.

RETURN: Change of direction of a pipeline or a wall by 90°.

RETURN BEAD : A bead having a quirk on its either side in a corner.

RETURNED END : A moulding end having the shape of the profile of the moulding.

RETURN PIPE : In a water-heating system, the pipe which comes out of a hot water cylinder and returns to the boiler after transferring heat.

RETURN WALL : A short wall at right angles to a long wall at its end.

REVEAL : The outer part of a jamb of a door or window opening. This is visible from outside.

REVEAL LINING : Lining a reveal with decorative stone slabs or rendering a reveal by plastering with ornamental works.

REVERBERATION : The echo of a sound of a certain frequency to improve the musical quality of sound.

REVERBERATION TIME : The time period in seconds required by the reverberation to diminish after the sound source is silenced. The reverberation time decreases with the increase in sound absorption capacity of an enclosure.

REVERSE CURVE : A serpentine curve as shown in illustration used in road alignments to avoid physical barriers.

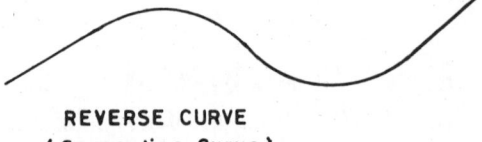

REVERSE CURVE
(Serpentine Curve)

REVERSE TEMPLET : A templet having a shape reverse to that of a moulding, used for checking the accuracy of a mould.

REVERSE OSMOSIS PLANT : A system consisting of a semi-permeable cellulose acetate membrane module in a pressure vessel. The driving force to extract dissolved solids from a waste is the reverse osmosis pressure. A schematic arrangement of a reverse osmosis plant is shown in illustration.

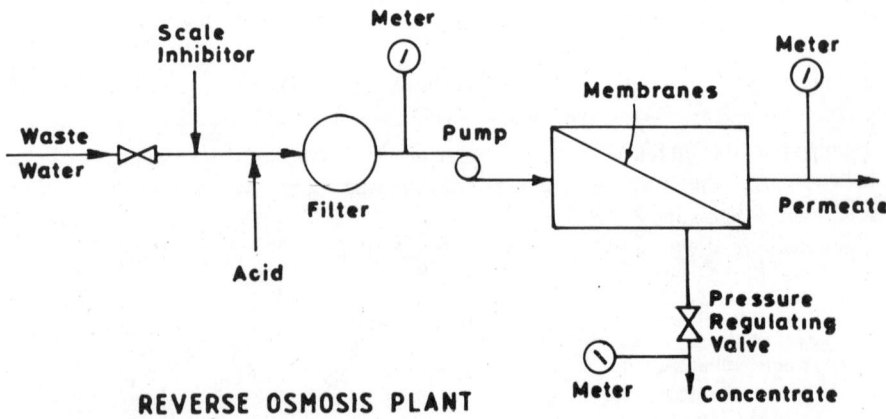

REVERSE OSMOSIS PLANT

REVERSIBLE LEVEL : A levelling instrument, in which the telescope can be rotated through 180° without disturbing the instrument, for checking the correctness of line of collimation.

REVERSIBLE LOCK : A door lock in which the hand of the lock is reversible i.e., the latch can be removed and reversed.

REVETMENT : A protective covering or lining given to a sloping bank of a river, stream or channel to protect the slope from scouring action due to wash water or due to water waves. See illustration.

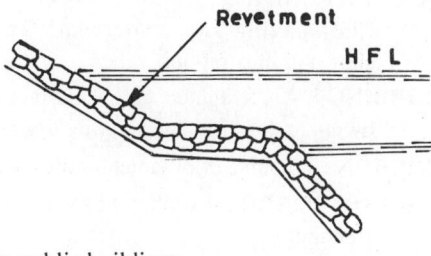

REVOLVING DOOR : A door having four leaves at right angles to each other, the leaves being pivoted on a central post. which rotates. This type of door is used in public buildings.

REVOLVING SCREEN : A trash rack comprising a drum or a belt moving mechanically. This is used as a screen to arrest floating matters in water or sewage.

REYNOLD'S GATE : The gate as shown in illustration is about 3 m. long and 2.5 m high, connected by means of a chain-pulley system with a counter weight in a well at the back of the gate. It operates automatically on the principle of buoyancy.

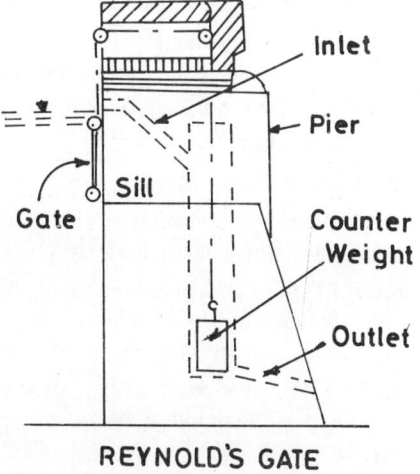

REYNOLD'S GATE

REYNOLD'S NUMBER : The non-dimensional factor governing frictional resistance. It is given by Re = vd/n ; where v is the relative velocity between surface and the fluid, d is the diameter of the pipe, and n = co-efficient of viscosity of fluid/absolute density of fluid. The Reynold's number of 2000 to 2500 holds good for fluid flow through circular pipes for all fluids. If the number is greater than 2500, the flow is turbulent.

REYNOLD'S CRITICAL VELOCITY : The velocity at which the flow of a fluid changes from laminar to turbulent condition is known as the critical velocity, which was found out by Reynold from his experimental results.

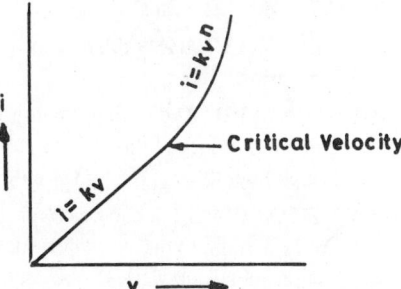

RHEOLOGY : The study of flowing streams and rivers.

RHOM BRICK : A special brick having a shape like a rhombus. The bonding of bricks in laying is very easy and the continuity of a vertical joint can be avoided easily.

RHONE : An eaves gutter at the edge of a sloping roof.

RIB : A very thin straight or curved beam to stiffen a flat slab or an arched roof or a domed roof

RIBAND: A timber piece running along the head of the struts under a R.C. girder, required for fixing up the formwork to the sides of the girder.

RIBBED FLOOR : A reinforced concrete floor having ribs underneath as shown in illustration.

RIBBED FLOOR

RIBBING : A corrugated timber surface owing to the shrinkages during drying of the timber.

RIBBON : A course of ornamental tiles or slates, which is used seldom.

RIBBON BOARD : A timber beam at ceiling level to support the floor beams in a baloon framing.

RIBBON COURSE : Alternate courses of tiles or slates laid to produce long and short depths.

RIBBON GRAIN : Interlocked grain in a quarter sawn timber which produces alternate light and shade strips due to reflection of light.

RIBBON RAIL : A metal rail connected to the metal balusters and covered by a wooden handrail, used in a stair or verandah railing.

RIBBON SAW : A band saw of very narrow width.

RIBBON STRIP : A ribbon board.

RIBBON STRIPE : Ribbon grain.

RIB HOLES : In tunnelling or shaft sinking, these holes are made at the sides of a tunnel or a shaft which are fired last i.e., after firing the cut holes and relief holes.

RICH LIME : Fat or pure lime having high calcium content, which slakes vigorously with evolution of heat. It forms a thin paste with water. Its hydraulic property is not good. It makes, very weak mortar and hence it is not suitable for important constructions.

RICH MIX : A mixture of ingredients having more quantum of cement than usual for preparation of mortar or concrete.

RIDDLE : A coarse screen or sieve.

RIDER SHORE : The topmost raking shore of short length resting on the back shore on top of the raking shore from ground, required to support a tall structure.

RIDGE : The apex of a sloping roof.

RIDGE AND FURROW : A method adopted in land irrigation.

RIDGE AND FURROW

RIDGE BOARD : The horizontal board placed at the meeting points of the sloping rafters i.e., at the ridge line.

RIDGE BOARD

RIDGE CAPPING : A cap or cover given all along over a ridge. This is usually done by placing sheet metal, clay tiles, A.C. sheets or roofing felt.

Common rafter

RIDGE COURSE : The topmost course of tiles or slates next to the ridge line in a sloping roll.

RIDGE ROLL : A 'Hip rool'.

RIDGE STOP : A flexible metal flashing, usually lead sheet, dressed over the ridge and up the wall, at the junction of a ridge and a wall rising above it.

RIDGE TILE : A burnt clay or concrete tile used to cover a ridge.

RIDGING : Capping or covering a ridge with tiles or sheets.

RIFT : Roughly perpendicular cleavage planes found in rocks which facilitate easy splitting.

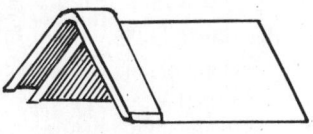

RIDGE TILE

RIFT SAWN : Quarter-sawn timber.

RIGID ARCH : An arch having fixed ends and fixity all throughout.

RIGID FRAME : A frame having rigid joints (fixed joints).

RIGIDITY : The resistance to shearing, bending or twisting.

RIGIDITY MODULUS : Modulus of rigidity. It is the ratio of shear stress to shear strain by applying Hooke's law.

RIGID PAVEMENT : A pavement made of concrete slabs.

RIM LATCH : A door locking latch operated by a knob. Also called 'Rim Lock'.

RINDGALL : An extra growth in a tree. A callus.

RING : A circle or semi-circle formed by placing segments of pressed steel, wrought iron or cast iron in lining a large sewer, Aquedet, tunnel and shaft.

RIGN COURSE : The course of bricks or stones next to the extrados of a multi-coursed arch.

RING-POROUS WOOD : A hardwood having pores in form of rings.

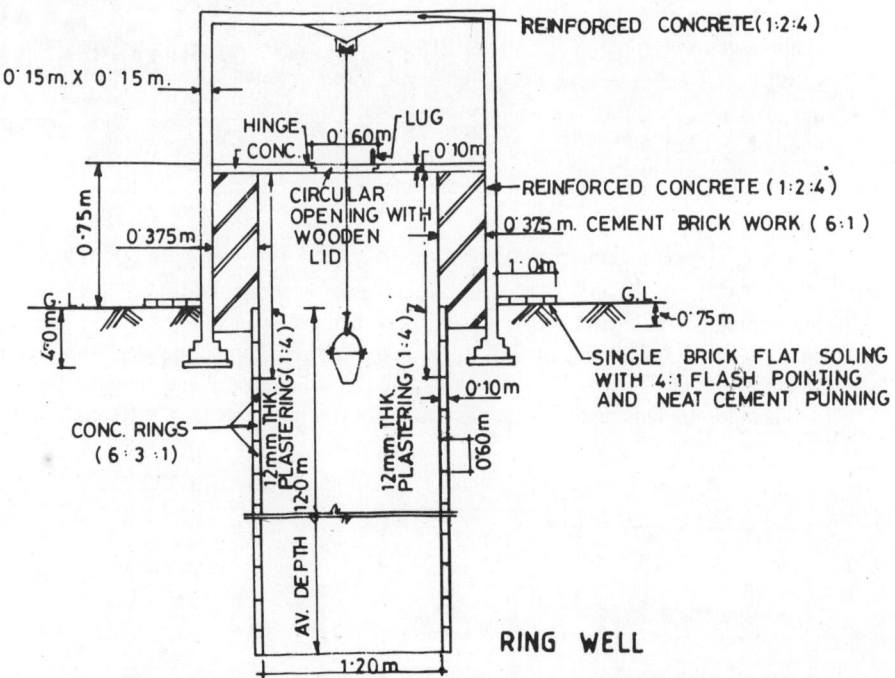

RING WELL

RING SHAKE : A 'cup shake' or 'wind shake', a defect in a timber.

RING TENSION : 'Hoop tension' occuring in the circular wall of a pipe carrying a fluid under pressure or in the wall of a circular tank containing liquid.

RING WELL: This type of dug well is useful in remote rural areas for drawing ground water manually for domestic use. For supply of potable water to a group of houses in a rural area, such type of well is constructed. Its diameter ranges from 1 m to 3 m. and the depth varies from 6 m to 12 m. Sometimes, a wind mill or a small pump is fitted to the ring well for drawing water mechanically. For drawal of water manually, pulley and bucket system is provided. See illustration.

RIP : To cut a timber parallel to the grain.

RIPARIAN : The bank area of a stream or canal.

RIPPER : (i)   A rip saw

       (ii)   A light rooter

       (iii)   A long thin cranked blade used by slate layers in cutting the slate nails by introducing it under the slate and pulling back.

       (iv)   A concrete breaker.

RIPPLE : (i)   A beautiful figure obtained in a sawn timber, which has formed due to buckling or twisting of the tree during its growth.

       (ii)   Beautiful figures formed on sea bed or sea shore due to the action of waves.

RIPPLE FINISH : Uniform wrinkled finish in a painted surface by stoving.

RIPPLE MARK : A standard feature in all shallow water bottoms in sand and alluvial material due to water currents.

RIP-RAP : Stones for bank revetment for protection against scouring.

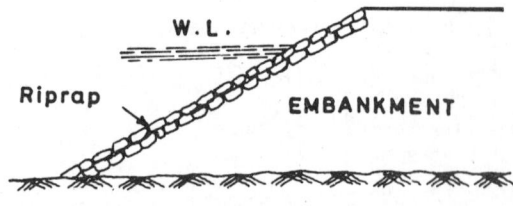

RIP SAW : A hand saw used by carpenters to cut a timber parallel to its grain.

RISE : (i)   The vertical distance i.e., height from the springing line to the soffit of an arch.

       (ii)   The vertical distance of the crown of a road from its lowest points at sides.

RISE AND FALL METHOD : The R.L. of each point is computed from that of the one preeceding it. The difference in readings between each pair of consecutive points is entered in 'Rise' column or in 'Fall' column, so that the difference between summation of the values of 'Rise' and 'Fall' should give the difference in level between the first and last points. This method is very popular for finding out R.L. of different points.

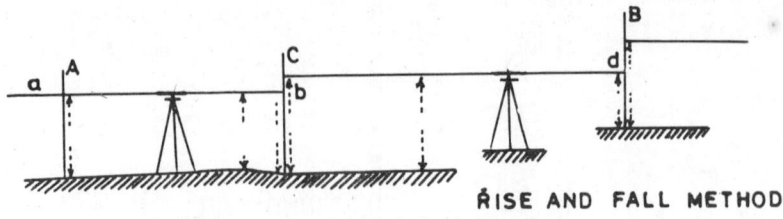

RISE AND FALL METHOD

RISE AND FALL TABLE : A bench or table for operation with a circular saw. The table can be raised up or lowered accordingly as required during sawing a timber.

RISE AND RUN : A certain vertical height for a particular horizontal run used in carpentry.

RISER : (i)   The upright face of a step, usually of height 125 mm to 175 mm.

(ii)   A deep stone thicker than other stones in a layer, which may act as a bond stone.

RISINT BUTTS : A special type of hinge which raises a door by 12 mm during opening. The hinge is made with a helical bearing surface.

RISING MAIN : A water supply line through which water is pumped to an elevated reservoir i.e., from a lower level to a higher level by pumping.

RIVET : A bolt-like mild steel piece with a head and a stem, but without any thread on the stem. The stem is heated to redness and is inserted into the hole through the parts to be joined together and the stem end on the other side of the hole is formed into a head by pressure. This forms a permanent fastening.

RIVET HEAD : Various forms of rivet head are used. These are snap, pan, cup, conical, spherical, etc.

RIVETING MACHINE : A hydraulic or pneumatic riveter, which forms the rivet head by pressure.

RIVET SET : Rivet snap.

RIVET SNAP : A punch having a rivet head like a recess in its head. A rivet head is formed by hammering this punch.

RIVET TEST : Testing of rivets by hammering to check whether a rivet is tight or slack.

ROAD : A track for communication. This may be an earthen road of stabilized soil or bituminous pavement or concrete surface.

ROAD BED : The ballasts carrying the sleeper and rails in a railroad i.e, in a permanent way.

ROAD FORMS : Wooden planks set on edge to form the sides of a road.

ROAD PANEL : Formation of a road surface by laying concrete slabs in panel with a view to breaking the continuity of the slab.

ROAD ROLLER : A power-driven roller weighing from 1/2 to 12 tons, used in rolling the road surface.

ROAD SURFACE : The wearing course or road carpet i.e., the topmost layer of a road.

ROCK ASPHALT BREAK WATER : A sloping type break water, in which the seaward slope consists of a mixture of large stones and asphalt placed hot under water in large batches.

ROCKER AND ROLLER: See illustration.

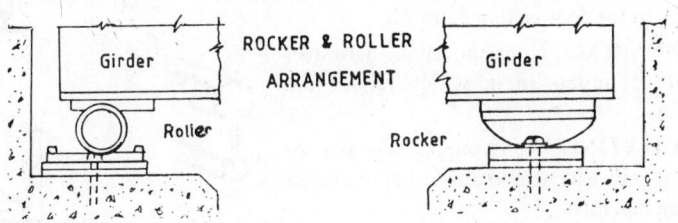

ROCKER BEARING+: A type of bearing at the support of a bridge or truss, which is allowed to rotate, but not to move horizontally.

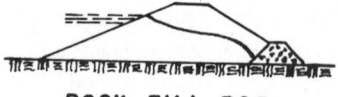

ROCK FILL DAM

ROCK FACE : The surface texture, of a stone.

ROCK FILL DAM : An earthen dam filled with broken rocks and well-compacted.

ROCK FILL TOE : See illustration.

ROCK FLOUR : Finely crushed rock.

ROCKING FRAME : A vibrating or oscillating frame on which moulds are placed for placing and compacting concrete.

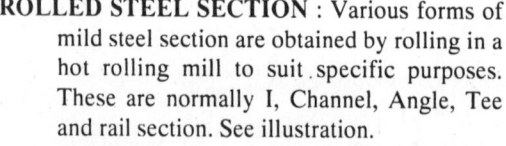

ROCK FILL TOE

ROCKWELL HARDNESS TEST : A method of testing hardness of metals by measuring the depth of penetration of a conical diamond point or a steel ball by hammering.

RODDING : Cleaning out drains by introducing and pushing rods into the drains.

RODDING EYE : An access eye in a drain, through which rods are pushed in for cleaning out the drain.

ROD FLOAT : A wooden rod designed to float vertically with most of its length submerged under water due to the weight attached to its bottom, required to measure the average velocity of flow of a stream throughout the submerged depth.

ROLL-CAPPED RIDGE TILE : A form of ridge tile having a cylindrical projection at top.

ROLLED ASPHALT : A base course in road making, consisting of coarse aggregate (stone ballasts) and asphalt laid hot uniformly and rolled until an even surface is formed.

ROLLED STEEL JOIST : R.S.J. or I-section of mild steel formed in a hot rolling mill.

ROLLED STEEL SECTION : Various forms of mild steel section are obtained by rolling in a hot rolling mill to suit specific purposes. These are normally I, Channel, Angle, Tee and rail section. See illustration.

ROLLED-STRIP ROOFING : Roll roofing with felt.

ROLLER : A heavy cylinder 1/2 to 12 tons in weight is rolled by a heavy-duty, slow-moving vehicle, used for rolling road surface or for compacting earth fill.

ROLLER BEARING : A bearing having hard steel smooth cylinders in it, which reduce the friction.

ROLLER COATING : Applying paint, enamel or distemper by a hand-operated roller or a roller coating machine.

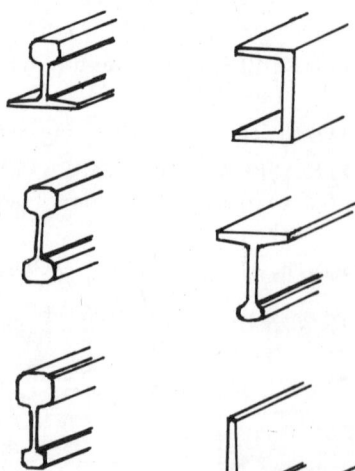

ROLLER GATE : A hollow cylindrical crest gate (provided for spillways in a dam) which is supported on toothed wheels meshing with inclined racks on either side. During opening or closing of the gate, it rolls over the rack.

ROLLING LIFT BRIDGE : A bridge that can be lifted up by means of a rolling system.

ROLLING LOAD : A moving load e.g., the load transferred by the rolling wheels over a bridge or girder.

ROLLING RESISTANCE : The tractive resistance i.e., the resistance caused by the friction between the rolling wheels and the track.

ROLLING SHUTTER : A shutter that can be taken up or let down by rolling from a roller carried by a lintel over the opening. This is commonly used for large door openings like a door of a garage or a shop.

ROLLING-UP CURTAIN WEIR : A framed weir which remains upright in position, but can be lifted up by means of chain and pulley system to open the weir.

ROLLWAY : An overflow spillway in a dam, over which the excess water overflows by rolling down.

ROMAN BRICK : The brick measuring 50 mm x 100 mm x 300 mm.

ROMAN CEMENT : Pozzolana, the impalpable powder of volcanic ash.

ROMAN MOSAIC : A terrazzo pavement or tessellated pavement with uniform distribution of 12 mm square marble chips. The chips are placed by hand to form some design or pattern.

ROMAN TILE : A roofing tile having either one waterway or two waterways divided by a central roll. These are burnt clay tiles.

RONE : An eaves gutter.

ROOF : The top cover of a room or a building, either flat or sloped.

ROOF BOARDS : Boards with tongued and grooved joints and nailed to the common rafter to form the base for roofing felt or flexible metal sheets in a pitched roof.

ROOF CLADDING : Covering the roof with tiles, slates, sheet metals, A.C. sheets, C.G.I. sheets, etc.

ROOF DECKING : For quick roofing, light-weight panels of strawboard or wood-wool slab are made with a covering of roofing felt or sheet metal.

ROOFING FELT : Fibre mats treated with bitumen, asphalt, tar or a water-proofing compound.

ROOFING TILES : Tiles made of burnt clay, concrete or cement-asbestos, having projections for interlocking.

ROOF LADDER : A cat ladder.

ROOF LIGHT : A skylight.

ROOF SPACE : The space between roof and ceiling.

ROOF TRUSS : A wooden or steel frame to support a roof.

ROPINESS : A painted surface with brush marks left on it.

ROPEWAY : See aerial ropeway. A device to transport material from one place to the other by means of a moving overhead cable or rope having containers hanging from it.

ROSE : A decorative cover, plate or boss used in door handles and electrical fixtures.

ROSE BIT: A countersunk bit used in timberwork.

ROSE NAIL : Wrought iron nails used in carpentry.

ROSIN : Colophony, a natural resin used for making varnishes, size, etc. This is the residue obtained from distillation of turpentine.

ROSSI-FOREL SCALE : A scale to measure the intensity of earthquake evolved by Rossi and Forel. In the scale, grade 1 is mild and grade 10 is catastrophic.

ROT : Decay of timber. This may be dry rot or wet rot.

ROTARY CUTTING : A method of cutting veneers from the log of wood taken out from the cooking vat. The log is set in a large powerful lathe machine and is rotated. A knife then cuts a continuous sheet of veneer.

ROTARY DRILLING : A method of drilling deep holes of 150 mm to 450 mm diameter for oil or water. Hollow shafts are screwed together to form the length, with the cutting bit rotating at the tip. The drilled material is taken out through the shaft.

ROTARY EXCAVATOR : A machine used for making large diameter holes - shafts for coal mines or tunnels.

ROTARY METER : A current meter, used to measure the current in a flowing river or stream.

ROTARY PUMP : A pump with gearing arrangment which deliver large volume of water at a low head.

ROTARY METER

ROTARY SCREEN : A screen consisting of perforated drums (cylinders) which are rotated for sizing the broken stones or gravel.

ROTARY SNOW PLOUGH : In cold climates, this is used to remove the snow off the road surface by ploughing with a rotating blade.

ROTATIONAL SLIDE : A sort of failure of a clay in slope by slipping of the earth along a curved surface. This is also called cylindrical slide.

ROTATION RECORDER : An instrument that counts and records the number of rotation. This is employed in various measuring appliances, e.g., current meter, water meter, rotary meter, etc.

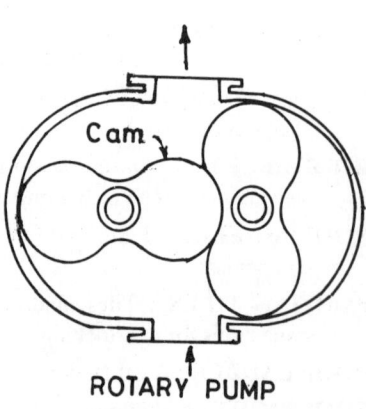

ROTARY PUMP

ROTTEN KNOT : An unsound or dead knot.

ROUGH ARCH : A relieving arch.

ROUGH ASHLAR : Roughly dressed stones.

ROUGH ASHLAR : A bracket under a wooden stair.

ROUGH CARRIAGE : A carriage under a wooden stair.

ROUGH CAST : Rendering a rough surface on a plastered wall with coarse stuff. Sometimes, fine-grained gravel, moorum or shingles with hot hydraulic lime is thrown on the surface to form a decorative texture.

ROUGH COAT : (i)   The base coat of plastering on a wall surface.

(ii)   The base coat of a paint on a surface.

ROUGH CUTTING : Cutting of a brickwork for facing.

ROUGH FLOOR : (i)   A floor with rough finish.

(ii)   A flooring base, a sub-floor.

ROUGHING IN : Carrying out the preliminary or primary work of plastering or plumbing.

ROUGHING OUT : Shaping a work piece roughly prior to its finishing stage.

ROUGH STRING : A carriage under a wooden stair.

ROUGH WORK : Brickwork or stonework that will be covered eventually by plastering or facing.

ROUND ABOUT : A circular traffic island provided at the centre of a road junction where four or more roads meet. This facilitates in guiding traffic flow.

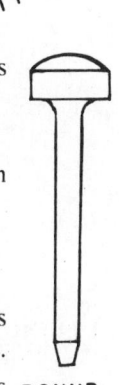

ROUND ABOUT

Traffic Island

ROUND END STEP : A bullnose step whose end is rounded.

ROUND HEADED BUTTRESS DAM : A buttress dam consisting of parallel buttresses made of mass concrete, touching each other at the ends. It looks like a multiple-Arch dam .

ROUND LOG CONSTRUCTION : Construction of houses in forest, with round logs.

ROUND SPIKE : A form of spike used in railway engineering to hold the rails on wooden sleepers. See illustration.

ROUND TIMBER : Freshly cut tree logs prior to conversion into timber.

ROUND TOPPED ROLL : A joint of sheet metal formed over a wood roll in flexible metal roofing.

ROUTER : A plough plane used in carpentry.

ROWLOCK : (i)   A course of brick on edge.

(ii)   A decorative brickwork used in paving, which consists of a ring of headers encompassing a frame of stretchers.

ROWLOCK ARCH : A relieving arch made of concentric courses of headers placed on edge.

ROUND SPIKE

ROWLOCK-BACK WALL : A wall facing backed with bricks laid on edge.

ROWLOCK CAVITY WALL : A wall made with rat-trap bond.

R.S.J. : A rolled steel joist i.e., an I-section used as a beam or column.

RUBBED BRICK : A soft brick without any frog, rubbed to required shape for a gauged brickwork.

RUBBED FINISH: A concrete surface finished by rubbing down with carborundum.

RUBBED JOINT: A joint formed between two smooth narrow boards by applying glue on both the surfaces and rubbing against each other to drive out the air bubbles and excess glue.

RUBBING FENDER : The original type of fender consisting of vertical and horizontal timbers fixed directly on the the quay wall, used to prevent minor damage to the moored vessel and the quay wall.

RUBBING VARNISH : A flatting varnish.

RUBBING PULLEY : A gin block.

RUBBLE : (i)   Roughly dressed squared stones for walling.

(ii)   Broken bricks, stones, etc.

RUBBLE ASHLAR : A wall with ashlar facing and rubble backing.

RUBBLE CONCRETE : A concrete in which large size stones, about 150 mm size, are used as coarse aggregate. This concrete is used in massive work like construction of a dam.

RUBBLE DRAIN : A blind drain in form of a trench filled with stones to permit flow through it.

RUBBLE WALLS : These are either random rubble or snecked rubble i.e.,either coursed of varying depth of layers or uncoursed.

RUBBLE MOUND BREAKWATER : See illustration. A form of breakwater built of heavy stones of large size by dumping on top of each other.

Sea Side   Harbour Side

RUBBLE   MOUND   BREAKWATER

RULE : (i)   A straight edge with graduations for measurement.

(ii)   A straight edge used by masons for levelling a plastered surface or concrete surface.

RULLING GRADIENT : In highway or railway engineering, it is the limiting gradient i.e., the maximum allowable gradient. It is 1 in 30 in plains and 1 in 20 in hilly areas.

RUMMEL : A term for 'soakway' used by Scots.

RUN : (i)   The stretch of waterline or sewer line.

(ii)   The distance between the supports of a rafter or a stair string.

(iii)   The layout of cables or pipelines in a building.

(iv)   A narrow strip of paint or varnish flowing down on a vertical surface with a teardrop at the lower end.

RUNG : A horizontal bar in a ladder used as a step.

RUN LINE : A line painted with the help of a straight edge or a lining tool or by stencilling.

RUNNER : (i) The guide in front of a plough or roller or cutter.

(ii) A horizontal piece of timber supporting the beams of a formwork (shuttering) under a concrete slab.

(iii) The rotating part of a turbine or water wheel.

RUNNING BOND : A stretcher bond.

RUNNING GROUND : Sand or silt in a soil flowing like a liquid. Such a situation is dealt with by artificial cementing of the soil, by close timbering, by making well points or by any other method of draining out water from the soil.

RUNNING MOULD : A metal or wooden templet having the shape of the profile of a moulding, used in a moulding work.

RUNNING OFF : Giving the finishng coat to a moulding.

RUNNING PLANK : A wooden plank in a barrow run. Also called 'gang plank'.

RUNNING RULE : A rule or straight edge nailed to the floating coat under a cornice which is to be moulded by sliding a horsed mould along the rule.

RUNNING SAND : The flow of subsoil sand along with flow of subsoil water which occurs, when the subsoil sand is very fine, it is saturated with water and there is a flow of subsoil water from one place to the other. Also known as 'Quick sand'.

RUNNING SCREED : A plastered band used for running a moulding, instead of using a running rule.

RUNNING SHOES : Metal pieces on a horsed mould to facilitate easy sliding of the mould.

RUNNING SYPHON TRAP : Now-a-days, this is not used, because it has no arrangement for cleaning and ventilation. See illustration.

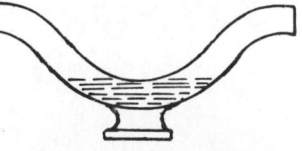

RUNNING TRAP : Similar to Running Syphon trap, but it has facility for cleaning. See illustration.

RUNNING SYPHON TRAP

RUN-OFF : (i) The overland flow of water derived from precipitation. This is also known as surface run-off, storm run-off or flood run-off.

(ii) The flow of water gathered into rivulets, streams or rivers.

RUNNING TRAP

RUN-OFF-CO-EFFICIENT : The quantum of run-off depends upon the intensity of rainfall, area of watershed and the run-off co-efficient. The run-off co-efficent, c, varies depending upon the imperviousness of the area, nature of surface and its moisture content.

RUSTICATED ASHLAR : An ashlar work in which rough faced alternate stones project about an inch from the joints.

RUSTIC BRICK : A facing brick whose surface is made rough by impressing it with a pattern.

RUSTIC JOINT : The sunk joints in a rusticated ashlar work.

RUSTIC SLATE : Random slates.

RUST JOINT : A rust-proof joint made with iron cement between cast iron pipes.

RUST POCKET : An access eye or door provided at the base of a ventilating pipe or duct. Accumulated rusts are removed by opening out this door.

RYBATE : A jamb stone forming a rebate in a door or window opening.

S

SABIN : A unit of sound absorption equivalent to absorption through one sq. ft. of an opening.

SABLE BRUSH : A pointed brush of sable hair used for making decorative painting.

SACRIFICIAL ANODE : The anode used in cathodic protection of pipes against corrosion.

SACRIFICIAL PROTECION : The protection of iron surface by coating with zinc, cadmium, etc.

SADDLE : (i)   A bolster.

        (ii)   A flexible metal piece used in dressing the lines of interscetion of sloping roofs.

     (iii)   A roll provided on a flat roof dividing it into bays.

     (iv)   A block of steel provided over the tower of a aerial ropeway or suspension bridge, over which the steel ropes pass.

     (v)   A steel strap with lugs on either end is placed over a pipe and screwed at the ends for fixing the pipe.

SADDLE-BACK BOARD : A narrow board with chamfer at both ends fitted under a door with a projection to clear the carpet, when the door opens.

SADDLE BAR : A metal bar fixed horizontally to stiffen a leaded light.

SADDLE BEAD : A bead for fixing glass.

SADDLE JOINT : Also known as 'water joint'. A joint between stones in a cornice or between cornice mouldings to form the shape of a horse-back (saddle) so that rain water is thrown away from the joint.

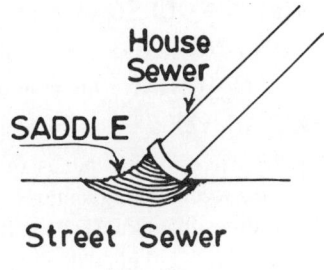

SADDLE PIECE : A piece of flexible metal used as a saddle.

SADDLE ROOF : A sloping roof with gables.

SADDLE SCAFFOLD : A scaffold formed over a ridge for repair work at the top of a building or a chimney.

SADDLE STONE : A stone used at the top of a gable.

SAFETY ARCH : A relieving arch to ensure safety of the main arch.

SAFETY BELT : A belt to be worn by a worker for his safety while working at a height more than one metre.

SAFETY FACTOR : Factor of safety used in design of a structure.

SAFETY GLASS : A 'Georgian glass' which is reinforced with chicken wire mesh.

SAFETY LINTEL : A load-bearing lintel provided to protect an ornamental arch or lintel over an opening.

SAFETY RAIL: A check rail or guard rail used on railway curves fitted outside the inner rail to keep the inner wheel in position on the rail and to reduce wear and tear of the outer rail due to the action of centrifugal force.

SAFETY VALVE : A spring-loaded valve used to release steam pressure in a boiler for safety. It alarms the workmen by producing sound while releasing steam pressure. See illustration.

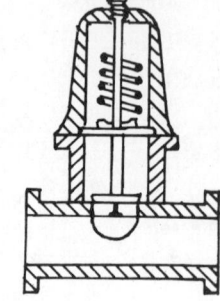

SAG BAR : An anti-sag bar provided in a truss.

SAG CURVE : A vertical curve with concavity upwards as shown in illustration.

SAG CORRECTION : A correction applied to the length measured by a tape (holding the tape ends by hands). The tape held by hands is liable to sagging and thus the measured length becomes greater than the actual length in field.

SAFETY VALVE

SAGGING : (i)   Bending with concavity upwards.

(ii)   Curtaining in painting.

SAGGING MOMENT : A bending moment which causes sagging of a beam.

SAG PIPE : A pipe to carry canal water under a road or railway to connect two canals.

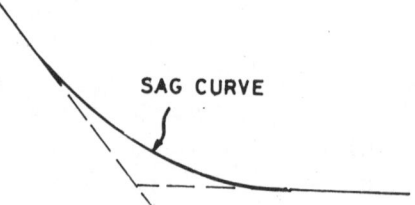

SAG CURVE

SAILING COURSE : A string course used in a building at each floor level.

SAINT ANDREW'S CROSS BOND : Engllish cross bond.

SAL AMMONIAC : Ammonium chloride which is used as a flux in soldering.

SALT GLAZE : A glaze given to stoneware articles and drain pipes by salt vapour.

SAMPLER : A continuous or intermittent sampling device for taking samples from a channel or steam for testing purpose. See illustration.

SAND : Granular material of quartz origin found in river beds, seashore and pits. Its grain size varies from 0.06 to 2 mm. It is cohesionless.

SAND BLAST : Throwing sand particles on to a surface through a compressed air jet to make a surface clean and smooth or to etch a surface for decorative work.

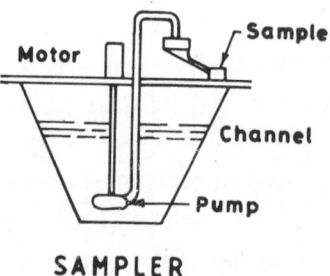

SAMPLER

SAND BOX : Dry sand packed in boxes used at the bottom of the posts supporting a centering for casting concrete. On curing, the centering can be quickly taken out by slipping out one side of each box. The sand falls down and the posts are removed.

SAND CATCHER : A hydrographic instrument used as a sand-grain meter through which water flows out and sand particles get deposited within the instrument chamber. On bringing the instrument to the surface, the quantum of sand deposited is measured. This gives a measure of sand particles flowing with the water in a stream.

341   **SAND PUMP**

SAND DRAIN : With a view to increasing strength of soil by consolidation with the help of accelerated drainage in relatively impervious deposits, vertical drains made

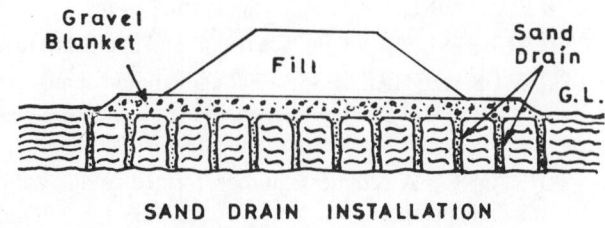

SAND DRAIN INSTALLATION

of sand columns are installed as shown in illustration. With the increase in load, water is squeezed out of soil into the drains and thus escapes.

SAND DRY SURFACE : A dry surface to which no sand can stick, which is suitable for painting.

SANDED BITUMEN FELT : A roofing felt coated with sand-bitumen mixture.

SANDER : A sand-blowing machine.

SAND-FACED BRICK : A brick coated with sand grains to form a rough decorative surface.

SAND FILTER : A filter bed made of graded sand, the finer particles being at top. It is used for filtration of water in small scale.

SAND GRAIN METER : See 'Sand catcher'

SAND HUMP : A hump of sand with a rising gradient at the dead end of a siding track in railway as shown in illustration. It stops a vehicle

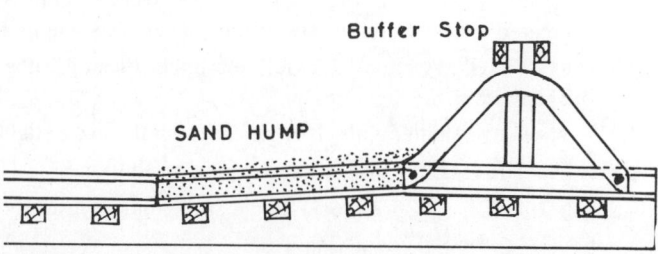

and prevents vehicles from running off the track.

SANDING : Smoothing a surface by using sand as an abrasive.

SANDING MACHINE : A power-driven sander.

SANDING SEALER : A hard sealing coat used on a wood surface for filling depressions, cracks, holes, etc. after which the surface is to be sanded for preparation of the surface prior to application of a paint.

SAND-LIME BRICKS : Bricks made of a mixture of sand and slaked lime by pressing them and afterwards maturing in a steam oven. Their crushing strength may be as high as 6000 psi.

SAND PAPER : An abrasive paper used for cleaning or smoothing a surface by rubbing.

SANDPAPER SURFACE : A road surface or a concrete surface, from which sharp pieces of aggregate do not protrude more than 5 mm.

SAND PILES : Piles of sand used for compaction of soil to increase the bearing power of soil. Deep holes are made in the ground, sand is poured into it and rammed for penetration into the soil.

SAND PUMP: A bailer or sludger pump to extract mud, silt, sand and other particles from a borehole.

SAND PUMP DREDGER: A suction-pump dredger.

SAND RUBBING : Applying a layer of sand over a bitumen impregnated roofing felt.

SAND SAUSAGE BREAKWATER : Giant sand-sanusages built to serve the function of breakwater, by placing sand-filled cylindrical synthetic-material bags on the sea bed. This is suitable for use on sandy coasts without rock.

SAND STONE : A sedimentary rock formed by deposition of quartz sand with other minerals in layers.

SAND TRAP : A trap as shown in illustration is used in a drainage system to arrest sand and grit.

SAND TRAP

SANDWICH BEAM : See 'Flitched beam'.

SANDWICH CONSTRUCTION : A composite construction with aluminium, light alloys and plastic which is strong, durable, light in weight and ornamental.

SANITARY SEWER : A conduit used as a sewer to carry sanitary sewage.

SAP : The moisture content in a freshly cut tree or freshly quarried stone.

SAPONIFICATION : A phenomenon by which a paint containing oil may be destroyed due to action of alkalies on the oil.

SAP WOOD : The wood in the outer part of a tree which is normally weaker than the heartwood (Core), light in colour and prone to decay more easily. It contains more sap compared to heartwood. Its absorptive power is also more than that of heartwood.

SARKING FELT : A bituminous felt laid under tiles or slates over roof boarding in a pitched roof.

SASH : A window shutter with glass panes for admittance of light.

SASH AND FRAME : A sash window with a cased frame.

SASH BALANCE : A spring operating a sash window instead of using pulley, cord and sash weights.

SASH BAR : A glazing bar to hold the glass panes in a window or door.

SASH CHAIN : A chain moving over a cogged pulley for heavy shutters, instead of using cord and ordinary pulley.

SASH CORD : A strong cord or rope passing over a pulley is held tight by means of a sash weight. This system is required to slide the shutter up and down.

SASH DOOR : A door whose upper half or 1/3 rd of height from top is glazed.

SASH FAST : A sash lock or fastener.

SASH PULLEY : Pulleys with cords and sash weights provided at each side of a sash window.

SASH RIBBON : A steel tape attached to a spring instead of using sash weight, required to hold the sash window up.

SASH RUN : A pulley stile of a sash window.

SASH WINDOW : A window having a cased frame, into which the opening lights slide up and down.

SATURATED AIR : An air containing maximum amount of water vapour at a given temperature.

SATURATED ROOFING FELT : A roofing felt saturated with bitumen.

SATURATION CO-EFFICIENT : The ratio of volume of water absorbed by a material and its volume of pore space.

SATURATION LINE : The water table in a saturated soil.

SAUCER DOME : A small dome like an inverted saucer which is glazed and used as a roof light.

SAUCLLER DRAIN : A bow shaped shallow roadside surface drain.

SAW : A toothed steel blade used to cut timber, plastics, commercial boards and light metals. This is either power-driven or hand-operated.

SAW BENCH : A long Bench through which a circular saw passes for sawing.

SAW FILE : A fine-cut file.

SAWDUST CONCRETE : A concrete made with sawdust, cement and plastic chips.

SAW HORSE : A four-legged stool used for handsawing.

SAW PIT : A pit in which a man stands for pitsawing.

SCABBLING : (i)   Dressing stones roughly by a pick hammer.

(ii)   Laying of stone flakes under stone pitching for revetment.

SCAFFOLD : A temporary frame made of timber or steel to support men and materials during a construction.

SCAFFOLD BOARDS : Wooden planks used for making platforms in a scaffold to support men and materials.

SCADFFOL POLES : Timber poles used as 'Standards' in a scaffold.

SCAGLIOLA : A plasterwork made with Keene's cement, coloured chips and pigments and polished by rubbing with pumice stone and linseed oil to resemble imitation marble.

SCALE : (i)   A graduated rule, 6" or 12" long, made of a piece of timber, steel, aluminium, plastic, ivory, etc.

(ii)   The ratio between dimensions of a drawing to represent an object.

(iii)   A thin film of paint which sometimes comes out from a painted surface after somedays.

(iv)   Black iron oxide formed on the surface of iron or steel during hot working, which requires to be removed prior to painting.

SCALLOPS : Withies of short length tied at the verge of a sloping roof to hold the thatch.

SCANT : A piece of timber, short of one dimension.

SCANTLE : A wooden gauge stick having two nails projected from it, used to mark the position of hole in a slate.

SCANTLING : A small piece of square-sawn timber.

SCARCEMENT : The reduction in thickness of a brick wall as it rises up.

SCARIFIER : An implement with downward projecting tines, attached to a roller, for breaking the old road surface.

SCARF JOINT: A joint made for lengthening a timber as shown in illustration.

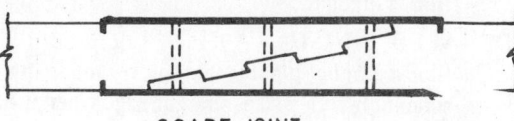

SCARF JOINT

SCENT TEST: A drain test carried out by pouring a strong scent into a drain and plugged. The next manhole is opened out and investigation is made whether there is any smell of the scent coming out of the manhole. This is to ensure the flow through the drain stretch.

SCISSOR'S CROSSOVER : A combination of two crossovers over lapping each other in form of scissors as shown in illustration. This is provided between two parallel tracks where shunting operations are frequent and enough space is not available for two separate crossovers. Also known as 'double crossover'.

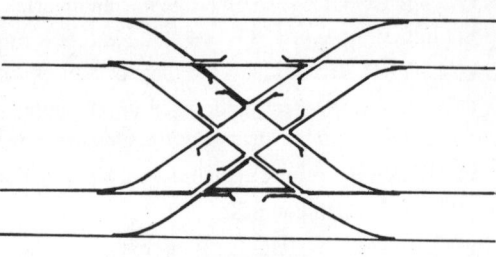

SCISSORS CROSS

SCISSOR'S TRUSS : A truss with four members framed in such a manner that it has a scissors-like appearance. There are two rafters from wall plates to the ridge and the other two members start from wall plates and meet the middle of the opposite rafter. These two members cross each other at the mid-span. Thus, a good headroom is available at the centre of the span.

SCOTCH BLOCK : A block made of wood or steel as shown in illustration is placed in position on the rail to make an obstruction to wagons to move beyond the dead end of siding track.

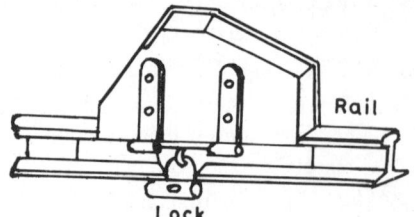

Rail

Lock

SCOTCH BLOCK

SCOTCH BOND : American bond used in brick or stone masonry.

SCOUR : Erosion of river bank or bed by wave action of flowing water.

SCOURING SLUICE : A sluice provided at the lower part of a dam to expel out deposited sand, silt, gravel etc. through the sluice.

SCOUR PIPE : Similar to scouring sluice, but it is ineffective since it gets blocked.

SCOUR PROTECTION : For protection of river bank against scouring due to wave action and high velocity of flow, there are many devices that can be adopted. One of those is shown in illustration which is made by dumping sand bags (bag work) and erecting a wall by tremie concrete.

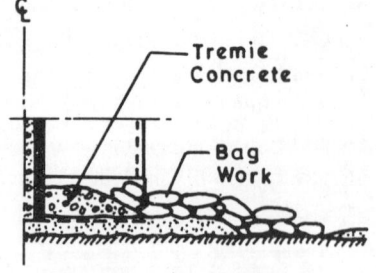

Tremie
Concrete

Bag
Work

SCOUR PROTECTION

SCOW : A dumb barge to carry dredged material from a dredger.

SCRAPED FINISH : A finishing coat of lime-cemeng plaster or stucco made by scrapint it with a steel straight edge when it starts hardening.

SCRAPER : A scraping tool or machine or a scraper plane.

SCRAPER BUCKET : The bucket with a cutter attached to a scraper loader, which is usd in excavation.

SCRAPER EXCAVATOR : A multi-bucket excavator used in excavation.

SCRAPER LOADER : Bowl scraper.

SCRATCH COAT : The first coat of plaster on a lath.

SCRATCHER : A nail float or drag used in plastering.

SCREED :(i)   A long wooden templet, each end resting on forms, is drawn over a green concrete surface by one man at each end for levelling and finishing the surface. It is also known as 'screed rail' or 'screed board'.

(ii)   A layer of mortar used as a bed for laying marble slabs or floor tiles.

(iii)   A band of plaster laid and finished to a correct level, which serves as a guide for the screed rail for levelling and finishing a concrete surface.

SCREEDERS : The persons who draw the screed board for levelling and finishing a concrete surface.

SCREED RAIL : See 'screed'.

SCREEN : It may be made of fine wire mesh for sieving fine materials of bars for screening coarse materials and large size floating matters in a flowing liquid.

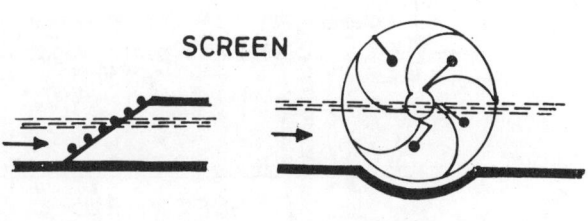

SCREEN

SCREEN ANALYSIS : Sieve analysis which is a mechanical analysis usually carried out for determining the particle size distribution in a soil or any other material.

SCREENINGS : Rejected materials by screening.

SCREW : A nail-like metal piece having a head with a slot, the shank being tapered towards its tip and threaded all along. This is used in temporary fastening with the help of a screw driver.

SCREW ANCHOR : A sleeve used in a bottomless hole for anchoring a screw.

SCREW AUGER : An auger.

SCREW CLAMP : A handscrew used in joinery work.

SCREW CONVEYOR : A helical conveyor.

SCREW DOWN VALVE : A cock or valve which is operated by screwing down a jumper to give a high resistance to the flow of a liquid.

SCREWED PIPE : Gas or water pipes made of wrought iron or dead mild steel having external thread at one end and internal thread at the other end. The pipes are galvanized and are available upto 6" (150 mm) diameter. The pipes are joined by screwing.

SCREW EYE : A screw having a head in form of a loop, which is used in door and window fastening.

SCREW GAUGE: A number varying with the diameter of screws which indicates the screw size.

SCREW PILE: A form of steel pile as shown in illustration is used for support in soft soils. The shafts of screw piles may be of cast iron or mild steel. The helix consists of two half turns. The pile is driven by fixing to the head, a capstan with a power-driven motor. The spacing of screw piles is about twice the diameter of the helix.

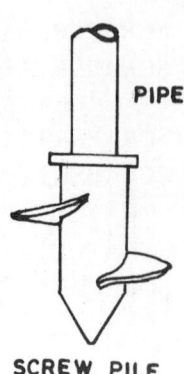

PIPE

SCREW PUMP : A non-clog type sewage pump equipped with screw-feed mechanism, which macerate solids before they enter the pump chamber. See illustration.

SCREW SHACKLE : A 'turn buckle' or a 'tension sleeve' which is a long nut having internal threads right-handed at one end and left-handed at the other. It is used in connecting two rods having threaded ends. When the nut is tightened, the rods are in tension.

SCREW PILE

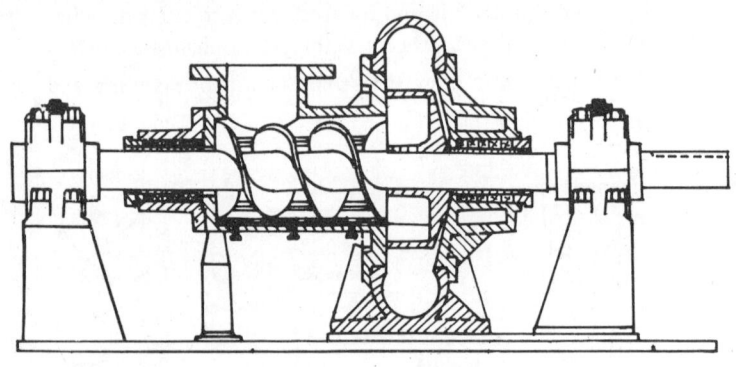

SCREW PUMP

SCREW SPIKE : A coach screw which is used as a rail fastening. It is driven into the sleeper by screwing.

SCREW THREAD : A thread cut in form of a helical ridge round a shank of a screw or bolt-nut, which is required to make a male-female joint.

SCRIBED JOINT : A joint between two mouldings.

SCRIBER : A pointed tool like a divider used to mark the outline of a templet on a surface.

SCRIBING PLATE : A templet used for marking out scribing lines.

SCRIM : Cotton or canvas or rust-proof fine wire mesh used to bridge gaps in joints.

SCRIMMING : Setting a scrim over a joint between plaster boards or timber boards to bridge the gap prior to painting or papering.

SCREW SPIKE

SCRUB BOARD : A skirting board.

SCRUB PLANE : A plane with a rounded cutting iron, which facilitates in removing an appreciable quantum of shavings at a time.

SCUMBLE GLAZE : A transparent paint used in modifying the colour of a finishing coat in painting.

SCUMBLING : Softening the first coat or base coat in painting to bring uniformity of colour and its modification.

SCUMBOARD : A board or thin baffle which is partly submerged in a liquid to prevent scum flowing out along with the flow of the liquid.

SCUTCH : A small hammer with a cross peen on both ends of its head i.e, having no striking head, is used by masons for cutting bricks during laying.

SCUTCHEON : An escutcheon.

SEAL : (i)   A tight joint to prevent flow of fluid through it.

     (ii)   The water in a trap in a drainage or sewerage system to prevent coming out of gas or foul air from the drain or sewer.

     (iii)   An air-tight cover to preserve a material.

SEALANT : (i)   A compound of plastic consistency spread over a surface to fill up the pores, cracks, crevices and gaps in joints so that the surface will not permit the penetration of water or any other fluid through it. Some of the sealants are bitumen, synthetic resin and building mastics. Amongst building mastics, thiokol polysulphide seems to be the best one.

     (ii)   A fluid bitumen laid cold or hot over a roofing felt to seal the laps.

     (iii)   Epoxy resin or polyurethane painted on a surface to form a water-repellant durable plastic coating.

     (iv)   Bituminous emulsion applied to a freshly cast concrete surface to prevent loss of water during setting and curing. This is known as liquid membrane curing compound.

     (v)   Sodium silicate solution applied to a set concrete surface, which prevents disintegration or dislodging of aggregates.

SEA-LEVEL CORRECTION : If 'R' is radius of earth at sea level and 'l' is the measured length of a line at an altitude 'h' above the sea level, then the value of 'l' at sea level is found out by deducting an amount 'x' from 'l' and the value of 'x' is found out as follows :

$x = (h/R).l$

SEALING COAT : (i)   A thin coat of bitumen, tar or an emulsion applied to a road surface.

     (ii)   A base coat applied to a timber.

SEALING COMPOUND : A 'sealant'.

SEAM : (i)   A joint made between two flexible metal sheets in roofing by turning the edges vertically upwards, doubling them over and dressing them.

     (ii)   An endless joint.

SEAM WELDING : A type of resistance welding.

SEASHORE CONFIGURATION: See illustration.

SEASONING: (i)   Hardening of freshly quarried stone by drying it.

     (ii)   Driving out moisture content or sap from a freshly cut timber tree. It is either 'natural-seasoning' or 'kiln-seasoning'. Natural-seasoning may be air-seasoning or water-seasoning.

SEAT OF SETTLEMENT : The thickness of soil mass underneath a foundation, within which about 75% of the settlement takes place.

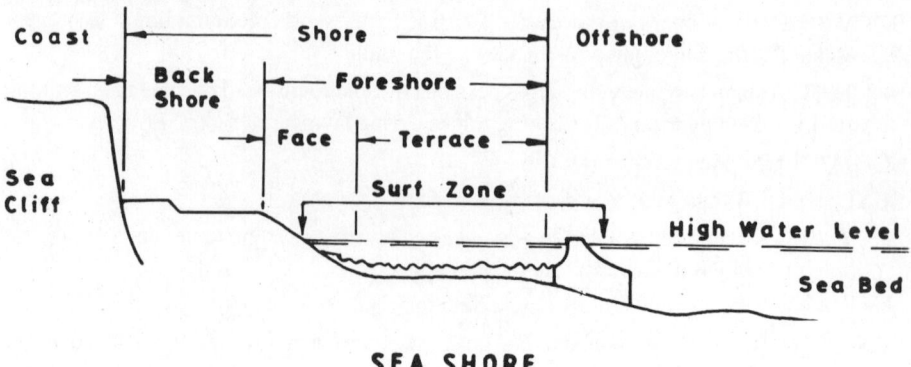

SEA SHORE

SECANT MODULUS OF ELASTICITY : For materials having a variable modulus of elasticity (E), it is the slope of the secant to the stress-strain curve. For elastic materials the secant and tangent to the stress-strain curve coincide within the elastic limit.

SECONDARY BEAM : A beam that is supported by a main beam.

SECOND FIXINGS : Fixing of wall linings, skirting boards and other joinery works including plumbing and electrical works which are carried out on completion of plastering.

SECOND-FOOT : One cusec which is a unit of flow.

SECOND-FOOT-DAY : A unit of volume, which is one cusec flowing for a day.

SECOND-GROWTH TIMBER : The timber trees which have grown later in a virgin forest.

SECOND MOMENT OF AREA : Moment of moment of area. Also known as moment of inertia.

SECRET DOVETAIL : A mitre dovetail.

SECRET FIXING : Fixing of joinery in such a manner that they do not appear on the surface.

SECRET GUTTER : A closed valley gutter in which the lead is hidden by the tiles.

SECRET NAILING : Slant nailing through tongue or rebate made in boards such that this can not be seen on the surface.

SECRET TACK : Tack welding at back.

SECRET WEDGING : Inserting wedges into a blind mortise during driving of a tenon into the mortise by pressure.

SECTION : The transverse section of a member speaking of its shape and cross-sectional dimension.

SECTIONAL ELEVATION : The elevation of an object drawn by sectioning the object by a plane parallel to its vertical back or front.

SECTIONAL VIEW : A view obtained by removing the sectioned part of an object.

SECTION MODULUS : The modulus of section obtained by dividing the second moment of area of a section about its neutral axis by the depth of neutral axis from the flange. It is commonly denoted by 'Z'.

SECTION MOULD : A template for moulding the desired section of a member.

SECTION PROPERTIES : The properties of the section of a member e.g, cross-sectional area, second moment of area, radius of gyration and section modulus.

SECTOR GATE : A roller gate, the roller being a sector of a circle.

SEDIMENT : Suspended particles settled at the bottom of a liquid.

SEDIMENTATION : A phenomenon by which particles suspended in a fluid settles down.

SEDIMENTATION TANK : A reservoir in which sedimentation of particles (suspended in a liquid) takes place. Its shape and size vary with the nature of suspended particles, desired degree of clarification, type of coagulant used, detention time in the tank and other factors.

SEEDINESS : A rough, sandy surface produced by a paint film. This occurs due to a faulty paint.

SEEPAGE : Percolation of small quantum of water through a permeable material.

SEEPAGE LOSS : Loss of water due to seepage through the reservoir walls or canal embankments.

SEEPAGE PIT : A lined excavation in ground, into which septic tank effluent is discharged and the effluent seeps through its bottom and sides. Also known as 'soak pit'.

SEGMENTAL ARCH : An arch in form of a segment of a circle as shown in illustration.

SEGMENTAL SLUICE GATE : A radial gate or sector gate comprising a pair of sectors placed one at each end of the span. The sectors revolve about the horizontal axis passing through the vertex of the sectors.

SEGMENT SAW : A circular saw of speicial type used to cut very thin slices of wood.

SEICHE : A tide in a lake water due to wind.

SEISMIC DESIGN : The design of a structure by considering seismic load that may act on the structure during its life span.

SEISMIC LOAD : The load on a structure due to vibratory motion of earth - during earthquake or blasting operation or volcanic eruption or underground explosion.

SEISMOGRAPH : An instrument, which records time and amplitude of earthquakes or earth shocks.

SEISMOLOGY : The subject dealing with the study of earthquakes.

SEISMOMETER : A meter that transforms vibrations of earth shock into electrical energy for recording time and amplitude in a seismograph.

SELENITIC LIME : Lime having 5 to 10% of plaster of paris.

SELF-ANCHORED BRIDGE : A suspension bridge in which the cables are tied to the ends of the girders and thus it requires no anchorage.

SELF-CLEANSING GRADIENT : A gradient given to a channel or a pipe such that the flow will carry solids with it.

SELF-CLEANSING VELOCITY : A non-silting velocity of flow which facilitates in keeping a drain or sewer free silt deposition.

SELF-DOCKING DOCK: A floating dock which is built in parts and one such part can be docked on the others for repair, as and when needed.

SELF-FACED STONE: Flagstones which split neatly and do not require any surface dressing.

SELF-LOAD : Self weight of a structure.

SELF-PRIMING PUMP : A pump that does not require priming by pouring water prior to pumping. Sometimes, the suction pipe end is fitted with a foot valve to avoid the trouble of priming.

SELF-PROPELLED HOPPER : A dumb barge.

SELF-READING STAFF : A levelling staff used in level survey, which can be directly read by looking through a telescope.

SELF-STRESSING CEMENT : A hydraulic cement which expands during its setting and enables reinforced concrete to prestress themselves after casting.

SELF-SUPPORTING WALL : A self-load bearing wall which carries no other load.

SEMAPHORE SIGNAL : A stop signal consisting of a vertical post on which movable arm is fitted at top. This signal is placed on the left hand side of a railway track facing in the direction of movement of the train. The side of the arm is painted red with a white vertical band. The horizontal position of the arm indicates 'stop' or 'danger' and the inclined position at 45° indicates 'proceed'.

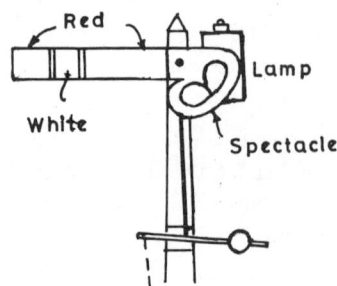

SEMAPHORE SIGNAL

SEMI-CIRCULAR ARCH : Most commonly used arch of semi-circular shape.

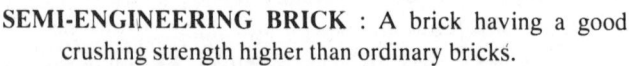

SEMI-DETACHED HOUSE : A double house speaking of duality. Two identical houses built with a common wall (party wall).

SEMI-ELLIPTICAL SEWER : A special sewer section as shown in illustration, used as a trunk sewer.

SEMI-ENGINEERING BRICK : A brick having a good crushing strength higher than ordinary bricks.

SEMI-HYDRAULIC LIME : A lime having properties which are in between eminently hydraulic lime and high calcium lime.

SENSIBLE HORIZON : The visible horizon which is apparent.

SENSITIVITY RATIO : The ratio of unconfined compressive strength of an undisturbed soil sample to that of the soil when remoulded. It gives a measure of sensitivity of a soil

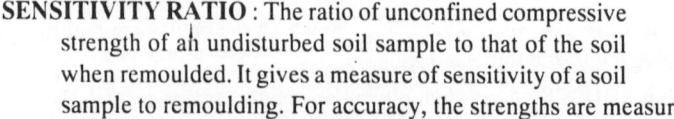

SEMI-ELLIPTICAL
SEWER

sample to remoulding. For accuracy, the strengths are measured at equal strains. For a sensitive clay it varies from 4 to 8.

SEPARATE SYSTEM : A system of sewerage in which sanitary sewage and storm water are conveyed through two separate sewer lines.

SEPTIC TANK : A tank with compartments and piping arrangement for inflow, outflow and ventilation as shown in illustration, is used in treating domestic sewage from individual house, by gassification and liquefaction and sedimentation of sludge at the bottom.

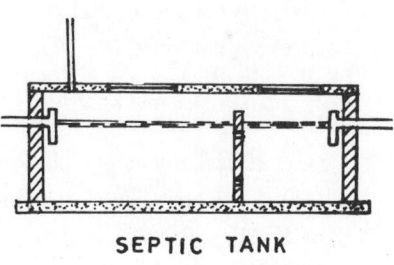

SEPTIC TANK

SERIES-PARALLEL TRICKLING FILTER : A process of treatment of sewage by two-stage trickling filters with series-parallel arrangement; Its B.O.D. removal efficiency is 90 to 95%.

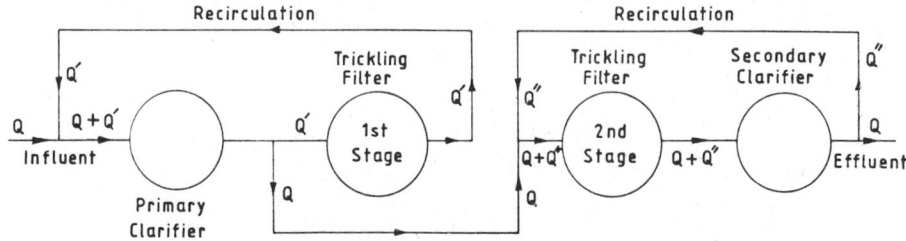

SERIES – PARALLEL TRICKLING FILTER

SERVICE CABLE : The power line joining the supply authority's main to the house cut out board.

SERVICE CORE : The mechanical core of an enclosure or a building.

SERVICE ELL : An L-shaped pipe fitting with external thread at one end and an enlarged socket with internal thread at the other, used in pipeline works.

SERVICE PIPE : The pipeline from a street water main or gas main to house premises for supply of water or gas.

SERVICE PRIVY : A toilet with a bucket or tub or a container at the bottom as shown in illustration to receive the human excreta which is replaced when full and is disposed off in a trenching ground or night soil digestion tank for production of manure and gas. This is a premitive type of toilet used in rural areas.

SERVICE RESERVOIR : A water tank in a house, which is fed by a gravity main or sometimes by a force main for distribution of water to different units within the house.

SERVICE RISER : A rising gas pipe to supply gas to an upper floor of a building.

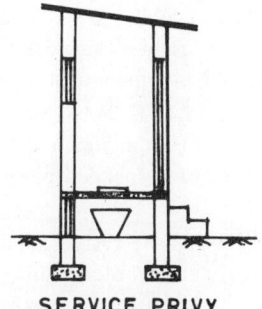

SERVICE PRIVY

SERVICE ROAD : A road of small width parallel to the main road to avoid obstruction to through traffic.

SET :   (i)   A nail punch.

(ii)   A bend in a metal piece.

(iii)   A deformation.

(iv)   The penetration of a pile for each blow of the hammer.

(v)   The direction of water flow.

(vi)   The hardening of plaster, mortar or concrete.

(vii)   The non-flowing state of a paint film after application on a surface.

SET BACK: The shifting of a building line in the upper floors above a certain level of a tall building, to a distance of about one-storey height back for each floor, as per the building laws of a municipal corporation.

SET SCREW : A small screw like a grub screw to be fitted into a tapped hole.

SET SQUARE : A drawing equipment of triangular shape with one angle 90° and the two other angles are either 45° or 60° and 30°. This is chiefly used in ingineering drawing.

SETT : (i)   A hammer-shaped cutting chisel used by blacksmiths.

(ii)   Stone blocks used for paving a road.

(iii)   A follower used in pile driving.

SET SCREW

SETTER OUT : A long-experienced skilled joiner or carpenter.

SETTING : (i)   Initial hardening of glue, paint, mortar, plaster and concrete.

(ii)   Laying of tiles, bricks, stones and timber blocks.

(iii)   Masonry work to enclose a furnace.

SETTING COAT : The finishing coat of a plaster.

SETTING OUT : Marking on ground by driving pegs for a building layout, ambankment or cutting, alignment of curve, etc.

SETTING STUFF : A finishing coat of plaster with a paste of hydrated lime and sand.

SETTING UP : The gelling of a paint during storage, which can be thinned and made uniform by stirring.

SETTLEMENT : Subsidence i.e., downward movement of a structure which occurs, when the load of a structure exceeds the bearing power of soil.

SETTLEMTNT CRATER : The crater of settlement i.e, transformation of a level surface to a basin shape due to settlement of a uniformly loaded soil, which causes differential settlement of structures.

SETTLING BASIN : A basin in which settling of suspended particles takes place.

SETTLING TANK : A tank either rectangular or circular, in which suspended particles settle down and the clarified liquid flows out from top. See illustration.

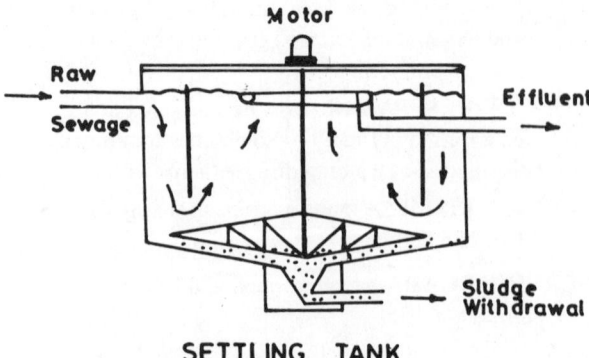

SETTLING TANK

SETTLING VELOCITY : The velocity at which a particle settles down.

SET UP : To caulk a spigot & socketed joint by driving lead in, with the help of a blunt chisel and a hammer.

SEWAGE : Foul water carrying human excreta and filthy matters. It is usually domestic sewage. Industrual effluent is called industrial sewage.

SEWAGE DISPOSAL : Disposal of sewage by discharging it to sea or a flowing river or stream, by sewage farming (land treatment) and other methods to safeguard against health hazards.

SEWAGE FARMING : A method of disposal of sewage by land irrigation especially in sandy soil which is caspable of taking hydraulic load. The nutrients present in sewage increase the fertility of soil. Land areas should be irrigated with sewage alternately with a gap in between, with a view to avoiding sewage sickness of land.

SEWAGE GAS : Bio-gas containing about 65% of methane and 35% of carbon dioxide and other gases obtained by digestion of sludge. The gas can be used for heating, lighting, cooking and generating electrical power by running gas engines.

SEWAGE PUMP HOUSE : A pump house close to a sewage treatment plant or outfall to pump the sewage from a lower level to a higher level. This may be a dry well type or wet well type. The illustration shows a wet well type pump house.

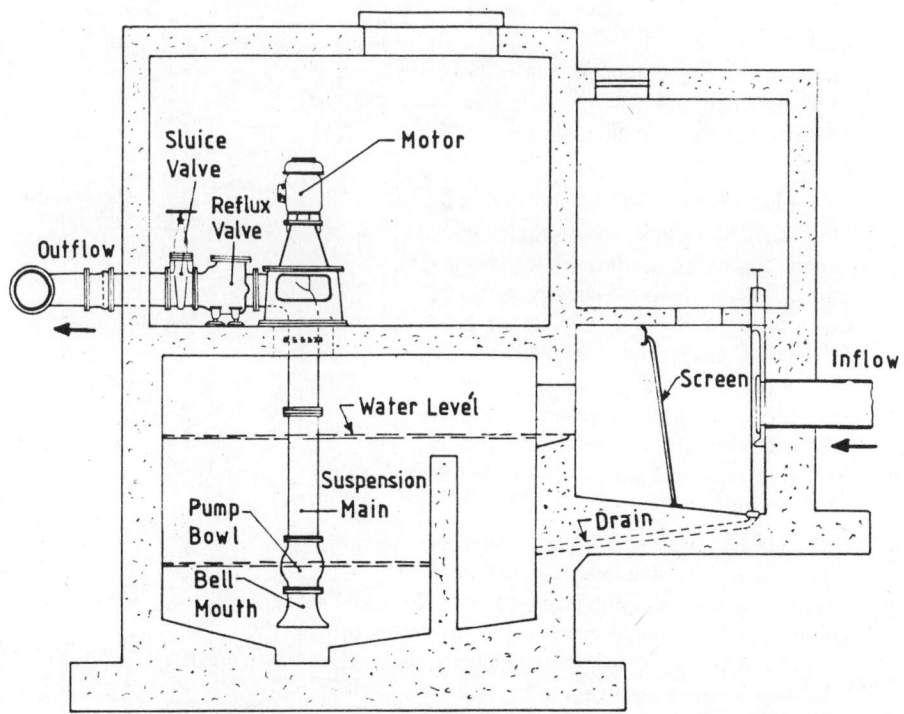

SEWAGE PUMP HOUSE

SEWAGE LIFT STATION : A pumping station installed in a sewer network to lift the sewage from a lower level to a higher level, with a view to avoiding too deep excavation for laying sewers. Such a lift station is required to be installed when a sewer invert reaches 6 metre or more below the ground surface. See illlustration.

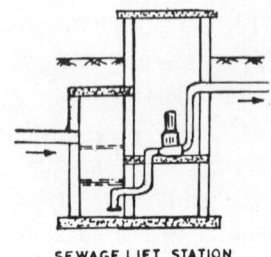

SEWAGE LIFT STATION

SEWAGE TREATMENT: The treatment of sewage to reduce the B.O.D. load and settle the suspended solids by various techniques. The methods usually adopted for sewage treatment are : Bio-filtration, Activated sludge process, Aerated lagoon, Oxidation ditch, Anaerobic pond and Stabilization ponds.

SEWER : A pipe or conduit or duct, built undeground to carry sewage by gravity.

SEWERAGE : A system of sewer network for transportation of sewage.

SEWER BRICK : Engineering brick having low absorptive power and high resistance to abrasion.

SEWER CHIMNEY : When a street sewer is at a great depth, a vertical pipe of smaller diameter with branches encased in concrete, called a chimney is used to facilitate house sewer connection as shown in illustration. The house sewer is connected to the branch at the top of the chimney.

SEWER CLEANSING : Small diameter sewers are cleaned by rodding or by dragging some sewer-cleansing appartaus or device through them. Medium size sewers can be cleaned with the help of movable scrapers. Plows and scrapers are dragged through sewers to loosen sludge banks within the sewer line. Large diameter sewers can be cleaned manually or mechanically by means of cable and windlass or hydraulic sand ejector or turbine machine.

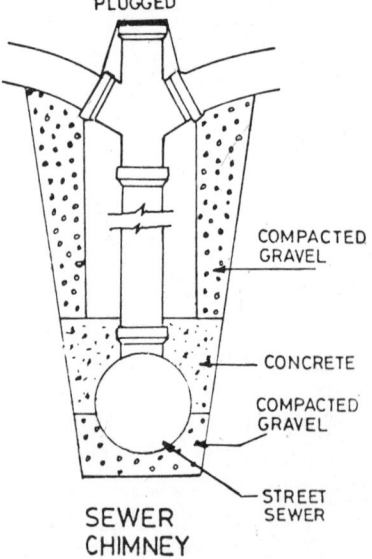

PLUGGED

COMPACTED GRAVEL

CONCRETE

COMPACTED GRAVEL

STREET SEWER

SEWER CHIMNEY

SEWER CORROSION : The corrosion of sewers takes place due to H_2S oxidation. In fact, H_2S gas evolved gets mixed with CH_4 and CO_2 coming out from sludge decomposition, which is corrosive to pipes. Again, H_2S can be biologically oxidised to H_2SO_4 which is corrosive to sewers.

SEWER JUNCTION : A typical sewer junction of three sewers meeting at a place is shown in illustration. Difficulties arise in providing supports for the partial arches whose supports are cut away in making the junction. It is therefore

SEWER CORROSION

necessary to build a masonry chamber enclosing the entire junction and with a vault roof spanning all the sewers. The bottoms of the junction are the mathematical intersections, executed in masonry. Bell mouth arch is used to span the opening.

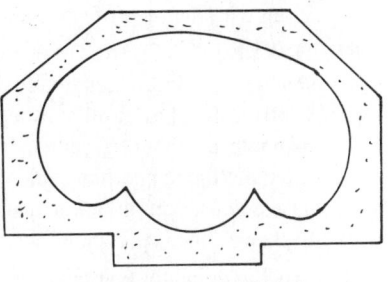

SEWER JUNCTION

SEWER LAYING : The laying of a sewer is shown in illustration. On completion of digging and preparation of bed roughly by dresing, sight rails or batter boards are placed across the trench at intervals supported on both sides on uprights as shown. Then, the centre line of the sewer is marked on the sight rail by nailing an upright cleat. A cord is stretched between two such cleats. Then, the sewer grade is checked by resting an adjustable bonding rod over the invert of the sewer grade is checked by resting an adjustable bonding rod over the invert of the sewer and keeping it vertical.

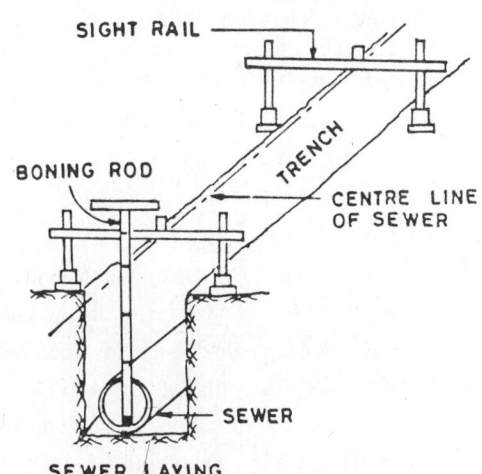

SEWER LAYING

SEWER NOMENCLATURE : The disignation of different sewers in a sewer network is shown in illustration. The lateral sewers feed a branch sewer and the branch sewers discharge into a trunk sewer. The intercepting sewer receives sewage from trunk sewers and lead to either an arterial sewer or an outfall sewer.

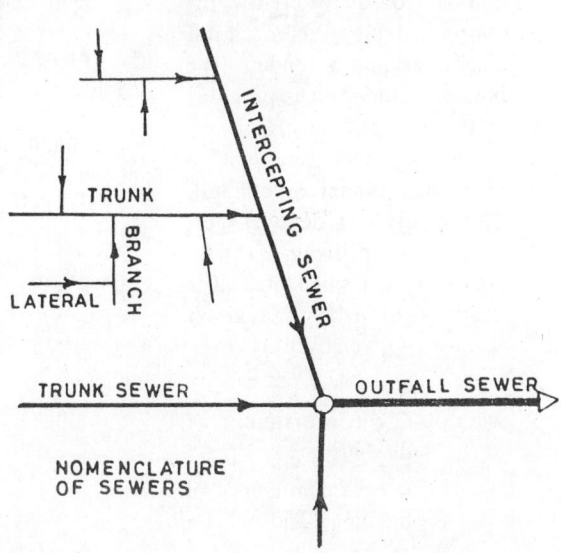

NOMENCLATURE OF SEWERS

SEWER PILL: A ball-shaped skeleton frame almost of same diameter of sewer is allowed to pass through the sewer along with the flow of sewage. During its travel it cleans the sewer walls.

SEWER VENTILATION: The ventilation of sewers is sometimes made by building a ventilating shaft and connecting it to the manhole chamber as shown in illustration to drive out the dangerous explosive or corrosive gases. Thus, accumulation of such gases within the sewer line is prevented.

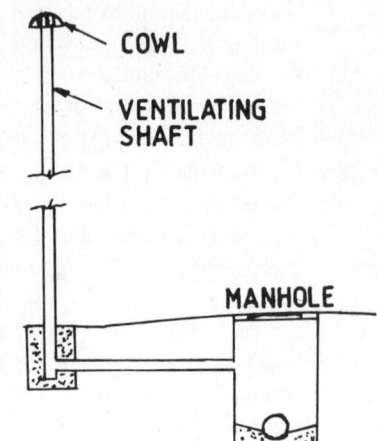

SEXTANT : A handy compact instrument used to measure angles. 'Box sextant' is commonly used in survey work. There is also 'nautical sextant' which is similar in principle to the box sextant. The adjustments required in a 'nautical sextant' are similar to 'box sextant'.

SHACKLE : (i) The lifting ring on a crane hook.

(ii) A coupling of chain and hook to connect wagons forming a train.

SHAKES : Fissures and cracks as found in a timber.

SHAFT : (i) A duct for ventilation.

(ii) A pipe to be sunk underground or under water.

(iii) A vertical passage for disposal of wastes down below.

(iv) A deep hole down below the ground surface.

SHAFT PILLAR : The pillar of coal left in a mine to support the shaft.

SHAFT SINKING : Excavation of a shaft down below the ground.

SHAFT SPILLWAY : Drop inlet spillway as shown in illustration, consists of an overflow lip supported on a riser which discharges into a conduit. The bottom of the shaft is provided with a right angle bend connecting the shaft with a horizontal tunnel or conduit. This is used in deep gorges, where the spillway is to be provided in a small area. The discharge through a shaft spillway is given by

$$Q = C L H^{0.67}$$

where C = Co-efficient of discharge,

L = Circumference of the funnel and

H = Head of water over the crest of the funnel.

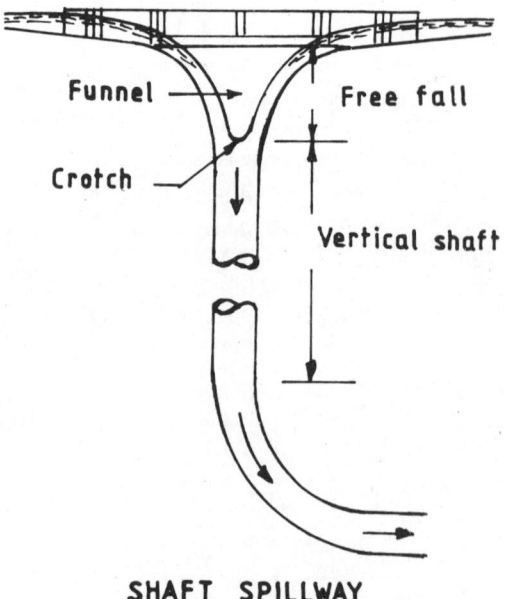

SHAFT SPILLWAY

SHAKING TEST : A simple and quick field test to determine whether a fine-grained soil sample is silt or clay without use of any apparatus. The wet soil sample is held in the palm of the hand, the top surface of the sample is smoothened with a knife and then it is shaken by tapping the hand.

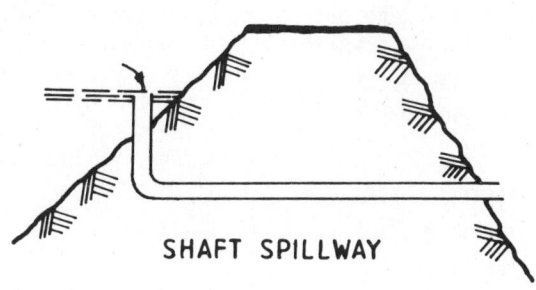

SHAFT SPILLWAY

Due to shaking, the top surface begins to glisten with water. When the sample is squeezed, the water disappears from surface and enters into the soil. This property of silt is called 'dilatancy', which is not found in a cohesive soil like clay.

SHALE : A compressed laminated clay or silt, often found in coalmines. It splits along its bedding plane and it is in form of a hard clay or slaty rock.

SHALLOW MANHOLE : An inspection chamber in a drainage system.

SHALLOW WELL : A shaft of shallow depth.

SHARP COAT : A coat of oil-based white lead suitably thinned with turpentine.

SHARP PAINT : A quick-drying paint having strong pigment used for sealing or priming.

SHARP SAND : Angular sand grains containing no clay and dirty matters.

SHAVE HOOK : A tool used by plumbers for shaving lead pipes prior to soldering.

SHEAR FORCE : The load acting across a member. For a simple beam subjected to a symmetrical loading, the maximum shear is half the total load on the beam. In case of cantilever, it is the total load acting on the cantilever.

SHEAR GATE : A gate as shown in illustration is used for controlling flow of water into or from open basins or streams or rivers.

SHEARINESS : The opalescent oily surface produced when a greasy surface is painted or there is lack of compatibility in the paint. This is a sort of failure of a painted surface.

SHEARING : Failure of a member due to the action of shear force.

SHEAR LEGS : Two poles tied at top and a pulley is hung from the lashing with the help of which loads are lifted up.

SHEAR MODULUS : Modulus of rigidity, which is the shear stress divided by shear strain.

SHEAR PLATE : A connecting plate held by bolts.

SHEARS : (i)   Shear legs.

(ii)   A device for cutting steel plates.

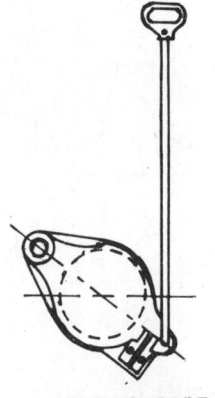

SHEAR GATE

SHEAR SLIDE : Land slide; sliding of a mass of earth in block.

SHEAR STRAIN : The angular deformation of a member due to shear force.

SHEAR STRENGTH : The strength of a material to resist shear.

SHEAR STRESS: The shear force divided by the shearing area.

SHEAR TESTS: Shear strength of soil samples is determined by carrying out shear tests like 'Vane test' or 'triaxial compression test'.

SHEATH : A covering.

SHEATHING : (i)   Covering with a sheet metal or paper or felt.

(ii)   Close boarding fixed to a timber frame.

SHED : A roofing.

SHED ROOF : A pent roof or lean-to roof.

SHEEN : A gloss.

SHEEPSFOOT ROLLER : A cylindrical roller with projections in form of sheep's foot arranged in rows round the cylinder. This is used in compacting soil.

SHEET : A very thin plate, less than 10 mm. thicknes, of tin, steel, aluminium, copper, asbestos-cement, timber, etc. used as a covering material and for various types of jobs.

SHEET GLASS : Thin plates of glass.

SHEETING : (i)   Holding timber planks against the sides or trenches and excavations by means of struts and walings.

(ii)   Covering a roof with sheets.

SHEET METAL : Metal sheets.

SHEET METAL WORK : Works made with sheet metal, such as ventilator, chimney, hopper, duct, etc.

SHEET PAVEMENT : Road surfacing with asphalt or concrete which is thinner than stone block pavement.

SHEET PILES : Piles made of thin sheets of steel, concrete slabs, wooden planks, etc. closely set and driven underground to retain earth or water.

SHEET PILE WALL : A wall made of sheet piles to withstand the thrust of water and earth.

SHEET STEEL : Sheets made of steel.

SHELF ANGLE : An angle section usually made of mild steel, used in structural steel work.

SHELF LIFE : The time period upto which a glue or paint can be stored without any deterioration.

SHELF NOG : A timber piece built into a wall and projecting out to hold a shelf.

SHELF RETAINING WALL : A retaining wall having a relieving platform or shelf at its upper part.

SHELL : A structure of any desired shape and size made of very thin curved plates.

SHELLAC : An incrustation formed by female 'Coccus lacca' an insect, on the twigs of certain trees. It contains about 90% resin (soluble in spirit) and about 5% wax and other colouring matters. It is the natural resin used in varinshes and polishes.

SHELL AND AUGER BORING : The primitive method of making exploratory bore holes by turning the auger manually. A rope is used for the shell and a boring rod for turning the auger. Shear legs are used for carrying the tools and rods and for lifting up the excavated materials.

SHELL GIMLET : An old form of gimlet having a hollow shank.

SHELLING : Crazing in plastering.

SHELL PERM PROCESS : A method adopted in reducing the flow of water into an excavation by closing the pore spaces by injecting bituminous emulsion into the soil.

SHELL PUMP : A sand pump used in making boreholes through sandy soil below the ground water table.

SHELL ROOF : A roof made of a shell structure.

SHELL SHAKE : A ring shake appearing on the surface of a sawn timber.

SHELL STRUCTURE :   See illustration. A structure made of very thin slabs.

SHERADIZING : Zinc coating given to iron and steel articles by heating them with zinc dust at about 350°C. This is more durable than galvanizing.

SHIELD : A steel hood used to protect the workmen engaged in driving a tunnel through soft soil. The shield is fitted with jacks all round its edge to push on the C.I. lining and thus the shield advances.

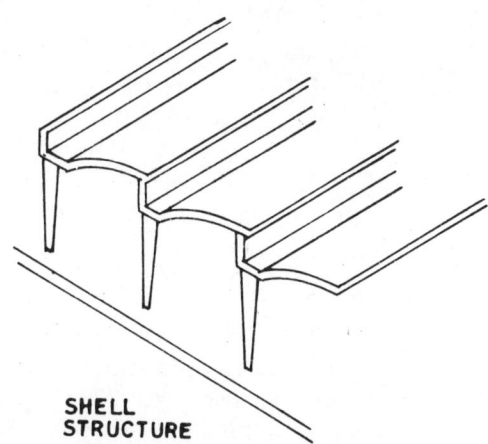

SHELL
STRUCTURE

SHIFT : (i)   A working period of 8 hours at a stretch.

(ii)   Change of position of instrument in surveying.

(iii)   The displacement from the circular shape required to form a spiral or a transition curve from a circular curve.

SHIM : (i)   A very thin slice of timber inserted in veneer.

(ii)   A thin insertion between two surfaces acting as a gap-filling material.

SHIN : (i)   A fish plate used in railway.

(ii)   The edge of a mould board which can be replaced.

SHINGLE : Rounded stones of various shapes and sizes found in river beds.

SHINGLING HATCHET : A lath hammer-like hatchet having a notch for drawing out nails.

SHIP CAISSON : A floating box made of steel and reinforced concrete, used for closing the entrance to a dock or a lock. When the entrance is open, it fits into the recess in the dock wall and in closed position it fits into the grooves made in the walls. Also, known as 'sliding caisson'.

SHIPLAP BOARDS : Weather boards having a rebate on each edge, which fits into the corresponding rebates on the adjacent boards.

SHIP SPIKES : Heavy-duty spikes made from a square bar for fixing large size timbers. Also known as 'boat spikes'.

SHIP WORM: Marine borer, which attacks timber and form holes in the wooden body of boats and ships.

SHOCK FORCES: The illustration depicts the ventilated shock, hammer shock and compression shock on break waters due to wave action.

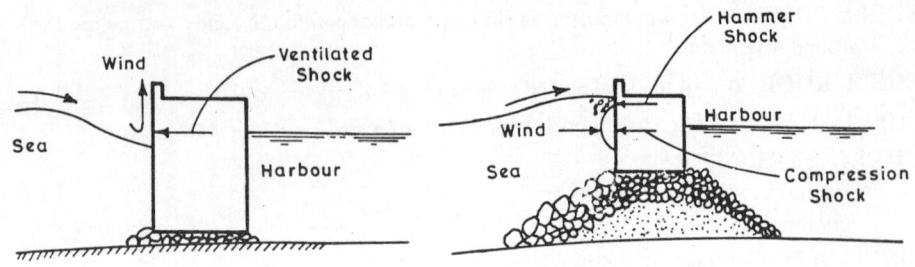

SHOCK FORCES IN BREAKWATER

SHODDY WORK : Inferior quality of work.

SHOE : (i)   A pile shoe.

(ii)   A cutting curb.

(iii)   A rainwater shoe at the foot of a rainwater pipe.

(iv)   A metal socket enclosing the end of a rafter or end of a ranging pole or ends of legs of a tripod stand.

(v)   A base shoe.

(vi)   A fitting to hold the lower end of a glazing bar on to the member of a roof.

SHOOK : A small piece of sliced veneer.

SHOOTING BOARD : A board framed to hold another board steadily during planing of the edge of the later.

SHOOTING PLANE : A jointing plane.

SHOP DRAWING : A drawing used for manufacturing an article in a factory or workshop.

SHOP RIVET : A rivet used in riveting at workshop.

SHOP WELD : A weld made at workshop.

SHOP WORK : Manufacturing or fabrication at a workshop.

SHORE : A horizontal or vertical or inclined support or prop of temporary nature.

SHORING : Providing a temporary support to a structure during its repair or renovation work.

SHORELINE CONFIGURATION : See illustration. It shows how a shoreline configuration changes due to deposition of sand and formation of a new shoreline.

SHORE RAMP : A slope given to a shore as

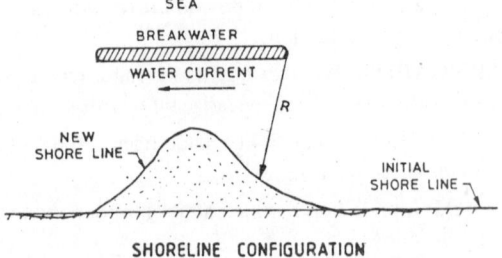

SHORELINE CONFIGURATION

shown in illustration to facilitate landing of ships, unload- ing and load- ing.

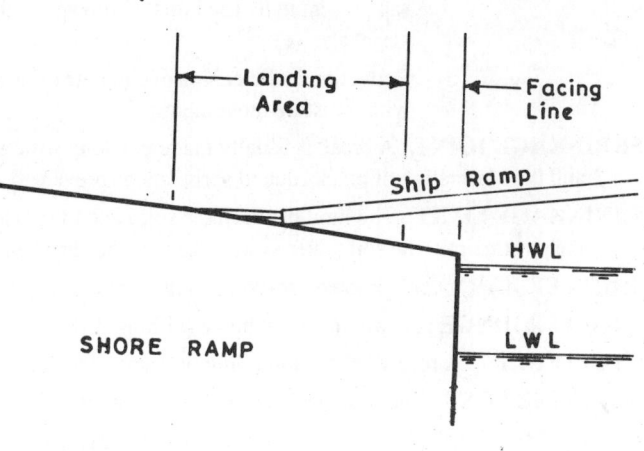

SHORT COLUMN : A column having slenderness ratio less than 32 or length to diameter ratio less than 8. Such a column does not fail by buckling but by direct compressive stresses.

SHORT GRAIN : Caused due to fungus in a timber, which makes the timber brittle.

SHORT OIL : A low ratio of oil to resin present in a varnish.

SHORT-OIL ALKYD : An alkyd resin containing oil upto 40%.

SHORT-OIL VARNISH : A varnish containing 1.25 parts of oil to 1 part of oleo-resin by weight.

SHORT TON : American ton weighing 2000 1bs.

SHORT BLASTING : Cleaning a steel surface by throwing steel shorts on to it with the help of steel impellers prior to metal coating or painting.

SHORT FIRING : Blasting with explosives to break rocks.

SHOT HOLE : A worm hole in a timber.

SHOT SWAN : A building stone sawn by chilled shot to produce a sommot surface.

SHOULDER AND HOUSED JOINT : See illustration.

SHOULDERED ARCHITRAVE : An architrave round a door frame, with widened top.

SHOULDERING : (i)   Splay cutting of top corners of slates.

(ii)   A bed of mortar placed under the head of each slate in roofing.

(iii)   Cutting small diagonal pieces from opposite corners of a slate.

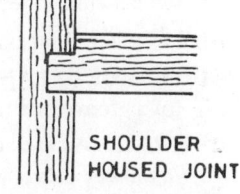

SHOULDER NIPPLE : A small short piece (pipe fitting) having a space of 20mm at middle between threads at the ends.

SHOVED JOINT : A joint in a brickwork in which the last brick is buttered and pushed against the last but one brick laid.

SHOVEL : (i)   An excavator.

(ii)   A hand tool used for mixing mortar, concrete, etc.

SHREAD HEAD: A jerkin head.

SHRINKAGE: (i) Reduction in dimension of a timber during its drying by loosing the sap present in it. The radial shrinkage is about half of the tangential shrinkage.

(ii) Reduction in dimension of concrete during its setting and hardening with moisture movement.

SHRINKAGE JOINT : A break is usually made in a long structure to allow for shrinkage and thus formation of cracks due to shrinkage is prevented.

SHRINKAGE LIMIT : The limit between the solid and the plastic states of a clay. Below the plastic limit, the soil-water system reaches the shrinkage limit.

SHUTTER BAR : A bar pivoted for fixing shutters over a window.

SHUTTER HINGE : A parliamentary hinge (H-hinge).

SHUTTERING : A formwork for moulding or casting concrete.

SHUTTING POST : The post against which a gate shuts.

SHUTTING STILE : The lock stile or meeting stile carrying locking arrangement.

SIALKOT TILE : A special form of roofing tile.

SIDE BOARD : A timber sawn excluding the heart centre of the wood.

SIDE CUT : A 'side board'.

SIDE-ENTRANCE MANHOLE : A deep manhole in a trunk sewer having access through the side of the inspection chamber.

SIALKOT TILES

SIDE FLIGHTS : Double-return stairs.

SIDE FORMS : Road forms used in construction of concrete roads.

SIDE GUTTER : A gutter provided down a roof slope at the intersection of a vertical surface with the sloping roof.

SIDE HOOK : A bench hook required for joinery work.

SIDE LIGHT : A flanking window.

SIDE POND : A pond or reservoir made by the side of a lock chamber for storage of water with a view to reducing loss in lockage.

SIDE POST : 'Princess post' in a long span 'queen post truss'.

SIDE RABBET PLANE : A small plane with cutting iron at its side, used for planing walls of a groove or a rabbet.

SIDE RAIL : A check rail or a guard rail which is a third rail fixed outside the inner rail on a railway curve, which reduces wear on the outer rail.

SIDE SWAY : Sideways movement of a framed structure due to unsymmetrical loading or wind force.

SIDE TELESCOPE : An auxiliary telescope sometimes fitted to a surveying instrument.

SIDE TREE : A post holding the head tree and side board.

SIDE WAY : A footway provided by the side of a road for use by pedestrians.

SIDING : A wall cladding with metal or A.C. sheets.

SIENNA : A pigment of yellow-brown to orange-brown colour, obtained from hydrated iron oxide and used in preparation of paints.

SIEVE : A screen to separate grain sizes.

SIEVE ANALYSIS : A screen analysis to determine grain-size distribution.

SIGHT DISTANCE : The distance for which an object is clearly visible to an observer on a road surface, both being at the same height.

SIGHT RAIL : A horizontal board set on ground for fixing the invert level of a drain or sewer, with the help of a boning rod.

SIGHT RULE : An alidade used in plane table surveying.

SIGHT SIZE : The dimensions of an opening in a building for admitting light.

SILICA BRICK : A variety of refractory brick containing about 90% of silica bonded with lime. This brick can withstand a very high temperatue (upto 1700°C) and is used in lining furnaces.

SILICATE COTTON : Slag wool obtained from furnace.

SILICATE INJECTION : Injection of sodium silicate solution followed by jetting of calcium chloride solution into a soil to make it water-tight.

SILICATE PAINTS : Non-inflammable paints containing sodium silicate (water glass).

SILICATES : Common rock-forming minerals containing SiO_3.

SILICON CARBIDE : An important constituent of 'carborundum' — an abrasive.

SILICONE : Chemically inert and highly heat-resistant silicone resins containing an appreciable quantity of silicon.

SILICOSIS : A disease of lungs due to breathing of silicon dust.

SILL :   (i)   The horizontal part at the base of a window.

        (ii)   The floor of a lock chamber against which the gates rest when closed.

       (iii)   The horizontal flow line over a weir or notch or spillway.

      (iv)   The horizontal member at the bottom of a framed partition.

SILL ANCHOR : A plate anchor at the sill level to hold a frame.

SILL BEAD : A deep bead at sill level.

SILL BOARD : A window board.

SILL BRICK : A specially-moulde, brick as shown in illustration is used in forming an ornamental sill.

SILL BRICK

SILL COCK : A hose cock.

SILL DRIP : A moulding used at sill level.

SILO : A tall tower of cylindrical shape or other forms used for storage of fine-grained materials.

SILT : A fine–grained soil whose individual particles can not be differentiated by the naked eye. A silt may be organic or inorganic. Organic silts contain inorganic silts with certain percentage of decomposed organic materials or colloids. The particle sizes range from 0.06 to 0.002 mm. Inorganic silts are mostly coarser than 0.06 mm. and contain mineral grains.

SILTATION : Deposition of silt by sedimentation.

SILT BOX: A box kept at the bottom of a gulley pit for collection of silt which is disposed off by taking out the box and emptying it.

SILT DISPLACEMENT: A method of tunnel driving through silty-clay almost in fluid state, in which a shield is pushed forward and the silt comes out through rectangular openings.

SILT EJECTOR : A hydraulic ejector. It is used as a device for controlling silt at canal headworks. It consists of a number of piers of suitable height constructed at equal intervals along the width of the main canal with bends to divert the water through 90°. The piers are covered with slabs forming under tunnels. The water with coarse silt passes through the under tunnels and the clear water passes over the slab. See illustration.

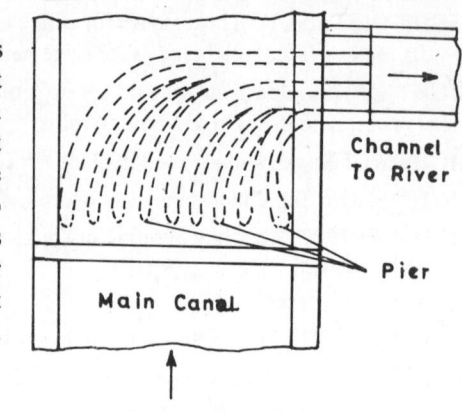

SILT EJECTOR

SILT EXCLUDER : See illustration. It is provided with under tunnels through which water with coarse silt passes. The approach velocity through the tunnels is very much reduced and thus the silt gets deposited at the lower level. The clear water passing over the tunnels is drawn by the head sluice.

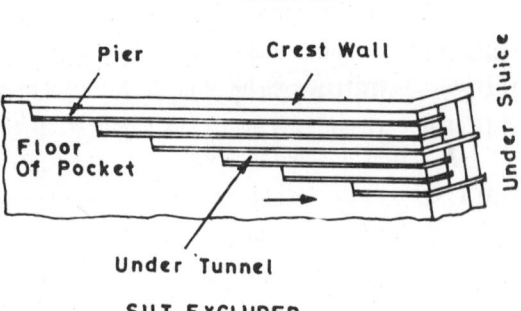

SILT EXCLUDER

SILT GRADE : A material of grain size as that of silt.

SILTING : Deposition of silt by sedimentation.

SILT VANES : Silt vanes with curved wings constructed in the bed of a parent channel at the head of an off-taking canal act as a device to exclude silt from an off-take. This is suitable, when the off-taking canal has a discharge of more than 1/3rd of the discharge of the parent channel. See illustration.

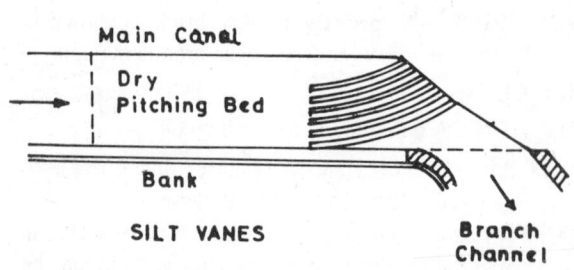

SILT VANES

SILVER BRAZING : Brazing copper tubes and articles to form neat joints with the help of an alloy of copper, silver and phosphorous.

SILVER LOCK BOND : A 'Rat-trap bond'.

SIMPLE BEAM : A beam simply supported at its ends.

SIMPLE BENDING : Bending of a simply-supported beam.

SIMPLE CURVE : An arc of a circle joining two straights as shown in illustration.

SIMPLE FRAME WORK : A perfect frame with minimum number of members.

SIMPLY-SUPPORTED BEAM : A simple beam which is freely supported at its ends.

SIMPLE CURVE

SIMPSON'S RULE : A rule used for estimating the area of a plot of land of irregular shape. The plot is divided into an even number (say, n) of parallel strips of width, d. Then, the lengths of ordinates h_0, h_1, h_2 ..... h $(n-1)$, h_n are measured. The area is given by-

$$A = d/3 \{h_0+h_n+2(h_2+h_4+\ldots h_{n-2})+4(h_1+h_3+\ldots h_{n-1})\}$$

SIMPSON'S RULE

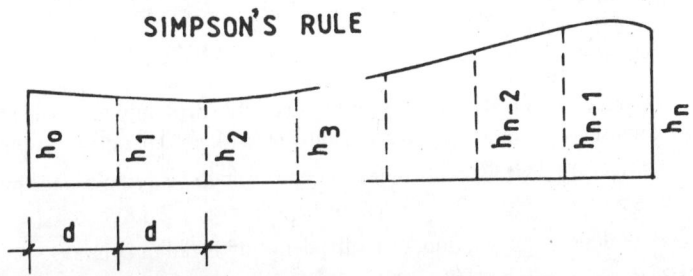

SINGLE ACTING PUMP : A reciprocating pump driven by steam engine or compressed air, in which every second stroke of the engine is a power stroke and at every stroke the pump discharges fluid.

SINGLE BRIDGING : Herring bone strutting at the middle of the floor joists in timber flooring.

SINGLE-CUT FILE : A file that cuts in one direction only, while filing.

SINGLE-FLEMISH BOND : A bond in brickwork which is used in facing of a wall, English bond being used in the core of the wall.

SINGLE FLOOR : A wooden floor, in which the joists span from one wall to the other.

SINGLE-HUNG WINDOW : A single window-shutter hung by hinges on one side only.

SINGLE LAP TILES : Curved tiles used in roofing such that they overlap the tiles of the next course.

SINGLE-LOCK WELT : A cross welt used in plumbing.

SINGLE PASS SOIL STABILIZER : A soil-stabilizing machine which stabilises soil rapidly by pulverizing the soil to a measured depth, mixing it with a liquid binder and intermingling the soil with a solid binder in one operation.

SINGLE-PITCH ROOF : A pent roof or lean-to roof having slope in one direction.

SINGLE ROOF : A close-couple roof or collar-beam roof in which common rafters carry the roof and no roof truss with purlins & principal rafters is used.

SINGLE-SIZED AGGREGATE: An aggregate of size between two consecutive sieves, to be used in making a concrete.

SINGLE SLING: A sling with a hook at one end and a ring at the other, used for lifting materials.

SINGLE SLIP : An arrangement made in a diamond crossing in railway tracks to enable trains to change the tracks appro-aching from one direction only. See illustration. This type of railway junction consists of two acute-angle cross-ings, two obtuse -angle crossings, four check rails and switches.

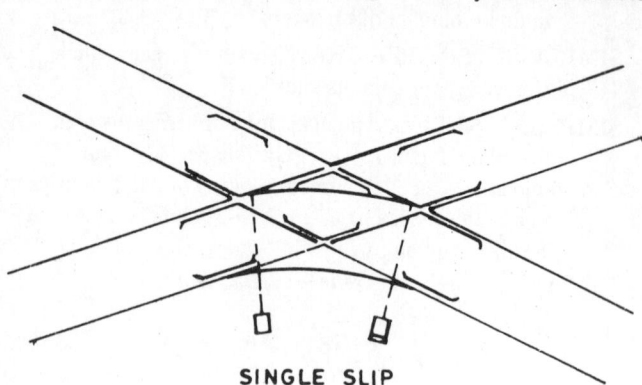

SINGLE SLIP

SINGLE STACK SYSTEM : A single-pipe system used in plumbing, in which one vertical pipe carries both sanitary sewage and sullage, with no separate anti-siphonage pipe.

SINGLE-STAGE PUMP : A centrifugual pump is said to be a single stage pump, when it delivers a liquid upto a 30 m head in one stage.

SINKAGE : A depression in a floor or a subsidence of a wall or column.

SINK BIB : A bib cock provided in a kitchen sink.

SINKER DRILL : A hand-held rock drill of large size, used for shaft sinking.

SINKING : A recess cut below a surface for door and window fittings and fixtures.

SINKING IN : Loss of gloss of finishing coat of paint due to absorption of vehicle by the undercoat.

SINKING BUCKET : A kibble used in mining.

SINKING PUMP : A power-driven or pressure-operated robust pump used for keeping a shaft dry during sinking. In-built protection is provided against damage due to external causes.

SINTERED CLAY : Expanded-clay aggregate used in light-weight concrete.

SINTERED FLY ASH : A kind of light-weight aggregate prepared by pelletizing and sintering fly ash.

SINUOUS FLOW : Turbulent flow.

SIPHON : A full-flowing pipe, a part of which rises above the hydraulic gradient.

SIPHON AQUEDUCT : An arrangement to pass the drainage water through an inverted siphon under an irrigation canal as shown in illustration. A siphon aqueduct is

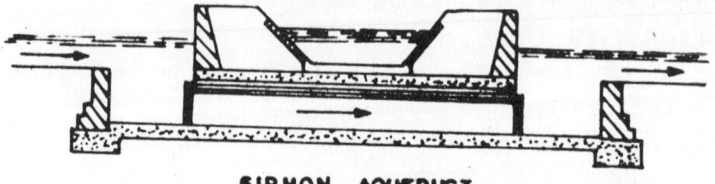

SIPHON AQUEDUCT

constructed either with vertical drops at the two ends of the tunnel or with sloping approach and exit.

SIPHON SPILLWAY (i) In sanitary engineering, it is an arrangement of diverting excess flow of sewage from a combined sewer by siphonic action as shown in illustration. The siphonic action continues till the mouth of air pipe remains submergd in sewage.

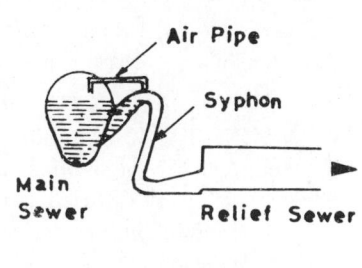

SYPHON SPILLWAY

(ii) A spillway built as a siphon in a dam over its crest such that when the water level rises to the crest of the siphon it starts flowing of water automatically and the flow of water stops when the level of water goes below the inlet of the siphon.

SITE : A plot of land for construction or for any other purposes.

SITE WELD : Welding at construction site.

SIZE : (i) A liquid sealing compound used in preparation of a surface prior to painting or varnishing.

(ii) A binder used in distempers.

SIZE DISTRIBUTION : Grain size distribution by sieve analysis.

SKELETON : (i) A network of lines obtained by a triangulation survey.

(ii) A framework of a structure.

(iii) A schematic representation of a pipeline network.

SKELETON CONSTRUCTION : A steel or reinforced concrete framework of columns, girders and beams for a building.

SKELETON CORE : The internal frame of a door or wooden partition wall, which remains hidden;

SKELETON STEPS : Concrete or metal treads without resers.

SKEW BACK : The top surface of a springer in an arch.

SKEW BACK SAW : A special hand saw with its back curved.

SKEW BRIDGE : A bridge that spans obliquely.

SKEW CORBEL : A gable springer.

SKEW FLASHING : An oblique flashing made between the gable coping and the roof.

SKEW NAILING : Driving nails obliquely to secure a joint.

SKEW TABLE : A kneeler in gable coping.

SKID ROAD : A road made by embedding round logs for use by sledges or stone boats.

SKIDS : Short lengths of rough round timber or pipe or steel rails set under a heavy machine or any other heavy object for its movement without sinking into the ground.

SKIFFLING: Rough dresing of stones in the quarry.

SKIMMER: An excavating bucket with which the excavated material moves away from the machine and slides along an excavator jib.

SKIMMING : (i)   Dressing a soil surface by removing irregularities.

(ii)   Removing froth from the liquid surface of a primary clarifier in sewage treatment.

SKIMMING COAT : A very thin finishing coat in plastering.

SKIMMING TANK : A long trough-shaped tank as shown in illustration with provision of diffused aeration and separation of oil & grease by skimming from the top.

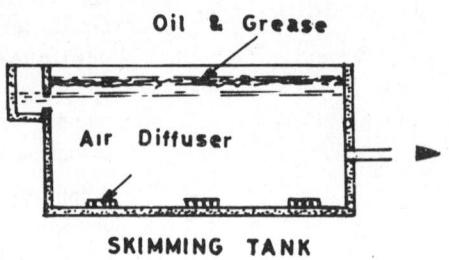

SKIMMING TANK

SKIN FRICTION : The frictional resistance offered by the underground soil surrounding a pile during its driving into the ground. It increases with the depth of penetration into the soil.

SKINNING : Formation of a skin (a thin film) on a painted surface due to oxidation of the drying oil of the paint.

SKIP : A carrier in an aerial ropeway or a kibble or hoisting bucket used in shaft sinking.

SKIRTING : A special treatment given round the base of a wall to protect it from damages.

SKIRTING BOARD : Scrub board or mop board or wash board used in skirting.

SKIRTING BLOCK : An architrave block.

SKY CLEARANCE FACTOR : A factor expressed in per cent of the days clear in a year, used in the design of an oxidation pond which is a non-conventional sewage treatment plant.

SKY FACTOR : Daylight factor.

SKYLIGHT : A dead light provided in a roof.

SKY SCRAPER : A multi-storied high-rise building.

SLAB : A flat piece of timber or concrete or stone .

SLABBING : Squaring a log and cutting slabs out of it.

SLAB FLOOR : A floor, made of R.C.C. or stone slabs.

SLACK-LINE CABLEWAY : A cableway with a track rope between a low tower and a high tower. The tension of the track rope is adjusted accordingly as required. A bucket moves on carrier wheels along the track rope.

SLACK-WATER NAVIGATION : A navigation in still water or with a low current by keeping the water level with weirs and dividing the stretches with provision of locks so that required draught is available all throughout.

SLAG : The glass-like molten impurities flowing off above the molten metal in a furnace. It is used in making slag cement, expanding cement, supersulphated cement and slag bricks or slag concrete.

SLAG BRICKS : Bricks made from a mixture of crushed furnace slags and lime. Also known as sand-lime bricks.

SLAG CEMENT : A cement made by pulverizing blast furnace slag and mixing it with powdered lime or ordinary portland cement.

SLAG STRIP : A wood strip fixed to the edges of a roofing felt to keep the gravels in position and to produce a neat appearance.

SLAG WOOL : Filaments of furnace slags used in thermal insulation and fire resistance. Also used for sound absorption.

SLAKED LIME : Hydrated lime made by slaking quick lime with water.

SLAMMING STILE : A shutting stile of a door or window shutter.

SLAMMING STRIP : A band provided at the edge of a shutting stile of a flush door.

SLAP DASH : Roughly cast.

SLASH GRAIN : The grain observed in a flat-sawn or slash-sawn timber.

SLAT : (i)   A slate made of stone or concrete.

(ii)   A thin wood strip in a louvre.

SLATE : Clay, silt, shales, etc that has been compressed underground for long years by earth movements.

SLATE BOARDING : Close boarding under tiles or slates in a roof.

SLATE NAILS : Durable 'Cast yellow metal' nails used for fixing slates. These may be made of brass or copper.

SLATE POWDER : Fairly opaque, dark, impalpable powder of slate used as an extender in paints.

SLEDGE HAMMER : A heavy double-headed hammer used by black smiths and also for breaking stones.

SLEEPER : (i)   A beam of wood, steel or concrete placed under the rails in a permanent way.

(ii)   A valley board placed over the rafters of a roof to replace the valley rafters and to carry the base of the jack rafters.

SLEEPER PLATE : A timber plate onto which floor boards are fixed or a wooden plate on a sleeper wall.

SLEEPER WALL : A honeycomb wall.

SLEEPINESS : Reduction of glass as a paint film dries.

SLEEVE : (i)   A short piece of pipe surrounding a pipe joint for protection.

(ii)   A short length of a thin-walled pipe built into masonry or concrete, through which a water line or gas pipe or an electrical conduit passes.

SLENDER BEAM : A beam which has a tendency to fail by buckling in compression flange, when overloaded. A concrete beam is said to be a slender beam when its length is greater than 20 times its width. In such a beam compressive load should be reduced accordingly and in proportion to its slenderness ratio.

SLENDERNESS RATIO: The raio of the effective length or height of a column to its minimum radius of gyration. It is given by L/K, where K is the minimum radius of gyration.

$$K = (I_{min}/A)^{0.5}$$

SLICED BLOCKWORK: Stone or concrete blockwork for construction of a break water in slope. Blocks when lowered by a crane, they slide along the slope and comes to rest in position.

SLICED VENEER : Machine-cut veneer slices from a cooked wet flitch.

SLICKENSIDED CLAY : A stiff-fissured clay.

SLIDING CAISSON : A ship caisson.

SLIDING DOOR : A door which shuts or opens by sliding.

SLIDING FORMS : A form work which can be raised simultaneously with the casting of concrete. This facilitates in rapid construction of a high and continuous wall without any construction joint. Also known as climbing form or moving form.

SLIDING GATE : A crest gate which shuts or opens by sliding on rollers to avoid high frictional resistance.

SLIDING-PANEL WEIR : A wood-panelled weir that slides between grooved uprights at the two sides.

SLIDING SHUTTER : A shutter that slides to shut or to open.

SLIDING WEDGE METHOD : The wedge theory adopted for determination of active or passive earth pressure on a retaining wall.

SLIDING SASH : A window sash that opens by horizontal movement.

SLING : A steel wire rope or chain hanger with a loop at one end to tie an object, the other end being tied to a crane hook.

SLIP : (i)   A sloping slab to form a slipway.

(ii)   A landslip.

(iii)   A parting slip.

(iv)   A fluid grout.

(v)   A fillet.

SLIP-CIRCLE METHOD : A method to determine the stability of slopes or failure of a mass of earth by slip or land-slide showing a circular rupture arc.

SLIP DOCK : A deck having a sloped bottom with a gate for keeping out water.

SLIP FACTOR : The co-efficient of friction between a friction grip bolt and the gripped member.

SLIP FEATHER : A tongue in a feather joint.

SLIP-FORM : A sliding formwork used in casting concrete.

SLIP-FORM PAVER : A concrete road making machine, which takes out road forms by pulling after casting concrete.

SLIP JOINT : A contraction joint made between two sections of a long wall such that the relative movement of the two sections can be allowed without causing any damage to the wall. The vertical tongue provided in one section fits into the groove made in the other section. Such a joint is made water–proof.

SLIP MORTISE : An open mortise used in joinery.

SLIPPER : A shoe on a horsed mould sliding along the running rule used during plastering.

SLIP SILL : A sill which is not built into the wall, but fitted between the jambs of an opening.

SLIP STONE : An oilstone slip for sharpening tools.

SLIP SURFACE : The surface or plane of failure of an earth mass by sliding-wedge action.

SLIP TONGUE : A tongue in a feather joint, which is a false tongue.

SLIP WAY : See 'Slip'.

SLOPE : The inclination or gradient of a surface.

SLOPE CORRECTION : A correction to be applied to a length measured by a tape or chain in slope in order to obtain the horizontal length. If 'h' is the difference in level between two points and 'L' is the apparent length (tapered length measured along the slope), then to yield the true length a correction,

$$C = \frac{h^2}{2L}$$ is to be subtracted from the apparent length.

SLOPE GAUGE : A measuring device to indicate depth with the help of a graduated staff laid at a slope.

SLOPE STAKING : Marking on the ground by driving stakes or pegs to indicate points for earthwork in cutting or filling.

SLOP MOULDING : Moulding of bricks by dipping the mould into a water tub each time.

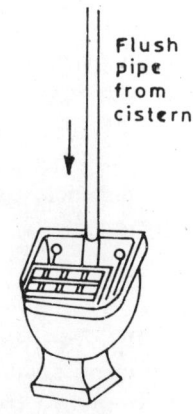

Flush pipe from cistern

SLOP SINK : A low-height large sink similar to a water closet, topped with a brasswire grate as shown in illustration is used as a housemaid's sink provided in a kitchen or bathroom.

SLOT MORTISE : An open mortise.

SLOT SCREWING : Screwing boards to the stiffening battens through slots to permit slight movement due to shrinkage.

SLOW SAND FILTER : A water-tight basin as shown in illustration, containing a filter media of sand and gravel with an underdrainage system, used for filtration of water. The rate of filtration is very low, about 2.5 to 7.5 1pm/sqm. A water having turbidity above 5 mg/1 should not be passed through the bed. This filter is highly effective in removing bacteria from water.

SLOP SINK

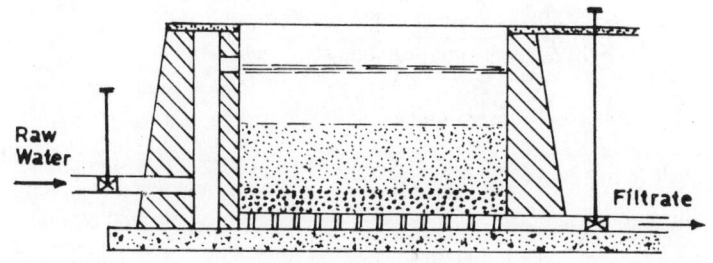

Raw Water

Filtrate

SLOW SAND FILTER

SLOW TEST : A drained shear test of a soil sample.

SLOUGH: (i) Sliding or breaking off a mass from a surface.

(ii) A secondary channel having sluggish flow of water.

SLUDGE: The solid matters obtained by settling a sewage. It may be inert or organic. Bio-gas and manure are obtained by decomposition of organic sludge.

SLUDGE AGE : The time a particle of suspended solids in sewage remains under aeration.

$$\text{Mathematically, sludge age} = \frac{\text{lbs of activated sludge in aeration}}{\text{lbs of S.S. load / day entering the aeration tank.}}$$

SLUDGE BANK : Deposition of sewage solids on the bed of a stream, waterway or open surface drain.

SLUDGE CAKE : Dewatered sludge obtained from a sludge drying bed, centrifuge, vacuum filter or sludge press.

SLUDGE DIGESTION : Digestion of sludge by anaerobic decomposition which produces sewage gas and reduces the volume of sludge.

SLUDGE DRYING BED : A bed of sand and gravel with under–drainage system as shown in illustration, used for drying digested sludge from a digester.

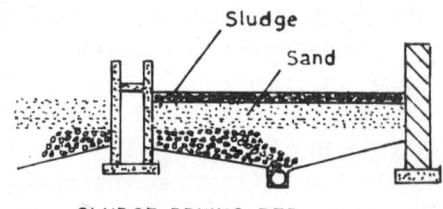

SLUDGE DRYING BED

SLUDGE GAS : Bio-gas containing about 65% methane and 35% Carbon dioxide and other gases obtained by sludge digestion. Also called 'Sewage gas'.

SLUDGER : (i)   A scraping tool used for taking out drilled material from a hole in rock prior to inserting explosive for blasting.

(ii)   A sand pump or a sludge pump.

SLUDGE SLURRY : A sludge containing a high moisture content as obtained from a clarifier or digester.

SLUDGE VOLUME INDEX : The volume in ml. occupied by aerated mixed liquor containing 1 gm. of dry solids after 30 minutes' settling. Also known as 'Mohlman index'. This is expressed as

$$\text{S.V.I.} = \frac{\text{settled sludge volume in ml for 30 mins.}}{\text{MLSS concentration in mg/l}}$$

When SVI exceeds 100, settling characteristic is poor and when above 200, bulking of sludge takes place.

SLUG : The slurry of oil-well cement of small quantum sandwiched between two masses of drilling fluid is injected through a pipe for cementing oil-well casing.

SLUICE GATE : A gate as shown in illustration is used to open or close openings into walls of a water reservoir. This is operated either manually or by hydraulic pressure. A light type of sluice gate is used to control flow in open conduits.

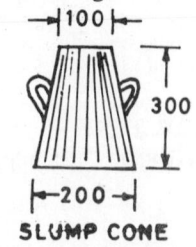

SLUMP CONE

SLUICE VALVE : A valve having a threaded spindle carrying a circular gate which fits into the groove made in the body of the valve. See illustration. This valve is used for controlling the flow of a liquid.

SLUMP CONE : A hollow truncated cone of sheet metal with top and bottom open, used in carrying out slump test. The diameters at top and bottom are 10 cm. and 20 cm. respectively. The height of the cone is 30 cm. See illustration.

SLUMP TEST : A common and popular test which gives a measure of consistency of concrete. A sample of mixed concrete is poured into the slump cone (mould) full to the brim and the cone is gently lifted up. The subsidence of the concrete cone in height is the slump. See illustration.

SLURRY : Very thin paste or semi-fluid state of a material, e.g. cement slurry, sludge slurry.

SLUSHED JOINT : A vertical joint in a brickwork into which a thin mortar is poured or by throwing in mortar by the trowel edge.

SLUSHING : Hydraulic filling as practised in Mine Engineering.

SMALL BORE SYSTEM : Small pipe system as used in central heating with radiators. Actually, the radiators are fed by small diameter copper tubes carrying hot water.

SMALL LIME : The residual small lumps of 'quick lime' in a kiln after taking out the best quality (pure) quick lime lumps by hand picking. The 'small lime' contains kiln ashes and it is used in inferior works.

SMALL TOOL : A curved small tool of different shapes used for finishing of mouldings of mortar or plaster.

SMELL TEST : A test for drains by using a strong scent.

SMITH WELDING : Forging in smithshop.

SMOKE TEST : A test for detecting leakage in drains and sewers by introducing smokes in the drains or sewers after plugging the upstream and downstream ends of a stretch. Smokes will come out through leaky joints.

SMOOTH ASHLAR : A smooth-faced squared stone.

SMOOTHING IRON : An iron tool which is heated for smoothing asphalt and other sealing compounds.

SMOOTHING PLANE : A small 'bench plane' used for smoothing timbers.

SMUDGE: (i) A mixture of lamp black and size.

(ii) Waste paint that can be used for coating a formwork.

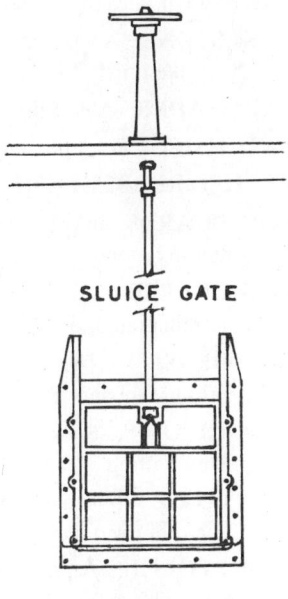

SLUICE GATE

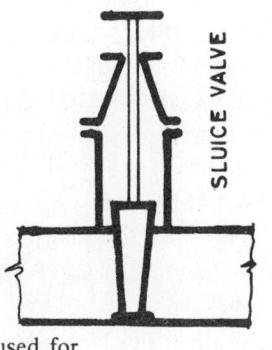

SLUICE VALVE

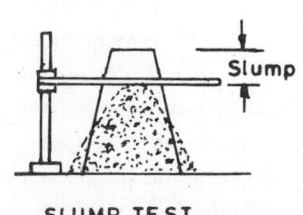

Slump

SLUMP TEST

SNAKE: A flexible long spring-like tool used for cleaning drains and small diameter sewers.

SNAKESTONE : A very smooth stone used for polishing keene's cement work to have a glass-like finish. Also known as 'scotch stone'.

SNAP HEADER : A half-brick which is used as a blind header.

SNATCH BLOCK : A block with a hook used for lifting purpose.

SNECKED RUBBLE : A rubble wall made of squared stones of different sizes.

SNOW BOARDS : In cold countries, wood slats nailed horizontally with gaps over a box gutter to facilitate melting of snow for draining away.

SNOW COURSE : A course identified on an area, along which snow sampling is done for determination of depth and density of snow.

SNOW DENSITY : The water content of a snow, which is expressed as the ratio of snow depths before and after melting.

SNOW GUARDS : Snow boards at least 100 mm high fixed above the eaves gutter to prevent falling of large masses of snow by slipping off the sloping roof.

SNOW LOAD : A live load due to deposition of snow over a roof is considered in the design of the roof in cold climates.

SNOW PLOUGH : A ploughing machine with a blade towed by a tractor for removing snow off a road in cold countries.

SNOW SAMPLE : A core of snow taken by a sampling pipe by introducing it into the snow depth for determination of snow density.

SNOW SAMPLER : It comprises a set of sampling pipes lightly jointed and a spring balance to read directly the water depth from the measured weight of snow. The balance is graduated accordingly.

SNOW SAMPLING : A sampling device to take snow samples from the snow core with the help of sampling pipes or tubes lightly conneected .

SOAKAWAY : A soak pit filled with brick bats or stone boulders or any other coarse material.

SOAKER : A flexible metal piece used for interlocking tiles in sloped roofing to make water-tight joints, especially at a hip or valley.

SOAP : A term used for a 'queen closer' in brick or stone measony work.

SOCKET : (i)   A groove to receive a pin.

(ii)   An enlarged end of a pipe into which a narrowed end (spigot) of another pipe is fitted.

(iii)   A coupling used in joining two pipes in plumbing.

(iv)   The mortise of a 'mortise and tenon' joint.

SODA-LIME PROCESS : A process for softening hard water by adding soda-ash and lime into water in a reaction basin. By this process both temporary and permanent hardness are removed. The chlorides and sulphates of calcium and magnesium are removed by reacting with lime and soda ash. .

SOFFIT : The under-surface of a roof, vault, stair, arch, etc. or the upper-surface inside a drain, sewer, duct or culvert.

SOFFIT BOARD : A board fixed underside of rafters to form the soffit of an overhanging eave in a sloped roof.

SOFT BURNT : Bricks or tiles are called 'soft burnt', when they are burnt at a low temperature.

SOFT CLAY : A clay that can be moulded in hand easily and can be excavated by a spade.

SOFT SOLDER : A solder that melts below red heat.

SOFT WOOD : Sap wood comprising broad annual rings, light in colour and weight and is liable to decay. This wood is found below the cambium layer of a tree.

SOIL : (i) The term 'soil' covers a very wide range of materials. In a broad sense, soils include all earth materials, organic and inorganic, occuring in the zone overlying the rock crust of the earth.

(ii) sewage carrying human excreta.

(iii) A smudge

SOIL BRANCH : A branch of a soil pipe.

SOIL-CEMENT : A mixure of good quality cohesive soil free from harmful ingredients, debris, etc. and cement in the proportion of 1:1 used as a binding/cementing material for mortar and concrete. A soil-cement construction is economical as compared to cement construction.

SOIL CLASSIFICATION: Different soils may be classified

(i) By soil types

(ii) By mineralogical composition

(iii) By origin of soil

(iv) By structure of soil

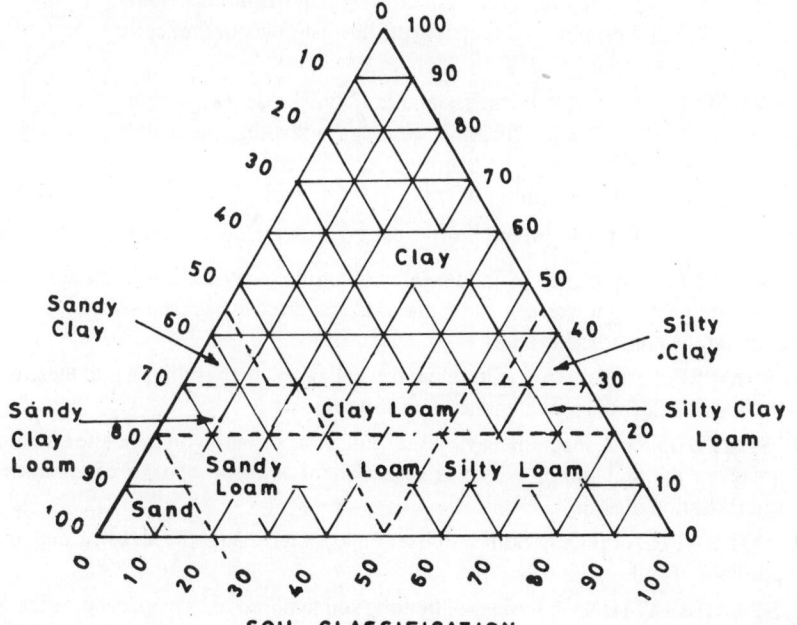

SOIL CLASSIFICATION

    (v)   By grain size

    (vi)   By Atterberg limits

    (vii) By texture of soil (MIT classification) and

    (viii) By triangular classification developed by U.S. Bureau.

A triangular classification chart is shown in illustration, which is widely used in the fields of highway engineering and agriculture. It reveals the grain size distribution in percentages. After the percentages of sand, silt, clay, etc. in a given soil sample are determined, the point on such a triangular chart representing the mechanical composition is located and the soil is given the name as shown.

SOIL CONSOLIDATION: The compression of soil, which is a gradual process of decreasing void ratio and excess hydrostatic pressure and increasing intergranular pressure in the soil. As the compression occurs, the pore water escapes. If the soil under compression has a low co-efficient of permeability, a long time would be required for the consolidation to take place. A honeycomb-structured soil with high porosity is more compressible than a dense structure. A soil of flat grains is more compressible than a soil having spherical grains. The more compressible the soil, longer the time required for consolidation. The more permeable the soil, shorter the time required.

SOIL DRAIN : A foul drain carrying domestic sewage or waste liquid from trade centres. Usually, this drain feeds a sewer.

SOIL MIXER : A mixing machine used to pulverise soil and to mix chemicals for soil stabilisation.

SOIL OFFSET : An offset piece of pipe as shown in illustration is used in soil pipe plumbing.

SOIL PIPE : A pipe used to receive sanitary sewage from individual toilets of a building and to convey it to the nearest sewer or the septic tank.

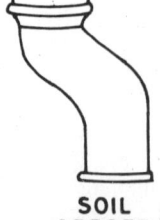

SOIL
OFFSET

SOIL PROFILE :   (i)   The thickness and horizontal extent of various soils, location of ledge and other such data shown on a profile. Also called 'stratigraphical profile'.

              (ii)   A ground profile shown in a vertical section.

SOIL SAMPLE : A specimen of soil taken by a sampling device for testing its properties. For certain tests, undisturbed soil samples are required to be taken out from a specified depth at the construction site.

SOIL SAMPLER : A sampling instrument or appliance which is driven into the ground to the specified depth to take undisturbed soil samples.

SOIL SHREDDER : A machine having two half drums rotating in opposite directions to break up the soil lumps which is essential for intimate mixing of chemicals for stabilitisation of soil.

SOIL SOLIDIFICATION : Soil stabilisation by increasing the density and binding property of soil.

SOIL STABILISATION : A process of treating soil to improve its physical properties. The various methods are :

(i)   Mechanical stabilization by mixing solids of different grain sizes and compacting it by rollers.

(ii)   Cement stabilization by mixing cement intimately with soil and compacting it.

(iii) Lime stabilization similar to cement stabilization.

(iv)   Bituminous stabilization by adding bituminous emulsions to bind the soil grains.

(v)   Chemical stabilization by injecting hygroscopic chemical solutions into the soil.

SOIL SURVEY : Examination of soil at site regarding its quality and strength.

SOIL WATER : Water or moisture content in soil, just below the ground surface, which has agricultural importance. Soil water is classified into three categories. See illustration, which indicates relative positions of soil water classes with equilibrium points.

SOIL WATER DIAGRAM : See illustration. It depicts the nature of water content below a ground surface viz. gravitational water, capillary water and hygroscopic water. It also speaks of maximum water capacity, field capacity, available water, wilting point, hygroscopic co-effecient and zero vapour pressure.

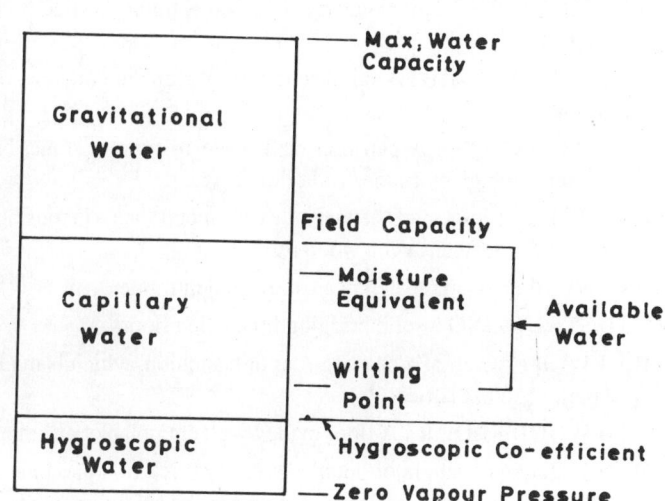

SOIL WATER DIAGRAM.

SOLAR REFLECTING SURFACE : A special finish given to a flat roof surface either by paint- ing it with an aluminium paint to reflect the sunrays or by coating it with asphalt and stone chips to reduce the heating effect due to sunrays.

SOLDER : A fusible alloy usually of tin and lead used in soldering in electrical engineering. This is a soft solder. A hard solder consists of copper and zinc and sometimes small percentages of tin and lead are also mixed with this.

SOLDERING IRON : A steel rod with a copper bit at one end and a wooden handle at the other. The bit is heated and used for making a soldered joint.

SOLDIER: (i)   An upright timber held by struts in timbering a trench without use of walings.

(ii)   An upright ground for fixing boards.

(iii)   A brick-on-end which is upright.

SOLDIER ARCH: A flat arch.

SOLDIER BEAM : A steel beam of I-section driven into the ground to resist the horizontal thrust of an earthen bank.

SOLDIER COURSE : A course of brickwork with bricks-on-end used in coping.

SOLE : The under surface of a plane or base of any engineering component.

SOLE PLATE : A base plate or mud sill or sole piece.

SOLID BRIDGING : Strutting wooden floor joists by short lengths of joists fitted tightly between the floor joists and at right angles to them.

SOLID DOOR : A single leaf flush door with a solid core. This door is sometimes plated with metal sheets.

SOLID FLOOR : A solid floor slab made of concrete or timber.

SOLID MAP : A geological map depicting the bedrock formations of an area.

SOLID MOULDING : A moulding made on a timber article by engraving and not by planting.

SOLID NEWEL STAIR : A spiral stair built of stone or concrete, the inner ends of which form a cylinder.

SOLID PARTITION : A partition wall made of timber, stone, brick-work or concrete without having any framed or hollow core.

SOLID ROLL : A joint made between flexible metal sheets in roofing by folding and rolling their ends together over a wood roll.

SOLID STOP : A stop rebated in a solid wooden frame.

SOLID STRUTTING : Solid bridging in wooden floors.

SOLID WEB : A web of a beam, girder or stanchion, which is made of a section (usually a plate) and not lattice.

SOLID WOOD FLOOR : A floor made by placing planks on edge.

SOLING : Laying a single or double layers of brick or stone in a foundation trench over which concrete is cast. The gaps between the bricks or stones are filled with sand.

SOLUBLE DRIERS : Driers which are soluble in hydrocarbon solvents or in drying oils at ordinary temperatures for use in preparation of a paint.

SOLUTION INJECTION   : Injection of chemical solutions into a soil mass for stabilisation of the soil.

SOLVENT : A liquid that dissolves other matters. Water is nature's universal solvent. Acetone dissolves solids. Volatile liquids dissolve binders in a paint, which evaporate during drying of a paint film.

SONIC PILE DRIVER : A vibrating pile driver used in pile driving without hammering.

SOOT DOOR : A clean-out or ash pit door provided at the base of a chimney for removal of soot.

SOOT POCKET : An extended part of a chimney fitted with a soot door below the level of smoke inlet.

SOPWITH STAFF : A telescopic self-reading levelling staff used in surveying.

SOREL'S CEMENT : Magnesia cement, the strongest hydraulic cement first made by 'Sorel', a French scientist in 1853 A.D. This cement is a mixture of magnesium oxide and magnesium chloride. Its production cost is very high. This cement is suitable for works in sea water.

SOUND BOARDING : Horizontal timber boards fitted closely over joists, carry slag wool, fibrous materials or coarse stuff for the purpose of sound insulation.

SOUNDING : A technique used to determine the depth of river or sea bed. It is also used in measuring depth of a bedrock.

SOUNDING LEAD : A lead weight attached to a lead line used in hydrography. Also known as 'hand lead'.

SOUNDING LINE : A lead line to which a lead weight is attached for use in hydrographic survey.

SOUND KNOT : A solid knot held tightly with the surrounding timber and having the same strength as that of the timber.

SOUNDNESS OF CEMENT : Various methods viz. Pat test, Accelerated test, Expansion test, etc are used for testing soundness of cement. Of all tests, Le Chatelier's test is acceptable quantitative method for determining expansion of cement. The apparatus consists of a split cylinder with long indicators, two glass plates, cold water tray, boiler and weights.

SOUND REDUCTION FACTOR : A value expressed in decibels indicating 'acoustical reduction factor' to be used for a wall. It gives a measure of reduction in intensity of sound through a wall.

SOUTHING : In surveying, a distance measured in southward direction from an east-west axis.

SOUTH-LIGHT ROOF : A north-light roof built at southern hemisphere such that its glazed slope faces south.

SOYA GLUE : A vegetable glue resembling 'Casein', obtained from soya bean after extraction of soya oil.

SPACE : A distance, gap or area.

SPACED SLATING : Open slating.

SPACE FRAME : A three-dimensional frame, which is stable against wind and other loads, without having any support from any other adjacent structure.

SPACE HEATING : Heating a space or area in a building.

SPACE LATTICE : A space frame built of lattice structures.

SPACING : The centre to centre distance.

SPAD : A surveyor's nail 'Spud'. The plumb bob is hung from the spad driven into a frame.

SPAD

SPADER : A grafting tool (narrow spade) used in digging hard soil.

SPALL : A gallet or stone flake used to fill spaces in stone pitching/revetment or in a rubble wall.

SPALL DRAIN: A blind drain in form of a trench filled with stones to allow drainage of water through them ensuring safety of the structure.

SPAN : The distance between supports of a beam, girder, slab, bridge, truss, arch, etc. The clear distance between supports is called 'clear span'.

SPANDREL : (i)   The triangular space between an abutment, extrados of an arch and the top surface in line with the crown of the arch.

(ii)   The triangular infilling made under the outer string of a stair.

SPANDREL STEP : A triangular shaped step made by carving a solid stone as shown in illustration.

SPANDREL WALL : A wall built on the extrados of an arch.

SPANISH MODULE : A rigid outlet module to maintain a constant discharge by a float device operated by a needle valve. When the water level increases in the distributary, the area of flow gets reduced by the needle valve and thus a constant discsharge is maintained.

SPANDRIL STEPS

SPANISH TILE : A form of roofing tiles having shape like a half cylinder with one end being wider than the other. The under tiles are slightly larger than the over tiles.

Guide

Channel

Float

Plug

SPANISH MODULE

SPAN PIECE : A collar beam in a truss.

SPAN ROOF : The common pitched roof.

SPAR : (i)   A common rafter used in a roof truss.

(ii)   A brotch pointed at both ends are driven into the thatch over runners or scallops, used in a thatched roof.

SPARGED TURBINE : An axial flow turbine as shown in illustration is used for aeration of sewage.

SPARGE PIPE : A perforated pipe used in flushing an urinal.

SPARROW PECK : (i)   A texture made in a plaster-work by pitting it with a stiff broom, while the plaster is green.

(ii)   A stone surface obtained by picking.

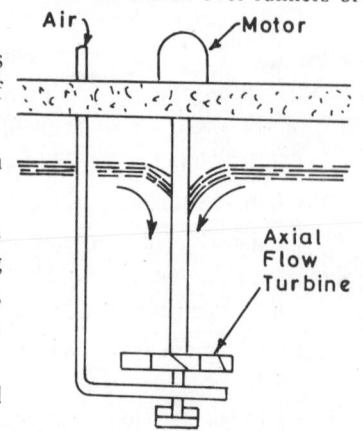

Air

Motor

Axial
Flow
Turbine

SPARGED TURBINE

SPAR VARNISH : A varnish consisting of tung oil or linseed oil and weather-resisting resins applied to boats.

SPATTER DASH : A thin but rich mix of cement and sand (1 : 1.5 or 1 : 2) applied on a smooth masonry or concrete surface by throwing it so that it provides a key on setting and hardening and forms the base for regular plastering.

SPECIALS : Pipe specials are bend, tee, reducer, etc. Special bricks are bullnose brick, plinth brick, cant brick, coping brick, corbel brick, cornice brick, mullion brick, squint quoin brick, etc.

SPECIFIC ADHESION : The chemical bond between two cemented or glued substances.

SPECIFICATION : A detailed description of a material , a machine, a mechanical or electrical unit, or an item of work speaking of activities and procedure of work.

SPECIFIC CAPACITY OF WELL : Used to indicate the yield of a well in cubic metre per metre of draw-down. The yield of a well varies directly as drawdown.

SPECIFIC GRAVITY : The weight per unit volume of a substance compared with the weight of same volume of water at a specific temperature. Thus, it is a pure number.

SPECIFIC RETENTION : The ratio of volume/weight of water retained by a soil against the gravitational force to its self volume/weight.

SPECIFIC SPEED : (i) The speed of a pump in rpm., at which an impeller of identical geometrical shape and suitable diameter will discharge 1 gpm. at 1 foot of head. It has bearing on the design of an impeller. In case of a single–stage centrifugal pump, it is given by

$$N_S = \frac{N(Q)^{0.5}}{H^{0.75}}$$

Where N_s is specific speed (rpm),

Q is discharge in gpm,

N is actual speed of impeller (rpm) and

H is head in feet.

(ii) The speed in rpm at which a geometrically similar runner of a water turbine of suitable diameter will develop 1 bhp. under 1 foot of head. The specific speed indicates the performance of a turbine. It is given by

$$N_S = N(\text{bhp})^{0.5} / H^{1.25}$$

Where N = Speed of turbine runner (rpm),

bhp = Brake hourse power and

H = Head in feet.

SPECIFIC SURFACE : It is a measure of fineness of a powder and is expressed in sq. cm/gm. It is the ratio of area covered by a powder to its weight or volume. This is required in determining the fineness modulus of cement and similar such materials.

SPELTER: Zinc with a very small percentage of impurities, which is used for galvanizing.

SPHERICAL TRIANGLE: A triangle which is formed upon the surface of a sphere by the intersection of three great circles as shown in illustration.

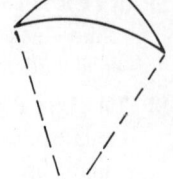

SPHERICAL TRIANGLE

SPIDER LINE : Spider web or cross hair provided in the diaphragm of the telescope of a surveying instrument.

SPIGOT : A pipe end that is introduced into the socket end of another pipe.

SPIGOT AND SOCKET : A pipe joint in which the spigot end of one pipe is fitted into the socket end of another pipe as shown in illustration.

Centre

SPIKE : A nail-like wedge-shaped or conical metal piece with a pointed end to be driven. Its stem is either threaded or plain with a head.

SPIGOT AND SOCKET JOINT

SPIKE KNOT : A splay knot in timber.

SPILE : (i)   A board with sharp edges.

  (ii)   A timber pile .

SPILING : Driving forepoles (spills).

SPILL : A forepole.

SPILLWAY : Waste weir or wasteway or an overflow weir over a dam to allow passing of excess water.

SPILLWAY GATE : A crest gate provided over a dam to control the flow of excess water.

SPINDLE : A short length of shaft or rod or a small axle used in valves, gates, paddle mixer, etc.

SPILL – THROUGH ABUTMENT

SPIRACTOR : An inverted conical tank, one half of its depth being filled with marble chips. The hard water with chemicals is forced under pressure in a tangential direction. The softeining of water takes place by the spiral motion of water, the chips being kept in suspension. The detention period in the tank is 10 minutes only. It combines flocculation, settling and sludge removal in one unit

SPIRAGESTER : A modified form of 'Imhoff tank' applicable to moderate size installations is shown in illustration. It has three compartments viz. flow-through chamber, digestion chamber and scum chamber. The sedimentation and digestion of sludge takes place in separate compartments in the same tank.

SPIRAL CURVE : A curve constructed upon a series of equal chords and the angle subtended by the first chord is made the common difference for the angles subtended by the succeding chords. It is a multi-compound curve with degree of curve progressing in arithmatical ratio from one chord to the other. Various spiral curves are shown.

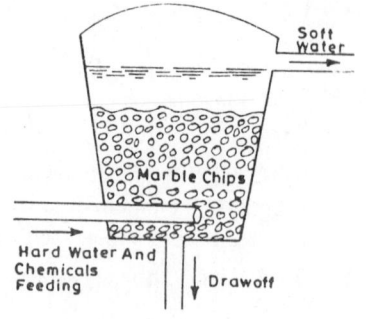

Soft Water

Marble Chips

Hard Water And Chemicals Feeding

Drawoff

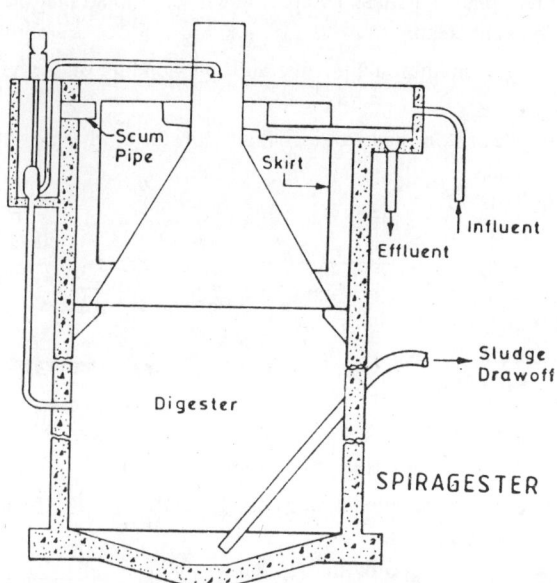

SPIRAGESTER

HYPERBOLIC SPIRAL

ARCHIMEDIAN SPIRAL

SPIRAL GRAIN : Fibres grown spirally round a tree trunk. It is difficult to work with such a timber.

SPIRIL STAIR : A circular stair in which all steps are winders. The inner ends of the steps form a continuous cylindrical solid newel. Such a stair is usually made of stone, timber, cast iron and reinforced concrete. This stair can be housed in a small space. Actually, it forms a helix and not a spiral. See illustration.

SPIRITING OFF : Finishing operation in French polishing, which involves removal of oil traces if any left on surface, by drawing quickly a cloth damped with methylated spirit.

SPIRIT LEVEL : A barrel-shaped level tube made of glass containing a liquid such as alcohol, chloroform or petroleum ether forming a bubble being occupied by air and spirit vapour. The central position of the bubble indicates the horizontal.

SPIRIT STAIN : Shellac and dye are dissolved in methylated spirit and is used for darkening the surface of a timber.

SPIRIT VARNISH : A varnish made by dissolving soft resins in methylated spirit, amyl acetate, wood naptha, acetone, ether or benzol. The drying of spirit varnishes occurs due to volatilisation of the solvent leaving resin on the surface. It does not stand weather.

SPLASH BOARD: (i)   Wedge-shaped weather moulding planted at the foot of the shutters of an outside door to splash out rain water.

LOGARITHMIC SPIRAL

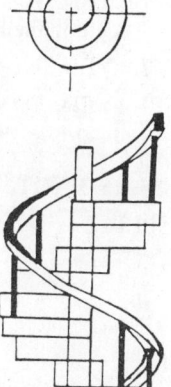

SPIRAL STAIR

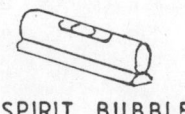

SPIRIT BUBBLE

(ii)   A board placed against a wall beside a scaffold so that the wall can be kept clean.

SPLASH LAP : The part of the overcloak of a sheet metal drip extending on to the next sheet on flat.

SPLASH PAD : A metal or concrete or stone pad provided just below the discharging mouth or pipe or slot so as to splash the fluid for its well spreading. This type of pad is found in distributing arm of a trickling filter, at sludge discharge point in a sludge drying bed and at point of inflow in an oxidation pond.

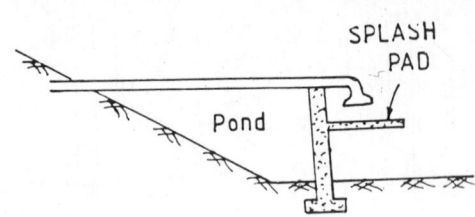

SPLAT : A cover strip over the wallboard joints.

SPLAY : A bend or a slope usually at 45°across the full width of a surface.

SPLAY BRICK : A cant brick which is bevelled at one end. This is a type of special brick used at plinth level.

SPLAY BRICK

SPLAYED COPING : A coping with feather edge (wedge-shaped) sloping out-wards.

SPLAYED GROUND : A base acting as a screed with a bevelled edge to provide a key for the plaster.

SPLAYED COPING

SPLAYED HEADING JOINT : A joint made between floor boards by cutting their edges or ends at 45° and by over-lapping each other.

SPLAYED SKIRTING : A skirting having its top edge bevelled.

SPLAYED, TONGUED & GROOVED JOINT : A joint between floor boards as shown in illustration.

SPLAYED HEADING JOINT

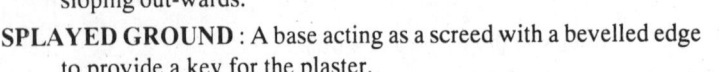

SPLAY KNOT : A knot cut parallel to its length.

SPLICE : (i)   A joint between two halved timber, which is made by covering them with wood or steel straps and bolting them together.

SPLAYED,TONGUED & GROOVED

(ii)   Increasing length of a short R.C.C. pile by exposing the reinforcement bars from the pile head and casting concrete after lapping additional bars of desired length. This is hardly practised. The concrete should be made with high-alumina cement.

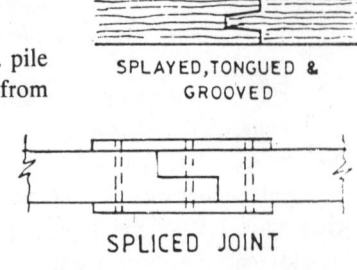

SPLICED JOINT

(iii)   A joint between two columns or stanchions which is made by fish plates as shown in illustration. The plates are used on the flanges and webs of the columns or stanchions. For unequal sections, additional plates are required.

(iv)   A feather.

(v)   A wood strip nailed down a timber sheet pile onto which the next pile fits.

(vi)   A strip of wood nailed to a window frame for glazing.

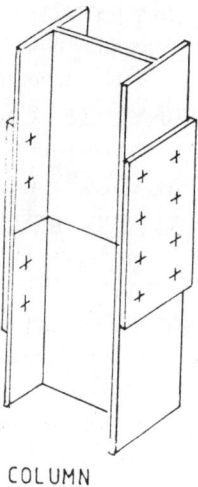

COLUMN
SPLICING

SPLICE BAR : A fish plate or rod.

SPLIT COURSE : A layer of bricks, cut length wise for reducing the thickness of a partition wall.

SPLIT HEAD : A tripod with an upright standard (notched at top) placed at four corners to carry two scaffold boards on edge to form a platform for workers.

SPLIT PIPE : A pipe cut lengthwise to form a channel.

SPLIT RING CONNECTOR : A form of timber connector, required in connecting two timbers by inserting it into the ring-shaped groove made in both the timber pieces.

SPLIT SHAKES : Wood shingles made by splitting and not by sawing the wood.

SPLIT SYSTEM OF HEATING : Heating a room by blowing hot air and using radiator simultaneously.

SPLOCKET : A sprocket or a cocking piece.

SPOIL : Wastes.

SPOIL BANK : Deposition of excavated earth in form of a bank.

SPOOL : A distance piece placed between two timbers.

SPOON BIT : A dowel bit used by carpenters.

SPOT BOARD : A gauge board of 1 metre square used in plastering.

SPOT GLUING : Gluing timber in separate small spots.

SPOT LEVEL : The elevation or depression of spots.

SPOTTING : Laying or dumping materials at regular intervals in a stretch along a road, railway embankment, canal, etc.

SPOTTING IN : Spot finishing of defective patches by rubbing down and coating.

SPOUT DELIVERY PUMP :A pump similar to a diaphragm pump, which delivers water through its spout with no higher head.

SPRAY BAR : A pipe fitted with jets for spraying binder onto a road.

SPRAYED ASBESTOS : Adhesive asbestos blown on to a surface to give a fine coating of asbestos, which makes it vermin-proof, rot-proof and incombustible.

SPRAYED CONCRETE : Thin slurry of concrete sprayed by guniting with the help of a compressor pump for filling up cracks and crevices in a construction.

SPRAYER : A spraying gun with a pressure tank.

SPRAY GUN : A gun with a nozzle for ejecting a fine mist of powder or paint operated by compressed air.

SPRAY LANCE: The pipe of a sprayer to be held by hand for spreading binding material onto a road surface.

SPRAY PAINTING: Painting a surface by means of a sprayer, which facilitates uniform painting by ejecting a fine mist of powder on a hot surface or a mist of paint on a surface at normal temperature.

SPRAY UNIT : An unit comprising a compressor, compressed airtank, paint container and a pistol with a jet (spray gun).

SPREADER : A machine used for spreading concrete mix or road metals with binder.

SPREADER BEAM : A beam with ropes or chains hanging from a crane hook, used for the purpose of lifting R.C.C. piles and fragile objects to avoid breakage during lifting and transfering it to a place.

SPREAD FOOTING : A column or wall footing gradually widened towards its base for distribution of load over a large area.

SPREADING BOX : A container having a device of receiving road materials and spreading it uniformly to the required thicknes on a surface.

SPREADING RATE : The covering power of a paint or any other liquid.

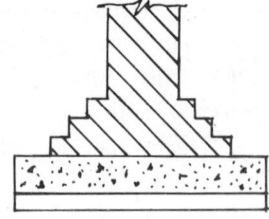

SPREAD FOUNDATION

SPREAD RECORDER : A recording device or instrument to measure the outward spread of a bridge abutment during its testing by aplying load.

SPRIG : A headless small wire nail.

SPRIG BIT : A bradwal.

SPRINGER : The first brick or stone laid on the springing line in an arch, which forms the skew back. This is also called a springing brick or springing stone.

SPRING-HEAD NAIL : A fixing screw for roofing sheets.

SPRINGING LINE : The horizontal line joining the springing points of an arch i.e, the points where from the intrados of an arch starts from either end.

SPRINGINGS : The intersections between the wall surfaces and the intrados of an arch on either side.

SPRINGING POINTS : See 'Springing line'.

SPRING SNIB : A spring-controlled sash fastener.

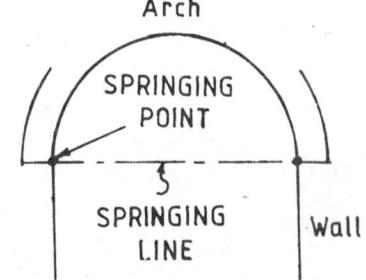

SPRING WASHER :A ring-like open ends washer, which is bent to a helical curve to produce the springing effect. This prevents a nut from unscrewing.

SPRING WOOD : The annual ring formed in a wood during autumn, which is usually pale in colour, soft and light in weight.

SPRINKLER SYSTEM : (i) A system of sprinkling water in a lawn or garden.

(ii) A fire-fighting or extinguishing system which consists of a network of pipes built inside the duct of the ceiling with projected pipes sealed with a metal plug or a plastic plug with a liquid. In case of any fire, the plug gets released by fusing or by bursting automatically when the temperature attained is 60°C. such installations are provided in multistoreyed buildings, offices, hotels, hospitals and auditoriums.

SPROCKET : A cocking piece.

SPROCKETED EAVES : Eaves tilted by sprockets.

SPRUCE : A kind of soft wood grown in North America, which weighs only 450 kg/cu.m. at 15% moisture content.

SPRUNG : A timber that has warped by spring.

SPUD : (i)　A dowel provided at the foot of a door post for fixing it to the floor concrete.

(ii)　A nail-like hook used by surveyors for hanging a plumb bob for marking a survey station.

(iii)　A steel post that can be lowered by a toothed rack under a dredger until it anchors the dredger.

SPUDDING : A technique of enlarging a pile hole by dropping and raising a spud of heavy weight and diameter larger than the pile diameter. This helps in reducing the friction during pile driving.

SPUR : A groyne used for controlling the direction of flow of a river water.

SQUARE : (i)　A measure of area for floors, roofs and walls.

(ii)　Any material having square cross section.

(iii)　A try square or a tool for setting out right angles.

SQUARE CROSSING : A railway crossing as shown in illustration is provided when one track crosses another track at right angles to each other. This type of crossing should by avoided on main lines.

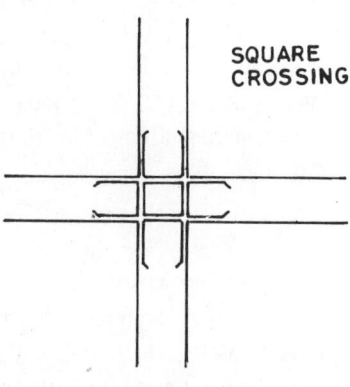

SQUARE
CROSSING

SQUARED LOG : A baulk or half-timber.

SQUARED RUBBLE : Squared stones of varying sizes used for rubble walling.

SQUARE-EDGED TIMBER : A squared timber without wane.

SQUARE MILE-FOOT : A measure of volume of water of 1 foot depth standing over an area of 1 square mile. Similar such measure of volume of water is acre-foot.

SQUARE JOINT : A joint made between two pieces by butting them against each other.

SQUARE ROOF : A sloping roof in which rafters are placed at 45° to the horizontal and thus the rafters from either end meet at right angles at the ridge.

SQUARE-SAWN TIMBER : A timber sawn to a square or rectangular cross section.

SQUARE SHOOT : A wooden down pipe used in early days.

SQUARE STAFF : A wood fillet used as an angle bead at the corner of a room.

SQUARE THREAD: A form of screw thread, square in section, is capable of transmitting thrust in both directions.

SQUARE-TURNED BALUSTER: A baluster of square section with mouldings on its faces.

SQUINT BRICK: A brick cut to a special shape as shown in illustration, used in making a squint junction.

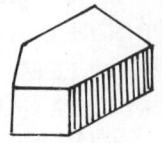

SQUINT BRICK

SUINT JUNCTION : See illustration.

SQUINT QUOIN : A corner of a room or building which is not a right angle.

SQUIRREL CAGE MOTOR : A polyphase induction motor resembling a squirrel cage. Under normal operating conditions for pumping water, it runs almost at a constant speed. This motor is suitable for use in waterworks because of its simple design, ruggedness of construction, trouble-free operation and other facilities.

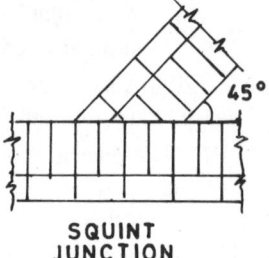

SQUINT JUNCTION

SQUIRREL TAIL PIPE JOINTER : A clip used for jointing pipes.

STABILIZATION POND : A pond used for stablization of organic wastes by utilising natural process of aeration, solar energy and symbiosis. A stablization pond may be aerobic pond or oxidation pond or facultative pond or anaerobic pond.

STABILIZED SOIL : A soil that is made compact and stable by the process of soil stablization.

STABILIZER : A material used for stabilization.

STABLE DOOR : A door shutter cut horizontally at half its height and the shutters are hung separately. Also known as 'Dutch door.

STABLE STRUCTURE : A balanced well-built structure, which is capable of resisting sway, sliding, tilting or overturning.

STACK : (i)   A chimney stack.

(ii)   A soil pipe or a rainwater down pipe.

STADIA HAIRS : Two horizontal lines in the reticule of a telescope of a theodolite, at equal distance apart above and below the line of sight.

STADIA ROD : A levelling staff of special type having graduations for stadia work.

STADIA WORK : The use of stadia hairs of the theodolite telescope and the stadia rods for determination of horizontal distances and altitudes of visible points by tacheometric surveying. It facilitates in quick mapping when used with plane table. Whatever may be the position of the visible points, the stadia rod is always held vertical with the help of a plumb bob. The apparent horizontal distance is given by

h^1 = 100 x staff intercept + additive constant.

The true horizontal distance is given by

$h = h^1 \cos^2 \alpha$, where α is the angle above or below the horizontal.

STAFF : Usually a wooden bar of rectangular section with graduations painted on it.

STAFF BEAD : A guard bead or an angle bead.

STAFF GAUGE : A graduated wooden bar or metal plate for indication of water level in a reservoir. Sometimes, graduations are made on a bridge pier or abutment to record the water level in a flowing stream or river throughout the year.

STAFFMAN : The man who carries and holds a staff in position.

STAFFORDSHIRE BLUE : A deep blue brick of Britain, which is very hard and dense having a crushing strength of about 16000 psi.

STAGE : (i) A deck or platform

(ii) A phase

(iii) A water level with reference to a line.

STAGGER : To array alternately the placement of bolts and rivets or decorative pieces of timber, stone or brick in any ornamental work.

STAGGERED COURSES : Courses of bricks, stones or tiles laid in staggered position.

STAGGERED JOINT : A joint used in railway tracks as shown in illustration.

STAGING : A scaffold for masons.

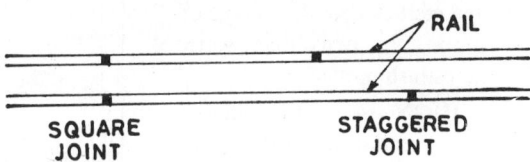

STAGING HEIGHT : The height of platform over which an elevated reservoir is built i.e. the base of the elevated reservoir.

STAIN : A solution of colouring pigment or dye in a vehicle, which penetrates a surface and gives the colouring effect. True stains are oil or spirit stains.

STAINERS : Coloured pigments ground in a vehicle, which have intense staining power and therefore sometimes a small amount of a stainer is added to a ready-mixed paint.

STAINING POWER : The amount of colour attributed by a certain quantum of coloured pigment to a white pigment.

STAINLESS STEEL : It is produced by adding over 12% of chromium to low carbon steel. In general, this alloy contains 18% chromium and 8% Nickel and is used for cutlery and surgical instruments. It is unaffected by acids, alkalies, salts and weathering actions.

STAIR : A series of steps with balustrades and handrail for safety and with or without landings from one floor to the other. Usually, a landing is provided in between two flights. See illustration.

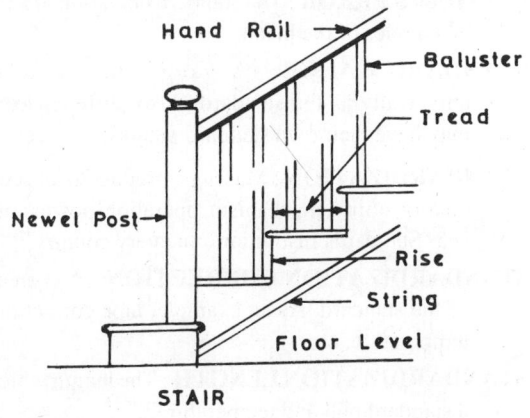

STAIRCASE : The space within which a stair is housed.

STAIR CLIP : A clip that holds a stair carpet in proper position.

STAIR HEAD : The top of a stair.

STAIR HORSE : A carriage for a stair.

STAIRWAY : A staircase.

STAIRWELL : A well kept in a building for provision of stairs.

STALK : The stem of a R.C.C. retaining wall.

STAKE : A peg or a pointed piece of timber for driving into the ground.

STALLBOARD : A window sill with its frame over the stall riser.

STALLBOARD LIGHT : A pavement light close to a stallboard.

STALL RISER : The polished vertical surface from the pavement upto the stall board.

STALL URINAL : A public urinal for a community as shown in illustration. Similar such urinal is provided in public places, auditoriums, hospitals, offices and market areas. Such type of urinal is provided with auto-flushing system.

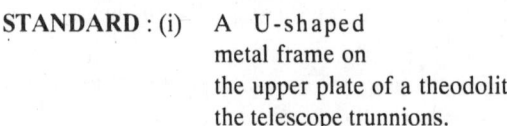

STANCHION : A steel column of heavy section, usually built-up as shown in illustration.

STALL URINAL

STANDARD : (i) A U-shaped metal frame on the upper plate of a theodolite to cary the telescope trunnions.

(ii) A vertical member of a scaffold.

(iii) A common measure accepted by all in a country or accepted universally.

STANDARD DEVIATION : The square root of the variance which gives a measure of spread of observations. It is given by $n=(a)^{0.5}$, where 'n' is the standard deviation and 'a' is the average of the squares of the deviations of all observations.

STANDARD ERROR : The standard deviation of a number of samples of the mean.

STANDARD GAUGE : A gauge that is accepted universally as the standard. Also, different countries may have their own standard gauges.

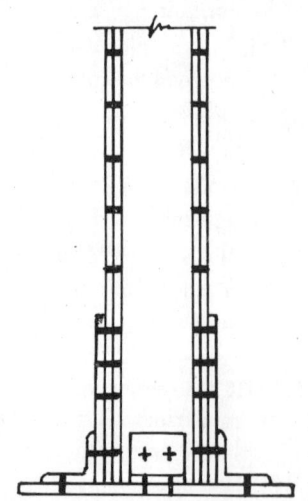

STANCHION

STANDARDIZATION : Making a product to an acceptable mark as regard its dimension, quality, utility, mechanism, operation, power consumption, etc. For this purpose, there is a 'Standards Institution' in every country.

STANDARDIZATION CORRECTION : A correction needed for bringing to the level of the standard. As an example, tape correction is needed in a surveywork for pull, temperature, etc.

STANDARDIZATION LENGTH : The length which is taken as the standard length under a standard pull and temperature.

STANDARDIZATION TEMPERATURE : The standard temperature at which dimensions and properties are standaridized.

STANDARD KNOT : A knot of 40mm diameter or less in a timber.

STANDARD PILE : A guide pile used in pile driving.

STANDARD SECTION : Rolled steel sections of standard dimensions (shape & size) and weights.

STANDARD SPECIALS : Standard pipe fittings.

STANDARD SPECIFICATIONS : Descriptions of standard materials and workmanship.

STANDARD WIRE GAUGE : Stanardized number given to specify the diameter of nails & wires and thickness of sheet metals.

STANDING DERRICK : A guyed mast or a pole held nearly in vertical position by means of guy ropes. A hoisting rope passes over a pulley fitted at the top of the mast for raising loads.

STANDING LADDER : A ladder having rectangular stiles.

STANDING LEAF : The leaf of a folding shutter which is bolted when closed.

STANDING PIER : A pier with spans on either side.

STANDING SEAM : A seam in sheet metal (flexible) roofing, starting from ridge to eaves.

STANDING TIMBER : A living tree.

STANDING URINAL : A bowl type or stall type urinal for use by gents. The bowl type should be set at a height not more than 20" (500 mm) above the platform level. The width of stall should not be less than 24" (600 mm) and it should be supported by cast iron brackets fixed at about 3" (75 mm) above the platform. Arrangement must be made for watering and drainage of waste water.

STANDING WATER LEVEL : The level of ground water remaining unchanged in an open well or pit.

STANDING WAVE : See illustration. This is formed at the downstream foot of the weir, when there is insufficient depth available at the downstream to form a hydraulic jump. No impact or energy loss is encountered in the formation of standing waves.

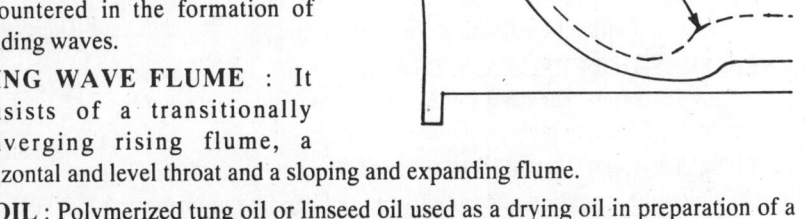

STANDING WAVE FLUME : It consists of a transitionally converging rising flume, a horizontal and level throat and a sloping and expanding flume.

STAND OIL : Polymerized tung oil or linseed oil used as a drying oil in preparation of a paint. The polymerization is done by heat treatment. In earlier days, polymerization of the oil used to be done by standing it in sun for a period of time and as such it was named so.

STAND PIPE: An elevated water column in a cylindrical shell built of steel or reinforced concrete, used to develop distributing pressures. Standpipes are usually located on a hill top. The volume of the tank is determined on the basis of the required capacity of storage. See illustration.

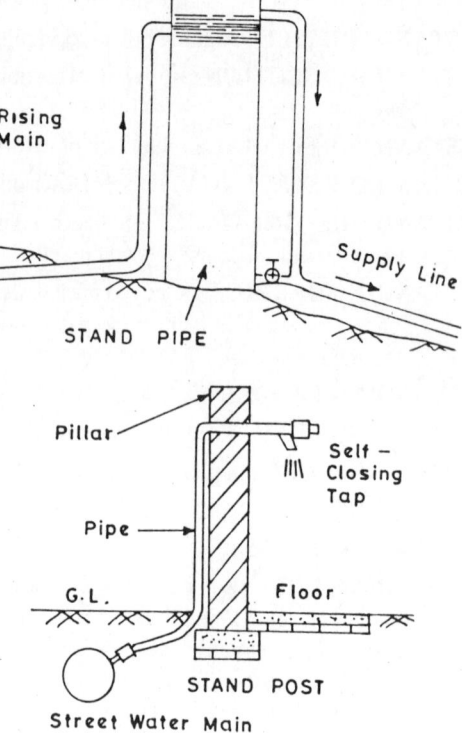

STAND POST : A short column with a platform and carrying a water tap by the side of a road. This is a street tap provided for water supply to a small community having economically weaker section of people. For every 100 persons, one such tap should be provided in a rural area or in city slums when the water supply is intermittent and the supply norm is 50 to 70 1pcd. See illustration.

STAND SHEET : See 'Dead light'.

STANK : A small coffer dam of timber made water-tight with puddled clay.

STAPLE : (i) A U-shaped spike with two points.

(ii) A U-shaped metal loop used for locking door.

STAR DRILL : A plugging chisel of star shape.

STARLING : Timber piles driven into the river bed upstream of a bridge pier to guard it from floating matters.

STAR SHAKE : A defect in timber showing several fine cracks or splits radiating from the pith or from the bark. See illustration.

STARTER BAR : A bar embedded on either side in a construction joint, which knits the adjoining concrete masses.

STATICALLY-DETERMINATE FRAME : A structural frame in which reactions, shearing forces and bending moments can be found out by using laws of statics.

STAR SHAKE

STATICALLY-INDETERMINATE FRAME : A hyperstatic frame in which the reactions, shearing forces and bending moments can not be determined simply by using laws of statics. Additional equations making use of strain energy or deflection are called for calculating forces in each and individual member.

STATIC HEAD : The difference of height between two points or the difference in two water levels in static condition.

STATIC LOAD : The self load of a structure or any load simply kept on a structure without any increment in weight or any movement.

STATIC MOMENT : The first moment of an area about an axis.

STATIC PENETRATION TEST : Soil tests like cone penetration test, penetrometer and plate bearing test, in which the testing device is gradually pushed into the soil by means of a measurable force.

STATION : (i)   A point of reference.

         (ii)   A starting, halting or finishing point in a railway track.

         (iii) A point at which an instrument is set in a surveywork.

         (iv) A measuring point.

STATIONARY DREDGER : A bucket-ladder dredger which discharges the dredged material into a barge or a pipeline for transportation.

STATION ROOF : A roof supported on columns in a line. Also known as 'Umbrella roof'. It is cantilevered to one or both sides.

STAUNCHING BEAD : A vertical gap left between successive bays of concrete during concreting a dam, which is concreted afterwards when shrinkage has taken place in adjoining concrete bays.

STAUNCHING ROD : A rubber rod which gets compressed between a crest gate and the structure to form a water-tight joint.

STAUNCHING WALL : A wall similar to a curtain wall is built to prevent leakage at the junction of earthwork and masonry walls. Actually, it increases the traversing length of creeping water.

STAVE : A ladder rung.

STAY : A guy rope, tie rod or a diagonal bar used to prevent movement.

STAY BAR : (i)   A bar holding together two opposite walls of a building.

         (ii)   A horizontal bar strengthening a mullion.

STAY LOG : A log of timber fitted to a veneering machine for cutting half-round veneers.

STAY PILE : A pile driven or cast into ground which serves as an anchor for holding back a sheet pile wall.

STEADY FLOW : A streamline or laminar flow expected in an open channel.

STEAM CURING : Curing of precast concrete products in a steam bath under pressure for hastening the maturation.

STEAM ENGINE : A reciprocating piston-driven engine worked by steam pressure.

STEAM HAMMER : A pile hammer operated by steam pressure.

STEAM ROLLER : A road roller driven by a steam engine which is now obsolete. Now-a-days, diesel engines are used.

STEAM SHOVEL : An excavator of early days driven by a coal-fired steam engine.

STEAM TURBINE : A turbine operated by the force of high-pressure steam onto the vanes for generation of electrical power.

STEEL : It is manufactured mostly by removing a portion of carbon content from pig iron and refinig it. Wrought iron can be converted into steel by adding the required quantum of carbon to it. Steel is extensively used in civil engineering constructions and in manufacturing mechanical and electrical instruments, appliances and equipment.

STEEL BAND : A band chain made of steel, used in surveywork.

STEEL BENDING SCHEDULE: A bar bending schedule for reinforcement in concrete.

STEEL CASEMENT : A window casement made of steel.

STEEL CORE : A steel member encased in other material.

STEEL DOOR : A door shutter made of sheet steel .

STEEL ERECTOR : A skilled worker with specilisation in erection of steel structures.

STEEL FRAME : A framework of steel sections.

STEEL GRIT BLASTING : A type of shot blasting.

STEEL LATHING : Metal lathing for plasterwork by using steel wire mesh or expanded metal.

STEEL PILE : Steel pipe piles filled with concrete and steel H-piles are used, when extremely long piles with high bearing capacity are required for the foundation of a bridge or a tall building.

STEEL RING : A pressed steel ring used in lining a circular section of a tunnel.

STEEL SHEET : Sheets of steel used in various fabrication works.

STEEL SHEET PILE : Sheet piles made of rolied steel sections having interlocking arrangement. On completion of sheet piling, excavation can be done safely, because sheet piles keep out the flow of soil.

STEEL SLEEPER : A railway sleeper made of steel as shown in illustration. It is made from 6mm thick steel plate, pressed in form of an inverted trough with folded ends. The rail seat on either side is given an inward slope of 1 in 20 for the required tilt of rails.

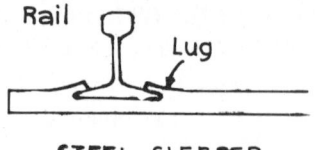

STEEL SLEEPER

STEEL SQUARE : A graduated L-shaped steel plate with the legs at right angles to each other. It is like a carpenter's square.

STEEL TAPE : A measuring tape made of steel, used in survey work.

STEENING : Lining of a well with bricks, stones or earthen rings laid dry. Also known as 'Steining'.

STEEPLEJACK : A tradesman who has skill in building and repairing long tower, chimney, lightning conductor, weather cocks, etc.

STELLITE : An alloy containing chiefly chromium, cobalt and tungsten with traces of carbon, silicon, manganese and iron. This alloy is harder than steel.

STEP : One unit in a stair, comprising a tread and a riser.

STEP FLASHING : A roofing felt or flexible sheet metal introduced into the joints of bricks between the wall and the sloping edge of the roof to make the joints leak-proof. This is also used at the junction of a sloping roof with a chimney.

STEP IRON : A heavy type malleable cast iron U-shaped staple, whose two ends are built into concrete or masonry wall of a manhole chamber, an inspection chamber, a sump or reservoir, at a vertical spacing of 225 mm to 300 mm (9" to12") to facilitate workmen to go down or to climb up.

STEP JOINT : A birdsmouth joint between a tie beam and a rafter in a wooden truss, which is formed by notching the tie beam to make a birdsmouth for receiving the rafter end.

STEP LADDER : A wooden ladder having stiles and treads and no rung.

STEPPED FOUNDATION : A benched foundation as shown in illustration is required, when the wall load is to be distributed over a large area. The width of foundation is increassed in layers with the increase in the depth of foundation.

STEPPED SKIRTING : An asphalt skirting provided at an inclined intersection.

STERIOMETRIC MAP : A map clearly depicting valleys and ridges of a hilly area.

STEPPED FOOTING

STERIOSCOPIC VISION : The three-dimensional vision as viewed by a person, in which the distance can be estimated. The vertical photographs can be seen sterioscopically by means of a plotting instrument.

STEVENSON'S FORMULA : A formula developed by stevenson to determine the height of waves formed due to travelling of gales over the water surface through a distance in nautical miles. It is expressed as

$$h = 1.5 \, (s)^{0.5},$$

where h=height of wave in feet, and s=distance in nautical miles.

STICK AND RAG WORK : Fibrous plaster work.

STICKER MACHINE : A machine that cuts and shapes mouldings out of solid wood.

STICKING : Shaping of a moulding with the help of a machine or manually by a plane.

STICKING BOARD : A board with a frame to hold wood blocks for making mouldings with the help of a plane.

STIFFENER : A member added transversely to a slender beam to prevent buckling. In a steel built-up girder, angle or tee sections are either riveted or welded to the web and flange of the girder.

STIFFENING GIRDER : In a suspension bridge, when the estimated deflection is found to be greater than one-three hundredth of the span, it is usually strengthened by providing a stiffening girder for uniform distribution of loads among the suspenders.

STIFF-FISSURED CLAY : Slickensided clay i.e, a clay which is stiff when dry, but contains a network of fissures which makes it porous.

STIFF FRAME : A perfect and rigid frame.

STIFFNESS : The resistance of a structural member to its bending, deflection or buckling. It is directly proportional to the modulus of elasticity of the material, second moment of area of the section and is inversely proportional to the length of the member.

S-TILE : A pan tile having a sharp curvature.

STILE : An upright member at the two ends of a door or window shutter, which is mortised to receive the rails and the panels.

STILLING POOL : A water cushion made at the downstream foot of a weir or a spillway by deepening the bed of the river.

STILLING WELL : A tank having communication with the main source of water through an inlet.

STILL WATER NAVIGATION : A navigation in slack-water.

STIPPLER :   (i)   A bristle used to break up the colour or texture of a coat .

(ii)   A brush of soft bristles in a flat stock used gently on a freshly-painted surface to remove brush mark and unevenness in painting.

STIRRUP: A binder of small diameter steel bar (usually 6mm.) used to hold the reinforcement steel in position. This also takes diagonal tension in a beam.

STIRRUP STRAP : A steel strap used for holding purpose.

STOCK : (i)   The body or handle of a tool, e.g. the wooden part of a plane.

(ii)   A tool to hold a die for cutting threads on a pipe.

(iii) A converted timber.

STOCK BRICK : A type of brick which is readily available and commonly used in a particular place.

STOCK BRUSH : A large brush with bristles arranged in a cylindrical form, used for wetting a wall prior to plastering or distempering.

STOCK LUMBER : A form of sawn and sized timber available in market.

STOCK RAIL : The outer rail fixed at a turnout against which the movable rail point is held.

STOKE'S LAW : A classic law of sedimentation of particles in a liquid formulated by stokes. The terminal velocity of a particle is given by

$$V_c = [g (P_s—P) d^2]/18 m$$

where g is acceleration due to gravity,

d = diameter of particle,

P_s = density of particle,

P = density of fluid and

m = viscosity of fluid.

For laminar flow conditions, stokes found the drag force Fd to be 9.42d mv, where v is the particle velocity. The settling of discrete, non-flocculating particles can be analysed by this law for design of sedimentation basins.

STONES : Derived from rocks composed of mineral earths like silica, alumina, lime, magnesia, etc. held by cementing materials. There are various types of stones varying in composition, colour, density and strength.

STONE BLOCK PAVEMENT : A pavement made by setting blocks of stones cut to rectangular shape and more or less of uniform size.

STONE BOAT : A boat-shaped tray made of wood and steel and set on sledge runners which is used to transport stones through a short distance over a road by skidding.

STONE DRAIN : A drain made of rubbles.

STONE LIME : A pure lime obtained by calcination of limestones (carbonate of lime), which is used in brickwork, flooring and lime terracing. This lime has good hydraulic property.

STONE SAW : A saw having a long reciprocating blade with no tooth, used for cutting stones.

STONE SLATE : A thin slab of stone used in flooring or roofing (sloped roofs).

STONE TOOL : A percussive tool used for dressing or carving stones manually by hand.

STONEWARE : Made usually from refractory clays or clays mixed with ground glass, stone dusts, ground potteries, silica, etc. Stonewares are burnt at a high temperature. These are usually hard, impervious, and close-grained resembling fire bricks.

STONEWARE ARTICLES : Acid jars, bombomes, paving tiles, sanitary fittings & fixtures, drain pipes, etc. are usually salt-glazed.

STONEWARE CLAY : It is composed of silica, alumina, iron oxides, magnesia, lime, soda and potash with water.

STONEWARE PIPE : Pipes made of salt-glazed stoneware are used in making a drain or a branch sewer. The diameter of these pipes range from 150 mm to 450 mm.

STONEY ROLLERS GATE : A common type of vertical lift gates. A roller cradle is hung at each end of the gate by means of chains fixed to gate bridge. The gate moves freely up and down with the cradle rollers which are in contact with the vertical sides of the gate.

STONING : Pressing a carborundum stone to a running circular saw, at right angles to the axis of saw.

STOOLING : Preparation of the upper surface of a concrete or stone sill to form the bed for masonry work over it.

STOP : (i)   A moulding planted (fixed) on a door frame, against which the shutter closes.

    (ii)   A bench stop.

    (iii) A door stopper.

    (iv)   A finishing touch to a stuck moulding.

STOP BEAD : A guard bead.

STOP COCK : A valve or cock provided in the gas or water supply pipe in an individual house for opening or closing the line completely and not for controlling the flow quantum. It is usually kept under the jurisdiction of the supply corporation.

STOP END : A stunt end of a moulding.

STOP LOGS : Precast R.C.C. beams, steel joists, wooden planks or baulks that fit between the vertical grooves made in walls, to close the flow through a channel or a spillway.

STOP MOULDING ; A planted moulding which ends at a stop and is not continuous upto the end of a member.

STOPPED CHAMFER : A chamfer which gradually merges into a sharp edge.

STOPPED MORTISE : A blind mortise.

STOPPER : A filler used in a paint or a hard stopper.

STOPPING KNIFE : A knife used by glaziers and painters for smoothing putty or for hard stopping into cracks and holes in a timber.

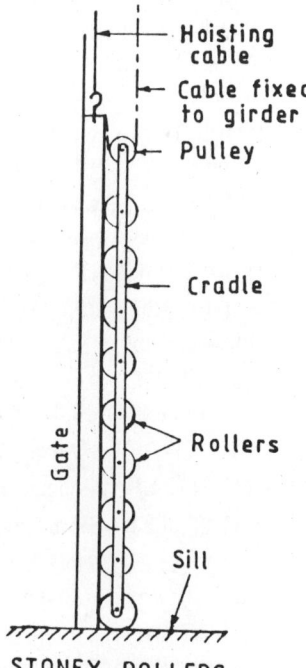

STONEY ROLLERS GATE

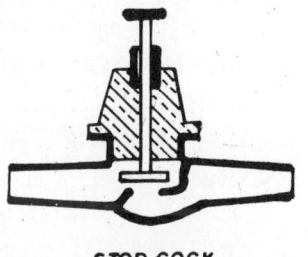

STOP COCK

STOP PLANK : See 'Stop Log'.

STOP VALVE : A stop cock.

STORAGE RESERVOIR : A reservoir used for storage of a liquid, the capacity of which is determined by the requirement of storage volume.

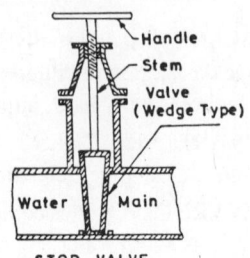

STOP VALVE

STORAGE TANK : A storage reservoir.

STOREY : The part of a building from one floor to the next floor level. For example, the second storey means the part from the second floor level to the third floor level. A two-storeyed building means it has two floor levels, ground floor and first floor.

STOREY ROD : A storey pole or batten or rod cut to the height of a storey with marks showing window-sill level, door and window height, height or cupboard, height of ventilators, etc.

STORM CELLAR : A cyclone cellar in which the residents of a house take shelter during a violent stormy weather or a cyclone.

STORM CLIP : In patent glazing, a metal clip of saddle shape is fixed to the glazing bar to hold the glass down against a violent wind.

STORM DOOR : An additional inner door provided to protect a room from the violent wind in a hard weather during winter.

STORM PAVEMENT : A sloping bank to a breakwater.

STORM-PROOF WINDOW : A window provided with protectionary measures against thunder shower.

STORM SHEET : A roofing sheet curved down at one edge to protect the eave from rainwater during a downpour.

STORM SEWAGE : Overland flow during a storm.

STORM SEWER : A duct or pipe laid underground to carry storm sewage.

STORM WATER : Storm sewage.

STORM WATER TANK : A retention tank for storm water, in which the solid matters settle down prior to discharge of the storm water into a stream or nullah.

STORM WINDOW : An additional external window provided with an air space between this and the inner window to protect the room against the cold waves during a violent wind in winter. The air space helps to keep the room warm.

STOVING : Drying a surface by heat radiation.

STRADDLE POLE : A sloping pole laid along a pitched roof in a 'straddle scaffold' from a 'standard' and it meets the other straddle pole at the ridge of the roof.

STRADDLE SCAFFOLD ; A saddle scaffold.

STRAIGHT ARCH : A flat arch.

STRAIGHT COURSES : Tiles, slates or shingles laid with their ends in a line.

STRAIGHT EDGE : A long piece of planed wood with parallel straight edges used by masons in brickwork and plastering for checking the straightness and level of the work.

STRAIGHT FLIGHT : A stair of single flight with no winders.

STRAIGHT GRAIN : Grains of a timber parallel to its length.

STRAIGHT JOINT : A butt joint.

STRAIGHT-RUN BITUMEN : Residual bitumen obtained after distillation of petroleum.

STRAIGHT TONGUE : One of the edges of a wooden board or plank is made thin by forming rebates on either side so that this tongue can be introduced into a matching groove made in another board or plank.

STRAIN : A ratio of change in dimension of a member (due to an applied load) to the original dimension of the member. A longitudinal strain is the increase or decrease in length divided by the original length. A lateral strain is the increase or decrease in diameter or in lateral dimension divided by the original diameter or original lateral dimension.

STRAIN AGEING : Increase in strength and hardness in iron and steel with time after cold working or overstressing.

STRAIN ENERGY : The energy stored in an elastic body subjected to a load. This is used in determining deflections based on the principle of conservation of energy.

STRAIN ENERGY METHOD : A method of analysis of structures based on the principle of strain energy i.e., based on the amount of energy stored in the loaded structure. This method is extremely versatile.

STRAINER : A perforated or slotted pipe with or without copper wire mesh is fitted at the bottom of a tubewell pipe so that only water will enter into the pipe excluding sand, silt, moorum and mud. There are different types of strainers, of which one is shown.

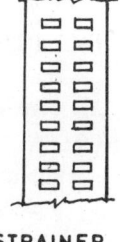

STRAIN GAUGE : A sensitive electrical instrument capable of measuring very small deflections in structures from which strains can be estimated. The electrical resistance varies with its deformation. The gauge is cemented to the test specimen, the strain in the specimen being measured as a function of the change in the electrical resistance of the wire element.

STRAINER

STRAIN HARDENING : Hardening of steel due to cold working.

STRAINING BEAM : A horizontal strut between the heads of queen posts in a truss.

STRAINING PIECE : A horizontal piece of timber bolted to the middle of a flying shore, from which sloping short struts at each end receive the thrust.

STRAINING SILL : A horizontal timber lying on and bolted to the tie beam of a queen post truss between the queen posts to keep them in position.

STRANDED CAISSON : A box type caisson.

S-TRAP : A trap of S-shape fitted to a W.C. pan for discharge of waste water by keeping a water seal of minimum 40mm depth. See illustration.

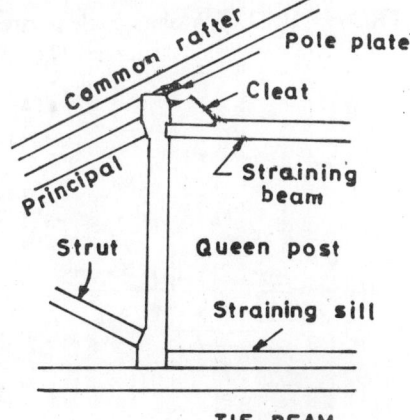

STRAP : A metal plate used for holding and binding parts together in a joint.

STRAP ANCHOR: A steel plate joining two joists butting against each other over a support.

STRAP BOLT : A metal strap having holes at one end and a threaded rod at the other end, used for fastening members in a timber truss or frame.

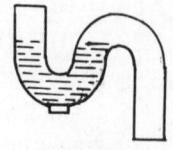

S-TRAP

STRAP HINGE : A cross garnet hinge.

STRAPPED ELBOW : A drop elbow used in plumbing.

STRAPPING : Marking common grounds for lathing and plastering.

STRATIGRAPHICAL PROFILE : A soil profile.

STRAW BOARDS : Boards made by cooking and compressing straws, which are used for sound insulation.

STREAMLINE FLOW : A laminar flow i.e., a flow in form of sheets and not turbulent.

STREET GULLEY : A receptacle or a chamber with an entry point with grate and an outlet pipe connecting the street sewer. This is built at the edge of a road and under the road surface to receive the discharge from gutters and to convey it to the sewer. See illustration.

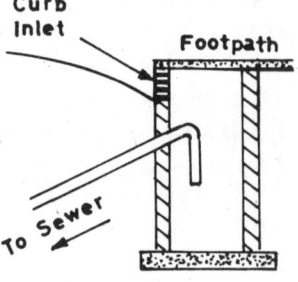

STRENGTH OF MATERIALS : The ability of materials to resist stresses due to shearing force, bending, torsion, tension and compression.

STRESS : The load per unit area of a structural member subjected to tension, compression or shear. In

STREET GULLEY

general, the stresses are either tensile or compressive or shear. The stress due to a direct load is called 'direct stress'. The stress due to bending is called 'bending stress'. The stress due to shearing force is 'shear stress'. There are also other forms of stress.

STRESS ANALYSIS : The determination of load per unit cross- sectional area of each member of a loaded structure either graphically or analytically.

STRESS CIRCLE : Mohr's circle of stress drawn to a scale to give a visual interpretation of two-dimensional stresses. The results are obtained graphically. A mohr's circle for

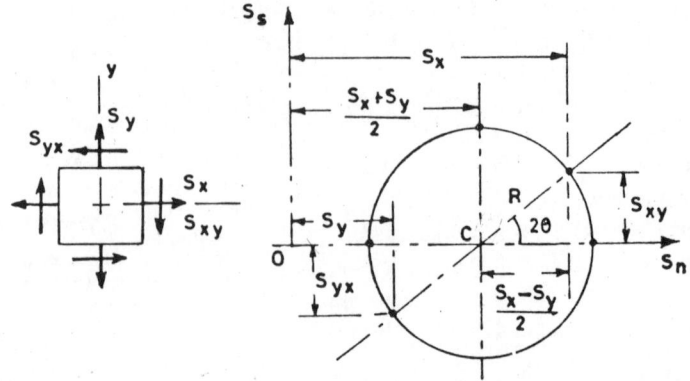

MOHRS CIRCLE FOR GENERAL STATE OF STRESS

general state of stress is shown in illustration. The centre, C is the average of the normal stresses. The normal and shearing stress components are represented by the co-ordinates of points whose position shifts around the circumference of mohr's circle. S_x, S_y and S_{xy} are known constants defining the specified state of stress.

STRESS CONCENTRATION : Concentration of stress (rapid increase of stress) due to abrupt change in cross section of a structural member, where the sectional area gets reduced due to notching and making holes. This effect can be detected by photo-elastic analysis.

STRESSED SKIN CONSTRUCTION : A geodetic construction or sandwich construction, which is a composite construction with structural material used at surface or skin, the core being framed or filled with materials of low strength.

STRESS-GRADED TIMBER : For use of timber in structural constructions, converted timbers are classified on the basis of the annual growth rings, grains, colour, texture, and presence of knots, shakes, wanes, etc. Thus, different timbers are graded according to their ability of withstanding the bending stresses and a reduction in strength is assessed for any particular defect in the timber.

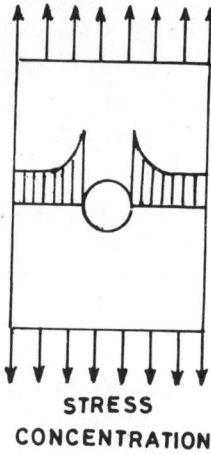

**STRESS
CONCENTRATION**

STRESS-NUMBER CURVE : A curve showing the range of stress in a material plotted against the no. of cycles to failure, as obtained in fatigue-testing of the material.

STRESS RELIEVING : Relieving internal stress in steel by heating it to a temperature below the critical point and then cooing it gradually at a slow rate.

STRESS-STRAIN CURVE : A curve obtained by plotting the stress v. strain in a test specimen usually under

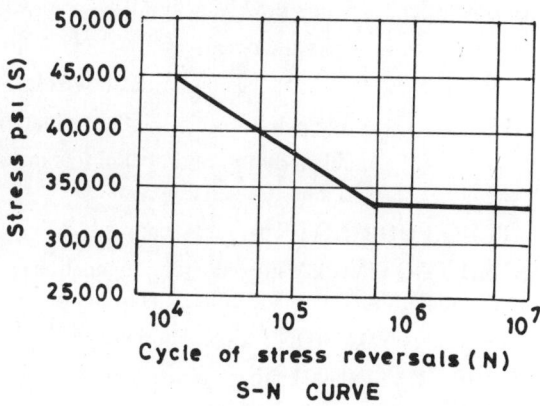

S-N CURVE

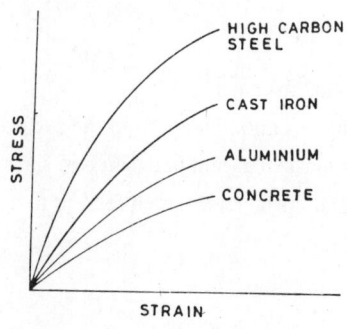

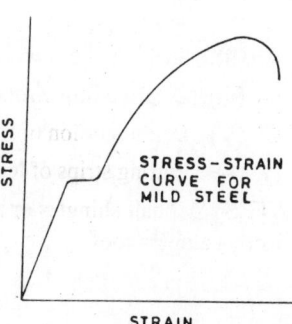

tensile test. The illustration shows stress-strain curves for mild steel, high carbon steel, cast iron, aluminium and concrete.

STRETCH: A particular length of a road, canal, stream, pipeline, etc.

STRETCHER : A stone or brick laid with its length along the wall.

STRETCHER FACE : The long face of a brick or stone when laid as a stretcher in a wall

STRETCHING BOND : A brickbond in which all the bricks are laid as stretcher, the wall thickness being half a brick.

STRETCHING COURSE : A layer of stretchers.

STRIDING LEVEL : A sensitive bubble tube fitted at each end of a theodolite telescope required for cross levelling of the instrument. The bubble tubes are fixed on the top of the legs standing on the telescope trunnions.

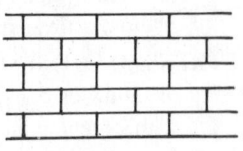

STRETCHING BOND

STRIKE : The horizontal line in a sedimentary rock at right angles to the dip.

STRIKING OFF LINES : Lines marked on a wall or ceiling for fibrous plaster work or for any other ornamental works.

STRING COURSE : A thin horizontal course of brick or stone or concrete projected outside the wall of a building at sill level or floor level.

STRINGER : A sloping board at the ends of treads in a stair to support the treads and risers.

STRINGING MORTAR : A bulk of thin mortar spread on the bed joint to facilitate laying of several bricks at a time.

STRING PIECE : The tie beam of a 'Belfast truss'.

STRIP : (i) A thin piece of short width of a material.

(ii) Lumber–a form of timber of thickness less than 50mm. and width less than 200 mm.

STRIP FLOORING : A parquet strip.

STRIP FOOTING : A strip-like long foundation of a wall with short projection on either side of the wall, made of brick masonry or concrete.

STRIP FOUNDATION : A strip footing.

STRIP HEATING : Heating a timber joint by a hot metal strip for gluing with resins.

STRIPPING : (i) Removing old paint by a stripping knife or with the help of a blow lamp.

(ii) Striking a formwork.

(iii) Clearing brushwood or turf from a site.

(iv) Disintegration of road metals and binders from the surface of a road.

(v) Putting strips of lead or glass in a joint between two slabs.

STRIP SLATES : Asphalt shingles or n.ineral-surfaced bitumen felt cut to slate size and laid in strips along a roof.

STRIP SOAKER : A strip of waterproof roofing sheet placed under each layer of shingles at a roof valley.

STRUCK CAPACITY : The rated capacity of a dredger bucket. The capacity of a bucket, skip, kibble or container calculated by considering upto a certain depth or height, beyond which it is imagined that the material will be struck off.

STRUCK JOINT : A weather joint in a masonrywork.

STRUCTURAL ANALYSIS : Analysis of a structure for determination of forces and moments coming on each member of the structure, prior to the design work.

STRUCTURAL CLAY TILE : Different forms of well-burnt clay tiles having good structural strength, used in various works. Some of these are hollow block tiles, furring tiles, tiles with horizontal or vertical cavities, etc.

STRUCTURAL DESIGN : The design of a structure based on the structural analysis such that each member is capable of taking the loads and moments coming on it and on the whole the structure is made stable against bending, sway, overturning, subsidence, etc.

STRUCTURAL GLASS : Reinforced glass blocks or tiles used in making partitions and in facing walls.

STRUCTURAL LUMBER : Stress-graded pieces of timber which are structurally sound.

STRUCTURAL STEELWORK : Fabrication of steel members and steel structures with rolled steel sections and their erection by joining the parts together either by welding or by riveting.

STRUCTURE : A frame or skeleton work for a building, bridge, trestles, etc.

STRUT : A column or a member subjected to compression.

STRUTTING : (i)   Bridging for a timber floor.

(ii)   Use of struts in timbering a foundation trench.

(iii)   Provision of temporary supports such as fly shore, raking shore and dead shore.

STRUTTING PIECE : (i)   A joist piece used for solid bridging in a timber floor.

(ii)   A straining beam.

STUART'S KEY : A type of key made of a steel plate bent in form of letter E, which is used as a wedge and introduced at the ends to keep it tight against the rail web and the outer jaw of the rail chair. See illustration.

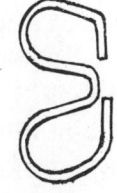

STUB : The nib of a roofing tile.

STUB TENON : A tenon to be introduced into a blind mortise.

STUC : Plasterwork resembling stone surface.

STUCCO : A smooth plasterwork consisting of an undercoat and a finishing **STUARTS** coat . The undercoat is made with a mix of 1 part of cement, 3 parts of **KEY** lime putty and 10 parts of sand. The finishing coat is made with the similar mix, excepting that 12 parts of sand are added in place of 10 parts of sand. Stucco plaster is more durable and weather-proof.

STUCK MOULDING : A moulding that made out of a solid by cutting and carving.

STUD :   (i)   A headless bolt i.e., a spindle or rod having threads on either end.

(ii)   Intermediate vertical member in a framed wooden partition.

STUD PARTITION : A framed wooden partition built of studs.

STUMP FOUNDATION : A brush block foundation made for bungalows or buildings with timber. In this foundation, round upright timber poles rest on timber sole plate in a pit, the pit being back-filled with soil after erection of the timber frame.

STUMP VENEER : A veneer obtained from the butt of a walnut tree exhibiting fine figures.

STUNT END : Vertical shuttering placed temporarily across a concrete wall or floor slab at the end of the day's casting so that it forms a construction joint.

SUB-BASE : A course of material laid under a road base and over the sub-grade.

SUB-BASEMENT : A second storey below the basement floor.

SUB-CASTING : A blind casting.

SUB-FLOOR : The load-bearing floor of wood or concrete, over which a finished floor is laid. Also known as 'Blind Floor'.

SUB-FRAME : A frame that is attached to or built into a main frame for fittings and fixtures.

SUB-GRADE : The formation level or ground surface below a road.

SUB-IRRIGATION : Irrigation by raising water level close to the plant roots.

SUBMERGED FLOAT : A subsurface float.

SUBMERGED WEIR : A weir completely under water used to raise the upstream water level.

SUBMERSIBLE PUMP : A centrifugal pump which remains submerged under water and is driven by an electric motor from top. This is used for pumping water or sewage.

SUBSIDENCE : Settlement or downward movement of the ground surface due to consolidation.

SUB-SILL : An additional sill provided at the outside of a window sill for throwing off rain water.

SUB-SOIL : The soil below the formation level or the top soil.

SUB-SOIL DRAIN : An agricultural drain under the top soil for drainage of water. This drain is built with open-joint clay pipes or tiles surrounded with gravel or coarse aggregates.

SUB-STRUCTURE : The foundation of a structure or the part of a structure below ground surface.

SUB-SURFACE EROSION : The underground movement of soil.

SUB-SURFACE FLOAT : An underwater float which is tied to a surface float.

SUB-SURFACE WATER : Water beneath the ground surface, whose occurrence is broadly divided into zones of aeration and saturation. The aeration zone consists of interstices occupied partially by water and partially by air. In the saturation zone, all the interstices are filled with water under hydrostatic pressure. See illustration.

SUBTENSE BAR : A bar of known length held at the distant point in the subtense method of tacheometric surveying. The distance is computed from the angle subtended by the bar at the observer and the known length of the bar.

SUBWAY : An underground passage for pedestrians in a busy area with a view to avoiding congestion on surface. See illustration.

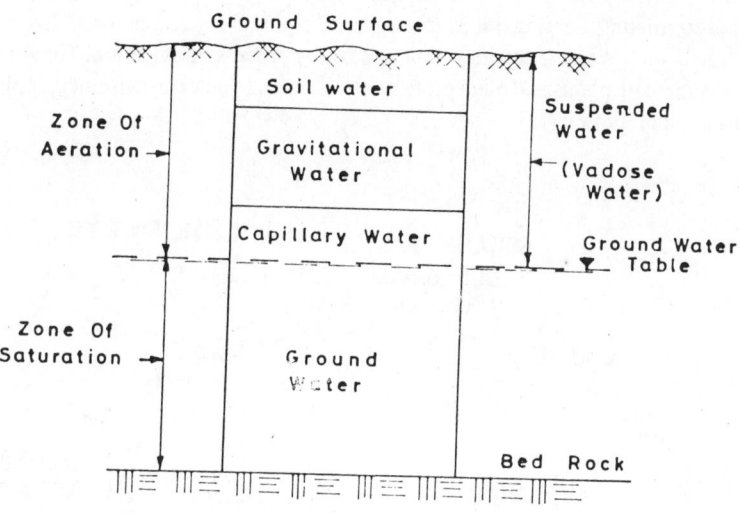

SUB-SURFACE WATER

SUCTION-CUTTER DREDGER : A suction-type dredger with a rotating cutter provided at the end of a suction pipe. This is used in digging stiff clay.

SUCTION DREDGER : A heavy-duty suction-type dredger without any excavating buckets. It dredges out mud and water mixture by suction pumps.

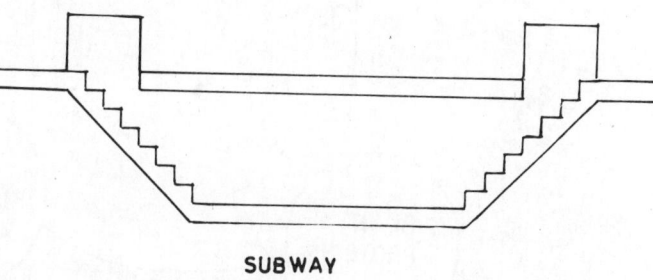

SUBWAY

SUCTION HEAD : The height through which a liquid is to be raised at the suction side of a pump. See illustration. This head in practice is restricted to a height of 6 m (20 ft.)

SUCTION VALVE : A check valve used at the suction end of a pipe.

SULLAGE : Silt deposited by a waste water coming out from bathroom, Kitchen, floors, etc.

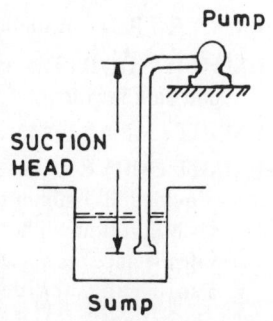

SULPHATE BEARING SOIL : A soil containing sulphates. If a soil contains more than 0.1% SO_3, it is necessary to take precaution in a foundation. If it is more than 0.5%, high alumina cement should be used.

SULPHATE-RESISTING CEMENT : A cement which resists the action of sulphates on it.

SULPHUR CEMENT : A portland cement containing powdered salmoniac and flour of sulphur, used for making joints in C.I. tanks.

SULPHUR CYCLE : One of the cycles of decomposition of organic wastes. The organic sulphurous matters coming out from plants and animals are decomposed to hydrogen sulphide, which is converted into sulphates by the process of oxidation. The sulphates are consumed by plants and plant proteins are formed, which are consumed by animals to form animal proteins.

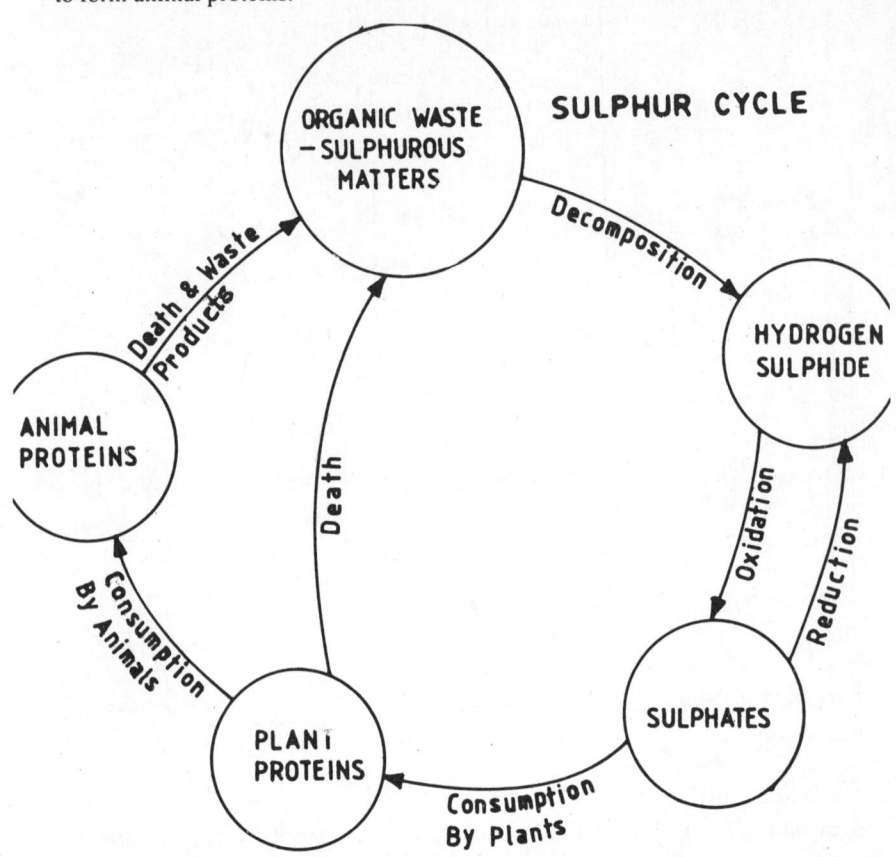

SUMMER TREE : A timber beam.

SUMMER WOOD : The layer of wood formed in a tree during summer. This wood is dense and dark in colour.

SUMMIT CANAL : A canal that crosses a summit by means of pumping water to it.

SUMMIT CURVE : A vertical curve provided at a summit such that visibility from either end at a safe distance may be achieved to avoid an accident. See illustration.

SUMP : A pit or underground chamber into which water or sewage is collected and then pumped to a higher head for further transportation of water or sewage.

SUMP PUMP : A pump provided for emptying a sump.

SUMP SIZE : The sump size is determined by the rate of inflow of a liquid and the detention time that can be allowed in the sump. The minimum dimensions that should be kept are shown in illustration. In case of a sewage sump, the maximum allowable detention time should be restricted to 30 minutes.

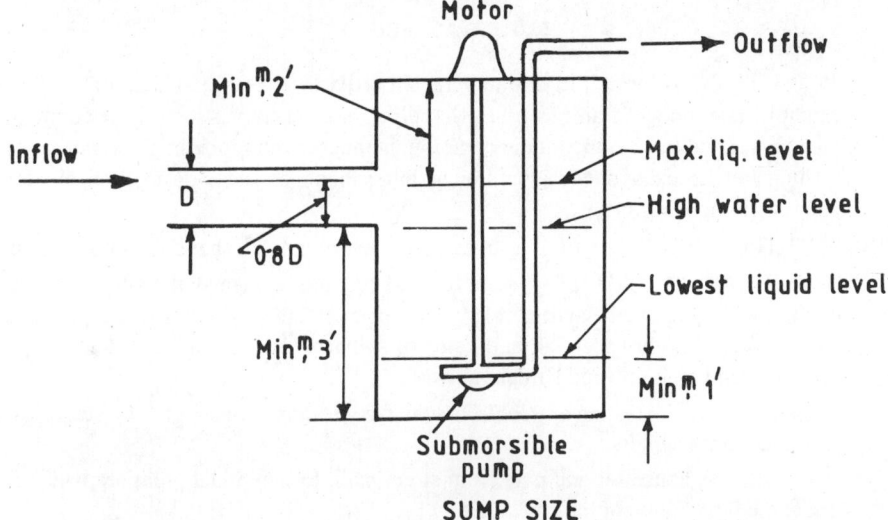

SUMP SIZE

SUNK DRAFT : A margin of a stone kept below the wall face.

SUNK GUTTER : A gutter which is provided below a roof surface by sinking it down and keeping it hidden.

SUPERELEVATION : A tilt given to a road surface or a railroad in a horizontal curve to counteract the effect of centrifugal force, as shown in illustration. The road is elevated at the outer curve.

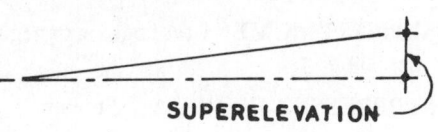

SUPERFICIAL COMPACTION : Compaction of soil in layers not exceeding 150 mm (6") by means of pneumatic-tyred rollers, light-weight sheep–foot rollers, hand rammers, vibrators, etc.

SUPERFICIAL MEASURE : Surface measurement of a solid object.

SUPER HARDBOARD : A hardboard which is most dense and water–resistant, used in boat building and floor finishing.

SUPERIMPOSED LOAD : A live load which is imposed in addition to other loads on a structure. Also known as 'super load'.

SUPER PASSAGE : A construction as shown in illustration, in which the natural drainage is carried over a canal. The bed level of the canal should be kept lower than the bed of the natural drainage channel so that this construction may be carried out without any obstruction to the flow in canal. A super passage is superior to a siphon.

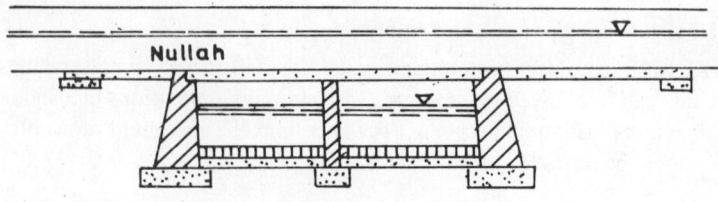

SUPERPASSAGE

SUPERPOSITION: A principle adopted in simplifying structural calculations; For example, a redundant frame is first splitted into perfect frames with some common members. Loads will come on the common members from more than one system of loading and the loads for each common member should be added up algebraically for determination of stress.

SUPER STRUCTURE : The part of a structure above ground i.e., above the substructure.

SUPERSULPHATED CEMENT : A metallurgical hydraulic cement with high percentage of SO_3, which resists the action of sulphates on a structure. This cement is prepared by grinding blast furnace slag and calcium sulphate and adding to it only 1% of portland cement. It sets and hardens slowly.

SUPPY LINE : A pipeline along a street from which connections for supply of water or gas are given to consumers.

SUPPLY PIPE : A house service pipe from street main to individual premises which is under the jurisdiction of the consumer.

SUPPORTED RAIL JOINT : A joint between two rails in a railway track, in which adjoining rail ends rest on a single sleeper as shown in illustration. Normally, three wooden sleepers abutting against each other are placed under the joint.

SUPPORTED RAIL JOINT

SUPPORT MOMENT : A hogging or negative moment at a support.

SUPPRESSED WEIR : A weir without any end contractions used in measuring the quantum of flow.

SURBASE : An ornamental moulding, crowning a skirting board or a dado capping.

SURCHARGED WALL : A retaining wall subjected to a surcharge i.e., a load coming over the surface levelled with the top of the wall. This surcharge may be in form of an embankment or dumping of soil sloping up from the top of the wall or construction of a structure above the top of the wall and adjacent soil. This increases a considerable active earth pressure on the retaining wall, which may be decreased by constructing a relieving platform.

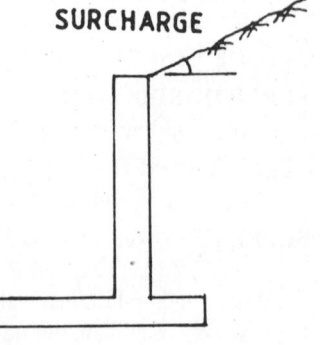

SURCHARGE

SURFACE ACTIVITY : A property of certain materials affecting the surface tension of a liquid and properties of soils.

SURFACE AERATOR : A mechanical aerator consisting of submerged or partially submerged impellers as shown in illustration, which is centrally mounted in an aeration tank for the purpose of aeration of sewage. The aerator agitates the sewage at the surface vigorously to entrain air in the sewage. Impellers are fabricated from steel, cast iron or non-corrosive alloys and reinforced fibre glass.

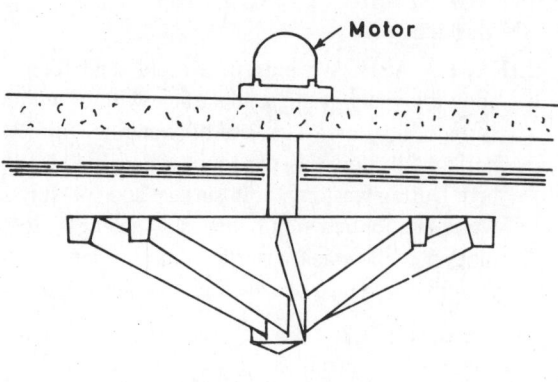

SURFACE ACRATOR

SURFACE AREA TEST : A test carried out on a cement to find out its specific surface which speaks of its quality. Ordinary portland cement has a specific surface around 2250 sq. cm. per gm., while high strength cement has 3250 sq. cm. per gm.

SURFACE DRESSING : (i)   Dressing of ground surface during site preparation.

(ii)   Application of a thin layer of bitumen-impregnated stone chipppings on a road surface and rolled to provide a wearing course.

SURFACE DRY : A stage when :

(i)   the surface of a timber is dry with sap-content inside.

(ii)   a paint is dry on the surface but wet beneath the surface.

SURFACED TIMBER : A dressed and planed timber.

SURFACE FLOAT : A float on a liquid surface.

SURFACE MEASURE : A superficial measure.

SURFACER : (i)   A tool or machine used for surfacing.

(ii)   A filler or a sealing compound used for making a surface even and smooth prior to painting.

SURFACE RESISTANCE : In thermal insulation of a building, it is the number of hours required for transmission of 1 BTU to 1 square foot of area from the surrounding air, when there is a temperature drop of $1^{\circ}F$ between the film of air and the surface area.

SURFACE RETARDANT : A liquid painted on the inner surface of a formwork to facilitate easy stripping of concrete from the surface after setting ; The surface is thus made rough by brushing so that the plaster will stick well with a good bondage with the concrete surface.

SURFACE TENSION : When two fluids of different density are in contact, a curve called 'meniscus' is formed at the surface of contact. This is due to attraction of molecules. There is also a little pressure difference between the fluids on either side of the surface. Thus the surface acts as an elastic skin which is in tension. This is the surface tension.

SURFACE TENSION DEPRESSANT : See
'Surfactant'.

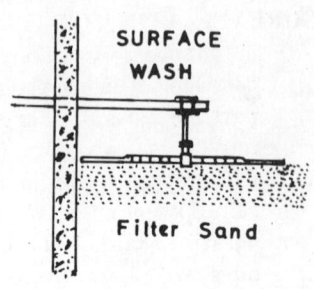

SURFACE WASH: Washing of a rapid sand filter by applying water at the sand surface with or without surface agitators. The jets of water are directed horizontally towards the four corners of the filter bed. During washing, both surface and sub-surface water are applied simultaneously causing swirling motion of the wash water to scrub the sand grains effectively against one another. See illustration.

SURFACE WATER : Surface run-off or overland flow from a paved, unpaved or a composite surface.

SURFACE WATER DRAIN : A drain, either open or conduit, to carry the surface water.

SURFACING : Surface treatment.

SURFACTANT : A material that disperses, dissolves, emulsifies or penetrates into other materials. In fact, this is a surface tension depressant.

SURGE : (i)   A horizontal force acting on a vertical structure at a high level.

(ii)   A sudden increase of pressure in a pipeline due to closing of a valve at its lower end.

SURGE PIPE : A stand pipe with open top provided at the lower end of a pipeline to release the surge pressure.

SURGE PRESSURE : The pressure in a pipeline due to a hydraulic transient caused by sudden opening or closing of a valve in the pipeline. Due to sudden closure of a valve, a high pressure wave travels back along the pipeline upto the point of stoppage, which causes the pipeline to vibrate. To counteract this phenomenon, a surge valve or pressure relief valve is used.

SURGE TANK : A tank provided at the top of a surge pipe to prevent loss of water by spilling during pressure surges.

SURPLUS ESCAPE : A device provided in an irrigation distribution system to allow water to escape as waste when the water rises above the full supply level in channels. This is

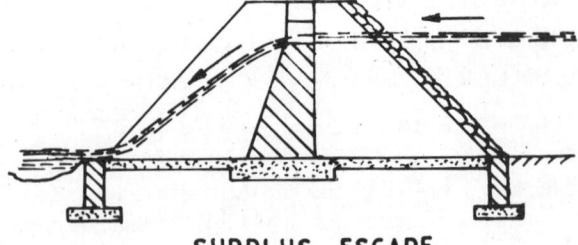

SURPLUS ESCAPE

provided at regular intervals along the length of a canal. The surplus escape consists of a weir wall. The capacity of a surplus escape should be 1/2 to 1/3 of the full supply quantity of the channel.

SURVEY : To take the view of a land with surveying instruments and equipment and to prepare a map of the land.

SURVEYING : The technique of conducting a survey work. There are numerous methods of surveying.

SUSPENDED CEILING : A false ceiling which is usually decorated.

SUSPENDED FLOOR : A floor which is hung from supports at its ends.

SUSPENDED-FRAME WEIR : A framed weir having its frame hung from a bridge and that is obviously above the water level during floods.

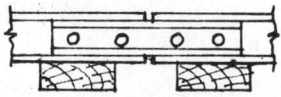

SUSPENDED RAIL JOINT : A rail joint in which both the adjoining rail ends remain projected beyond the shoulder sleepers as shown in illustration.

SUSPENDED RAIL JOINT

SUSPENDED SCAFFOLD : A hanging or projecting scaffold.

SUSPENDED SHUTTERING : A shuttering which is carried on the supports of a floor for casting floor concrete. It is not propped.

SUSPENDED SPAN : The freely-supported span of a cantilever bridge.

SUSPENDER : The hanger of a suspension bridge.

SUSPENSION BRIDGE : A bridge suspended from a pair of steel cables carried by two towers at the banks. The cable ends are anchored into a massive concrete or masonry. The road is carried by vertical rods spaced uniformly between road and cable on either side.

SUSPENSION CABLE : A steel wire stranded rope strong enough to carry a suspension road bridge. Two such ropes are needed for a bridge.

SUSPENSION CABLE ANCHOR : A massive concrete or masonry built behind the towers, into which the suspension cable ends are anchored.

SUTRO WEIR : A type of weir as shown in illustration, is used at the outlet of a grit channel to maintain a constant velocity of flow with the variation in depth of flow in rectangular channel.

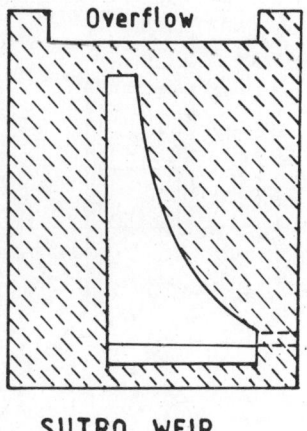

SUTRO WEIR

S V I : Sludge Volume Index, which is an empirical measurement, defined as the volume in ml. occupied by one gm. of mixed liquor solids (activated sludge) by dry weight after settling in 1000 ml. cylinder for 30 minutes. On this basis, the rate of sludge return and plant operation are controlled. In practice, it is the percentage volume occupied by the sludge in a mixed liquor after 30 minutes' setling.

S V R : Sludge Volume Ratio, which is the volume of sludge blanket maintaned at the bottom of a thickener divided by the volume of thickened sludge removed daily. The S V R normally ranges between 0.5 and 2 days. The lower values are required in hot climates.

SWASH BANK : The topmost part of the slope of a sea embankment.

SWATCH : A pile of samples of veneers or lionleum.

SWAY : (i)   The sideways movement or shifting of a structural frame.

(ii)   A tree sapling, about 20mm diameter and 1 metre long, laid horizontally across the rafters under the thatch to hold it down by means of ropes or cords.

SWAY BRACE: A brace placed diagonally in a frame to resist wind forces or other horizontal thrusts acting on the frame.

SWEAT: To join metal pipes by forcing one into the other and putting molten solder between them.

SWEATING : (i)   Sanding a dry painted surface to bring the gloss.

(ii)   Uniting pipes by forcing them together and putting molten solder at the joint.

(iii) Separation of vehicle and pigments in a paint so that the pigments appear on the surface.

SWEAT JOINT : A term used for capillary joint in USA.

SWEAT OUT : A plaster that appears damp and mushy on setting, which is likely to be caused due to use of dirty sand or imperviousness of brick surface.

SWEEP TEE : A 'Tee' is used in plumbing work for screwed pipes. The 'Tee' is perpendicular to the run with a gentle curvature.

SWELLING PRESSURE : The pressure exerted by an underlying layer of soil due to its swelling by absorption of water.

SWEPT VALLEY : A valley made of tapered slates or tiles to avoid the need of a flexible-metal valley.

S W G : Standard wire gauge.

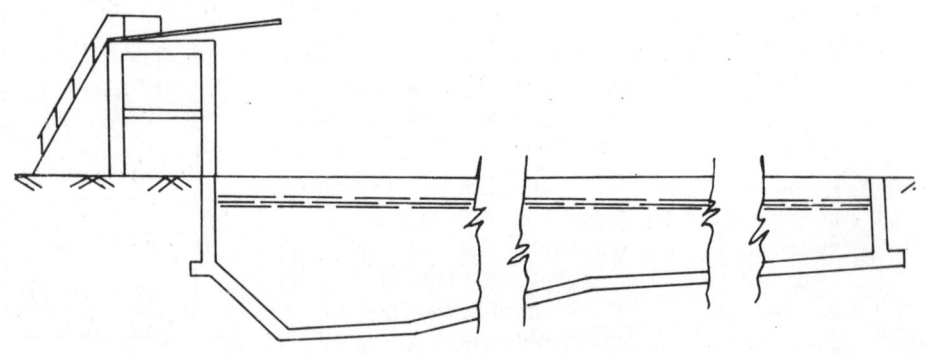

SWIMMING POOL

SWIMMING POOL : A pool of water in a tank of required shape and size for the purpose of practice of swimming as well as for the purpose of learning swimming. Care should be taken against physical injury and infection. In modern day swimming pools, recirculation of pool water is made with adequate treatment and disinfection.

SWING BRIDGE : A turn bridge which is pivoted at its centre and opens by swinging horizontally to allow water vessels to pass through. Such a bridge is provided when a road crosses a channel.

SWING DOOR : A door shutter fitted with a helical hinge such that it opens in both directions and closes automatically by spring action.

SWINGER : A pointed bar of about 1 metre length used to move runners in timbering a trench.

SWINGING POST : The hanging post of a gate.

SWING JIB CRANE : A crane with a horizontal boom which is capable of swinging through $360°$ with its counter-weight.

SWIRL : The grains of irregular form surrounding a knot in a timber.

SYLVESTER : A prop-drawing tool consisting of a steel rack with chain and lever system used for withdrawing steel or timber poles which were driven into the soil for a job.

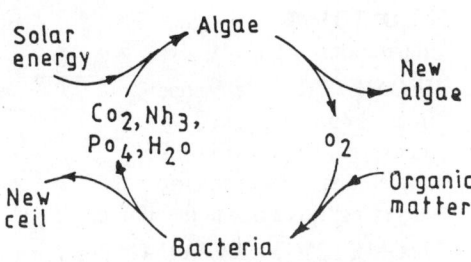

SYMBIOSIS BETWEEN ALGAE
AND BACTERIA

SYMBIOSIS : A schematic representation of the symbiotic relationship between the algae and bacteria as shown in illustration, speaks of the phenomenon 'symbiosis'. The oxygen released by algae is taken by the bacteria in degradation of organic matters present in sewage and carbon dioxide and nutrients obtained from degradation are consumed by the algae. This process goes on in cylic order in a stabilisation pond.

SYNTHETIC RESIN : Phenol formaldehyde, Urea formaldehyde and Melamine formaldehyde, which are highly water-resistant and are immune to attacks by moulds and bacteria.

SYNTHETIC RESIN CEMENT : A binding or cemeting material like synthetic resin glue. This is extensively used in plywood and lamin board works for buildings, automobiles and aircrafts.

SYNTHETIC STONE : A term for cast stone i.e., artificial stone.

SYSTEMATIC ERRORS : Cumulative errors which are always either positive or negative and not compensating.

SYSTEM BUILDING : A method of systematic construction following a module and the critical path method with least time and saving of materials, keeping the quality of work unaltered.

T

TABLE : (i)   A working top supported on legs; A furniture

(ii)   The flat part of a Tee section.

TABLED JOINT : See illustration.

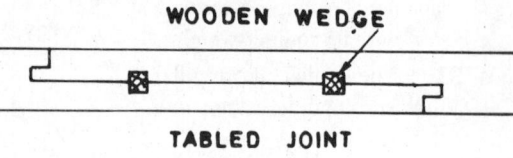

WOODEN WEDGE

TABLED JOINT

TACHEOMETER : A telescope in a theodolite used to measure distances by its stadia hairs in tacheometric surveying. It is a transit theodolite.

TACHEOMETRIC SURVEYING : Preliminary location survey and for filling in details on topographical maps.

TACK : (i)   A very small nail for fixing sheets.

(ii)   Stickiness of a drying paint film.

TACK COAT : A thin coat of tar or bitumen applied on a road surface to increase the adhesion of the top course (chipping carpet or wearing course).

TACK RAG : Cloth damped with slow-drying varnish used in removing dust from a rubbed or sand-papered surface.

TACK RIVET : A rivet used for convenience of fabrication or construction. This is not supposed to carry any load.

TACK WELD : A spot weld to hold the two edges together, which are to be welded afterwards as per requirement.

TACKY : A stage in drying of a paint film at which the paint is sticky.

TAFT JOINT : A finger-wiped joint used in lead pipe plumbing, which is made by wiping in plumber's solder instead of pouring fine solder in the joint.

TAG : A marking tally used in chains for measurement of lengths in surveying.

TAIL BAY : (i)   The part of a canal at the downstream of a tail gate.

(ii)   The end span of a timber roof.

TAIL BOLT : Bolts used for fixing the tail ends of roofing sheets, especially A.C. sheets.

TAIL ESCAPE : A weir wall with bed protections required to maintain water level at F.S.L. at the tail channel and to discharge the water to natural water course after serving the purpose of irrigation.

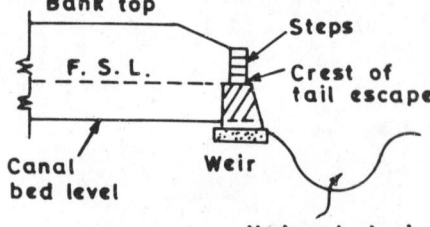

TAIL ESCAPE

TAIL GATE : The gate to the downstream of a lock.

TAILING IN OR TAILING DOWN : Fixing a cantilever (projected from a wall) by laying concrete blocks, bricks or stones on it (to develop the fixity).

414

TAILING IRON : A steel section embedded into a wall to hold down a cantilever at its point of projection from the wall.

TAIL JOIST : A timber joist resting on a tail trimmer.

TAIL PIECE : A trimmed joist used in timber flooring or in a frame.

TAIL RACE : A channel carrying water away from a turbine.

TAIL TANK : A reservoir having its own command area located near the tail end of a canal and is supplied with water from the canal whenever in excess of canal requirement.

TAIL TRIMMER : A trimmer joist to carry the floor joists in timber flooring.

TAIL WATER : The water in the tail race.

TAINTER GATE : A radial gate consisting of a pair of sectors placed one at each end of the span. A chain is fitted at the bottom of the gate at the ends. By pulling this chain, the gate is revolved about the horizontal axle passing through the vertex of the sectors and is opened allowing the flood water to discharge over the crest weir.

TAKE-OFF : Branching off from a pipeline.

TALBOT PROCESS : A process by which a coating of sand and bitumen is given on a cast iron pipe to protect the surface.

TALL BOY : A chimney hood about 1.5 m high made of galvanised sheet metal is provided over a chimney so that down draughts are prevented.

TALLY : A brass label with one notch cut in it for every 10 ft. (upto 40 ft. maximum) provided in a surveying chain to facilitate quick counting of a measured length. The tally for 50 ft. is of circular type with no notch in it. Also known as 'Teller' or 'Tag'.

10   20   30   40   50

or   or   or   or

90   80   70   60

TAG   OR   TALLY

TALUS WALL : A battered (slant) wall built to hold back an earth slope.

TALUS WEIR : A protection at the downstream end of a weir consisting of concrete or masonry blocks.

TAMBOUR : (i)   A drum-shaped stone in a circular column.

(ii)   A vestibule of cylindrical shape preventing draughts.

(iii) A circular wall carrying a dome.

TAMPER : A screed board used in concrete flooring or pavement.

TAMPING ROLLER : A sheep-foot roller used in construction of an embankment or dam.

TANDEM ROLLER : A road roller with rolls of same diameter behind each other.

TANG : The pointed back portion of steel tool like chisel, knife or file, which is driven into a wooden handle.

TANGENT DISTANCE: In setting out a road or railway curve, it is the distance measured from the point of intersection to the tangent point.

TANGENT POINT: A point on a curve at which it changes its direction or curvature.

TANGENT SCREW : A screw which moves the line of sight in a telescope through a very short distance. This is provided on both horizontal and vertical circles of a theodolite for fine adjustment.

TANGENTIAL SHRINKAGE : The shrinkage of timber parallel to the annual (growth) rings. This occurs in an unseasoned timber.

TANK : A reservoir, a container.

TANKING : Laying a thin waterproof layer of asphlat under a basement floor and up the walls with a view to making the basement watertight.

TANK SEWER : A storage tank in form of a detention tank to store sewage, when in an outfall system the high tide rises above the crown of the outfall sewer and it is not possible to discharge the sewage. This is built near an outfall.

TANK SPRAYER : A pressure tank required in a spraying job.

TAP : A bib-cock fitted to a water line for drawing water.

TAPE : A graduated ribbon of linen, invar, steel or linen interoven with wire (metallic) used in taking measurements in surveying.

TAPE CORRECTION : The following corrections may be required during measurement of lengths with the help of a tape in surveying. The corrections are for tension, sag, slope, temperature, alignment and reduction to sea level. Since field tapes are liable to suffer permanent elongation due to continual tension to which they are subjected, it is essential to compare its length with the length of a standard tape.

TAPERED AERATION : A modified form of activated sludge process in which compressed air is supplied at a higher rate at the inlet end of the aeration tank and is gradually decreased towards the outlet end of the tank. Thus, the rate of aeration is tapered. The tank is partitioned into three or four compartments.

TAPERED FLANGE BEAM : A R.S.J. section with the inner surfaces of its flanges tapered at 98° to the web.

TAPERED PARAPET GUTTER : A flexible metal sheet box gutter provided at the back of the parapet.

TAPERED WASHER : A bevelled washer to make a joint water-tight.

TAPER FILE : A fine-cut triangular file used for sharpening saw teeth.

TAPER PIPE : A reducer or enlarger used in plumbing a pipeline.

TAPER THREAD : Pipe thread which ensures water-tight or steam-tight joints in a pipeline.

TAPING STRIP : A strip of roofing felt used to cover joints between roof slabs, over which roofing felt is laid with a sealing compound.

TAR : A deep black viscous liquid. It may be coal tar, wood tar or mineral tar. Coal tar is a heavy, strong smelling black viscous liquid used chiefly as a preservative. Wood tar is obtained from resinous wood. Mineral tar is obtained by distillation of bituminous shales.

TARE : The weight of a lorry or wagon in unloaded condition i.e., the self weight.

TAR EMULSION : A bituminous emulsion.

TARGET ROD : A levelling staff.

TAR-GRAVEL ROOFING : A roofing felt with sand or small pea-size gravels embedded on the surface with the help of hot bitumen.

TARK SCREEN : A very fine screen with mechanical cleaning device. The sewage passes through the perforated surface of a drum rotating slowly with its long axis normal to the direction of flow. The sewage which enters into the drum, leaves it through one end and at right angles to its original direction of flow. The screenings collected at the top of the drum are swept by brushes mounted on an endless chain.

TAR MACADAM : A road material consisting of stone metals coated with tar or bitumen (applied hot).

TAR MACADAM PLANT : A plant for preparation of tar macadam.

TARPAULIN : A thick waterproof canvas used for protection against rain.

TAR PAVING : Surfacing a road with tar macadam.

TEAK : A close-grained hard and durable wood of very good quality used for manufacturing furnitures, doors and windows.

TEE : (i)   A rolled steel section having T-Shape.

     (ii)   A short pipe of T-shape used for branching a pipeline at right angles.

TEE BEAM : A tee-section used as a beam. In R.C.C. construction, it is a rectangular beam monolithic with the slab, the slab being the flange of the Tee.

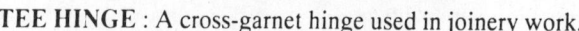

TEE

TEE HINGE : A cross-garnet hinge used in joinery work.

TEE IRON : A Tee section.

TEGULA : Italian under-tile used in roofing with over-tile.

TELEMETER ROD : A staff required for stadia work.

TELESCOPIC CENTERING : A formwork of steel sections made in such a way that it can be built rapidly by fitting them into each other telescopically. Also known as 'collapsible Pans' or 'Self- centering formwork'.

TELESCOPIC WEIR : Used for continuous sludge removal from pyramidal-bottomed sedimentation tanks, sludged under hydrostatic head.

TELLER : A tag or tally used in surveyor's chain.

TELPHER : A monorail (rolling on a single overhead rail) used in factories hung from roof girders for transporting machine and material from one end to the other.

TEMPER : (i)   To harden and toughen non-ferrous metals by the process 'annealing'.

     (ii)   To reheat a hardened steel to a temperature below the critical temperature and to cool it rapidly. The greater the quenching temperature, harder is the steel.

TEMPERA : A method of mural painting by using a mixture of gum, size, egg and water.

TEMPERATURE CORRECTION: The length of a steel tape increases as temperature rises and it decreases due to fall in temperature from the standardization temperature (normal). Therefore, correction for temperature variation is to be made accordingly. The temperature correction is equal to 0.00000625 foot per foot length measured per degree farenheit difference of temperature from the normal temperature.

TEMPERATURE GRADIENT : A change in temperature per unit length of a material.

TEMEPERATURE LOGGING : With a view to analyzing the subsurface conditions of earth, a vertical traverse measurement of ground water temperature in a well with the help of a resistance thermometer is made. The temperatures indicate the flow of water from different aquifers intersected by a well, presence of gas in a deep well and contamination from adjacent areas.

TEMPERATURE STEEL : Additional steel reinforcement provided in a concrete, especially to prevent cracks due to shrinkage and due to variation in temperature depending upon the climatic condition. It is usually 0.1% of the cross-section in any direction in addition to the normal reinforcement required in a R.C.C. member.

TEMPERATURE STRESS : The stress developed due to variation in temperature in a structural member. If 'a' be the co-efficient of expansion or contraction of the material used, 't' the difference in temperature and 'E' the modulus of elasticity of the material, then the temperature strain is 'at' and the temperature stress = temperature strain x modulus of elasticity $= a.t.E$.

TEMPLATE : A wooden or metal pattern used to form the shape of a moulding, plaster or concrete member.

TEMPORARY ADJUSTMENT : Initial adjustment of a surveying instrument (theodolite, level, etc.) by levelling or centering, during setting up the instrument at a survey station.

TENACITY : Tensile strength of a material.

TENDER : An offer from a contractor or consultant to do a certain construction work or planning and design for a job alongwith a price bid from the contractor or the consultant.

TENDERING : The procedure of inviting tenders by sending out drawings, specifications and bill of quantities to contractors. In case of a consultancy job it is the procedure to serve or publish a notice inviting tenders in newspapers stating the nature of work to be done and the 'terms of reference' including general terms & conditions, scope of services and payment schedule.

A tender may be 'open' or 'limited'. In case of a open tendering, anyone can submit his offers. But, in case of 'limited tendering', tenders are issued to the panelled or short-listed contractors or consultants for some specialised jobs. The empanelment is done on the basis of the prequalifying papers submitted by the contractors or consultants related to specialised jobs of similar nature.

TENDON : A prestressing cable bar, stranded wire or rope.

TENEMENT : An apartment with common toilets or privies for economically weaker section of people.

TENON : A narrowed projected part of a member that is introduced into the mortise of another member to make a joint.

TENON SAW : A mitre saw of about 12" length stiffened with a folded steel plate along its back as shown in illustration.

TENON SAW

TENSILE FORCE : A pulling force that causes tension in a structural member.

TENSILE STRENGTH : The internal force developed in a member under tension to resist the tensile force applied externally on the member.

TENSILE STRESS : The stress i.e., the internal force per unit area of the section of the member developed due to application of a tensile force.

Tensile stress = Tensile load applied/Cross-sectional area

TENSILE TEST : A test carried out on a standard specimen to find out the tensile strength of the specimen. On the basis of the test results, stress-strain curve or load-extension curve is drawn.

TENSIOMETER : An instrument used in field to measure the moisture content in a soil. It consists of a porous ceramic cup filled with water and connected to a mercury mano-meter. The cup is introduced into the soil and the pressure given by the soil moisture is indicated by the manometer.

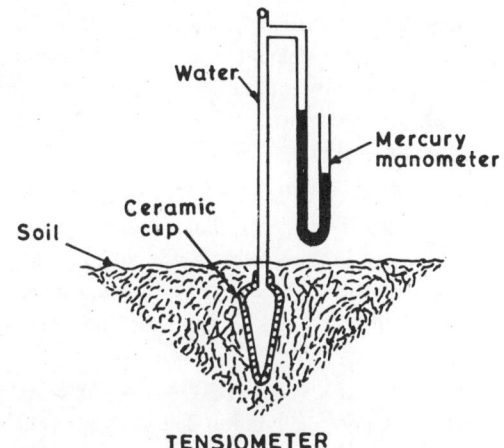

TENSIOMETER

TENSION : A member is said to be in tension when it is subjected to a tensile load.

TENSION CARRIAGE : A movable frame (wheeled) with belt-pulley system, used for tensioning the traction rope of an aerial ropeway or for tensioning guy ropes.

TENSION CORRECTION : The correction to be applied to a tape, when it is subjected to tension in a base line measurement during surveying. If the difference in tension between the pull applied and the standard pull at which the tape is made be P , the cross- sectional area of the tape is A, length of tape is L and modulus of elasticity of the tape material is E, then correction is given by $X = PL/AE$

If the applied pull is more than the standardization pull of the tape, X is to be added to the measured length. On the other hand, X is to be subtracted if the applied pull is less than the standardization pull.

TENSION FLANGE : The flange of a beam which is subjected to tension.

TENSION SLEEVE : A screw shackle used in structural connection

TERMINAL : (i)   An end point;

(ii)   A decorative end.

TERMINAL VELOCITY : The free falling or settling velocity of a particle that is attained by it in a medium through which it falls.

TERMITE SHIELD : A metal sheet used as a shield against damp and against climbing up of termites inside a house. This is a must in wooden constructions in temparate climates and tropical countries to prevent the timbers eaten away by termites.

TERRACE: A flat roof surface or a raised platform.

TERRACED ROOF: See illustration. Lime terracing made over the roof tiles supported on wooden burgahs. This type of roofing was used in earlier days. Now-a-days, lime terracing is done over concrete roofs.

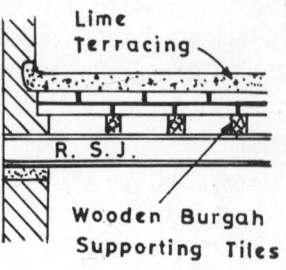

Lime Terracing

R. S. J.

Wooden Burgah

Supporting Tiles

TERRACED ROOF

TERRA COTTA : A strong and durable earthenware made from a refractory brick clay or from a selected clay mixed with ground glass, pottery and sand. It is used as a substitute for stone in ornamental parts of buildings to make architectural composition.

TERRAZZO : A venetian mosaic made of stones of different colours laid in cement mortar over a concrete surface. Designs are also made in mosaic. When set, the surface is polished by abrading to give a bright appearance.

TERREPHRAGM : A method of construction of an underground diaphragm wall.

TERRESTRIAL REFRACTION : Owing to the refraction, the line of collimation of a surveying instrument is not a straight line and hence correction is needed. The value of the angle of this refraction is not constant and it varies in different localities and at different times of a day depending upon the climatic conditon. The maximum value occurs in morning (5 to 6 AM) and decreases upto 10 AM. and again increases from 4 pm.

TERTIARY TRIANGULATION : A sub-division in a triangulation survey. It comes under minor triangulation. The governing chain used in tertiary triangulation is known as tertiary chain.

TESSELATED PAVEMENT : A pavement made of Roman mosaic.

TEST CUBE : A 4" (100 mm) cube of mortar or a 6" (150 mm) cube of concrete after 7 days or 28 days curing is tested in a compression testing machine, where the compressive strength is found out by crushing the cube. Test cubes are made of different mixes and a suitable mix is selected for a job on the basis of the cube test results.

TESTING MACHINE : There are a number of testing machines to determine the various properties of engineering materials. In a universal testing machine quite a number of tests can be carried out. These tests are for tension, compression, torque, impact and fatigue.

TEST PIECE : A test specimen made to a standard size for testing purpose.

TEST PIT : A trial pit to reveal the soil strata and also to conduct tests at different depths underground.

TETRAHEDRON : A four-faced solid bounded by four equilateral triangles.

TETRAPOD : An equiangular heavy solid figure acting as a tripod with its fourth leg (vertical) rising from the intersection of the other three. Such a heavy block of four legs are extremely stable and are used in 'break waters'. The blocks are piled up which break the waves of water.

TEXTURE : A face, fine-grained or coarse-grained, uniformly distributed or at random, smooth or rough, even or uneven and with or without designs.

TEXTURE BRICK : A rustic brick used in face work of a wall.

TEXTURE FINISH : Finishing a wall surface with cement-sand plaster or grout or concrete or paints to have different forms of texture.

THATCH : A covering in a sloping roof made of leaves, straws, reeds, etc. commonly found in village houses.

THEODOLITE : An instrument having a telescope fitted with horizontal and vertical circular graduated plates, carried on tribrach having three levelling screws for adjustment. This is used by surveyors for measuring horizontal and vertical angles. This instrument is chiefly employed in a traverse survey. There are various forms of theodolite available in market.

THERMAL BORING : A method of boring holes in concrete by means of a lance (a steel burning tube filled with steel wool through which an oxy-acetylene mixture is introduced to ignite the end of the tube).

THERMAL MOVEMENT : Movement caused by expansion or contraction of a material due to variation in temperature. Also known as 'temperature movement'.

THERMAL TRANSMITTANCE : The co-efficient of heat transmission from air to air. It is the U-value.

THERMIC BORING : See 'Thermal Boring'.

THERMIT : A mixture of aluminium powder and iron oxide required in alumino-thermic welding of steel parts which forms a trouble- free electrical and mechanical bond.

THERMOCOUPLE : A couple made by welding two wires of different metals (chosen suitably) at one end and the other two ends are connected to a circuit such that if there is a temperature difference between the two ends of the couple, there will be a flow of current. A pyrometer works on this principle.

THERMOLUX : A plyglass used for heat insulation.

THERMO-OSMOSIS : A natural phenomenon of movement of moisture from warmer part of a soil to the colder part.

THERMOPLASTIC : Cellulose, shellac, synthetic resins which soften on heating and become hard on cooling.

THERMOPLASTIC PUTTY : A putty which is made plastic by addition of tallow to it. This is used in glazing.

THERMOPLASTIC TILES : Flooring tiles made of thermoplastic resins, cellulose, shellac, asphalt and asbestos fibre.

THERMO-SETTING PLASTIC : Plastics which are progressively hardened when heated and which when moulded are permanently set. The common thermosetting plastics are bakelite, phenol formaldehyde and urea formaldehyde.

THERMO-SETTING GLUE : See 'Thermo-setting resin'.

THERMO-SETTING RESIN : A film glue (resin) used in joining plywoods. It hardens on heating and it does not soften again when reheated.

THERMO-SIPHON : A flow or circulation by gravity.

THIMBLE : (i)   A sleeve.

(ii)   A ring formed of sheet metal built into the end of a steel rope or stranded wire.

T-HINGE: A cross-garnet hinge used in joinery works.

THINNER: A solvent ot volatile liquid which is used in a paint or varnish to lower the viscosity so that it can flow easily. White spirit and turpentine are good thinners.

THINNING RATIO : The proportion of a thinner in a paint medium for good workability.

THIN SURFACING : A thin layer of bituminous carpet.

THIXOTROPIC FLUID : (i)   A thixotropic paint does not have running flow and the pigments settle in the paint very slowly.

          (ii)   Thixotropic clay gets weakened when in fluid state during moulding and attains strength when they are kept undisturbed.

THIXOTROPY : A property possesed by a soil which is related to the physical nature of structural strength, bond, congealing, orientation of molecules and adsorbed water films between particles.

THREE-COAT WORK : Rendering a surface by applying three coats of plastering, white washing or painting.

THREE-DIMENSIONAL DRAWING : A pictorial drawing which shows all the three dimensions of an object speaking of its shape. The drawing may be oblique, isometric, axonometric or perspective.

THREE-HINGED ARCH : An arch which is hinged at the two supports and at the crown. Each half may sink relatively to the other, without causing any damage to the arch.

THREE-LEG SLING : A sling made of three chains or ropes hung from a thimble or ring, used in loading and unloading of materials.

THREE-PEG TEST : A test used to be carried out for adjustment of the axis of the telescope bubble parallel to the line of collimation. This test is now discarded.

THREE-PINNED ARCH : See 'Three-hinged arch'. It is used where a differential settlement is likely to occur.

THREE-POINT PROBLEM : Used to locate on the plan the position of the instrument in the field by means of taking observations of three well-defined points, the positions of which are already delineated upon the plan (drawing).

THREE-QUARTER BAT : A brick cut across so that one-quarter of it is removed.

THREE-QUARTER HEADER : A header brick or stone of three-quarter wall thickness.

THREE-WAY STRAP : A T-shaped steel plate used in joining three members at a junction point of a timber truss.

THRESHOLD : A horizontal timber used as a sill of an outside door.

THROAT : (i)   An undercut in a coping, cornice or a drip course.

      (ii)   A gap through which wood shavings come out of a plane.

      (iii)   A flue gathering.

THROUGH BONDER : A bond brick or stone.

THROUGH BRIDGE : A bridge whose lower chord or deck carries a roadway or railway.

THROUGH LINTEL : A lintel of full wall thickness.

THROUGH STONE : A bond stone.

THROUGH TENON : A tenon that passes beyond the mortise.

THRUST : A horizontal or inclined force acting on a wall or support.

THRUST BLOCK : A heavy concrete block used at bends and at the dead end of a pipe to resist the thrust of water as shown in illustration.

THRUST BORER : An equipment to make a hole below an embankment or railway or to bore underground by applying thrust. It is a jack push.

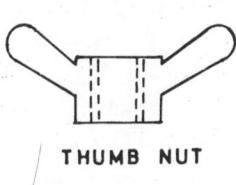

THRUST BLOCK

THUMBAT : A wall hook to fix lead sheet.

THUMB NUT : A nut similar to a 'wing nut' is provided with two conical rods to the cylindrical body of the nut to facilitate in turning the nut. See illustration.

THUMB SCREW : (i) A metal screw passing through the meeting rail of one sash screws into the meeting rail of the other sash for fastening them together.

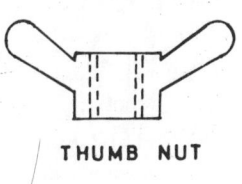

THUMB NUT

   (ii) A wing nut which is turned by the thumb for unscrewing or screwing.

THUMB TACK : A drawing pin for fixing a paper onto a drawing board.

TIDAL DOCK : A dock having no lock gate. Therefore, the water level inside and outside the dock is always same.

TIDAL LAG : The lag between high tide/low tide of an estuary and the highest/lowest resulting ground water table in the neighbouring area.

TIDE FLAP : A flap which closes against a seating on frame that is arranged for bolting to the faces of walls of an outfall structure. This is used to prevent back flow of flood water.

TIDE GATE : A lock gate provided at the mouth of an outfall structure to prevent the backflow of tide water.

TIDE GAUGE : A recording device to note the direction and speed of currents of tide water including the measurement of its level.

TIE : (i) A member which is in tension in a structure.

   (ii) A clip for fixing flexible metal sheets in roofing.

TIE BEAM : A beam which acts as a tie between two members. In a roof truss, it is the lowermost horizontal member between two supports. See illustration.

Principal rafter

TIE IRON : A wall tie made of iron or steel.

TIE LINE : A line that joins opposite corners of a four-sided figure for checking by triangulation.

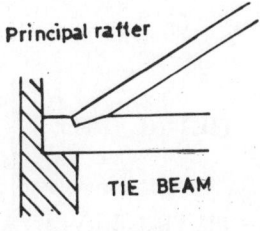

TIE BEAM

TIER : A leaf of half-brick thick wall forming one half of a cavity wall.

TIE ROD : A steel rod used as a tie. Sometimes the rod is threaded and is tightened by nuts.

TIGHT CESSPOOL : A water-tight cesspool from which the sludge slurry is taken out by pumping.

TIGHT KNOT : A firm knot around a timber.

TIGHT SHEATHING: Fixing matchboards diagonally by nailing them to rafters.

TIGHT SIZE : The size of a rebated opening for fixing glass with an allowance.

TILE : A thin burnt-clay or concrete plate commonly used for flooring or roofing. Roofing tiles are made of various shapes and sizes with fixing arrangement. Now a days, mosaic tiles with ornamental works are used in flooring and skirting. Glazed tiles are used in walls of bathrooms and toilets. Drain tiles are used for drainage in agricultural fields.

TILE BATTEN : A batten for fixing tiles.

TILED VALLEY : A valley covered with specially made valley-tiles at the junction of two sloping roofs.

TILE FILLET : A fillet made by cutting tiles and setting them at an angle to the wall over a roof.

TILE HANGING : Vertical tiling on a wall to throw out the rainwater. Also known as 'weather tiling'.

TILE PINS : Small wooden pegs used instead of nails in fixing tiles.

TILL : A term used for boulder clay. A till is a heterogeneous mixture of soil and rock fragments of varying sizes.

TILT : In air survey, the angle between the vertical and the optical axis of a camera is kept within 1° for taking vertical photographs.

TILTED PLATE SEPARATOR : A parallel plate separator used for removal of free oils from refinery effluents. It consists of corrugted plate modules installed at 45° to the horizontal. See illustration.

TILTING FILLET : (i)   An eaves board or fascia usually of triangular cross-secti on nailed

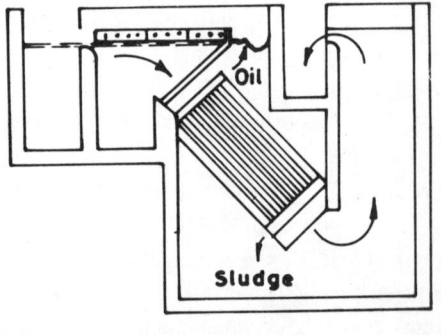

TILTED PLATE SEPARATOR

to the bottom ends of rafters or roof boarding.

(ii)   A sloping strip of wood fixed along the joint between a roof and a wall or at a valley gutter where two slopes meet.

TILTING GATE : A gate for spillways provided at the crest of a dam such that it opens by water pressure when the water reaches a certain level and it closes automatically when the water level goes down.

TILTING LEVEL : A levelling instrument with a bubble tube fitted on the telescope such that the axis of rotation of the telescope need not be vertical.

TILTING MIXER : A concrete mixing machine (power-operated or hand-operated) tilts when discharging mixed concrete.

TIMBER : Wood sections converted from logs (tree trunks) by dressing, baulking and sawing, which are used in constructional works.

TIMBER BRICK : A fixing brick.

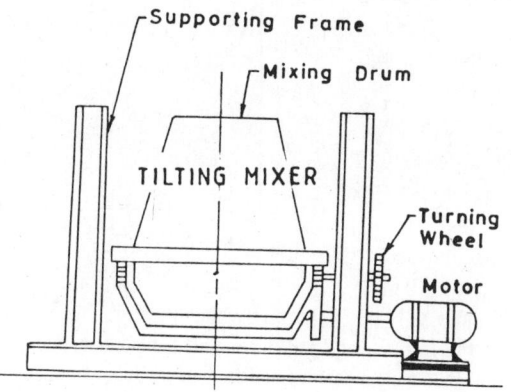

TIMBER FENCING : A fencing with timber pieces.

TIMBER FLOOR : A wooden floor formed by fixing timber boards on a timber frame.

TIMBER FRAMING : Making frames with timber sections for different types of roof truss, partitions, sloping roofs, flooring, etc.

TIMBER GRILLAGE : A grillage made of hard wood used in foundation of buildings in earlier days. This is replaced by steel grillage.

TIMBERING : Protection given to the sides of an excavation or a foundation trench with the help of supporting timbers.

TIMBER SCAFFOLD : A scaffold made of wood to facilitate construction of structures overhead. The frame consists of wooden standards, putlogs and ledgers fastened with ropes. Now-a-days, this is replaced by tubular steel scaffold.

TIME-AREA GRAPH : A graph which shows the impervious area con-tributing at any moment of time after commence-ment of a storm. In case of an absolutely regular impervious area it would be a straight line sloping upwards from zero acre at zero minute to total acres and time of con-

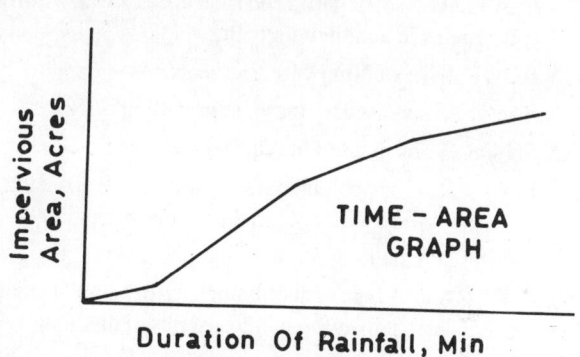

centration. The greater the deviation of the curve from the straight line and steeper the curve, the greater is the run-off expected.

TIME OF CONCENTRATION : It is the time required by an overland flow to reach the inlet point of a drain or sewer from the start of a rainfall plus the time of flow from the inlet point to the node at which the time of concentration is required. Thus, time of concentration = Inlet time + Time of flow.

TIME-SETTLEMENT CURVE: A curve presenting settlement predictions of a structure over a period of several years. The nature of such a curve is shown in illustration.

TIMESING COLUMN: A column in a proforma for quantity surveying, showing how many times the same quantity has been taken.

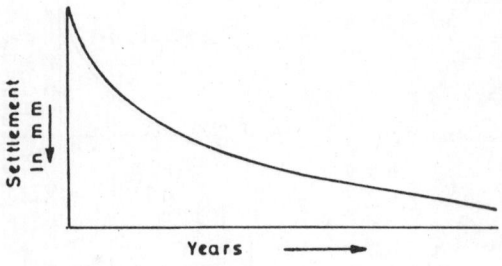

TIME- SETTLEMENT CURVE

TIN : A very soft and malleable white metal with slight tinge of blue. It can easily be flattened by hammering. The metal is rarely used alone. It is used for tin plating, lining lead pipes and for preparation of alloys and solder. It is one of the metals not attacked by water. Commercially pure tin is used for making infusion pots, evaporating basins, stills, etc. Tin is also used as a protecting coat for copper and iron utensils. Steel sheets coated with tin are used for making cans for milk, food and fruit industry.

TINE : (i)   A prong or tooth of a rake.

(ii)   A tooth of a dragline bucket or excavator bucket required for excavation.

TINGLE : A narrow strip of lead, zinc or copper used for fixing glass panes in patent glazing, or for fixing slates or for stiffening the edge of a flexible sheet metal.

TINNING : Coating iron or steel articles with a thin film of tin which resists corrosion or rusting.

TIN PLATE : A sheet of tin which is actually an iron sheet covered with a thin protective film of tin.

TIN PLATING : Coating iron and steel articles with a thin protective film of tin by dipping the articles in a molten tin bath.

TIN ROOFING : Roofing with tin sheets.

TIN SAW : A saw used by masons for cutting bricks.

TIN SNIPS : Scissors used for cutting metal sheets.

TINT : (i)   A light colour made by mixing more quantum of white pigment with a small amount of coloured pigment in a paint.

(ii)   A staining fluid.

TIPPER TRAY : A type of distributor used in a small percolating filter for sewage treatment which doses tank effluent into a series of distributor channels, whereby the effluent is spread over the bed.

TIPPING LORRY : A lorry which can unload the contents backwards by tipping the container.

TIPPING TROUGH : Same as 'Tipper Tray'.

TIPPING WAGON : A small railway wagon capable of unloading materials by end tipping.

TITAN CRANE : A heavy-duty crane in which the portal frame carries a swing jib crane capable of commanding an appreciable distance from its portal frame.

TITANIUM WHITE : Titanium di-oxide (TiO_2), a white pigment with good opacity is used in paints.

TOE : (i)   The blind end of a blasting hole.

(ii)   The free end (away from the retained material) of the base of a dam or retaining wall.

(iii) The lower part of a door stile.

TOE BOARD : A scaffold board which is fixed to the side of a scaffolding to prevent the dropping of tools and materials.

TOE FILTER : A graded filter provided at the toe of an earthen dam to protect it from piping due to seepage of water.

TOE LINE : A toe level i.e., the level to which the tipped ends of piles are driven.

TOE NAILING : Skew nailing used in carpentry.

TOE WALL : A vertically downward wall built at the toe of a retaining wall.

TONCAN IRON : A corrosion-resistant metallic substance.

TONER : An organic dye of strong colour.

TONGUE AND GROOVE JOINT : A joint between two boards used in flooring, ceiling, etc. as shown in illustration. The tongue of one board fits into the groove of the other board.

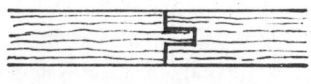

TONGUED & GROOVED

TONGUED, GROOVED AND METRED JOINT : A joint between two pieces of timber as shown in illustration. It is a secured joint.

TOOTH : Projected bricks or mouldings, e.g. dentiles.

TOOTHED PLATE : A bulldog plate used as a connector in woodwork.

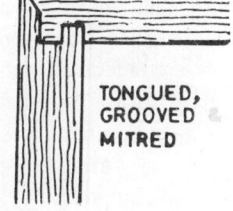

TONGUED,
GROOVED
MITRED

TOOTHING : Projections left in alternate layers of a brick work for bonding with future work.

TOP BEAM : A collar beam in a timber truss.

TOP BOOM : The top flange of a girder.

TOP COURSE TILES : The tiles used in the course next to the ridge tiles in a sloping roof.

TOP HUNG WINDOW : A window which is hinged at its top edge and opens outwards.

TOP LIGHTING : Lighting arrangement from overhead by providing skylight or borrowed light.

TOP LOG : The topmost log obtained from a tree trunk.

TOPOGRAPHICAL SURVEYING : The land survey plotting the natural features of a country, such as hills, woods, streams, rivers, lakes and artificial features like roads, railways, channels, canals, housing, etc.

TOPPING COAT : A floating coat in plastering.

TOP SOIL : The soil layer of 6″ to 12″ thickness at the ground surface that supports vegetation. This soil is usually composed of silt and humus.

TORN GRAIN: A strip of wood chipped off or torn below the finished surface, of a timber. This may occur due to defective cutting tool.

TORQUE: The twisting effect of a force acting tangentially on a body.

TORQUE STEEL : Stranded rods of steel used for reinforcement in concrete. This increases the bond strength.

TORSHEAR BOLT : A high strength friction-grip bolt.

TORSION : Twist or torque.

TORTUOUS FLOW : Turbulent flow.

TORUS ROLL : A horizontal wooden roll provided at the line of intersection of two slopes in a roof.

TOUCH DRY : The stage in drying of a painted surface when the paint film does not exhibit stickiness if touched by fingers softly.

TOUGHNESS INDEX : The ratio of plasticity index to flow index.

TOWER BOLT : A large size barrel bolt made of steel or brass, used in doors and windows.

TOWER CRANE : A swing-jib crane placed on a tower which facilitates in working in a congested area. The tower base may be mounted on wheels.

TOXICITY : The killing strength of a certain organic and inorganic compounds. The toxicity of waste waters is evaluated by 'Bioassay tests'.

TOXIC SUBSTANCES : Arsenic, cadmium, chromium, cyanide, lead and selenium are considered to be toxic substances, when their concentrations exceed a certain limit. The maximum alowable concentrations of these substances in water are presented in the illustration.

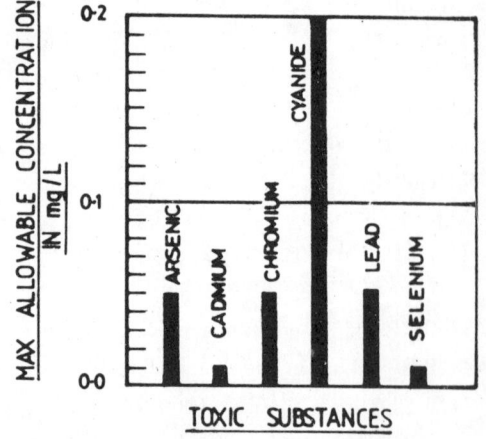

TRACK : (i) A path.

(ii) A single rail or a pair of rails over which a monorail or a train moves.

TRACK BOLT : A coach bolt or chair bolt for fixing rails to the sleeper.

TRACK CABLE : The cable of an aerial ropeway on which the container wheels move. This is made of steel wire rope.

TRACKING : Lines of wear marked on a road surface due to running of vehicles following the same track.

TRACK SPIKE : (i)   A spike of square cross-section made of steel driven into a wooden sleeper for fixing flat-footed rail.

(ii)   A hardwood piece inserted into a hole of a concrete sleeper to facilitate driving of a spike for fixing rail or cast iron rail chair.

TRACK STRINGER : A piece of timber 100mm x 250 mm placed under each rail in place of sleepers on soft ground.

TRACTION ROPE : The haulage rope used in a cableway or aerial ropeway.

TRACTIVE FORCE : The required pull to draw a locomotive over the rails, depending on the weight of locomotive and the co- efficient of friction. On upgrades, the tractive force is reduced by an amount equal to the loco-weight multiplied by the gradient and is increased by the same amount on downgrades.

TRACTIVE RESISTANCE : The frictional resistance offered to a tractive force and is given by the ratio of the locomotive draw bar pull to the weight of the locomotive. It comprises rolling resistance between wheels and rails, resistance due to bends and grades and bearing friction. Also known as 'Co-efficient of Traction'.

TRACTOR : A self-propelled vehicle used for ploughing or towing a rooter, bowl scraper and grader.

TRACTOR SHOVEL : A shovel mounted on crawler tracks or wheels used for loading with a maximum lift of 3 m. It has tipping arrangement.

TRAFFIC CAPACITY : The maximum number of passenger cars that can pass a given point on a roadway in one direction in an hour.

TRAFFIC DENSITY : The volume of traffic per metre width of a carriageway per day.

TRAFFIC INTENSITY : Weight of traffic in tons per metre width of a carriageway per day.

TRAILLING CABLE : A flexible, rubber-insulated electric cable (conductors) used for supplying power to a monorail, crane, loader, dragline, conveyor, etc.

TRAINING WALL : A wall built to guide the flow path of a river.

TRAINING WORKS : Spurs, groynes and dykes are examples of training works for a flowing stream or river. These are required to control the scour and siltation and also to regulate the flow path of a stream or river.

TRANSIT : (i)   The culmination of a star.

        (ii)   Change from one phase to the other.

        (iii)   A theodolite whose face can be changed.

TRANSITION CURVE : An easement curve of varying radius which negotiate a curve with a straight line. This facilitates in running a vehicle smoothly from a straight path to a curved path.

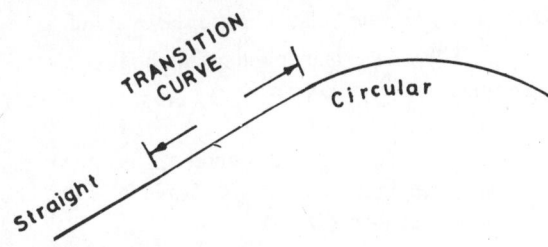

TRANSITION LENGTH : The length of a transition curve.

TRANSIT MIXER : A truck mixer which facilitates in mixing ingredients for a mortar or concrete during transit from the materials store to the work site.

TRANSLUCENT GLASS : Obscured glass which maintains privacy with entry of light.

TRANSMISSION LENGTH : In prestressed concrete it is the grip length of a tendon required for prestressing.

TRANSOME: (i)   A fanlight.

            (ii)   A horizontal piece of timber or stone separating the fanlight from a door or separating lights in a window.

TRANSPORTER BRIDGE: A lattice girder bridge spanning between two towers, along which a crab travels and a container carrying vehicles is hung at road level from the crab. Thus, vehicles are transported from one end to the other.

TRANSPORTER CRANE : A crab travelling along a lattice girder resting on two towers, carries materials by means of a grab and transfers the same to the desired place. This is used for loading and unloading ships and trains.

TRANSVERSE LOADING : Loading at right angles to a structure, e.g. a beam loading.

TRAP : A device provided in a waste water pipeline or in a drainage system to arrest solid wastes and to prevent foul gases coming in. There are quite a number of different types of trap to serve specific functions. These are grease trap, drip trap, slush-mucks trap, silt trap, P-trap, Q-trap, S-trap, etc.

TRAPEZOIDAL RULE : A rule used to determine the area of an irregular shape of land by dividing it into 'n' number of strips of equal width (common distance) 'd'. If O_1 be the first ordinate and O_2 be the second ordinate, then the rule is :

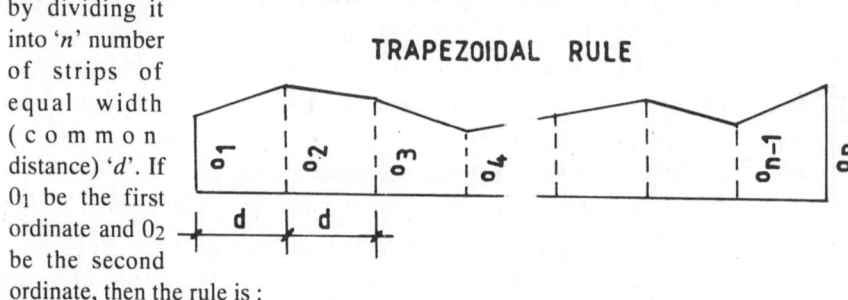

TRAPEZOIDAL RULE

Area = $d(O_1/2 + O_2 + O_3 + \ldots\ldots O_{n-1} + O_n/2)$

TRAP POINTS : In railway engineering, points are provided on running rails to avoid incorrect switching of a train.

TRASH RACK : A screen made of parallel bars provided across a stream to arrest floating matters as well as large heavy solids carried by the stream.

TRASS : (i)   Burnt clay clinker used by grinding i.e., surki

(ii)   A volcamic ash resembling pozzolana.

TRAUTWINE'S FORMULA : An empirical formula used for fixing the thickness of an arch with cut stone work.

$t = (R + 0.5S)^{0.5}/4 + 0.2$; Where t is the thickness of arch, in ft., R = radius of the arch at crown, in ft. S = span in ft. r = rise in ft.

$R = [(0.5S)^2 + r^2]/2r$

TRAVEL : The projection to be given to a flue wall by corbelling.

TRAVELLER : (i)   A single beam or a pair of beams mounted on wheels run on the crane rails. The beams carry the travelling crab of an overhead travelling crane.

(ii)   The middle 'boning rod' which is moved along the ground between the end boning rods for the purpose of checking the ground level.

TRAVELLER GANTRY : A stationary gantry on which a crane crab travels over the rails.

TRAVELLING FORMS : Moving forms, usually a built-up formwork for casting concrete walls. The formwork is supported on wheels or rollers so that it can be moved from one position to the other with the progress of the work and without dismantling and rebuilding it. Travelling forms are similar to sliding forms with the difference that a travelling form moves horizontally and a sliding form moves vertically.

TRAVELLING GANGER : A moving or walking ganger.

TRAVELLING GANTRY : A gantry with a crab or hoist is built on wheel and it travels on the rails.

TRAVELLING SCREEN : A moving trash rack or a drum type rotating screen.

TRAVEL MIXER : A self-propelled mixer which receives soil at its one end from a pile, mixes up at optimum moisture content and discharges through the other end. This is used in soil stabilization work.

TRAVERSE : (i)   A survey in which a number of lines are connected with measured lengths and bearings. It may be a open traverse or a closed traverse.

(ii)   A dressing iron.

TRAVERSE TABLES : Tables made to find out the differences of latitudes and departures for different angles.

TRAVERSING   :   The illustration depicts how traversing is done in a traverse survey.

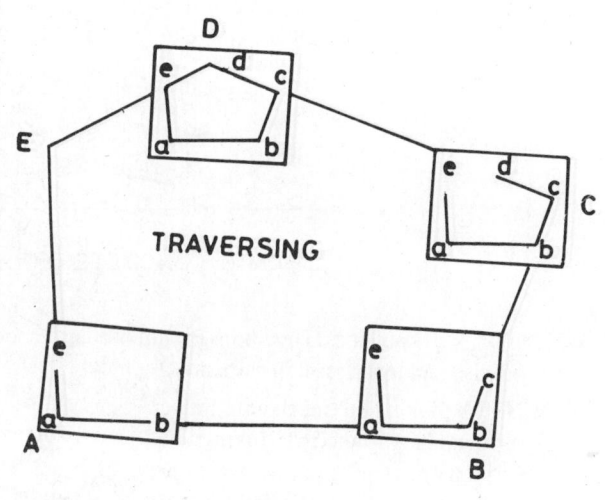

TRAVERSING

TRAVERSING BRIDGE : A bridge across a stream or channel that moves to allow a vessel to pass.

TRAVERSING SLIP-WAY : A slipway on which ships of light-weight are hauled and it is traversed sideways to transfer the ship to another berth for repair.

TRAVIS TANK : The first dual purpose tank incorporating sedimentation and sludge treatment, installed in England in 1904. See illustration.

TREAD : The horizontal part of a step.

TREE CUTTER : A tree-dozer or a tractor with a sharp toothed blade ahead, used in cutting a tree.

TREE GUARDS : Fencing made from old drums and bamboos or made of brickwork or concrete rings to protect young trees by allowing free circulation of air and sunlight.

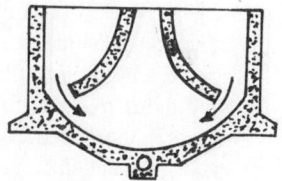

TRAVIS TANK

TRELLIS WINDOW : A lattice window usually made of timber battens or strips of steel.

TRELLIS WORK : Wooden jalli made of battens by placing them at right angles or diagonally.

TREFOIL ARCH: See illustration. The use of this type of arch is found in temples, mosques and churches. It produces architectural effect.

TREMIE : A water-tight steel pipe long enough to touch the bottom of a river or stream with its upper end above the water level. The upper end is provided with a hopper for pouring concrete.

TREMIE CONCRETE FOUNDATION : A break water constructed with tremie concrete foundation as shown in illustration. Placing of concrete by tremie prevents segregation of the ingredients.

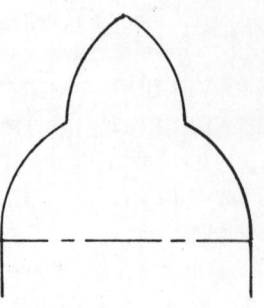

TREFOIL ARCH

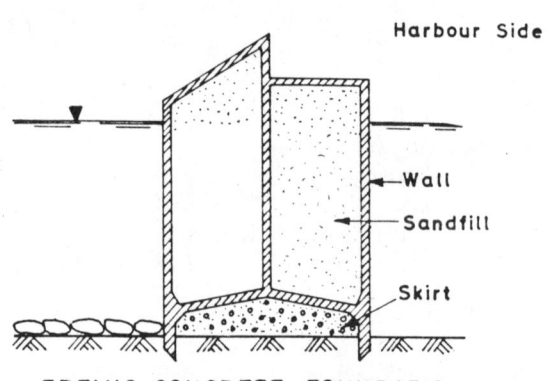

TREMIE CONCRETE FOUNDATION

TRENAIL : Also known as 'Drawbore Pin'. It is a hardwood pin driven into a hole across a mortise and tenon joint in carpentry.

TRENCH : A foundation trench with timber supports is shown in illustration. Timbering is essential in deep foundation trenches, especially in soft or loose soil.

TRENCH EXCAVATOR : A self-propelled trenching machine on a crawler track operated by a bucket-ladder excavator for digging trenches quickly. In normal digging, the speed is 1 ft./minute. It makes a clean-sided neat trench upto a depth of 12 ft. with 4 ft. width.

TRENCH HOE : A backacter.

TRENCHING GROUND : Sewage sludge and night soil are sometimes deposited in shallow trenches in porous soil and covered with good soil. The sludge and night soil become converted into manure by anaerobic decomposition.

TREPAN : A tool used in shaft sinking by dropping it through 1 ft. in each stroke and turning slightly simultaneously.

TRESTLE : A tower usually made of steel frame which can be erected quickly and dismantled as and when required without any damage of individual members. This is chiefly used for electric transmission lines, aerial ropeways and also for sending radio waves.

TRESTLE BRIDGE : A lattice girder bridge resting on trestles at its ends. This type of structure is light in weight.

TRIAL PIT : A pit or hole dug to find out the nature of soil strata or to reveal the underground utility lines.

TRIANGLE : (i)   Set square.

(ii)   In railway engineering, the arrangement made in form of a triangular track for changing the directon of engines as shown in illustration. This type of junction has three acute angle crossings, three pairs of switches and 6 nos. of check rail.

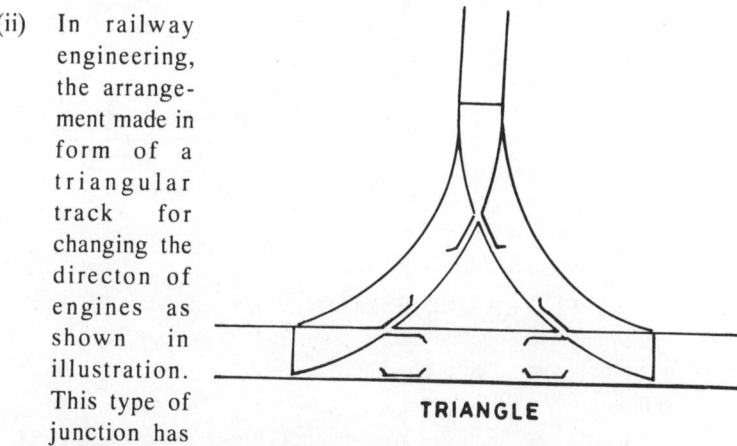

TRIANGLE

TRIANGLE OF ERROR : The most frequently used method for determining the correct orientation of the table at a survey station in plane table surveying. First, the table is oriented as correctly as possible and clamped. The straight edge is laid on the paper to intersect 'A' the line of sight directed to A. Similary two more rays are drawn in the direction of B and C. If the rays are concurrent, the point of intersection is the required station point, 0, i.e., the table is correctly oriented. A triangle of error is formed if the table is not correctly oriented. The solution is made graphically as three-point problem.

TRIANGULAR NOTCH : A Vee notch or weir to measure the discharge of a liquid through it. If the angle of the V-notch is $90°$ and h is the depth of liquid in ft. above the tip, then discharge $Q = 2.5\,h^{2.5}$ cusecs.

TRIANGULAR WEIR : See 'Triangular notch'.

TRIANGULATION: In topographical or photographic survey, the whole area is covered with a network of triangles for accurate measurement of the area. A base line is selected judiciously and measured accurately. The angles of the triangles are measured correctly and the lengths are computed by applying trigonometry. Large triangles are formed in primary triangulation, which are divided to form secondary triangulation and they are further subdivided into tertiary triangulation.

TRIAXIAL COMPRESSION TEST: The confined compression test of a soil sample contained in a rubber bag, the pressure being given to the sample from all directions. By this test, measurement of sheering strength of a soil sample is done. During the test, the deformation against load applied is recorded.

TRIBRACH : The base of a theodolite carrying three foot screws, which is essential for levelling the instrument. It facilitates quick adjustment of the instrument.

TRICKLING FILTER : The sectional view of a trickling filter (Bio-filter) is shown in illustration. It consists of a bed of stone blocks or brick bats over which settled sewage is sprinkled. As the sewage trickles through the bed, B. O. D. is removed due to biosorption by the slimes formed on the surface of the stones or bats. Natural aeration is employed in this process. The sewage is sprinkled over the bed uniformly by means of a distrubutor and the effluent goes out through the underdrain.

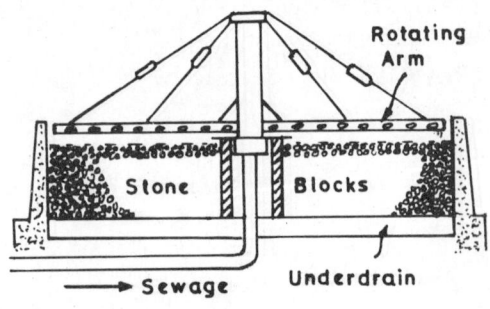

TRICKLING FILTER

TRICKLING FILTER UNDERDRAIN : The underdrain is made of hollow concrete blocks forming a false bottom. The underdrain should be open at both ends for the purpose of ventilation and flushing.

TRIEF PROCESS : A process of making blast furnace slag cement by wet grinding the furnace slag and mixing it with ordinary portland cement.

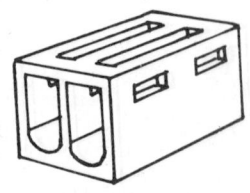

TRICKLING FILTER
UNDER DRAIN

TRIGONOMETRICAL STATION : A survey station used in triangulation survey.

TRIGONOMETRICAL SURVEY : A survey by triangulation to obtain the highest degree of precision in measuring areas and plotting the field data in a map.

TRIM : Dado,skirting Architrave, etc.

TRIMMED JOIST : A joist trimmed at an opening and is supported by a trimmer joist.

TRIMMER ARCH : An arch spanning from a chimney back to a trimmer joist carries a fireplace in a timber floor.

TRIMMER JOIST : A header joist of short length carried on the trimming joist, by tusk tenon through them. The trimmer joists carry the common joists.

TRIMMING : (i)   Final dressing up of a surface.

(ii)   Framing round an opening to strengthen it.

TRIMMING JOIST : A heavy joist parallel to the common joists, carries a trimmer joist (header joist). Thus, the load carried by the common joists is tranferred to the trimming joist through the trimmer joists.

TRIMMING PIECE : A camber slip.

TRINITROTOLUENE : TNT, which is highly explosive.

TRIPOD : A three-legged support used for placing a survey instrument.

TRITURATOR : A type of grinder used for shredding sewage screenings .

TROCHEAMETER : An instrument to determine the distances in expeditionary or route surveys by measuring and recording the number of revolutions of a wheel.

TROMMEL : A rotating screen.

TROUGH GUTTER : A box gutter.

TROWEL : A steel plate with a wooden handle used by masons in brick laying, flooring and plastering. Its shape and size vary with the nature of work.

TROWELLED FACE : A plastered surface finished with a trowel.

TRUCK MIXER : A transit mixer for making mortar or concrete. The mixing machine is mounted on a truck or lorry so that mixing is done during transport from the store house to the worksite.

TRUE BEARING : The horizontal angle made by a line with the true north.

TRUE MERIDIAN : The geographical north-south plane.

TRUNNION AXIS : The horizontal axis about which a thedolite telescope can rotate.

TRUSS : A frame of timber or steel used in roofing, bridging and making partitions.

TRUSSED ARCH : An arch made of rolled steel sections.

TRUSSED BEAM : A beam stiffened by a camber rod or tie rod.

TRUSSED PARTITION : A framed partition, the frame being covered by lath.

TRUSSED ROOF : A roof supported on trusses.

TRY PLANE : A truing plane, usually a bench plane used after shaping by a jack plane.

TRY SQUARE : A simple instrument as shown in illustration is commonly used by the carpenters and fitters to check the evenness of a surface and also to check whether the two adjacent planes or edges are at right angles or not.

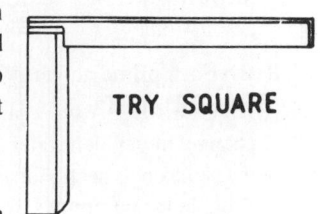

TRY SQUARE

TUBE : A pipe of very small diameter.

TUBE RAILWAY : An underground railway through a cylindrical tunnel. This is built either by cut and cover method or by tunnelling.

TUBULAR BRICK : A hollow brick as shown in illustration.

TUBULAR SAW : A hole saw.

TUBULAR SCAFFOLDING : A scaffolding made of steel tubes or tubes of light alloys. This can be erected or dismantled quickly.

TUBULAR SECTIONS : Hollow rolled sections-circular, square or rectangular.

TUBULAR BRICK

TUCK : A recess kept in a horizontal mortar joint in brickwork which is filled with putty with projection. This is also known as 'Tuck pointing'.

TUCK IN: The part of a bitumen felt sheet in roofing or skirting, which is bent into a chase in the wall.

TUCKING BOARD: A thin board placed horizontally in a 'poling frame'.

TUCKING FRAME : A poling frame used in timbering a trench, in which the walings support the poling boards.

TUCK POINTING : A protective as well as decorative pointing work in brick wall. See 'tuck'.

TUMBLER : The part of a lock to hold the bolt in place for locking.

TUMBLING BAY : A back drop in a manhole.

TUNG OIL : China wood oil which dries quickly and is water- resistant; It is used in paints.

TUNNEL : An underground passage.

TUP : A drop hammer.

TURBIDITY ROD : For measuring turbidity of a raw water, this rod is lowered into the water and the depth at which the platinum needle ceases to be seen under standard light indicates the turbidity of water.

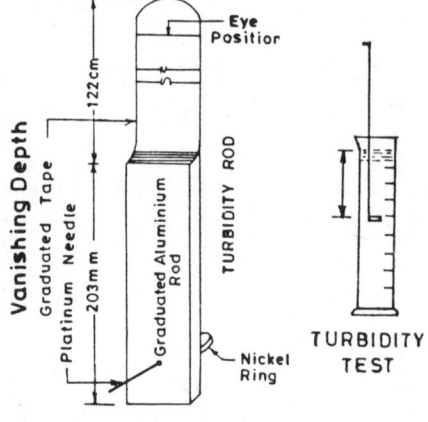

TURBIDITY TEST : A simple test to measure turbidity of water by immersing a platinum needle of 1 mm. dia. into a turbid water contained in a graduated glass jar. The depth at which the needle becomes invisible, gives the measure of turbidity from the reading of the graduations.

TURBINE : A prime mover driven by gas, steam under high pressure or falling water.

TURBINE PUMP : A type of a centrifugal pump in which the channel into which water flows on leaving the impeller has the same cross sectional area throughout. This pump is not suitable for pumping sewage.

TURBINE SEWER CLEANER : It consists of a set of cutting blades which are revolved by a hydraulic motor. The turbine is pulled through a sewer by means of a cable.

TURBULENCE : A flow condition when a liquid gets disturbed by eddies.

TURBULENT FLOW : A tortuous flow of a liquid at a velocity above the Reynold's critical velocity. This is opposite to streamline flow.

TURBINE PUMP

TURFING : Covering a ground with growing grass taken out along with soil from another site. This helps to prevent erosion in slopes of earthen embankments.

TURN BRIDGE : A swing bridge.

TURN BUCKLE : A screw shackle used in holding and tightening guy ropes, tie rods, etc.

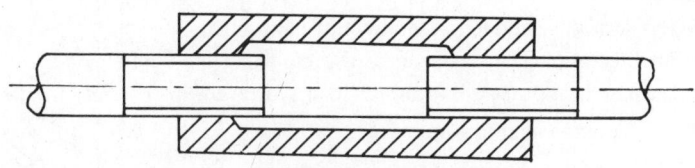

TURN BUCKLE

TURNED & BORED JOINT : A type of pipe
 joint as shown in illustration.

TURNED BOLT : A tight-fitting bolt which is
 made of exact dimension as that of the hole
 by means of turning it in a lathe machine.

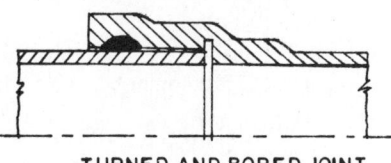

TURNED AND BORED JOINT

TURNING BAR : A chimney bar.

TURNING PIECE : A chamber slip or a trimming piece.

TURNING PIN : A tampin used in plumbing.

TURNING POINT : A survey station or a change point.

TURNING SAW : A bow saw.

TURN OUT : A railway line branching
 out from a main track running
 straight as shown in illustration.

TURN SCREW : A screw driver used
 by iron mongers.

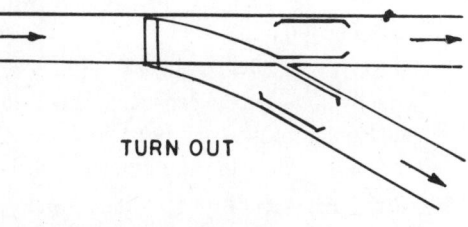

TURN OUT

TURN TABLE : A round table
 centrally pivoted such that it can
 revolve. It is provided with rails
on top and it has locking arrangement. This is used in diverting locomotives to any
desired direction. The locomotive is placed on it by running and it is sent off to the
desired direction by revolving the turn table and matching the railway tracks in the
desired direction.

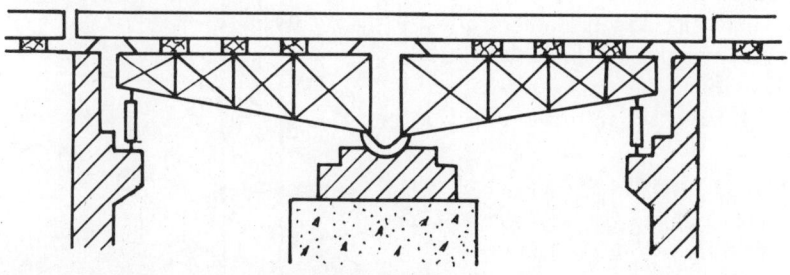

TURN TABLE

TURPENTINE : A good solvent used in paints. This is obtained by distillation of oleo resin
 of pine trees.

TURRET STEP : A triangular step used in a spiral stair.

TUSK NAILING: Skew nailing i.e, nails are driven inclined.

TUSK TENON: A mortise and tenon joint which is keyed as shown in illustration. The tenon of a trimmer joist is introduced into the mortise of a trimming joist and keyed.

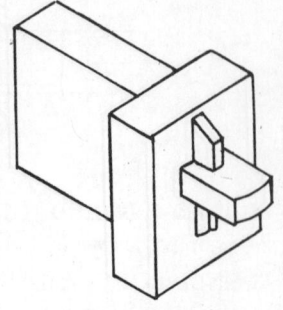

TWIN TENON : A tenon having two prongs.

TWIST : Spiral warp found in timber.

TWISTED FIBRES : Inter-locked fibres or grains as found in a timber.

TWIST GIMLET : A gimlet with a helical groove which facilitates in removing wood shavings.

TWITCHER : A trowel used in finishing a margin in plastering.

TUSK TENON

TWO-COAT WORK: Application of two coats of plastering, painting, whitewashing, etc.

TWO-HINGED ARCH : Two-pinned arch i.e, an arch-shaped structure provided with hinges at two supports.

TWO-LEG SLING : A sling made of two chains, each hanging from one thimble and with hooks from their lower ends.

TWO-LIGHT FRAME : A window frame with a central mullion dividing equally into two lights.

TWO-PEG TEST : A test carried out on a moderately level site by setting the thedolite exactly midway between two pegs at a convenient distance apart, the objective being adjustment of the telescope bubble which is one of the permanent adjustments of a theodolite. Thus, the axis of the bubble is made parallel to the line of collimation.

TWO-PINNED ARCH : A two-hinged arch.

TWO-PIPE SYSTEM : A house drainage system in which two vertical waste water pipes are provided, of which one carries sullage from bathroom & kitchen sink and the other carries night soil as shown in illustration. The two down pipes have separate vent pipes.

TWO-POINT PROBLEM : When in plane tabling one or two points are found either inaccessible or unsuitable for instrument station due to local attraction by magnetic influences, the instrument is fixed at the desired location by accurate orientation of the

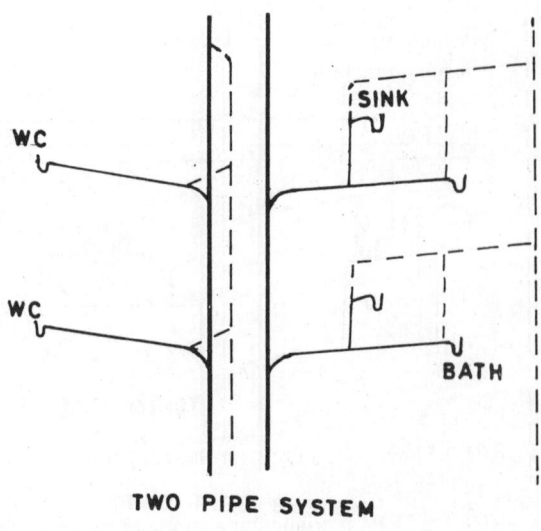

TWO PIPE SYSTEM

table by the compass and by intersection of rays. This method is known as 'Two point problem'. The figure illustrates an example in which A is unsuitable for instrument station and it is required to locate O and C in plan.

TYROLEAN FINISH: A surface made by throwing plaster onto a wall and left rough as it is.

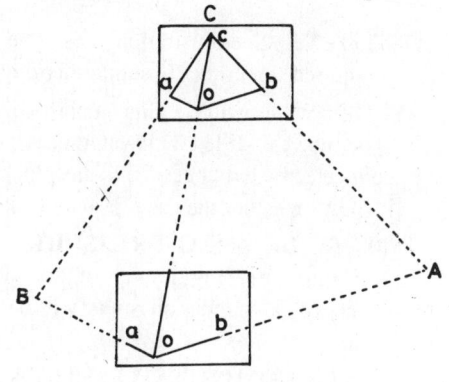

TWO POINT PROBLEM

U

U-ABUTMENT : See illustration. This type of abutment is required sometimes to support a bridge at its ends.

U-GAUGE : Water gauge ; This consists of a U-Shaped glass tube half-filled with water, one end of which is connected by a rubber tube to gas pipes to be tested. It shows whether the gas pipes are leaky or not.

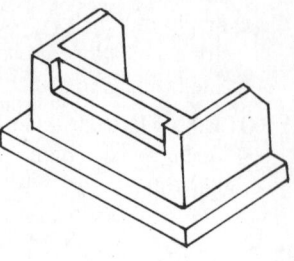

U – ABUTMENT

ULTIMATE BEARING PRESSURE : It is the maximum bearing power of soil which when exceeded, a foundation sinks without increase of load.

ULTIMATE COMPRESSIVE STRENGTH : This is the maximum compressive stress which a material can withstand. If this stress is exceeded the material crushes.

ULTIMATE STRENGTH : Breaking strength. The maximum stress which a material can withstand before breaking.

ULTIMATE TENSILE STRENGTH : The load at which a specimen under tensile test breaks, divided by its original cross-sectional area.

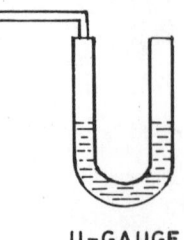

U-GAUGE

ULTRAMARINE : This is a precious blue pigment with good alkali resistance. Natural ultra-marine costs as much as gold.

UMBER : Raw umber is a brown pigment consisting of hydrated iron-oxide, obtained from naturally coloured clays originally found in Umbria (Italy). The best umber comes from Turkey. Burnt umber, which is rich and deep reddish-brown in colour, is produced by calcining raw umber. To produce stone colour, this is to be mixed with white lead.

UMBRELLA ROOF : A station roof.

UNAVAILABLE MOISTURE : The moisture in between the hygroscopic co-efficent and the wilting co-efficient, which can not be normally used by plants.

UNBUTTONING : Demolition of steel structures by chipping of the rivet heads.

UNCLOSED TRAVERSE :

The traverse in which the starting point and finishing point do not coincide i.e. the traverse does not form a complete circuit. This is required to survey a long narrow strip of country coast line, meandering river, etc.

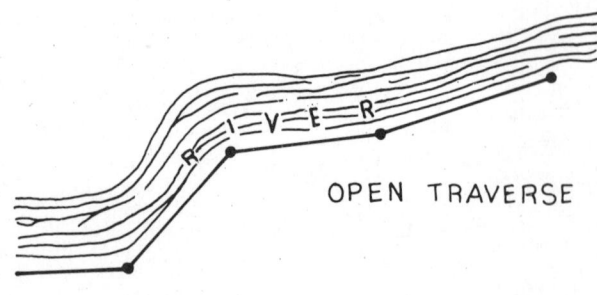

OPEN TRAVERSE

UNCONFINED COMPRESSION TEST : The simplest and quickest test to determine the shear strength of a soil. This is carried out on a clay sample of cylindrical shape, 3.8.

cm. dia. and 7.6 cm. high without lateral restraint. Half the unconfined compressive strength of a soil is its cohesion or shear strength.

UNCONFINED COMPRESSIVE STENGTH : This is a measure of the consistency of clays and is defined as the load per unit area, at which an unconfined cylindrical specimen of soil will fail in a simple compression test.

UNDER BRIDGE : A road passing under a bridge carrying another road along the top of the embankment.

UNDERCLOAK : A layer of slates or tiles at eaves or verges placed under the surface tiles in a roof.

UNDERCOAT : This relates to the filling properties, opacity, colour and base for the finishing coat. This is highly pigmented to cover most of the surface irregularities. Heavy undercoat should be avoided.

UNDERCURING : Insufficient hardening due to too short a hardening period or low temperature (in case of glue), or insufficient water (in case of concrete).

UNDERDRAINAGE : It is a necessary part of sewage-treatment by land filtration . It is the drainage of the subsoil which is important to the public health Engineer as well as structural Engineer . Underdrainage for building development is frequently constructed.

UNDERFILTER : See illustration.

UNDERFLOW : Movement of subsoil water under a structure.

UNDERFILTER

UNDERGROUND KILN : Bull's trench kiln, circular or oval in plan with concentric brick walls and one or two chimneys. This is required to burn bricks.

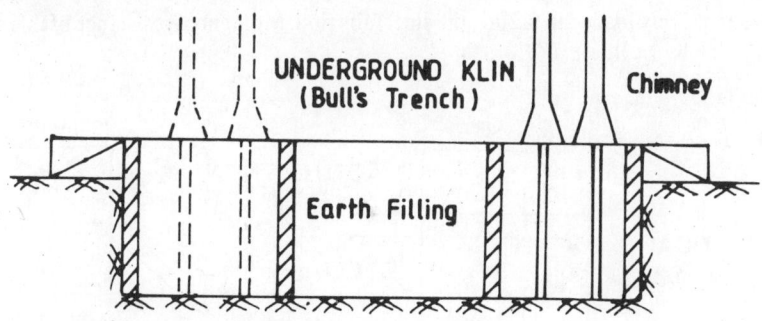

UNDERGROUND RAILWAY : Tube railway laid below street level of a city.

UNDERGROUND SURVEY : Mine survey.

UNDERLINING FELT : Sarking felt.

UNDER MINING: The rise of water level at the up-stream side of a weir increases the flow through subsoil . If the force exerted by the sub-soil water to flow in its direction is greater than the retaining force of the soil in front, the water will carry with it the soil particles by making them loose. Thus, undermining of the foundation of a weir takes place. The undermining effect starts from end and it may cause either sinking or washing away of the structure itself.

UNDERPIN: To provide a new permanent support beneath a wall or column or any other structure without removing the existing superstructure or without creating any disturbance to the existing structure. The underpinning may be carried out in concrete or brickwork. This is required when the load coming on a wall or column exceeds the predetermined value or there is any unequal settlement due to low-bearing power of soil, or when it is required to make a large opening in an existing wall.

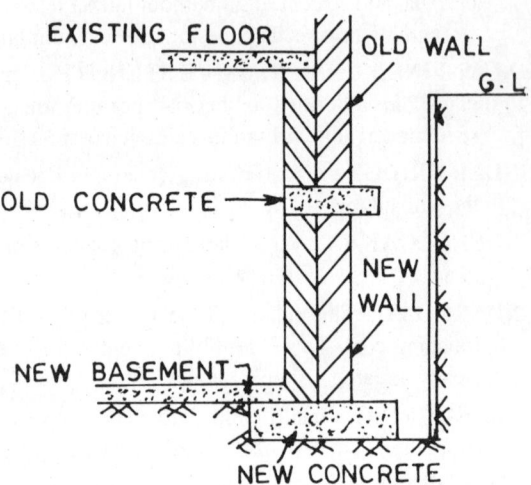

UNDERPITCH GROIN : Intersection of two barrel–vaults of different rise.

UNDERPLANTING : A contract on which the machinery is over loaded, is said to be underplanted i.e. more machines are to be employed.

UNDER-REAMING : To introduce sufficient gravel outside and around the screen at the bottom of a wall for straining effect, under-reaming or hydraulic jertting is resoted to. Under-reaming may be done by the use of expanding reamer blades.

UNDER-REINF RCED BEAM : If a reinforced concrete beam has steel less than the necessary amount of steel, the beam is said to be under-reinforced. such a beam when progressively loaded to failure, the steel fails first. Moment of resistance of this section is given by M.R. $= t.A_t(d-n/3)$.

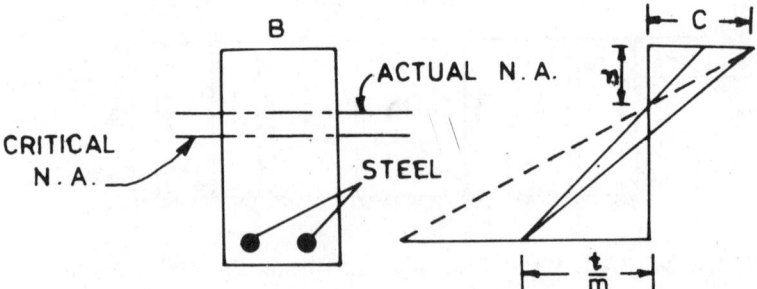

UNDER-RIDGE TILES : Tiles set in the top course of a roof under the ridge tiles.

UNDERSLUICE : This is constructed at the flank of the weir near the head regulator . Its function is

    (i)   to scour silt accumulated in the u/s of weir in front of the head regulator and

    (ii)   to reduce the highest flood level during flood season.

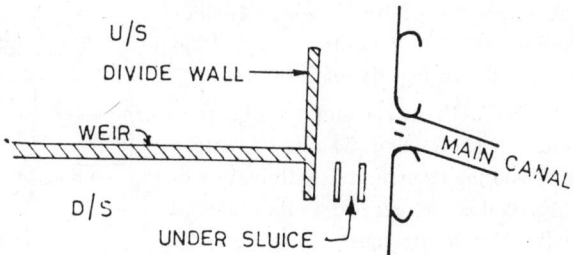

UNDER TILE : See 'under-ridge tile'.

UNDERTONE : The colour obtained by mixing a large quantity of white pigment with the coloured pigment.

UNDISTURBED SOIL SAMPLE : A sample of cohesive soil from a bore hole or trial pit obtained by driving soil sampler into the ground . This is done such that the soil structure and its properties are not disturbed during sampling .

UNDRAINED SHEAR TEST : A 'Quick test'; Tri-axial compression test of a sample of soil.

UNDRESSED TIMBER : A timber sawn, but not planed.

UNEQUAL ANGLE : An angle section having legs of unequal lengths.

UNFRAMED DOOR : A batten door; Ledged & braced door.

UNGAUGD LIME PLASTER : Plaster made with lime, sand and water, lime having compressive strength of at least 100 psi.

UNIFIED SOIL CLASSIFICATION : This calssification was introduced by cassagrande in U.S.A. in 1942. This is also termed the Airtield classification (A.C). The coarse soils are classified by their grain size and the fine-grained soils are classified according to their indices.

UNIFLOW SETTING TANK : A specially designed settling tank to reduce the detention period and cost of construction by decreasing the depth gradually towards the effluent end without affecting the uniform velocity through the tank. This is provided with link-belt sludge removal equipment. This tank can be used for primary as well as secondary sdimentation. See illustration.

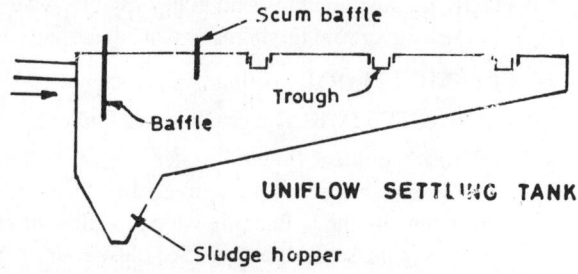

UNIFORMITY CO–EFFICIENT : The ratio between the grain diameter larger than 60% by weight of the soil particles to that diameter larger than 10% by weight of the soil particles in a soil sample . It is expressed by D_{60}/D_{10}. D_{10} is called the 'effective size'.

Uniformity coefficient of a uniform soil is less than 3.

UNIFORM FLOW : It is the steady flow in a stream.

UNIFORM SAND · A sand, most of the grains being of uniform size and it is not a graded sand.

UNION : A screwed pipe fitting which enables water or gas pipes to be easily connected and disconnected. It is usually made of brass.

UNIT HYDROGRAPH : The unit hydrograph is a curve that shows that the rate of surface run-off resulting from rainfall within a unit of time. From this, the peak and other rates of run-off from a particular basin can be determined.

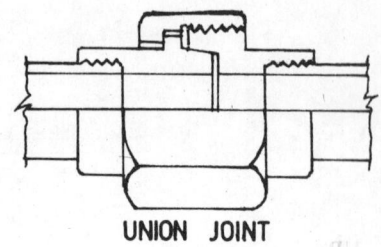

UNION JOINT

UNIT STRESS : Unital stress. load per unit area, or internal force.

UNIT WEIGHT : The weight of unit volume of a material.

UNIVERSAL JOINT : See illustration . This has the advantage of not requiring the pouring of a molten compound. It does not require skilled labour.

UNIVERSAL MOTOR : A motor which can work on A.C. or D.C. This is usually of less than 1 H.P.

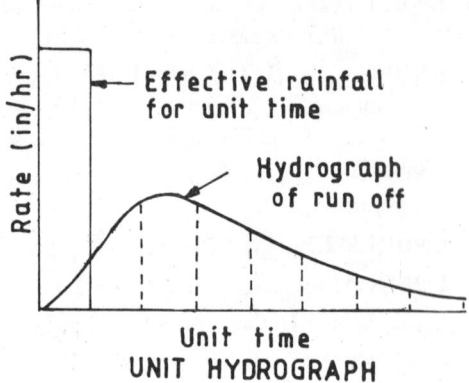

Unit time
UNIT HYDROGRAPH

UNIVERSAL PLANE : A combination plane; This is provided with various types of cutting irons to cut different mouldings rebates, beads, grooves, etc.

UNSOUND KNOT : A knot which is softer than the wood round it. It is a defect of timber.

UNSTABLE : A structure is said to be unstable, when it is liable to fail by overturning, sliding or other defects in construction.

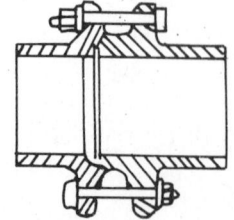

UNTRIMMED FLOOR : A floor simply-supported by common rolled steel joists.

UNWROUGHT TIMBER :See undressed timber.

U-PUMP : Intermittent flow of water in a pump results in vibration, shock and loss of energy. This is partly overcome by the U-Pump in which the flow of water is always in the same direction . See illustration.

U-SHAPED SEWER : A sewer whose cross-section is like U. This is used in combined system of sewerage. See illustration.

UPCAST SHAFT : A ventilating shaft to pass the vitiated air in an upward direction.

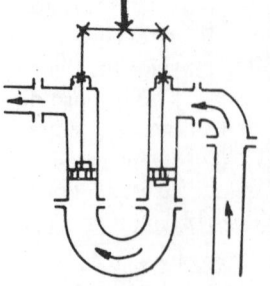

UPLAND CATCHMENT : In some areas water supplies are dependent on upland catchments . The run-off per acre from upland catchments is generally very high compared with lowland catchments . Water from upland catchemnts is of excellent quality, free from suspended matters, free from animal contamination and sewage and it is soft in nature.

UPLIFT PRESSURE : An upward force due to the presence of water in earth, which finds its way through cracks or pores into interior of a dam or any other structure. This may cause the formation of quick sand and it may be prevented by draining the water. This is also called 'Pore pressure'.

UPPER TRANSIT : The upper culmination of a star or the sun.

UPSETS : Rupture of tissues found in a tree. These are formed, if an young tree is injured by crushing due to violent wind.

UPSTREAM BLANKET : See illustration. This gives protection to a dam.

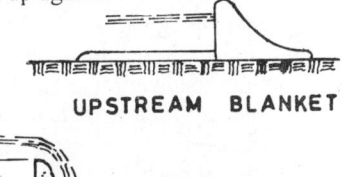

UPSET

UPTURN : The part of a lead flashing which is dressed up against a wall face.

UPTURN BUCKET : This is constructed to protect the down stream bed of a dam from the high velocity of water flowing through deep set sluices. At the bottom of the dam, an upturn bucket is provided.

UPSTREAM BLANKET

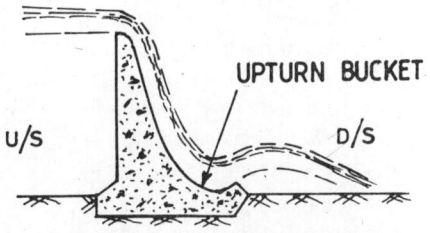

UPTURN BUCKET

U/S   D/S

URINALS : These are of two types : (i) slab type consisting of a back of white glazed fireclay and (ii) stall type (a serires of urinals) required for public buldings.

U.S.B.R. CLASSIFICATION : The United states Bureau of Reclamation modified the Airfield soil classification and adopted it in the year 1946. This was again revised in the year 1952 and was designated as Unified soil classification system.

U-TIE : A wall tie made of heavy wire bent into a U-shape or a top-hat shape.

U-VALUE: It is air to air heat transmission co-efficient. This value is obtained from experimental results and it speaks of the amount of B.Th.U. that will pass through 1 sft of the wall or roof, when the temperature of the air on one side of the wall or roof is 1oF higher than that on the other side.

V

VACUUM-CLEANING PLANT : This is usually installed in the basement of a building for cleaning the floor with the help of exhaust fan and filter . The air velocity is 50 to 80 ft/sec and the suction is 2.5 lbs. per sq.in.

VACUUM CONCRETE : This concrete reaches its normal strength in 10 days and its crushing strength is 25% higher than that of ordinary concrete . The vacuum mat fitted into the formwork sucks out excess air and water from the green concrete (newly poured) and makes the concrete dense and well-shrunk.

VACUUM FILTER : A cylindrical drum filter having a filter medium of synthetic fibre-woven cloth or wiremesh fabric enveloping the drum. The drum dipped into a vat of sludge slowly rotates and a part of the circumference is subject to an internal vacuum, by which the sludge is drawn to the filter medium. This is used for dewatering a sludge.

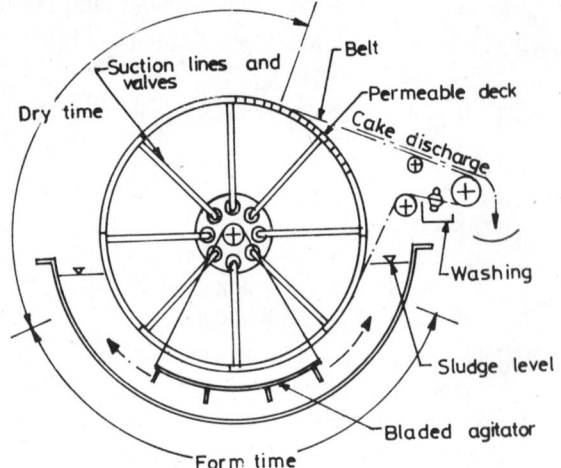

VACUUM HEATING : The heating system for buildings in which a vacuum pump works to remove the condensate and air from the radiators and helps in returning the water to the boiler feed tank.

VACUUM LIFTING : The raising of concrete slabs through the suction attachment to the sling and crane hook. The suction attachment is just like a vacuum mat.

VACUUM MAN : The man engaged in vacuum lifting .

VACUUM MAT : A flat metal screen which is faced by a linen filter and its back is kept under a partial vacuum. It is used to make vacuum concrete .

VACUUM METHOD OF TESTING SAND : Triaxial test of a sand sample by creating a partial vacuum in the rubber tube, the outside of the tube being kept at atmospheric pressure.

VACUUM PUMP : A pump to extract air from a space, the object of whcih is to maintain pressure in the space below atmospheric. This is also known as 'air pump'.

VALAMOID : Black putty; water repellant putty.

VALENTIAN SERIES : The lowest series of rocks in the sillurian series.

VALLEY : A valley is formed by the intersection of two roof planes sloping downwards towards their juntion. See illustration.

VALLEY BOARD : A board fixed on and parallel to the valley rafter, used for supporting slates or tiles.

VALLEY-GUTTER : A gutter lined with flexible metal and placed along the valley line of a roof to drain away the rain water.

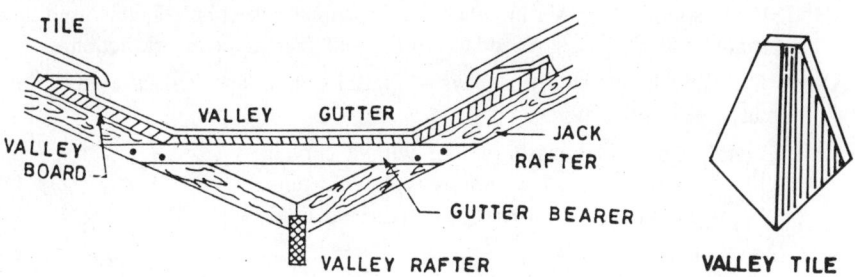

VALLEY-JACK : A jack-rafter which fits onto a valley rafter. See illustration.

VALLEY-RAFTER : Along the valley-line is laid the valley-rafter to receive the ends of the jack-rafters and to carry the valley-gutter.

VALLEY-TILE : A tile of special shape, concave upwards to fit the valley of a tiled roof. The pattern of the tile varies according to pitch of the roof.

VALVE : A device to open or close or to control a flow.

VALVE CLOSET : See illustration. This type of closet is not used now a days due to its system complicacy.

VALVE TOWER : A hollow C.I. or masonry tower built within a tank, from which control valves are operated to draw off water at different levels.

VANADIUM : The hardest known metal used to prepare strong alloy steels.

VANDYKE PIECES : The lead scraps remaining after cutting out of a stepped flashing.

VANE : A weather cock. A disc attachment to a levelling staff, which provides a sliding target.

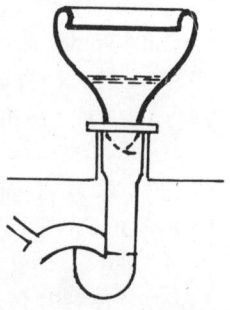

VANE ANEMOMETER : Used for measuring wind-velocities for meteorological purposes and air-velocities in large ducts.

VANE TEST : A four-bladed vane as shown in fig. is pushed slowly into a chunk of soil. The rotating motion is imparted to the vane until the soil shears. The applied torque is measured in the calibrated disc at the top of the torque rod.

VAPOUR BARRIER : An air-tight skin consisting of a thin coating of rubber-like paint or metal usefully employed on the inner, warm face of the insulation to prevent condensation.

VAPOUR HEATING : Water-vapour heating system, in which the condensate returns to the boiler feed tank by gravity.

VAPOUR PRESSURE: The pressure exerted by a vapour at a given temperature . This depends on the dryness-fraction of steam. The pressure of saturated vapour increases with the rise of temperature .

VAPOUR TENSION: See 'vapour pressure'.

VAQUEROS FORMATION : Strata of shallow-water origin and of lower miocene age.

VARNISH : A solution of resin in either oil, turpentine or alcohol. This is used to give brilliancy to painted surfaces and to protect them from atmospheric action.

VARVE CLAYS : Finely stratified clays of glacial origin deposited in lakes during the retreat-stage of glaciation.

VAULT : (i) An arched masonry roof with a curved soffit. There are various types of vaults having beautiful architectural compositions.

(ii) An underground chamber for preserving valuable goods safe from thieves.

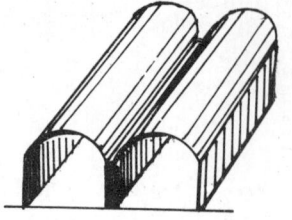

VAULT LIGHT : A pavement light .

VAULT PRIVY : Similar to 'pit privy'. The masonry vault is made water-tight, where the ground water table is close to G.L.

VAULT

V–BRICK : A perforated brick 9" x 9" x 3", with vertical cavities in it.

V–CUT : Wedge-shaped cut. See 'Vee-notch'.

VEE-GUTTER : A valley-gutter.

VEE-JOINT : See illustration. This is used in making a circular stone or brick masonry.

VEE-NOTCH : Triangular notch. See illustration.

VEE-ROOF : Valley formed, when two 'lean-to roofs' meet.

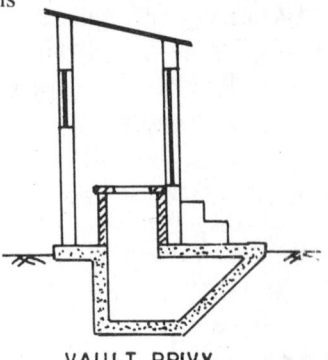

VAULT PRIVY

VEES : Soft earth occuring on the sides of a dyke.

VEE-TOOL : A parting-tool used in joinery works.

VEEVIE : Trade name for a paint solvent.

VEGITABLE GLUE : Starch-glue or protein-glue or glue obtained from ground nuts and rape seeds.

VEE JOINT

VEHICLE : The liquid part of a paint, usually oil, water or turpentine.

VELOCITY HEAD : Kinetic head, which is given by $H = v^2/2g$; where v = velocity in ft./sec. and g = acceleration due to gravity in ft./sec^2. It is the energy per unit weight of water due to its velocity, v.

VEE NOTCH

VELOCITY OF APPROACH : If the area of the channel at the approach of a weir is greater than the weir itself, the mean velocity on reaching the weir is called velocity of approach. This is taken to be uniform over the whole weir.

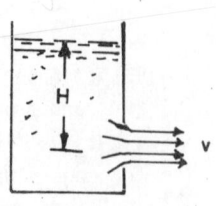

VELOCITY OF RETREAT : The mean velocity immediately downstream from the measuring weir constructed across the flow.

VENA-CONTRACTA : The cross-section of the 'nappe' or jet at which the streamlines first become parallel i.e., the narrowest point in the cross section of the nappe beyond the hole from which it issues. See illustration.

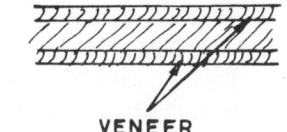

VENA CONTRACTA

VENEER : Timbers of inferior quality are glued together and faced with a thin layer of good timber. This is cheap in price and it looks like a solid timber of good quality. The thin layer is called veneer.

VENEER CUTTER : The man, engaged in cutting veneer from a log either by slicing or by rotary cutting.

VENEER SLICER : The operator of a veneer slicing machine.

VENEER

VENEERED STOCK : An early name for plywood.

VENEERED WALL : A wall having a facing which is attached to the backing. The facing may be of veneered brick, stone slabs, etc., This facing improves the appearance of the structure, but can not share load equally with the backing.

VENEERING : The mehtod of fixing decorative veneer to a backing of wood.

VENEER TIE : A tie for holdiong a veneered wall to its backing.

VENETIAN MOSAIC : Terrazzo.

VENETIAN RED : Oxide of iron (Fe_2O_3), a red pigment.

VENETIAN WINDOW : A window provided with venetian shutters.

VENT : An escape or outlet for air, a ventilating duct or pipe.

VENT PIPE : An anti-siphonage pipe which carries off foul gases and from a sanitary fixture leads into a ventstack.

VENT-STACK : A ventilation pipe which provides an escape for foul gases from sanitary fixtures and drains . Several vent pipes may be connected to a vent-stack.

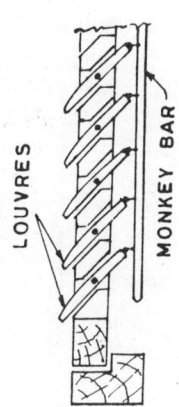

VENETIAN SHUTTER

VENTILATING BEAD : A deep bead.

VENTILATING BRICK : An air-brick.

VENTILATING JACK : A hood over the inlet to a vent pipe.

VENTILATION : Air change, air circulation; Arrarngement for the removal of polluted air and entry of fresh air.

VENTILATION PIPE OR VENT STACK OR VENT-PIPE : A soil-pipe must always be ventilated at its upper end by a pipe 2" dia. to expel the foul gases from the soil-pipe and drains.

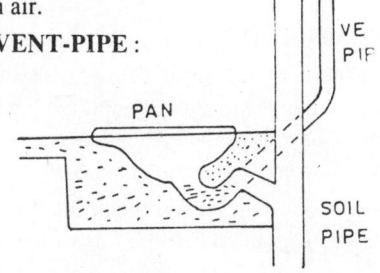

VENTILATOR OR VENT-LIGHT: A device for ventilating a room by replacing vitiated-air by fresh air.

VENTURI FLUME : The quantity of water flowing along a channel can be measured by such a control flume. This corresponds to the throat of a venturimeter, used for the measurement of flow in a pipe. See illustration. The differnece in level of the water surface gives a measure of the flow.

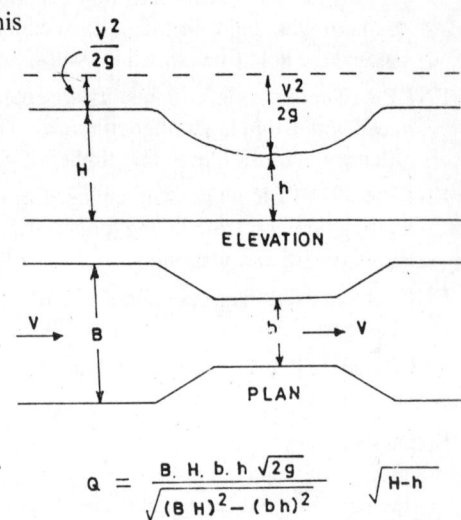

$Q = B.H. b.h. [2g(H{-}h)]^{0.5}/[(BH)^2 - (bh)^2]$

$$Q = \frac{B.\,H.\,b.\,h\,\sqrt{2g}}{\sqrt{(B\,H)^2 - (b\,h)^2}}\;\sqrt{H{-}h}$$

VENTURI-METER :
This is an instrument for measuring the quantity of liquid flowing through a pipe. It consists of a short length of pipe, tapering to a narrow throat in the middle. Tubes enter the pipe at the throat and at the enlarged end and pressure of water at these sections are measured. This may be employed as a water-meter.

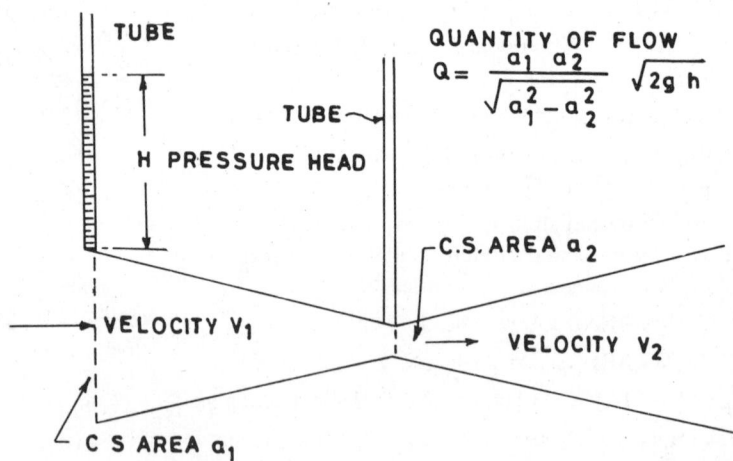

$$Q = \frac{a_1\,a_2}{\sqrt{a_1^2 - a_2^2}}\;\sqrt{2g\,h}$$

VENTURI TUBE : see venturimeter,

VERGE : A margin : (i) the edge of the tiles projecting over the gable of a roof; (ii) the unpaved portion of a footpath near road level, often of rough grass where trees are planted.

VERGE BOARD : Verge rafter. The board under the verge of gables . It is also known as Barge board.

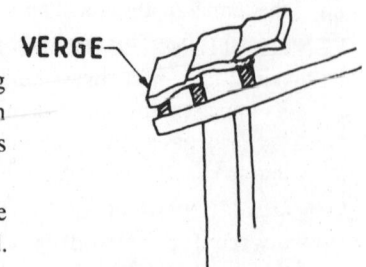

VERGE FILLET : A batten fixed on a gable wall to the ends of the roof battens.

VERGE TILE : A tile and a half tile used in alternate courses at the verge of a roof.

VERIBEST : Trade name for bituminous roofing felt, which is water–acid–and fume–proof.

VERMICULATION : A form of dressing of stone surface by making worm–shaped sinkings in the face of the stone.

VERMICULITE : Hydrous silicate. It occurs as decomposition product of Mica found in U.S.A and South Africa. When slowly heated, it expands and opens into long worm like threads. It forms a light insulating aggregate.

VERMICULITE GYPSUM PLASTER : For a 4-hour fire grading, this type of plaster is effectively used which reduces the slab and plasterweight.

VERMILION : A brilliant red pigment composed of mercuric sulphide (Hg S) which was first used in china. It is too expensive. The mineral is called 'cinnabar'.

VERMILIONETTE : A crimson or orange coloured pigment made from red lead stained with Eosine.

VERNIER : A small auxilliary scale attached to a main scale. The sacle can slide along the main scale to read fractional parts of the main scale.

VERNIER ARM : The arm of an instrument which carries vernier.

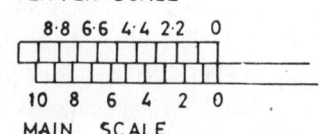

11 DIVISIONS ON MAIN SCALE = 10 DIVISIONS ON VERNIER SCALE

VERNON SHALE : It consists of red shales laid down under continental conditions .

VERONESE GREEN : Hydrated chromium sesquioxide, a transparent pigment also called viridian green.

VERTEX : Crown; Apex.

VERTICAL ALIGNMENT : Longitudinal section taken vertically through the centre line · of a road or railway.

VERTICAL CIRCLE : (i)   A great circle passing through the observer's zenith and cutting the horizon at right angles.

The graduated circular plate fitted to a theodolite for the measurement of the vertical angle of the Telescope.

VERTICAL CURVE : A curve, usually parabolic in nature, provided between two road– or railway gradients to have a gradual change from one gradient to the other.

VERTICAL INTERVAL : Con- tour interval . The vertical distance, apart which the contour lines are drawn.

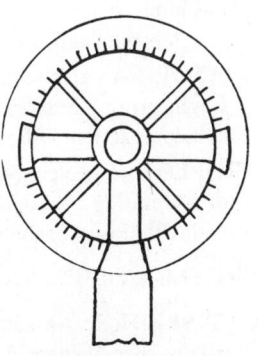

VERTICAL CIRCLE

VERTICAL SAND DRAINS : Installed for stabilisation of soft compressible soils beneath earth fills with a view to reducing the continued settlements after construction. Basic work in the design of sand drains is due to Terzaghi.

VERTICAL PHOTOGRAPH :
(AERIAL) A photograph of a
land taken from air for the
purpose of aerial survey,
which is chiefly employed
during war time. Thus, the
optical axis of the camera is
vertical or nearly so.

VERTICAL GRAIN : The edge
grain found in a quarter-sawn
timber.

VERTICAL SASH : A window
having vertical sash.

VERTICAL SHINGLING : Like tile hanging, shingles are hung on a wall; Hanging
shingling.

VERTICAL SPINDLE MOULDER : A spindle moulder who moulds vertically.

VERTICAL TILING : Tile hanging just like shingle hanging.

VESTIBULE : A small chamber just inside the entracce to a building, which seves as an
entrance room (waiting room) to a larger room. This is usually provided in office
buildings, Hospitals, Schools and colleges, Theatres and auditoriums.

VIADUCT : Just like an Aqueduct, a road or railway bridge over a valley .

VIBRATED CONCRETE : Concrete compacted and made homogeneous by means of
vibration. The vibration is given with the help of a mechanical vibrator.

VIBRATING PILE DRIVER : A device for driving piles was originated in U.S.S.R and
first used in 1949. On the top of a pile to be driven, two powerful motors ratate in
opposite directions and thus , strong vibration takes place. This pile driving is rapid
and noiseless.

VIBRATING ROLLER : A roller with a vibrated roll. The roller is self-propelled.

VIBRATION OF FOUNDATION : When heavey machines are installed in a building, the
building foundation becomes subjected to vibration due to the vibration of the
machines. This will cause compaction of subsoil and settlement of the foundation,
which is not at all wanted.

VIBRATOR : An electromechanical device used to perform various functions. In civil
Engineering, vibrators are used for (i) compcting concrete (ii) pile driving and (iii)
soil compaction. Different types of vibrators are available in market.

VIBROFLOT : A geotechnical process adopted for compacting sand and gravel. It consists
of a vibrating cylinder having water jets at both ends, which facilitates easy sinking
and withdrawal of the vibrator.

VIBROFLOTATION : See vibroflot.

VICAT NEEDLE : An apparatus used for testing the setting time of cement. The needle
has a special shape & size. See illustration.

VICE : (i)   A solid newel stair.

(ii)   A screwed clamp which is fixed to a working bench.

VICTAULIC JOINT : A Pipe joint. See illustration.

VICTAULIC PIPE : A special type of pipe with hydraulic joint which can be moved through several degrees, even after fixing and making water-tight.

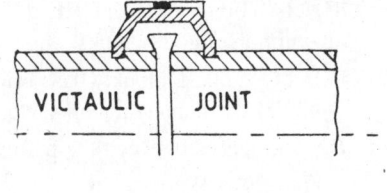

VICTAULIC JOINT

VIENNA WHITE : A paint base composed of pure white lead.

VIERENDEEL GIRDER : Open frame girder; A special type of girder formed of top and bottom booms rigidly connected by upright members without having any diagonal bracing. This type of girder is convenient to use when the diagonal bracings would be obstructive.

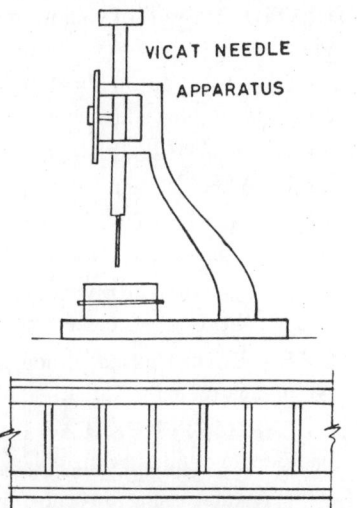

VICAT NEEDLE APPARATUS

VIGNOLES RAIL : A flanged-rail which is used in general in America and France.

VINYL RESIN : Thermo-plastic resin . These are odourless, tasteless, resistant to moisture and dilute acids and alkalies and are used for protective coatings, making gramophone records, water tubes, etc.

VIERENDEEL GIRDER

VIRIDIAN GREEN : See 'veronese green'.

VIRTUAL SLOPE : Hydraulic gradient. The slope shows the rate of head loss due to pipe friction in fluid flow system.

VISCOMETER : A device for measuring viscosity . There are various types of viscometer in which poiseuilles formula is used.

VISCOSIMETER : See viscometer.

VISCOSITY : The resistance of a fluid to flow due to internal friction caused by molecular cohesion in the fluid. Viscosity varies inversely with temperature.

VISCOUNTESS : A special building slate whose size is 18" x 10" (450 x 250).

VISCOUS FLOW : Streamline flow, steady flow or laninar flow. In this flow, there is a continuous steady motion of the particles of the fluid.

VISIBLE HORIZON : Sensible horizon or apparent horizon.

VISION-PROOF GLASS : Obscured glass.

VITREOSIL : Trade name for vitreous silica, which can withstand high temperature variation.

VITREOUS ENAMEL : Gloss coating to cast iron or steel articles which is more resistant to wear than enamel paint, but it cracks when struck by a hard blow.

VITRIFIED BRICK : Glazed hard-burnt bricks.

VITRIFIED TILE PIPE : Salt-glazed stoneware pipe.

VITRIOL : It is a mineral available in several colours such as blue, white and green. Sulphuric acid is the oil of vitriol.

VITROCIASTIC STRUCTURE : The structure of volcanic ashes produced by the disruption of glassy rocks.

VITROLITE : Glass block, Glass brick or Glass tile used for bathroom wall, facings of buildings, ornamental works, etc.

VOID : The spaces between the particles in a mass of granular material eg. sand, coarse aggregate, etc.

VOID RATIO : The ratio of volume of voids to the volume of solids in a sample of granular material.

VOLATILE : Literally means flying away. It descirbes fluids which boil at ordinary temperatures or at a temperature below the boiling point of water . The volatile fluids are to be kept air-tight.

VOLATILE ALKALI : An old name given to Ammonia.

VOLATILE CONTENT : It is obtained by dividing the wt. of volatile matter driven off, by the original wt. of the sample.

VOLCANTIC ASH : The product of explosive volcanic eruptions. When finaly ground, this may be used as a cementing, material.

VOLUME YIELD : This means how much volume of a material is obtained from a stated wt. of the original material.

VOLUMETRIC EFFICIENCY : This is the vol. of water entering a pump cylinder for each stroke divided by the vol. swept by the piston i.e. piston area x stroke.

VOLUTE : A spiral casing to a centrifugal pump, which reduces gradually the speed of fluid levaing the impeller.

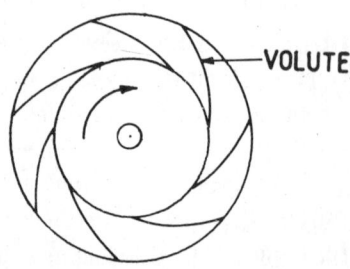

VOLUTE SIPHON : It consists of avertical down pipe or barrel bent at right angles at the bottom. The vertical down pipe has a bell mouth at the top. The inner surface of the mouth is provided with a number of volutes (curved blades) which guide and give spiral motion to water. The volute siphons are usually placed in batteries consisting of several siphon units.

VORTEX : A rotating fluid ; If the fluid is rotating freely without any external forces being impressed upon, it is called a free vortex.

VORTEX PUMP : A sewage pump having nonclogging type recessed impellers mounted outside the flow path of sewage between the pump inlet and discharge. Thus, solids equivalent to the pipe diameter

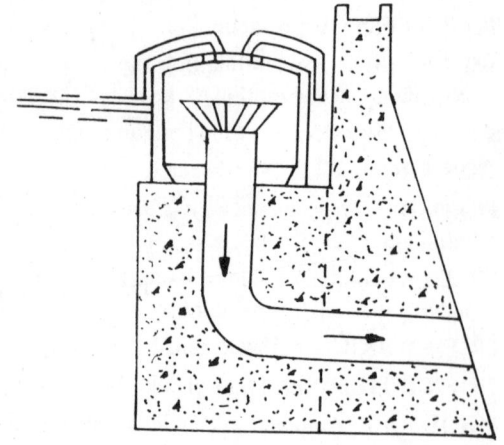

can be passed through it. See illustration.

VOUSSOIR ARCH :An arch made of wedge-shaped bricks or stones. Each such brick or stone is called voussoir.

VULCAN COUPLIG : A hydraulic shaft coupling to avoid torsional vibration.

VULCANISED FIBRE : A fibre made by treating paper pulp with zinc chloride solution used for low-voltage insulation. e.g., leatheroid.

VULCANITE: A name for fine grained igneous rocks.

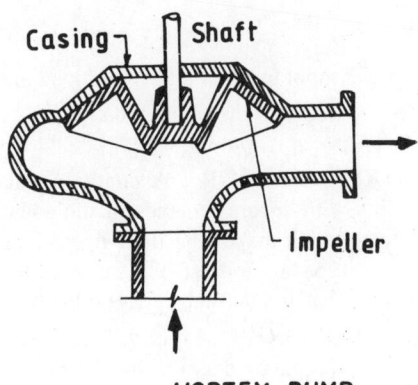

VORTEX PUMP

W

W : Symbol for total load, point load or concentrated load.

w : Symbol for distributed load; Weight per unit volume.

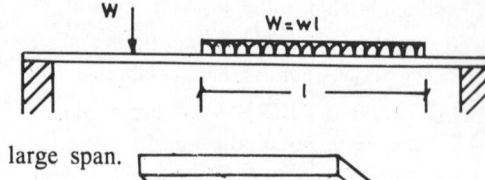

WAFFLE FLOOR : A concrete slab with square recesses in the soffit which makes the floor light. This type of floor slab is suitable for large span. Usually, the slab spans bothways.

WAGON DRILL : A rock drill mounted on a wheeled wagon, used for drilling deep holes in quarrying stones.

WAGON RETARDER : Braking bars which run parallel to the rails in a shunting yard. The signal man according to his decision can slow down the wagons by pulling the braking bar by means of a lever.

WAGON VAULT : Barrel vault.

WAGTAIL : A parting slip.

WAINSCOT : Wood panelling or wooden lining to the interior walls of a building upto the height of 'Dado'.

WAINSCOTING CAP : A moulding surmounting an wainscoting.

WAINSCOT OAK : Selected quarter-sawn oak for panelling.

WAIST : A narrowed down or constricted part; The least thickness of stair slab.

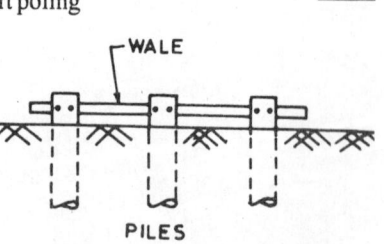

WALE : A horizontal timber used to tie together the piles driven in a row.

WALINGS : Beams which run horizontally to support poling boards in timbering trenches.

WALKING BEAM : A rocking beam for actuating the cable in cable drilling.

WALKING DRAGLINE : A moving dragline of large capacity. While walking, it rests on 6 metre dia discs.

WALKING GANGER : A travelling ganger who supervises the work of several gangs.

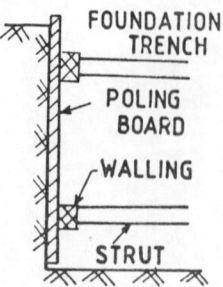

WALKING LINE : An imaginary line (setting out line) at 450 mm. from the centre line of the handrail of a stair, required to set out the winders of the stair.

WALK UP : A block of flats having no lift. The number of storeys is restricted to four.

WALK WAY : A permanent gangway provided with handrails, so that safe access may be given along a roof.

WALL : A construction to enclose a space, or to retain some materials, or to carry loads from floors. This also provides saftey and security.

WALL ANCHOR : A steel strap built into the brickwork and screwed to the common joists to ensure that the joists (beams) give lateral support to the wall.

WALL BOARD : Boards of laminated construction used for surfacing walls. These are used for the purpose of insulation as well as decoration.

WALL BOX : A box like cast iron support built into a wall to support the end of a timber beam.

WALL COLUMN : A steel or reinforced concrete column, a part of which forms the thickness of the wall.

WALL FRICTION : The friction between the back of a retaining wall and the retained material.

WALL HANGER : A cast iron or steel strap partly built into brickwork to support the ends of timber beams. In modern buildings, wall boxes are replaced by wall hangers.

WALL HOOK : An L-shaped heavy nail or spike used for the attachment of a pipe or a piece of timber.

WALL JOINT : A mortar joint running parallel to the wall face; The junction between two walls.

WALL PANEL : Brick wall in between the skeleton framework of a building. These panels of brickwork carry only their self–weights and they are restrained by the building frame.

WALL PAPER : A special paper used for decorating wall surfaces of a room.

WALL PIECE : A vertical piece of timber nailed to a wall which supports the end of a raking shore.

WALL PLATE : Timber, steel, stone or reinforced concrete blocks bulit into or placed over a wall to support the end of a beam or a rolled steel joist.

WALL PLUG : A pointed wooden peg introduced into a hole in a wall to facilitate easy driving of a nail or a screw in the wall.

Beam

Wall Plate

WALL STRING : A housed string placed against a wall to bear the inner ends of steps.

WALL TIE : A piece of twisted galvanised iron plate splayed in two limbs at either end, used to hold the two parts of a cavity wall. The splayed ends are built into the brickwork.

WALL TILE : Glazed burnt clay, terra-cotta, glass or terrazzo tiles used for making decorative and smooth surface of a wall. These are usually 100 x 100 and 150 x 150 in size.

WALLING : A general term for masonry walls.

WALLING BOARD : Vertical poling boards in a foundation trench are supported by waling boards placed horizontally and firmly secured by struts across the trench.

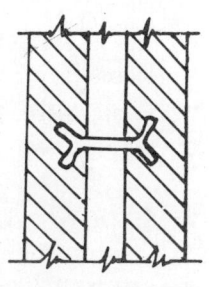

WALL TIE

WALLING MASON : A mason who sets stones or bricks for construction of a wall.

WALNUT: A decorative hard wood with a gray background and dark colour, used for carving and veneers.

WANE : Rounded surface under the bark of a tree.

WARE PIPES : Pipes made of salt glazed stone ware or china clay.

WARE RITE : Laminated plastic.

WARM-AIR HEATING : A central heating method of small capcity, heat source being gas or oil.

WARNER SIGNAL : This is a railway signal provided either away from semaphore signal or on the same post carrying the outer signal. Horizontal position of warner signal warns that the signal ahead is in stop position and the driver can go cautiously upto the home signal. When on the same post the semaphore is lowered but warner horizontal, the driver can move upto the next signal or station. When both the signals on the same post are lowered, the line is clear for running through without stopping.

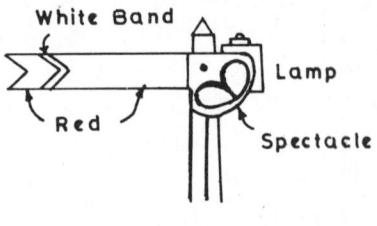

WARNER SIGNAL

WARNING PIPE : An overflow pipe fitted to a Cistern to discharge the overflow water into the open indicating the defects in plumbing and valves.

WARP : A permanent distortion in a timber due to the change in temperature and moisture content during the process of seasoning.

WARREN GIRDER : A form of girder, triangulated truss, having inclined members between horizontal top and bottom members without any vertical member.

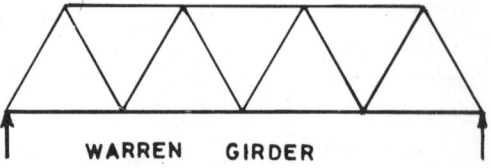

WARREN GIRDER

WARREN HILL SERIES : A group of volcanic rocks of pre-cambrian age.

WARRINGTON HAMMER : A Joiner's hammer.

WASH : A thin coat or film of a point. A weathered slope or weathering.

WASHABILITY : The property of retaining the paint and removing the dirts, when washed by water i.e., it is capable to withstand washing without damage.

WASHABLE DISTEMPER : Water paint ; A paint or distemper in which binder is insoluble in water, but the paint can be thinned easily by water.

WASH BOARD : A skirting board.

WASH BORING : Water-jet boring; Sinking a casing or drive pipe by a jet of water within it.

WASH DOWN CLOSET : This closet is most commonly used. It is made of glazed stone ware having a shorter cone. See illustration.

WASHED CLAY : An artificial imitation of natural marl made by mixing clay and chalk in wash mill.

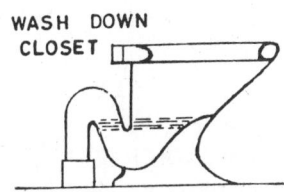

WASHER : A steel ring placed under a nut or a bolt head to distribute the pressure.

WASH LEATHER GLAZING : Bedding of glass in wash leather in place of putty, generally used for swing doors.

WASH OUT CLOSET : A Closet so shaped that a small quantum of water remains in it into which excreta falls and is flushed out over the edge of the basin into a trap below. See illustration.

WASH OUT VALVE : A scouring sluice; A valve introduced in a pipeline at the bottom of a valley.

WASTE : Spoil; A thing which is not usable; Building rubbish. Dirty or foul water from basin, kitchen, bath, latrine, etc. including night soil, Garbage, Screenings, Chippings, etc., and foul liquids from industries.

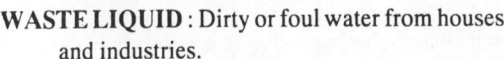

WASTE DISPOSAL UNIT : A unit to dispose off wastes; For easy disposal of rubbish, garbage, etc except papers, tins and glasses down the drain, a grinder driven electrically is installed near the kitchen into which all sorts of rubbish, garbage, etc. are ground and drained out through the drain.

WASTE LIQUID : Dirty or foul water from houses and industries.

WASTE PIPE : Soil pipe; A pipe carrying foul water or waste liquids from bath, latrine, kitchen, etc to drain.

WASTE PREVENTER : A cistern for flushing a water closet. Each time after use, the water closet is flushed with a definite amount of water either by pulling the chain or by lowering the lever attached to the cistern.

WASTE WATER : Waste liquid; Water containing foul matters.

WASTE WAY : Waste weir; A spillway.

WASTE WEIR : It is an escape device provided for flow of excess water over a weir from a water reservoir.

WASTER : (i)   A facing brick which is to be used as a backing brick due to its minor defects.

(ii)   A mason's chisel required for wasting prior to dressing a stone block.

WASTING : Prior to dressing a stone block, the method of removing excess stone with the help of a waster.

WATER BAR : A galvanised iron bar or lead bar introduced in the joint between the wood and sills of a window so as to prevent penetration of water. This is also known as weather bar.

WATER BEARING GROUND : A ground below the standing water level.

WATER BOUND MACADAM : A road surface formed of gravel or broken stones bound by soil and sand particles or hoggin by sprinkling water and rolling.

WATER BURNT LIME : Non-hydraulic limes form grains of a very slowly reacting lime after being slaked without excess of water. The grains swell, when combined with water.

WATER CARRIAGE SYSTEM : The system of disposing off waste matters involving the use of water to carry away the matters by water closets and through sullage water pipes.

WATER-CEMENT RATIO : The ratio of water to cement by weight is of great importance for the strength of concrete. Water-cement ratio determines the porosity of the hardened cement paste. In 1919, Duff Abrams established that when concrete is fully compacted its strength may be taken to be inversely proportional to the water- cement ratio.

WATER CHANNEL : A condensation groove provided in patent glazing to drain out the water.

WATER CHECKED CASEMENT : A casement in which stiles and mullions have grooves cut in the meeting edges to prevent water from getting through.

WATER CLOSETS : A closet connected to a drainage system for carrying human excreta and foul water from latrine.

WATER COLOURS : Pigments soluble in water which can also be ground up in a gummy medium.

WATER COLUMN : A column with a vertical pipe and a swan–neck swivelling pipe located at the ends of a platform for feeding water to steam locomotives, when the engine of a running train halts in a railway station. The column is provided with a foot valve for regulating the flow of water. A bag hose spout is fitted to the swan neck to direct the flow of water into the locomotive water tank without splashing.

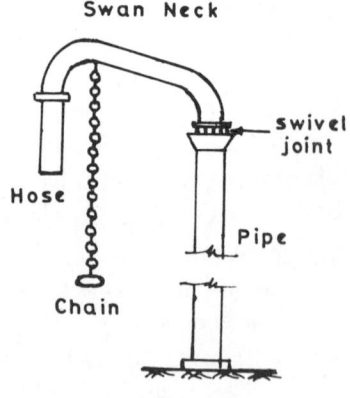

WATER COLUMN

WATER CONTENT : Moisture content; Quantity of water present in a matter; It is expressed in percent. The weight of water in a wet mass divided by the weight of the dry mass and multiplied by 100.

WATER-COOLED GREASE TRAP : A trap used for grease separation from sullage water. Cooling the interceptor aids grease separation. See illustration.

WATER CUSHION : A stilling pool of water.

WATER DIVINING : The selection of the proper location of the bore wells is known as water divining. Such a selection depends on the geological strata and condition of the underground water table and its surroundings.

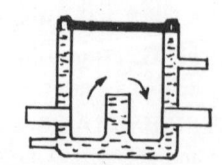

WATER-COOLED GREASE TRAP

WATER DIVINER : The person, expert in water divining.

WATER GAUGE : A U-shaped gauge filled with water adopted for measuring the pressure difference. The gauge having 24" (600 mm) of water indicates a pressure of 1 psi.

WATER GLASS : Sodium Silicate.

WATER HAMMER : The hammering effect of water with sound and high pressure developed in a pipe due to sudden closure of the flow of a liquid.

WATER
CUSHION

WATER JET BORING : Sinking of a tube-well by the action of water jet, which loosens the soil at the bottom of the bore facilitating easy sinking of the housing pipe. Also called 'Wash Boring'.

WATER JOINT : A saddle joint made in a water line.

WATER LEVEL : A simple instrument used for setting out levels at construction site. A rubber tube is connected to two vertical glass tubes filled with water maintaining the same level.

WATER LINE : The margin of water surface along the bank of a water pool i.e., a line joining the points where the water surface meets the bank.

WATER
GAUGE

WATER-LOGGING : Submergence under water due to inadequate facility of draining out water.

WATER LOWERING : Usually applied to the lowering of ground water table due to withdrawal of water from a well manually or mechanically.

WATER METER : A meter to read the quantum of water flowing through a pipeline. It is an integrating flow meter indicating the total quantity of water passed through the line.

WATER OF CAPILLARITY : Held water above the standing water level due to capillary action.

WATER PAINT : Washable distemper containing water-soluble binders. During application of the paint, when it becomes thick it can be thinned by water i.e., water is used as a thinner.

WATER PARTING : The dividing line between two catchment areas or watersheds i.e., a boundary line from which water flows in two different directions.

WATER PIPE : A pipe meant for carrying water. Mostly, these are pressure pipes.

WATER–PROOF CEMENT : The cement which has water-repelling power. It is used for construction of water-proof structures.

WATER PROOFING : A method of protecting walls and roofs from water absorption. This is usually done either by using water–proofing compound during construction or by applying a thin coat of water–repellant material or oil paint afterwards.

WATER–PROOF PAPER : A building paper waxed or impregnated in bitumen used in formwork for casting concrete with a view to retaining the water in green concrete and producing a smooth surface when the concrete is set and cured.

WATER–REPELLANT CEMENT: A hydraulic cement having resistance to penetration of water. This cement is used for works in damp places and to make a structure water-proof.

WATER REQUIREMENT: Water demand for a specific job. This may be for domestic, industrial, commercial, institutional and agricultural uses.

WATER RETENTIVITY : This is commonly defined as a property of agricultural soil (top soil) to retain the moisture content.

WATER SEAL : A seal of water provided in a trap in the waste water pipe system with a view to preventing the access of obnoxious odour.

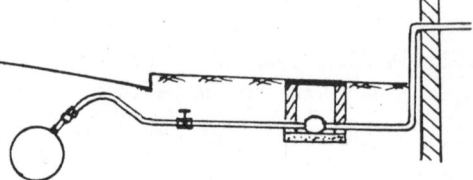

WATER SEASONING : A method of seasoning timber by immersing fresh cut logs under water for a fortnight or so and drying them afterwards in atmospheric air under a shed.

WATER SERVICE CONNECTION : Ferrule connection of a water line from the street water main for supplying water to individual houses.

WATER SHED : Defined catchment area from which run-off takes place. It is defined by the natural slope of a tract of land.

WATER SERVICE CONNECTION

WATER SHOT WALLING :
Dry walling with stones sloping towards the outside of the wall such that rainwater is thrown out and the wall is prevented from dampness.

WATER SOFTENER : A plant with a device to remove hardness of water. Usually, a zeolite bed is used. Ion-exchange process is employed in a water–softening plant.

WATER SOFTENING ACCELERATOR : A compact unit for accelerated water softening combining flocculation, settling and sludge removal. See illustration. Hard water is mixed with chemicals in the central portion of the tank and the sludge previously formed is churned by a rotating fan. The hard water with chemicals rises through the sludge blanket and the softening takes place in an hour. Sludge is removed intermittently.

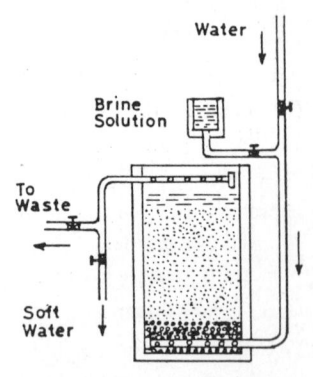

WATER SOFTNER

WATER SPOTTING : Pale spot marks on a painted surface caused by droplets of water. This usually occurs when painting is done outside a shed during drizzling.

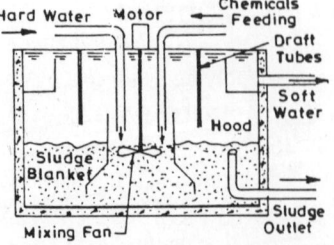

WATER SOFTENING ACCELERATOR

WATER STAIN : The discolouration of a converted seasoned timber on getting wet. The stain marks are found in the timber even after drying.

WATER SUPPLY : Specifically means a system for supplying potable water to a community. It covers intake of water, its pumping, treatment and distribution to the consumers in a locality.

WATER TABLE : The standing water level, overground or underground. In case of ground water, this is the saturation line.

WATER TEST : Hydraulic test for underground sewers and drains. The test is done in a stretch between two manholes with a water level in the manhole on the upstream side, the inlet and outlet ends being kept closed. The object is to find out leakage in the line, if any.

WATER TOWER : An elevated water-storage reservoir constructed for supplying water by gravity.

WATER TREATMENT : The treatment of water differs with the raw water quality and the desired quality of treated water. In general, the process involves removal of taste, colour, odour, turbidity, hardness and disease–causing organisms. For use in boilers, complete removal of hardness is required. For preparation of medicines, distilled water and deminerlized water are to be produced. For production of potable water, a process layout is shown.

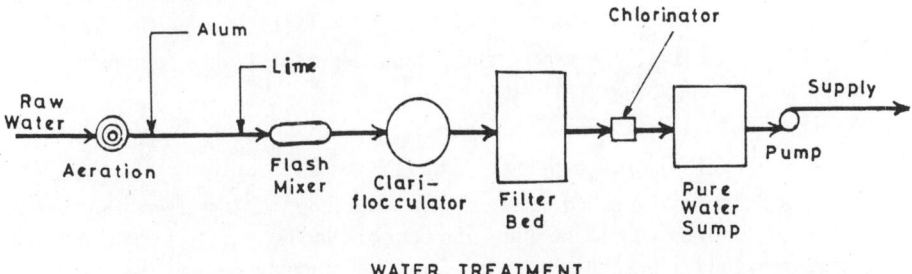

WATER TREATMENT

WATER –TUBE BOILER : A steam boiler having water and steam tubes in the furnace which make the boiler more efficient than a fire–tube boiler.

WATER TURBINE : Water wheel or pelton wheel which is turned by the force of water for generation of electrical power. The other well-known water turbines are Francis turbine and Kaplan turbine.

WATER WALL : A steel jacket filled with water required in a pulversied coal boiler.

WATER WASTE PREVENTER : The cistern provided in a toilet for flushing a closet after use.

WATER WHEEL : The earlier form of water–turbine which was not of high power and efficiency. The term is now used for overshot and undershot wheels.

WATTLE WORK : Hurdle work.

WAVE PRESSURE : The pressure of water waves on break-waters or marine structures.

WAVY GRAIN : A curly or wavy grain often found in certain timbers which is attracting.

WAY LEAVE: An approval for laying a cable or pipeline over a land.

WAYSIDE STATION : Also known as 'Non-junction station', which is provided with an arrangement for crossing of an uptrain and a down-train in case of a single railway track and for overtaking a slow-moving train by a fast-moving train.

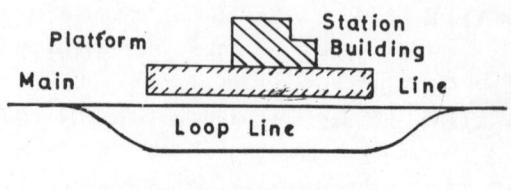

WAYSIDE STATION

WEARING COURSE : The road carpet i.e., the top layer of a metalled road.

WEATHER BAR : A term for water bar, which prevents draughts and water entering into the gap between sills of a casement during stormy weather.

WEATHER BOARD : Horizontal boards overlapping each other are nailed together on edge over the outside of a building with a view to keeping out rain.

WEATHER CHECK : A throat or drip used to keep out rain water.

WEATHER COCK : Also known as 'Weather Vane'. An ornamental finial which turns with the wind indicating the direction of blowing wind.

WEATHER POINTING : Weather-struck joints made in the brick masonry with a view to protecting the brick joints from weathering actions.

WEATHER JOINT : See 'weather pointing'.

WEATHER MOULDING : Weather board or a moulding fitted to the bottom rail of an outside door to throw off rain water.

WEATHER SHINGLING : Vertical Shingling.

WEATHER SLATING : Slate hanging by fixing the slates vertically.

WEATHER STRIP : Also called 'Air Lock' or 'Wind Stop'. A strip of wood, rubber or soft metal introduced into the joints of a door or window, such that no draught can pass through the joints. This is required for air-conditioning a room.

WEATHER-STRUCK JOINT : A 'Weather Joint'. A joint made in brickwork with cement mortar by smoothing off at the upper edge with the help of trowel such that rain water is thrown out of the brick face.

WEATHER TILING : Vertical tiling by hanging tiles.

WEATHER VANE : Weather cock.

WEB : The plate joining the flanges of a beam, girder, column or stanchion.

WEBBING : The wrinkling in a painted surface.

WEB STIFFENER : For stiffening the web of a plated girder, web stiffeners are required to be placed at intervals.

WEDGE : (i)   A tapered piece of wood required in timbering or centering for casting work or for shoring.

(ii)   A tapered piece of steel or timber used in a rail joint.

WEDGE COPING : A coping with a feather edge to protect a wall.

WEDGE CUT : A V-cut method used in tunnelling or shaft sinking, in which the cut holes slope towards each other like a wedge face.

WEDGE THEORY : The theory developed by coulomb in 1776 for analysis of the overturning force of a retaining wall. It was based on the weight of the earth of wedge shape at the back of a retaining wall, which would slide if the retaining wall fails.

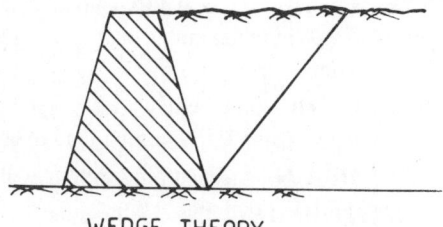

WEDGE THEORY

WEDGE-WIRE SCREEN : A screen made of wedge-shaped radial wires with the wide edges uppermost, is used for dewatering sand, grits and pebbles. This screen is of non-clog type.

WEEP-HOLE : (i)   Holes provided in any type of retaining wall with a view to allowing the escape of water from the retained material at the back of the retaining wall, which will reduce the pressure on the wall.

Weep Hole

(ii)   Small gaps left in the cross joints at the foot of a cavity wall for escape of water.

(iii)   A hole in a sill for allowing water of condensation to go outwards.

WEIGH BATCHER : A standard box used for measuring aggregates (fine & coarse) and the cement by weight for preparation of a dry mixture of concrete.

WEIGHT BOX : In joinery, it is the space for sash weights in a cased frame for window.

WEIGHTED AVERAGE : An average figure obtained by considering the weightage of individual values in a series.

WEIGHTED FILTER : Loaded filter.

WEIR : A dam-shaped structure constructed across a stream or river with a view to raising the water level, measuring the flow and storing water in the upstream during the dry period.

WEIR HEAD : The depth of water in the upstream of the weir from the bottom of the weir notch required for measuring the quantum of flow.

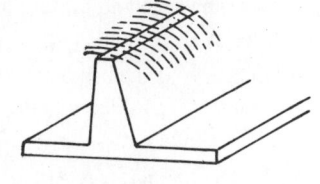

WEIR

WEISBACH TRIANGLE : In surveying with a theodolite, it is a triangle formed by a set up at the foot of a vertical shaft, from which both the plumb wires can be sighted. By this method, accurate observations can be taken.

WELD : A rigid joint made by welding pieces of metal or plastic.

WELDABILITY : The property of a material to facilitate welding.

WELDED FRAME : A structural frame made by welding its members.

WELL : A borehole in the ground to obtain water or any other fluid.

WELL BORER : A tradesman engaged in making shallow wells manually or in sinking tubewells with mechanical device.

WELL-CONDITIONED TRIANGLE : In surveying, this is a nearly–equilateral triangle formed, in which an error in measurement of an angle makes a very small error in length.

WELL CURB : A ring with wedge-shaped cutting edge, required for driving a hollow shaft into the ground by applying load over it.

WELL DRAIN : An absorbing well or soak pit.

WELL FOUNDATION : See illustration. Well foundation is used in soft loose soil. The well with the curb sinks under its own weight and the soil within the well is excavated. When the well rests on a hard strata, it is plugged by pouring concrete. While working under water, grab dedgers are used for excavating the soil.

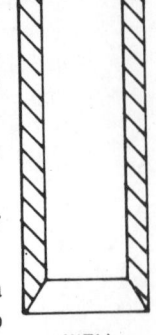

WELL
FOUNDATION

WELL HOLE : For quarrying stones, a deep borehole of about 150 mm diameter is made for blasting.

WELLINGTON FORMULA : 'Engineering News formula' used for driven piles.

WELL POINT : Ordinary shallow tubewell sunk by getting water into a soft water-bearing soil, the top being connected to a suction pump through a header pipe. A number of such well points are sunk around the periphery of an excavation with a view to directing the ground water flow towards the wells sunk away from the excavation. This helps in strengthening the soil and reducing pumping in the excavation.

WELSH ARCH : About 300 mm. opening bridged by a stretcher cut to wedge shape and resting on corbels on either side. Also known as 'Jack Arch'.

WELSH GROYNE : An underpitch groyne.

WELT : A seam made in a flexible metal sheet roofing.

WELTED DRIP : A drip formed by turning down the roofing felt at the verge and folding it back on the roof for sealing it.

Welted sheet

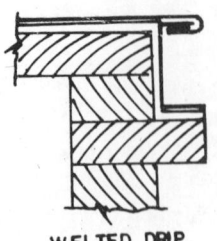

WELTED NOSING : The nosing made at the junction of a vertical sheet and a horizontal sheet in flexible metal roofing. The sheets are folded together at junction and dressed down at the top of the vertical edge.

WELTING STRIP : A strip of flexible metal with one end screwed to the sloping roof and the other end bent to hold the vertical sheet at the verge.

WELTED DRIP

WESTERN FRAMING : Platform framing in wood work.

WESTING : In surveying, it is the westward departure i.e., a co-ordinate measured westwards from a point.

WET ANALYSIS : Mechanical analysis of soil particles for determining settling characteristics. The soil sample is mixed in a measured volume of water and its density is checked by a hydrometer at intervals.

WET CUBE STRENGTH : The strength of a concrete cube in wet condition under a measured moisture content.

WET DOCK : A dock in which water is kept at high tide level and enable vessels to come in or to go out by opening the dock gate at high tide level.

WET DRILLING : Drilling hard rock by injecting water down the hole in the centre of the drill.

WET MIX : Mixing of dry ingredients for mortar or concrete with water in excess.

WET ON WET : Application of a special paint, the second coat being applied over the first coat when the later is still wet.

WET OXIDATION PROCESS : It facilitates wet oxidation of organic substances and raw sludge at an elevated temperature under high pressure, by feeding compressed air into the pressure vessel containing substances in liquid state.

WET ROT : Decay of timber by fungal growth due to alternate wet and dry conditions.

WET SAND PROCESS : A process for stabilisation of sandy soil with creosote oil, cut back bitumen and hydrated lime.

WET TIME : When the workmen report for work at site, but can not work due to bad weather. This idle time is wet time.

WETTED PERIMETER : The length of surface in contact with water measured across a channel or river. This is required for determining hydraulic mean depth.

WETTING CHANNEL : For consolidation of an earthen embankment, sometimes a dyke of low height is built parallel between the embankment and the river. During monsoon, water stands between the dyke and the embankment in form of a channel which helps in consolidation of the embankment and is known as wetting channel.

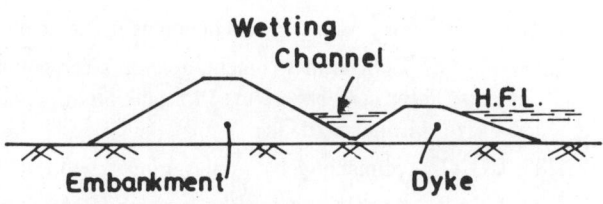

WET WELL : The sump of a pumping station from which water or sewage is to be pumped out.

WHARF : A berth of solid or open construction parallel to the waterfront. In solid construction, the wharf acts as a retaining wall by holding back all the earth behind it, while in open constructon no earth is retained and the wharf is supported on piles driven into the bed.

WHEEL BARROW : A container with a single or double iron or rubber-tyred wheels in front and two hand holds behind, by which it is pushed forward or pulled backward. This is used for transporting manually 2 to 6 cft. of material for a short distance. See illustration.

WHEELED TRACTOR: A tractor having rubber-tyred wheels like those of dumpers. It travels much faster than crawler tractors.

WHEEL BARROW

WHEEL GAUGE : In railway, gauge is the horizontal distance between the inner faces of the two rails. The 'wheel gauge' is measured as the horizontal distance from out to out of the wheel flanges.

WHEEL SCRAPER : A bowl scraper.

WHEEL STEP : Scots for winder.

WHETSTONE : An abrasive stone sometimes containing quartz rock or gritty slate, used for sharpening cutting tools.

WHIPPLE-MURPHY TRUSS : A bridge truss like a 'Pratt truss' made complicated. See illustration.

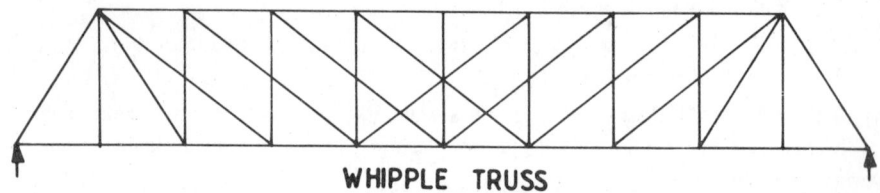

WHIPPLE TRUSS

WHIRLEY CRANE : A large revolving crane mounted on wheels or rollers or gantry.

WHITE CEMENT : Martin's cement, keating's cement and Keene's cement are white cements. These cements are used for making impervious, smooth and hard surfaces usually in skirting and dados.

WHITE COAT : A finishing coat in plastering.

WHITE LEAD : A basic lead carbonate which is a very brilliant white pigment used in white paint.

WHITE LIME : High calcium lime which is poorly hydraulic.

WHITE METAL : Anti-friction, tin-based alloy used for lining bearings and capping ropes.

WHITE PUTTY : The best putty prepared from white lead is used in plugging gaps exposed to weather.

WHITENING IN GRAIN : A streaky white appearance in a varnished or polished wood.

WHITE PIGMENTS : White lead, zinc oxide, titanium white, lead sulphate and antimony oxide.

WHITE SPIRIT : Spirit used as a thinner for oil paint, obtained by distillation of petroleum at 150° to 200° C. This is a substitute for turpentine.

WHITE WASH : Lime wash over a plastered surface.

WHITE WOOD : A soft, light wood, white in colour, obtained from pine, spruce etc.,

WHITING : Crushed chalk, the cheapest white pigment used for making glazier's putty, distempers, etc.

WHITNEY STRESS DIAGRAM : The stress-distribution diagram in R.C.C. beams designed by using ultimate load theory.

WHOLE-BRICK WALL : A wall of thickness equal to the length of a brick. It is usually 250 mm thick.

WHOLE CIRCLE BEARING : A bearing (used in surveying) that defines a horizontal angle from the true north.

WHOLE TIMBER : A squared baulk of timber.

WHOLE TIDE COFFERDAM : Full-tide cofferdam.

WICKET : (i) A sluice in a lock gate.

(ii) A small opening in a large gate. The large gate being kept closed, a man can pass through this opening in the gate which can also be closed, when required.

WICKET DAM : A movable barrier made of shutters that can revolve about an axis.

WIDE-GAUGE RAILWAY : A broad gauge railway, the length being 4'–8.5".

WIEP : A continuous cable made of willow to form the frame of a dutch mattress.

WIND BEAM : A beam used to resist wind force.

WIND BRACE : A wind beam.

WIND CRACK : Cracks developed in a timber due to wind force.

WINDER : A wheel step. A triangular or wedge-shaped tread changing the direction of a stair.

WIND FORCE : Wind load during violent wind blowing.

WINDING STAIR : A spiral stair winding round a shaft.

WINDING STRIP : Two short straight edges used for setting out a plane surface on wood.

WIND CRACK

WIND LOAD : Wind force.

WIND MILL : A wheel turned by wind pressure, used in pumping water and generating power.

WINDER

WINDOW : An opening in a room or in a closed space for circulation of air and entry of light.

WINDOW BACK : It is provided on the inner face of a window for hiding the lifting shutters.

WINDOW BAR : Glazing bar.

WINDOW BEAD : Guard bead.

WINDOW BOARD : A horizontal wooden board fixed at window sill level inside a window.

WINDOW EFFICIENCY RATIO : Daylight factor.

WINDOW FRAME : The outer frame of a window surrounding the casement.

WINDOW GLASS : Sheet glass.

WINDOW LOCK : Window sash fastener.

WINDOW STILE : Pulley stile.

WINDOW STOOL : Window board.

WIND PORTAL : A portal frame for resisting wind force.

WIND PRESSURE : Pressure of wind due to its velocity of blowing.

WIND SHAKE : Ring shake found in timber.

WIND STOP : Weather strip.

WINDROW : It is a long pile or stack with sloping surfaces as shown in illustration, used for composting solid wastes.

WIND TUNNEL : A hollow cone or cylinder through which air is blown at measured velocities for testing the wind effect on models of towers, bridges, etc.

WINDROW

WINDY : Pneumatic.

WING : A projected or extended part of a building.

WING COMPASS : A quadrant divider adjusted and held by a wing nut.

WIND DAM : A Spur.

WING LIGHT : Flanking window.

WING NUT : A Thumb nut or Fly nut; A nut provided with two wings which are gripped with fingers for turning.

WING WALL : A projected or extended wall from an abutment wall.

WIPED JOINT : A joint made in lead pipes with plumber's solder. One pipe end is widened and the other end tapered down to be inserted into it. After fitting, molten solder is poured on round the joint with a cotton wiping cloth. Fluid solder dropped out of the joint bottom is wiped back to the top.

WING NUT

WING WALL

WIRE CUT BRICK : In machine moulding, a long bar of clay moulded to a size of 4.5" x 3" is cut into brick lengths by wires. The production cost of these bricks is low. These bricks are not pressed and these do not have frogs.

Abutment

WIRED GLASS : Sheet glass reinforced with wire mesh. In case of breakage, the glass pieces are held in position.

WIRE GAUGE : Defining wire diameter by a number. The number for a thin wire is larger than that of a thick wire.

WIRE LATH : Expanded metal lath.

WIRE MESH : Wire net.

WIRE NAIL : Common nails made of steel wire.

WIRE ROPE : A rope made of steel wire.

WOBBLE SAW : A drunken saw used in wood work.

WOBBLE-WHEEL ROLLER : Pneumatic-tyred roller used in soil stabilazation.

WOOD : Standing timber.

WOOD BLOCK FLOOR : A floor made of wooden blocks. It is used in areas of cold climate.

WOOD BLOCK PAVING : Wood block flooring.

WOOD BRICK : A fixing brick.

WOOD CASING SYSTEM : An earlier system of electrical wiring through wooden grooves with cover.

WOODEN FORMWORK : A framework made of wood, required for casting concrete works. See illustration.

WOODEN RIDGE ROLL : See illustration.

WOODEN SLEEPER : Ideal sleeper for a railway track. These are made of sal or teak wood. Rails are fixed to wooden sleepers with the help of spikes.

WOOD FLOUR : Fine sawdust used in preparation of plastic wood, explosives and for other purposes.

WOOD MOSAIC : A decorative hard-wearing floor made of wooden blocks fixed with oxychloride cement.

WOOD RAY : Medullary ray that grow in a living tree from the outer wood towards the pith at the centre.

WOOD ROLL : A Piece of rounded wood fixed onto the roof boarding over which metal roof sheets are lapped.

WOODS CREWS : Screws suitable for fixing into wood.

WOOD SLIP : A fillet for fixing.

WOOD-STAVE PIPE : A pipe made of wood planks secured together with steel straps. Such a pipe was used earlier.

WOOD WOOL SLAB : Slabs of wood prepared from wood shavings compressed together. These look like fibre boards. These slabs are used as insulating boards, partitions, false ceiling and for decorative works.

WORKABILITY : Facility of working with a material.

WORKING CHAMBER : A chamber at the base of a pneumatic caisson where the workmen work in compressed air, the outside world being connected by a man-lock.

WORKING EDGE : The face edge of a timber.

WORKING FACE : The face end of a timber.

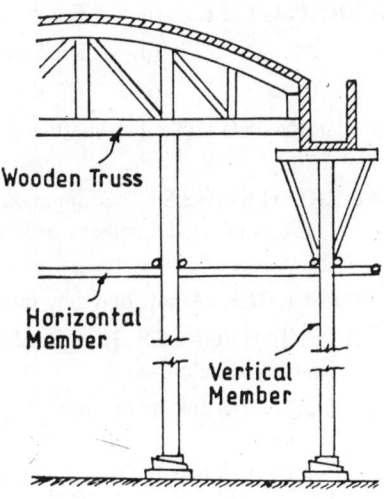

WOODEN FORMWORK

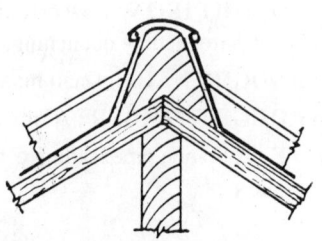

WOODEN RIDGE ROLL

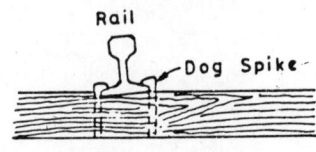

WOODEN SLEEPER

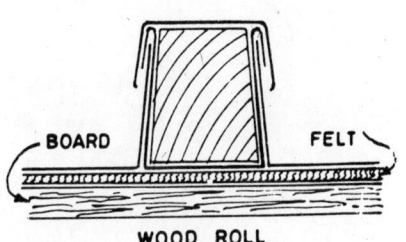

WOOD ROLL

WORKING LIFE :   (i)   The pot life of a glue.

                     (ii)   The longivity of a machine or system.

                     (iii) The durability of a material in use.

WORKING SHAFT : The shaft sunk to excavate a sewer and filled in after the sewer is laid.

WORKING STRESS : The safe maximum allowable stress to which a structural member can be subjected. In other words, it is the stress obtained by dividing the ultimate stress by the factor of saffety.

WORM HOLE : A hole bored by insects in timber.

WRACKING FORCES : Horizontal forces in the plane of a bent, which distort a rectangle into a parallelogram.

WREATH : The portion of handrail in a geometrical stair which is curved both in plan and in elevation.

WRINKLE FINISH : A finish with wrinkles made intentionally on a painted surface by stoving.

WROUGHT ALUMINIUM ALLOY : Cold-rolled, forged, drawn or pressed light alloy of aluminium.

WROUGHT IRON : Low carbon content, soft, malleable, fibrous pure iron which can not be hardened by quenching.

WROUGHT NAIL : A nail made of wrought iron.

WROUGHT TIMBER : A timber having planed surfaces.

WYE : A branch pipe leading off a main straight pipe. It looks like 'Y'.

X

X.P.M. : Expanded metal formed by cutting and expanding ribbed or plain sheets of steel.

XYLOL : It is a solvent (distilled from coal tar) for synthetic resins and gum.

XYLONITE : A thermoplastic material, like celluloid.

Y

Y : A branch pipe leading off a main pipe, usally at 45° to the main.

YALE LOCK : A Cylinder lock which is commonly used.

Y-ALLOY : It contains 4% copper, 15% Magnesium, 2% Nickel, 6% Iron, and 92% Aluminium. It is chiefly employed for casting engine piston.

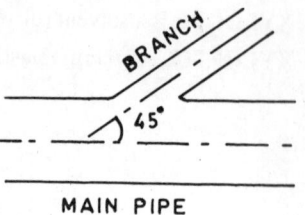

MAIN PIPE

Y-LEVEL : A kind of levelling instrument. The stage of the instrument carries two Y-supports in which the telescope is clipped.

YANKEE GUTTER : A wooden gutter built-up by metal sheathing. The metal sheet is held in position by an upright wooden block at the edge of the pitched roof.

YANKEE SCREWDRIVER : A special screw driver having a steep thread on the shaft which causes the screw to be turned easily, when the handle is pushed. This is used by carpenters to drive screws quickly.

YARD : The British unit of length (36 in. or 3 ft.); A lawn ; An area.

YARD TRAP : A gulley trap.

YEAR RING : An annual ring in a timber tree.

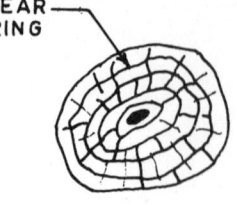

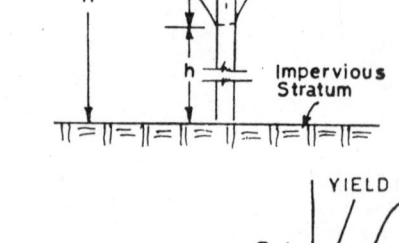

YIELD : The permanent set in a material when it is stressed beyond the elastic limit.

YIELD OF GROUND WATER : The maximum amount of water that can be drawn continuously from a well with a fixed 'draw down' (depression head). Yield of a well can roughly be calculated from the formula :

$$Q = K \ \frac{H^2 - h^2}{Log(R/r)}$$

where Q = yield in G.P.H.

$H - h$ = depression head in ft.

K = a co-efficient

R = radius of circle of influence in ft.

r = radius of well in ft.

YIELD POINT : The point at which suddenly increased yield occurs under slowly increasing load.

YIELD STRESS : The stress at yield point.

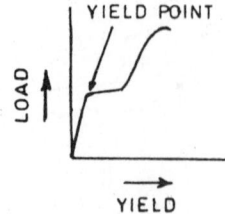

YOKE : (i) A spreader beam made of stout timber.

(ii) The outer framework of a 'moving form'. The forms proper are suspended from the yoke by lifting rods bolted to the cross bars of the yoke.

YOUNG'S MODULUS : The modulus of elasticity of a material. It is the ratio of longitudinal stress to longitudinal strain within the elastic limit of a material.

$$\text{Young's modulus} = \frac{\text{Stress}}{\text{Strain}}$$

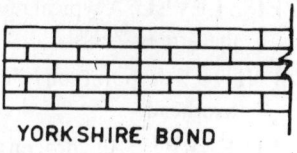

YORKSHIRE BOND : Modified Flemish Bond which shows on the face two stretchers and a header alternately in each course. This is also known as Monk Bond.

YORKSHIRE BOND

YORKSHIRE JOINT : The Capillary joint for copper tubes.

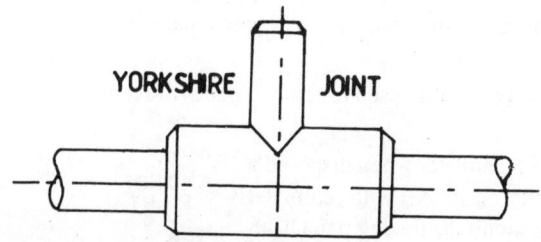

YORKSHIRE   JOINT

YORK : A slate having a curved cleavage.

Z

Z : Symbol for section modulus ; It is moment of inertia divided by depth of neutral axis.

$Z = \frac{1}{y}$. For a rectangular beam section Z is $1/6\ bd^2$ and for a circular section it is $3.14\ d^3/32$.

ZEISS LEVEL : A typical reversible tilting level introduced by the firm of Zeiss.

ZENITH : A point on the celestial sphere immediately overhead.

ZEO-KARB H : A hydrogen zeolite which removes all metallic ions present in water and replaces them with hydrozen ions by base exchange reactions.

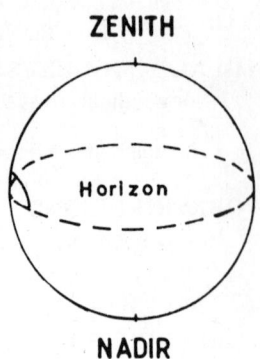

ZENITH

Horizon

NADIR

ZEOLITE : A chemical compound, found in green sand formation and also in certain deposits of marine origin.

ZEPHOLITE : A kind of zeolite used to remove traces of lead, zinc, copper and tin from water.

ZERO CIRCLE : A planimeter when disposed as in figure, describe an arc AB with centre P. If the arc be traced with the tracing point T, the measuring wheel wil not make any movement and thus the whole circle can be traced without any movement of the measuring wheel. This circle is called Zero Circle as shown.

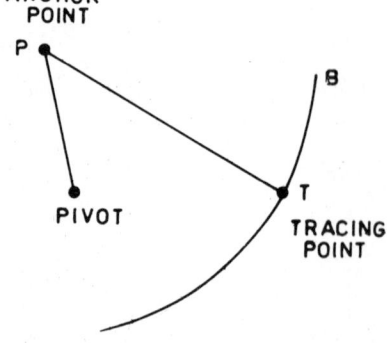

ANCHOR POINT

P

B

PIVOT

T

TRACING POINT

ZERO CIRCLE CONSTANT : It is the arc of the zero circle. When the pole of the instrument is inside the figure (whose area is to be found out), this constant is to be added to the measured area.

ZIG-ZAG BOND : Bricks are laid in a zig-zag fashion. This is similar to Herring–Bone bond, commonly used for brick pavings.

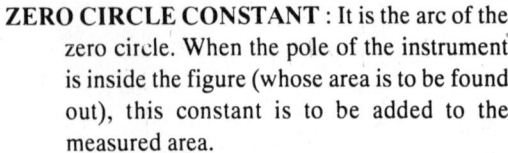

ZIG - ZAG BOND

ZIGERLI PROCESS : A modified process of re-aeration and re- activation of sludge.

ZINC DUST : Used for iron and steel surface painting, specially marine & port structures, and naval crafts which are in touch with sea water.

ZINC PAINT : It has the base of zinc oxide (ZnO). The paint is not affected by sulphur fumes.

ZINC-TANNIN PROCESS : Burnettizing process with some glue added to the zinc chloride solution. When the timber is well-soaked with zinc chloride solution, the timber is treated with tannin under pressure. This forms a leathery seal to the pores, which prevents water leaching out the zinc chloride.

ZIRCONIA : This is zirconium oxide (zio_2), a refractory used at very high temperatures.

ZONE FOSSILS : These fossils have characteristic of certain formations that assist the geologist in identifying the formations and determinging their positions in the geological series.

ZONE OF SATURATION: The ground below the water table which is saturated with water.

ADDENDUM

ACTIVATED AERATION : The waste activated sludge from an activated sludge tank is introduced to another tank and aerated there with sewage. The excess sludge is not returned, but concentrated and removed. In this system, air requirement is low.

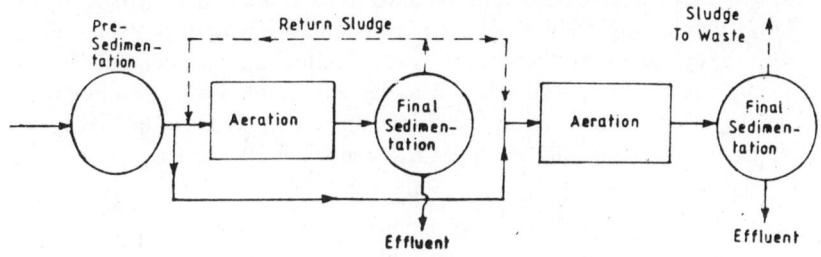

ACTIVATED AERATION

ALTITUDE VALVES : These valves are mainly used on the rising main (feeding line) to elevated reservoirs or standpipes. The valve closes automatically when the reservoir is full and opens when the pressure on the pump side is less than that on the reservoir side of the valve.

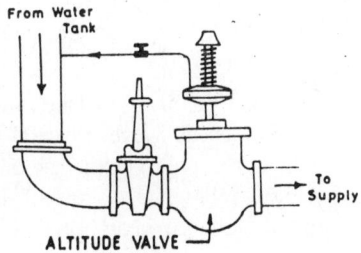

BALANCED VALVE : A type of valve as shown in illustration is used in controlling the flow of water with automatic devices. The most outstanding advantage of this valve is that it is opened and closed against the friction of the parts only, the water pressure on one disc being balanced by the pressure on the opposite disc.

BELLMOUTHED JUNCTION : This type of junction is used when two large sewers from different alignments meet at a point. The flow lines of the meeting sewers should coincide at the junction. A single large size arch is thrown over both sewers from each outer wall.

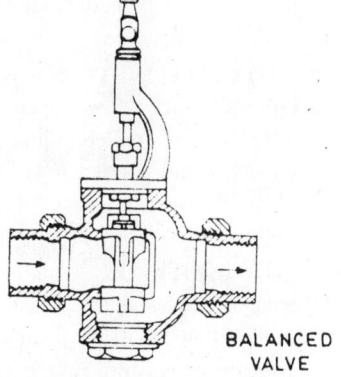

BALANCED VALVE

BIO-ACTIVATION PROCESS : It uses presedimentation followed by a 2' to 3' deep trickling filter, short-duration secondary settling, aeration tank and final sedimentation. It is actually a combination of trickling filter and activated sludge processes. This helps in absorbing high shock loads and removing high B.O.D. with only 1.5 hours aeration.

BLANKETS : These are composed tough workable highly impervious clay spread on the floor of a reservoir, under the fill or on the upstream from the dam to reduce seepage of water thickness of blanket $T = (2 + 0.02d)$ ft., where d is distance in ft. from upstream end of blanket to the dam.

CALIFORNIA BEARING RATIO : Abbreviated as CBR. The soil sample is compacted at optimum moisture content in five layers in a 150 mm dia. Cylinder by using 55 blows for each layer with a 10 pound rammer dropped through 450 mm. The sample with the cylinder is kept immersed in water for four days under a surcharge equal to the proposed surface load. The CBR of a soil sample is obtained by dividing the pressure required for 0.1 inch penetration (during CBR Test) by 1000 and is expressed in percentage. The CBR Test apparatus is shown in illustration.

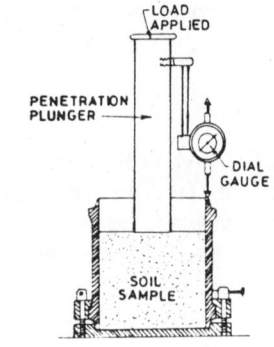

CARBONATE BALANCE : A state of water, at which it will neither dissolve nor deposit calcium carbonate. It is the most desirable condition of water from the view point of its corrosiveness or the possibility of clogging pipes. This can be determined by laboratory tests.

C BR TEST APPARATUS

CHECK VLAVE SLAM : Due to sudden closure of check valve, reversal of flow of water in a pipe gets accelerated. With a view to relieving the shock of slamming of the valve and air chamber is provided to the check valve. The illustration shows the provision of an air chamber to check valve or reflux valve.

CHEVRON-SHAPED TUBE : A tube having cross-section as shown in illustration. The V-groove in the bottom of the tube enhances the counterflow characteristics of sludge in a tube settler used for clarification of water.

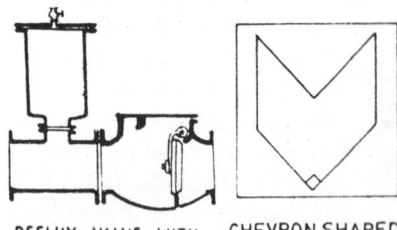

REFLUX VALVE WITH   CHEVRON SHAPED
AIR CHAMBER   TUBE

CHICAGO AERATOR : This is frequently combined with the final clarifier which is beneath a splash plate. The aerator has an impeller at the upper end which discharges mixed liquor to a vaned splash plate. This facilitates further agitation.

CONSTANT HEAD ORIFICE BOX : See illustration. It is a device to feed chemical solution to water at a constant rate. The float valve keeps the solution in the box at a fixed depth over the adjustable orifice for feeding the solution.

CONVECTIVE PRECIPITATION : It is exemplified by tropical rainstorms. Hot air masses near the earth's surface carry water vapour to higher altitudes under cold atmosphere at reduced pressure. On condensation of water vapour precipitation occurs.

COX BOLT GUN : A sampling gun adapted to obtain undisturbed samples of friable materials like weathered chalks, marls, shales, poorly-cemented sands and even soft rocks without rotary drilling. The gun comprises an outer barrel sliding on an inner barrel. An explosive charge is placed at the top of the inner barrel. At the base of the charge a plug is provided which is attached to the bolt. The gun is used to fire the sampling tube. When the plug breaks, the released energy drives the bolt through the inner barrel with a high velocity.

CYCLONIC PRECIPITATION : It is associated with unequal heating of earth's surface and build up of pressure differences. Differences in the earth's relative rotary speed between equator and poles deflect tropical air currents towards the poles. In planetary circulation of atmosphere between equator and poles, moisture-laden tropical air masses travel poleward, cooled and precipitate along their way. Cyclonic storms are eddies in the planetary circulation between equator and poles.

DIFFERENTIAL WATER PRESSURE : It is developed when the water level on the harbour side is lowered and it remains until the ground water level behind the wall reaches the same level. This is likely to build up on water retaining structures, water reservoirs, etc.

DIGESTED SLUDGE : The sludge obtained from septic tanks, inmhoff tanks and sludge digestion tanks. The volatile matter content of a well-digested sludge is 30 to 45%. It is dark brown in colour, homogeneous in texture having a tarry odour when wet. Its drainability is high and the solid particles separate easily from the liquid.

DRAFT TUBE AERATOR : An aerator using a draft tube, which generally requires maintenance of a constant level of mixed liquor in the aeration tank. The draft is downward in the tube and air pipes aspirate air into the liquor. Agitation at the surface also aids aeration.

DRIFTING SAND FILTER : This filter can be operated without any interruption for sand washing. The cleaning of sand proceeds with the operation of the filter.

EARTHQUAKE INERTIA : Earthquake cause both horizontal and vertical inertia forces. These forces are taken into account in design of dams and water retaining structures.

EDUCTOR PIPE : A pipe used in an air lift pump, at the bottom of which compressed air is released in lifting up water.

ELBOW METERS : Elbows in pipelines may be used for measuring the quantity of water flowing in the line. The pressure head difference between the inner and the outer edge of the bend, denoted by 'h', is measured by a differential gauge. The formula used is $v = kh^{0.5}$; where v is the mean velocity of flow and k is a constant varying from 7.0 for 1" pipe to 5.0 for 24" pipe.

FLOATING PIER : It is essentially a vessel and not a fixed structure. It consists of a pontoon kept in position by anchor chains and an access bridge connected to an abutment at the shore. A floating pier may be considered as moored vessel and as such there is risk, of its becoming loose. Berthing vessels should not be moored partly to this pier and partly to the fixed points.

FREEZING INDEX : This is expressed in Celsius-degree days required to measure ice thickness above water surface in a stream or river in cold climates. The empirical expression is

$$e = 0.035 \, a(s)^{0.5}$$

where *e* is ice thickness in m.

a is a co-efficient depending on local conditions,

and *s* is the freezing index.

FROST BOIL : It occurs during or after thawing due to softening of the subgrade combined with the effect of heavy traffic.

FROST HEAVE : Lifting of entire road surface with non-uniform severe cracking of the pavement due to frost action on coarse-grained cohesive soils.

FROZEN PIPE : The freezing of water in a pipe occurs due to long exposure of water in the pipe to a temperature below 0°C. The rate of freezing of water is static condition in a pipe is given by

$$t_1 = 0.3 \; \frac{Q}{K} \; \log \frac{T_2}{T_1}$$

where, t_1 = Time in hours for change in temperature,

T_1 = Original temperature difference between water and air,

T_2 = Final temperature difference between water and air,

Q = Quantum of water in kg in 1 m length of pipe and

K = Heat loss in Kg calories/hour for 1 m pipe length for a temperature difference of 1°C.

Water expands about $\frac{1}{12}$ th of its volume in a frozen pipe.

GEOPHYSICAL EXPLORATION : It provides required information on geological formations differing in their gravitational, seismic, acoustic, magnetic, electric and radioactive properties.

GRIFFITH PROCESS : Hays process of contact aeration as shown in illustration is widely used in sewage treatment. It provides 2 hours primary sedimentation, 1.2 hours primary aeration, 1 hour secondary sedimentation, 1.2 hours secondary aeration and 1 hour final sedimentation. See the flow diagram.

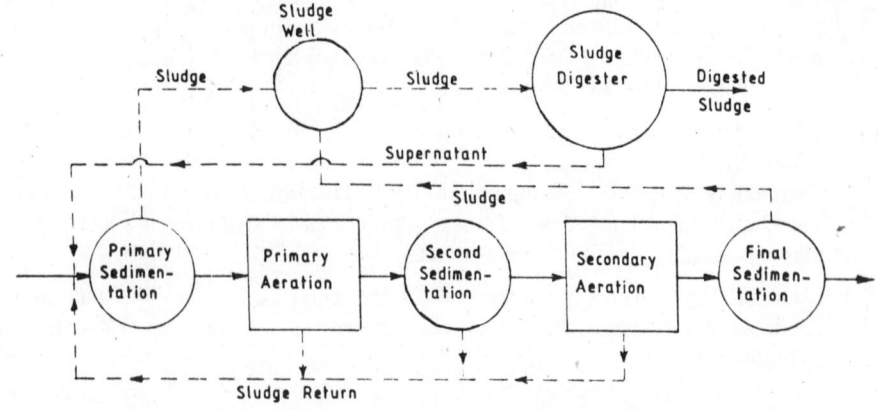

HAYS PROCESS / GRIFFITH PROCESS

GROUND KEY VALVES : These valves function as corporation cocks placed on house service pipes. See illustration. Multi-port ground key valves are used in filter control devices.

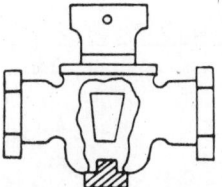

GROUND KEY
VALVE

HI-CONE SYSTEM : A mechanical aeration device, which has a vertical uptake tube with a vened cone or impeller at the top which discharges thin sheets of mixed liquor over the tank surface causing circulation of the mixed liquor with too much of agitation.

HIGH RATE DIGESTION : The rapid digestion of sewage sludge within a week or so in a sludge digestor is accomplished by keeping the digestor contents thoroughly mixed with continuous addition of thickened seeded sludge and preventing stratification of sludge into scum, supernatant and sludge solids. Mixing is done by recirculation of sludge gas which facilitates substrate to come in contact with the methane formers and thus digestion proceeds rapidly.

HYDRAULIC TRANSIENTS : Water transmission lines are subjected to transient pressures when valves are closed or opened and when pumps are stopped and started. Water hammer and surge are among such transient phenomena.

HYDRODYNAMICS : The subject dealing with fluids in motion and the application of its principles to water structures and water supply engineering. It includes the determination of quantum of flow, velocity of flow, change in pressures, loss of head and energy dissipation.

HYDROGEN ZEOLITE : Zeo-Karb H, which removes all metallic ions present in water and replaces them with H-ions by base exchange.

HYDROLOGICAL EQUILIBRIUM : It is expressed as :

$$\Sigma R = \Sigma D + \Delta S$$

where ΣR is various hydrological factors of recharge,

ΣD is various hydrological factors of discharge and

ΔS is the associated change in storage volume.

This is required to determine the safe yield of an underground aquifer.

ICE THRUST : In cold climates, effect of ice thrust is considered in design of dams and similar structures and also in operation of water intakes.

INSULATION JOINT : In a water transmission line this type of joint controls electrolysis by introducing resistance to the flow of stray electric currents along the pipeline. Rubber rings or gaskets are used in such joints.

KALMEIN PIPE : A pipe made of rustproof alloy of iron or steel (corrosion-resistant) for use in transportation of water.

LAMELLA SEPARATOR : A device recently developed for sedimentation of suspended particles in water. The influent enters at the top of the clarification basin and is then directed to flow downward through a series of parallel plates. The sludge settles at the bottom of the basin, while the clarified water rises to the top of the basin by return tubes. The plates are inclined at 30° to 45° to the horizontal and are spaced at 1.5 inches apart.

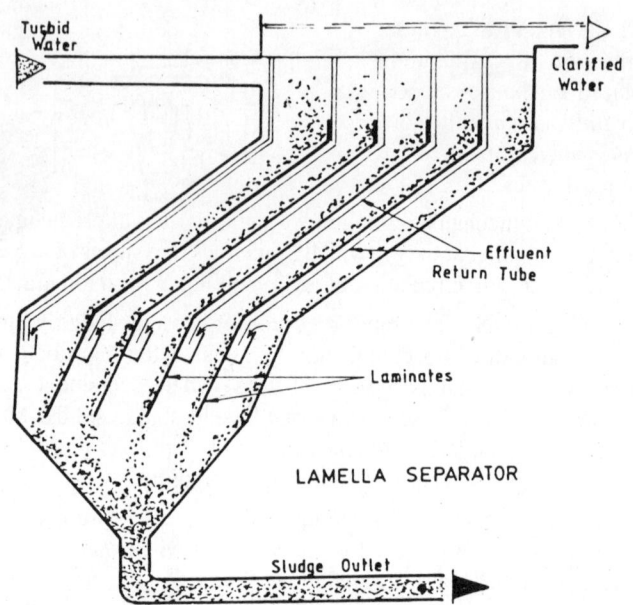

LAMELLA SEPARATOR

LANGELIER'S INDEX : An expression devised by W.F. Langelier to indicate the pH that a water should have to be in equilibrium with calcium carbonate.

LOCK-BAR JOINT PIPE : It is made of steel sheets bent into cylinders and locked by means of a bar along the longitudinal seams.

MAGNETITE SAND FILTER : A filter bed having magnetite sand made from magnetic iron ore, used in treatment of water or sewage with a view to removing finely divided suspended particles by mechanical straining, without using any coagulating agent.

MARSTON FORMULA : The load on a buried pipe can be determined by the formula

$$W = C \, \omega \, B^2$$

where C is a co-efficient depending on the nature of the filling material and the depth of fill to the top of the pipe.

        ω is weight of the filling material per cft.

and B is width of trench in feet at the bottom of the pipe.

McLAUGHLIN'S GRAPH : P.L. Mclaughlin plotted a graph as shown in illustration to determine the carbonate balance and hence, the corrosiveness of water. He proposed a laboratory procedure.

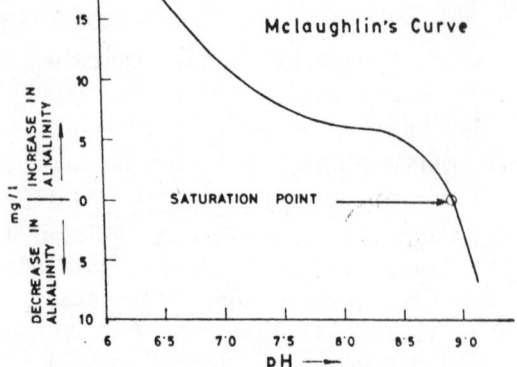

MEASURING WEIR : A device for

measuring quantum of flow of water,the weir head serving as an index of flow rate. It is usually a notch of rectangular, triangular or trapezoidal section.

MUD SCOW : A cylinderical sand bucket about 5 m long is attached to the bottom of the string of tools and is equipped with a flap valve to retain cuttings during drilling operation for a well.

NEUTRON MOISTURE METER : A measuring device to determine the soil moisture and density by nuclear radiations. The moisture meter is based on the fast neutron bombardment of the soil which produces reflection of slow neutrons acting on a deterctor of indium foil. The fast neutrons emitted by the source loose their energy due to collision with the H-atoms in molecules of soil water and the slow neutrons are absorbed by the indium foil making a change to radioactive indium. An equation relating amount of activity of the foil to the soil moisture content has been evolved. The component parts of the apparatus are shown.

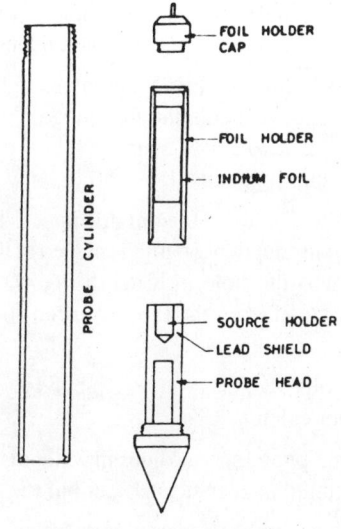

NEUTRON MOISTURE METER

NONEQUILIBRIUM FORMULA : A formula for drawdown based upon a number of assumptions. For an ideal aquifer, their expressed

$$S = \frac{114.6\,Q}{T} \int_{1.87r^2s/Tt}^{\alpha} \frac{e^{-u}/}{u}\,du$$

where s =   draw down in ft at any point in the vicinity of a well, pumping being at a uniform rate;

Q =   discharge of the well, in gpm;

T =   co-efficient of transmissibility, in gpd;

d =   distance in ft. from the pumped well to the point of observation;

s =   co-efficient of storage of the aquifer;

t = time of pumping of the well, in days

e = natural logarithmic base ie., 2.718

u = $1.87r^2s/Tt$

ORIFICE FORMULA : The quantum of flow thorugh an orifice is given by

$$Q = A\,C_1C_2\,(2\,gh)^{0.5}$$
$$= K_oA\,(2\,gh)^{0.5}$$

where A = Cross-sectional area of stream

C_1 = a Co-efficient of area reduction

C_2 = a Co-efficient of velocity

K_o = $C_1 \times C_2$ = Orifice co-efficient

g = acceleration due to gravity

h = velocity head.

OROGRAPHIC PRECIPITATION : When horizontal currents of moist air strike hills or mountains they get deflected upwards and on condensation precipitation occurs.

EARTH SYSTEM : Also known as PFT system named after Pacific Flush Tank Company, which consists of recirculating the sludge digester gas with the help of a compressor and releasing it to lower scum level to promote rapid digestion by circulating the actively digesting sludge.

PERCOLATION TEST : A test hole of 100 mm diameter is bored to the depth of a pipe trench and fine gravel to the depth of 50 mm is placed at the bottom of the hole. Clear water is then poured into the hole to a depth of 300 mm over the gravel. The percolation rate is computed from the time required for the fall of water level by 25 mm.

PERMUTITE : Sodium zeolite which is a precipitated synthetic compound used to exchange the sodium for calcium.

PIPE THAWING : Frozen mains and service lines in water distribution in cold climate are thawed by foreing a small diameter steam pipe into the frozen pipes.

PITCHER PUMP : A single-acting hand operated suction pump used to draw water from a well.

PITOMETER SURVEY : Waste survey by isolating a stretch of waterline by closing the valves, maintaining flow through one pipe only and measuring the rate of water flow by means of a pitometer inserted into the pipe (by boring a hole).

PM SPECTRUM : Most widely used PIERSON-MOSKOWITZ Spectrum representing fully developed waves in deep water.

PO : A symbol suggested by G.M. Fair to designate the odour scale. It is a unit of odour known as an *'olfactory'*. PO of an odour represents the logarithm to the base 2 of the odour threshold intensity.

PRESEDIMENTATION SLUDGE : Primary sludge or raw sludge obtained from presedimentation tanks. This sludge has offensive fecal odour. The volatile matter is about 70% of the dry solids.

PRESSURE REGULATING. VALVE : These are required to deliver water from a high

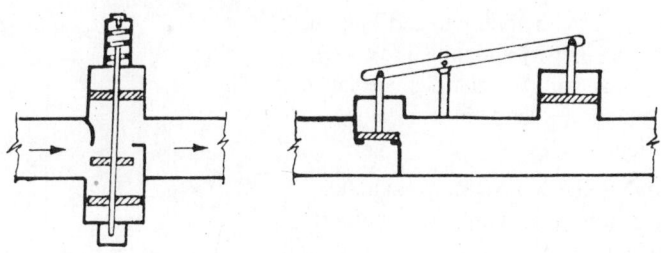

PRESSURE REGULATING VALVE

pressure to a low pressure pipeline. The two types pressure regulating device are illustrated. When the pressure below the valve, tends to decrease beyond the pre-determined valve, pressure in the chamber increased and the valve partly gets closed. If the pressure below the regulator becomes too low, the pressure on the flexible disc is relieved by opening the balanced valve in the regulator.

PUECH-CHABAL FILTER : Its use is almost exclusively confined to France. No coagulant is used in the raw water. Instead, the water is aerated between the stages of filtration. The water is passed through a series of sand filters at a progressively slower rate.

ROLLED FILL DAM : Rolled fills are constructed by depositing materials in layers and compacting it by means of sheep foot rollers at an optimum moisture content.

RORO CRAFT : Roll on Roll off (RORO) vessel has certain advantages over conventional water vessels. The vessel does not take a 360° turn. The propeller on

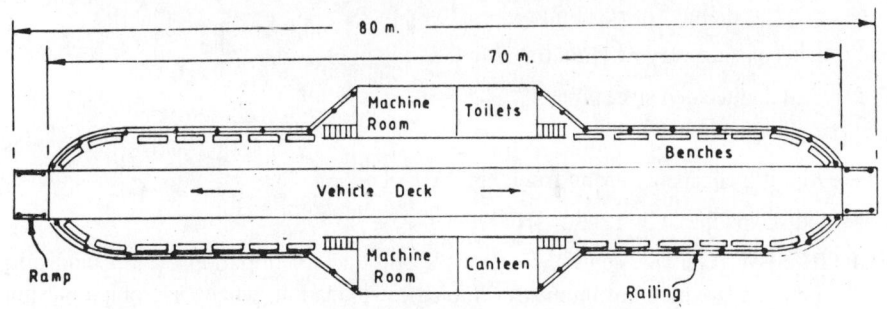

RORO VESSEL

retractable vertical shaft is capable of turning 180° facilitating equal propulsive efficiency for linear movement of the vessel either forward or backward. In shallow water, the shaft propeller unit can be lifted totally for its safety. The carrying capacity of RORO Vessel is more as compared to a conventional craft of same size. For the same speed, a larger conventional vessel will require more propulsion power.

SAND BOTTLE : With a view to carrying out tests on field density and moisture content of a granular soil, the samples are taken with the help of a sand bottle as shown in illustration.

SATURATION INDEX : An index to indicate the saturation of a sample of water with calcium carbonate. This is obtained as I_s = pH of water - pH of water at which it should be in equilibrium with $CaCo_3$.

A zero index means the water is in equilibrium with $CaCo_3$. A negative index indicates the water to be under saturated with $CaCo_3$ and a positive index indicates just the opposite.

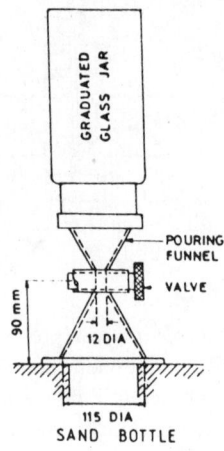

SAND BOTTLE

SCHMUTZDECKE : Its meaning is 'dirty skin'. A thin layer containing fine flocs bounded by zoogleal jelly formed on the sand surface of a filter bed, where biologic activities are at their highest. The successful operation of a filter with removal of bacteria is dependent on it. The formation of schmutzdecke is known as 'Ripening of a filter'.

SCOBEY FORMULAS : The formulas used for computation of head loss in various kinds of pipe.

For concrete pipes, $V = C_s H^{0.5} d^{0.625}$

where V is velocity is ft/sec.,

H is friction loss in ft/1000 ft.

d is diamter of pipe in ft.

and C_s is a constant varying from 0.267 to 0.345.

For riveted and welded steel pipes,

$V = H^{0.53} d^{0.58} k_s^{-0.53}$

where K_s is a constant varying from 0.32 to 0.48

For wood stave pipe , $V = 1.62\ d^{0.65} H^{0.55}$

SEPTIC SEWAGE : A sewage becomes septic due to formation of CO_2 and other compounds which are inimical to biological oxidation, when anaerobic digestion takes place due to stagnation of sewage in a sewer line or in a channel or clarifier with sludge accumulations or due to entry of a large volume of supernatant liquor from sludge digester to primary settling tank.

SEWAGE ACCELATOR : It combines air diffusion and agitation. The impeller of the accelator is placed close to the bottom of the tank. Air at the rate of 0.5 cft to 1.0 cft per gallon is mixed with the settled sewage entering at the bottom of the tank.

SEWER PASSAGEWAY : A shaft constructed with a passageway leading from the shaft through the sewer wall with a view to providing access into a large sewer from side as shown.

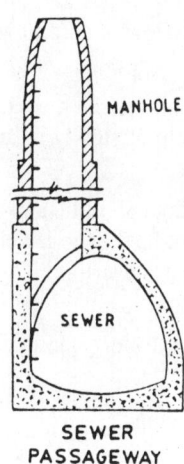

SEWER
PASSAGEWAY

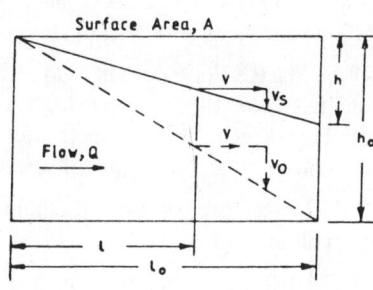

SHALLOW DEPTH SEDIMENTATION : The use of shallow depth sedimentation by providing additional horizontal trays at distances 'h' apart in a conventional sedimentation tank enables to increase the capacity of the tank and reduce the detention time of settling process to a few minutes. The particles with V_s less than V_o can be completely removed if false bottoms or trays are placed at intervals, h. As this interval is reduced, the size of tank decreases. The removal of settleable matters is a function of overflow rate and basin depth and is independent of detention period. See illustration.

SHALLOW SEWERAGE : With certain relaxations in technical standards it offers same level of user convenience as conventional sewerage. This is a cheap and effective means of sanitation, but it requires active participation of the EWS community living in a slum.

SHALLOW SEWERS : These sewers are designed to accept all household wastewaters for off-site treatment and disposal as an economical low cost sanitation measure in high density slums and squatter settlements in unsewered pockets of a city. These sewers need frequent flushing.

SHEAR BOX TEST : The test to ascertain the shear strength of a soil sample. The unconfined compression test apparatus of portable type is shown in illustration. An undisturbed soil sample is placed in position with no tension in coil spring. The load is applied gradually by rotating the handle at an even rate until the specimen fails either by bulging or by shearing at an angle to the vertical.

SLICHTER'S EXPRESSION : The velocity of flow of underground water was expressed as

$$V = \frac{KSe^2}{Y} \quad ; \quad \text{where} K \text{ is porosity constant,}$$

S is slope of hydraulic grade line,

Y is co-efficient viscosity of water, and e is the effective size of sand.

SLICHTER'S METHOD : A method suggested by Slichter for determining the velocity of flow of underground water by using a strong solution of an electrolyte into a well, putting two electrodes and measuring periodically, the flow of current with the help of an ammeter.

SLUDGE THICKENER : A hopper-bottom cylindrical tank in which concentration of primary and secondary sludge takes place with the help of slowly rotating vertical members and squeegees with rake blades.

SMALL BORE SEWERS : These are used to carry the effluent of septic tank and also the waste water from pour flush toilets in high density slums having no sewerage system. The small bore sewers are connected to the nearest sewerline. This system is marginally cheaper than conventional sewerage system.

SOIL STRIPPING : Removal of top soil containing about 2% organic matter from the entire reservoir area.

STABILITY NUMBER : It is obtained from a series of graphs to enable rapid calcula tion of factor of safety and is given by

$$N = \frac{C}{whF}$$

where C is cohesion per unit area,

w is weight of unit volume of soil,

h is height of slope and

F is factor of safety

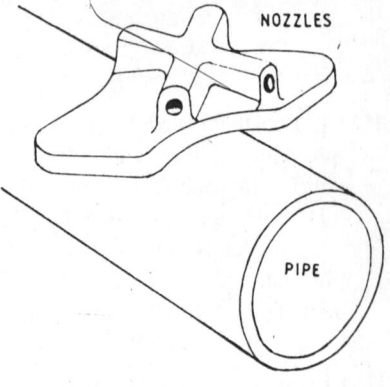

SPARJER NOZZLE

STEPHENSON'S FORMULA : The height of reservoirs or waves is given by

$$H = 1.5\,(D)^{0.5} + \{2.5 - (D)^{0.25}\}$$

where H is height of waves in feet and

D is length of water exposure in miles.

STORM-WAVE SPECTRUM : It has a typical skewed form with a larger high-frequency width than low-frequency width.

STOVEPIPE METHOD : A method of drilling originated in California. The casing consists of two sizes of lap-welded steel pipes, one of which just fits over the other.

SUPERNATANT LIQUOR : The liquid that leaves a digester by the overflow pipe when fresh sludge is introduced into it. The liquid is normally offensive in odour having suspended solids 0.05 to 0.10% and total solids 0.10 to 1.20%. It is discharged into the influent of primary clarifier. When the liquid contains more than 0.5% of total solids, it may cause floatation of sludge in primary clarifier.

TAYLOR'S CURVES : Taylor developed a method of investigation of the stability of homogeneous clay cuttings or slopes of regular section and prepared a series of curves by plotting stability number against angle of slope.

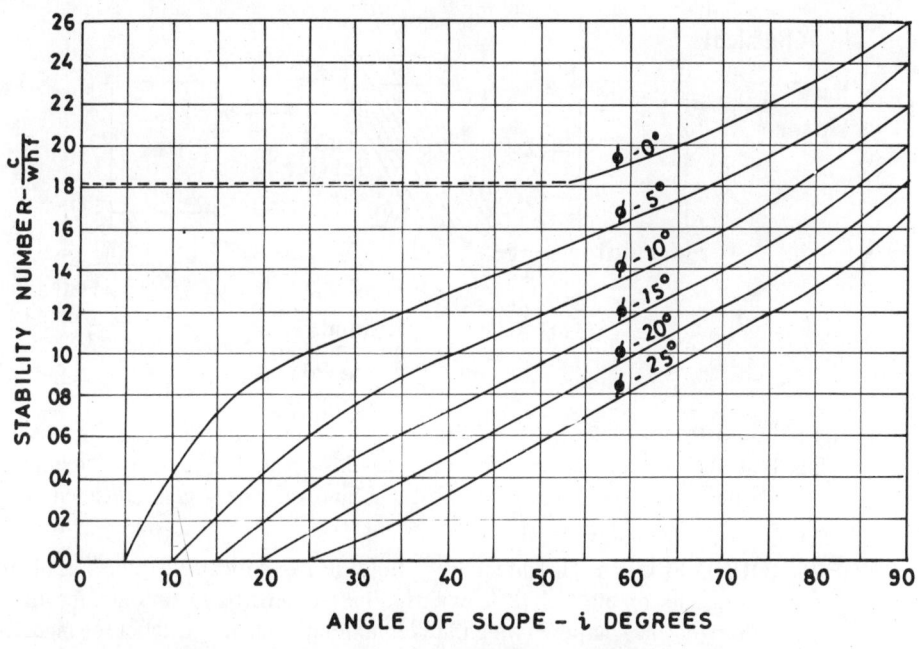

ANGLE OF SLOPE - i DEGREES

TAYLOR'S CURVE

THIEM'S FORMULA : The formula used for computing the co-efficient of permeability is

$$P = \frac{527.7 \, Q \log \frac{a}{a_1}}{t \, (d_1 - d)} \quad ;$$

where P is co-efficient of permeability,

        a is rate of pumping from well in gpm.,

       a, a_1 =   the distances in feet of two observation wells from the pumped well.

       d, d_1 =   the water depths in feet in the observation wells at distances a, a_1

and it is the average vertical thickness at a, a_1, in feet, of the aquifer.

TRICKLING FILTER SLUDGE : It is the humus obtained from film sloughing of a trickling filter. The volatile solids content is 45 to 70%. This sludge is digested with the primary sludge in digestion tanks.

TUBE SETTLER : A tube settler is based on the principle of shallow depth settling of suspended particles in water. It may be horizontal of inclined. In an inclined tube settler, the tubes are inclined at an angle of 45° to 60° to the horizontal. Sediment in tubes inclined at angles in excess of 45° does not accumulate but moves down the tube to eventually exit the tubes into the plenum below. A flow pattern is established in which the settling solids are trapped in a downward flowing stream of concentrated solids. The capacities of existing basin can be increased by 50 to 150% with improved effluent quality by providing striply inclined tubes with the clarifiers or horizontal sedimentation tanks. See illustration of a tube settler.

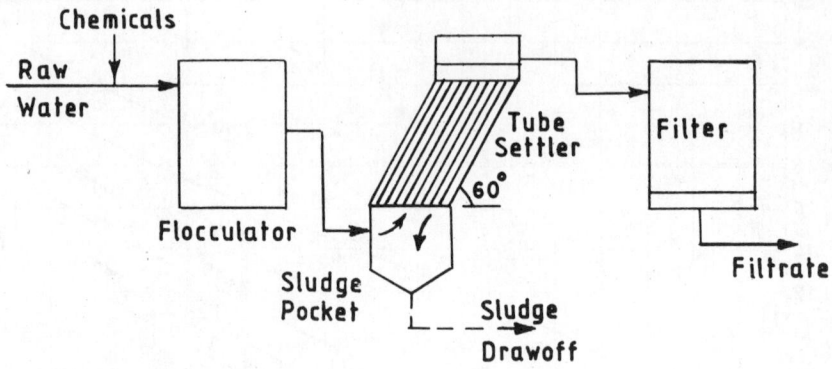

UMBRELLA SEPARATOR : In air-lift pumping, an umbrella-shaped plate is placed at the top of the eductor pipe, against which the rising stream of mixed water and air shrikes, the air escapes and the water gets collected in a tank.

UNDERGROUND FLOW : The direction of flow of ground water can be traced by using dyes, salts or electrolytic methods. Fluorescein and Eosine are the dyes best suited for the purpose which travel a long distance without change of properties. Sodium chloride is also used in tracing the ground water flow. Deep well current meters, salinity apparatus with water samplers are also in use for testing underground flow.

UNCONFINED COMPRESSION TEST APPARATUS : The required for conducting unconfined compression test of a soil sample is shown in illustration.

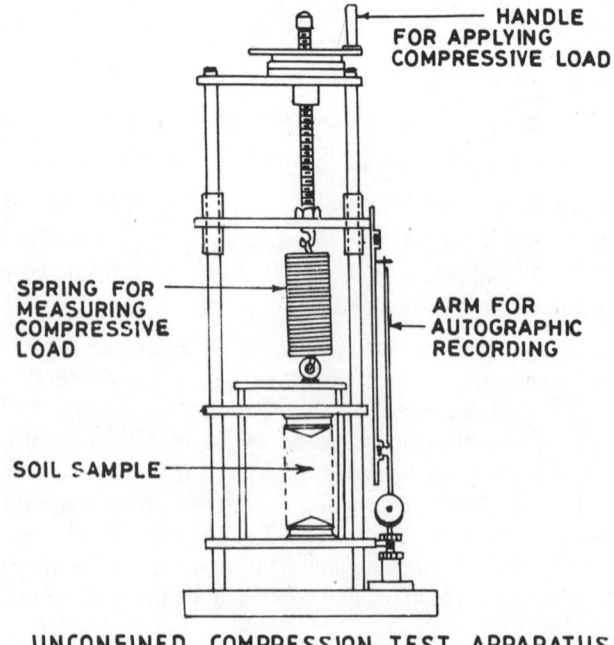

UNCONFINED COMPRESSION TEST APPARATUS

UNWIN FORMULA : A formula used in computing air friction through a fairly smooth iron pipe. This formula is applicable to low pressures and when the pressure drop does not exceed 20% of the initial pressure. It is given by :

$$Q = 60.4 \sqrt{\frac{hd^{0.5}}{L\{1+(3.6/d)\}}}$$

where Q is quantity of air in cft/minute,

       h is loss of pressure in inches of water.

       l is length of pipe in feet, and

      d is diameter of pipe in inches.

WATER LEAKAGE : In newly laid waterlines, the water leakage must be less than that obtained by the formula $L = \dfrac{ND\sqrt{P}}{1850}$

where L =   allowable leakage, in gph;

       N =   number of joints in the pipeline ;

       D =   nominal diameter of the pipe, in inches;

       P =   average test pressure during leakage test, in psi.

This formula gives 70 gpd of water leakage per mile length, per inch diameter of pipe, for 12' long pipes, at a pressure of 150 psig.

A test pressure of 50% above the normal operating pressure is recommended for at least 30 minutes.

WATER VALVE : An ordinary valve for a pump. This consists of a rubber disc resting across the ports through which water passes and the disc is held in position with the help of a coiled spring held by a threaded spindle into the centre of the spider. The valve opens by the pressure of water beneath it and closes when the water tends to flow back. The size of such a valve is restricted to 4" in diameter.

WAVE SPECTRUM : It depicts distribution of the variance density over the wave frequencies. The spectrum is conveniently obtained from wave recording by a Fourier analysis. Standard forms of wave spectra are used if the characteristic wave and period are available.

WAVE TRAIN : It is determined by its spectrum and it is considered a result of superposition of a large number of small, regular, sinusoidal waves, each of which is characterized by its height and frequency.

WEIR FORMULA : The discharge through weirs may be calculated by using the general form of weir formula $Q = CL(H)^{1.5}$

where Q is discharge through weir,

      C is weir co-efficient

      L is length of weir and

      H is total head of weir.

WELL LOG : In sinking a well, the valuable information of the materials encountered, their depth and characteristics of water bearing strata furnished by a log.

WHEELER FILTER BOTTOM : An arrangement in underdrainage systems of filter beds, comprising conical or pyramidal openings in the concrete bottom with the vertex of the cone or pyramid down. The openings are filled with heavy balls carefully placed to achieve the function of a strainer.